COLLEGE
ALGEBRA

HARCOURT BRACE COLLEGE OUTLINE SERIES

COLLEGE ALGEBRA

John L. Van Iwaarden

Department of Mathematics
Hope College
Holland, Michigan

Harcourt Brace College Publishers
Fort Worth   Philadelphia   San Diego
New York   Orlando   Austin   San Antonio
Toronto   Montreal   London   Sydney   Tokyo

Printed in the United States of America

5 6 7 8 9 0 1 2 3 4 074 13 12 11 10 9 8 7 6 5

ISBN 0-15-601520-X

PREFACE

Do not *read* this Outline—**use** it. You can't learn algebra simply by reading it: You have to *do* it. Solving specific, practical problems is the best way to master—and to demonstrate your mastery of—the theories, laws, and definitions upon which the study and use of mathematics is based. Outside the classroom, you need three tools to do algebra: a pencil, paper, and a calculator. Add a fourth tool, this Outline, and you're all set.

This HBJ College Outline has been designed as a tool to help you sharpen your problem-solving skills in college level algebra. Each chapter covers a unit of material, whose fundamental principles are broken down in outline form for easy reference. The outline text is heavily interspersed with worked-out examples, so you can see immediately how each new idea is applied in problem form. Each chapter also contains a Summary and a Raise Your Grades section, which (taken together) give you an opportunity to review the primary principles of a topic and the problem-solving techniques implicit in those principles.

Most important, this Outline gives you plenty of problems to practice on. Work the Solved Problems, and check yourself against the step-by-step solutions provided. Test your mastery of the material in each chapter by doing the Supplementary Exercises. (In the Supplementary Exercises, you're given answers only—the details of the solution are up to you.) Finally, you can review all the topics covered in the Outline by working the problems in the Unit and Final Exams. (Use the exam problems to diagnose your own strengths and weaknesses.)

Having the tools is one thing; knowing how to use them is another. The solution to any problem in mathematics requires six procedures: (1) UNDERSTANDING, (2) ANALYZING, (3) PLANNING, (4) EXECUTING, (5) CHECKING, (6) REPORTING. Let's look at each of these procedures in more detail.

1. **UNDERSTANDING:** Read over the problem carefully and be sure you understand every part of it. If you have difficulty with any of the terms or ideas in the problem, reread the text material on which the problem is based. (In this Outline, important ideas, principles, laws, and terms are printed in boldface type, so they will be easy to find.) Make certain that you understand what kind of answer will be required.

2. **ANALYZING:** Break the problem down into its components. Ask yourself
   - What are the data?
   - What is (are) the unknown(s)?
   - What equation, law, or definition connects the data to the unknowns?

3. **PLANNING:** Trace a connection between the data and the unknowns as a series of discrete operations (steps). This often involves manipulating one or more mathematical expressions to isolate unknown quantities. Once you have a clear, stepwise path between data and solution, take note of any steps that require ancillary operations, such as using special symbols or converting units. (Keep a sharp watch on units—they are often useful clues.)

4. **EXECUTION:** Follow your plan and execute the mathematical operations. It helps to work with symbols whenever possible: Substituting data for variables should be the *last* thing you do. Make sure you've used the correct signs, exponents, and units.

5. **CHECKING:** Never consider a problem solved until you have checked your work. Does your answer
   - make sense?
   - have the right units?
   - answer the question?

6. REPORTING: Make sure you have shown your reasoning and method clearly, and that your answer is readable. (It can't hurt to write the word "Answer" in front of your answer. That way, you—and your instructor—can find it at a glance, saving time and trouble all 'round.)

The path from concept to bound book is a long and time consuming one and we have not traveled it alone. I thank the editorial staff at Harcourt Brace Jovanovich for their expert work and helpful suggestions. I also thank Hope College for granting me a sabbatical leave to complete this work. I also thank my family for their patient support during these many months.

Hope College,
Holland, Michigan

JOHN L. VAN IWAARDEN

CONTENTS

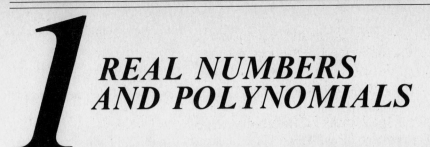

1 REAL NUMBERS AND POLYNOMIALS

THIS CHAPTER IS ABOUT

- ☑ **Sets and the Real Number System**
- ☑ **Properties of Real Numbers**
- ☑ **Mixed Operations on Real Numbers**
- ☑ **Polynomials**
- ☑ **Operations on Polynomials**

1-1. Sets and the Real Number System

A. Sets

A **set** is a collection (group, aggregate, assemblage) of objects called **members**, or **elements**, of the set. We'll use two methods of denoting sets:

(1) A set may be denoted by setting it equal to a *list of elements* separated by commas and enclosed in braces. Thus, the set C consisting of elements 1, 2, 3, and 4 can be written as $C = \{1, 2, 3, 4\}$. (It's customary, but not essential, to use capital letters to denote sets and numbers or lowercase letters to denote elements.)

(2) A set may be denoted by **set-builder notation**, in which the condition(s) of membership in the set are described. Thus, the set $C = \{1, 2, 3, 4\}$ may be written as

$$C = \{x \mid x \text{ is a natural number } < 5\}$$

which is read as "the set C of all elements x such that x is a natural number less than 5."

Some of the symbols used in set notation are shown in Table 1-1.

Sets follow mathematical rules:

- Two sets are equal if they contain the same elements.
- If every element of a set B is also an element of set C, then set B is a **subset** of set C.

TABLE 1-1: Definitions of Symbols	
$\{x, y\}$	the set having elements x and y
$\{x \mid \cdots\}$	the set of all x such that …
$X = Y$	set X equals set Y
$x \in Y$	x is an element of set Y
$X \subset Y$	set X is a subset of set Y
$X \cup Y$	the union of sets X and Y
$(\neq, \not\subset, \notin)$	(is not equal to, is not a subset of, is not an element of)

EXAMPLE 1-1

(a) Denote the set of denominations of common U.S. coins.
(b) Write the set $E = \{2, 4, 6\}$ in set-builder notation.
(c) Are the sets $A = \{1\}$ and $B = \{o, n, e\}$ the same set?
(d) If $X = \{t, o\}$, $Y = \{t, w, o\}$, $Z = \{t, o, o, t\}$, $W = \{t, o, w\}$, and $V = \{t, o, o\}$, which of these sets are equal?
(e) If $X = \{t, o\}$ and $Y = \{t, w, o\}$, is Y a subset of X?

Solution

(a) The common U.S. coins have denominations of 1, 5, 10, 25, and 50, so you can denote the set as C and write $C = \{1, 5, 10, 25, 50\}$.

(b) The elements in $E = \{2, 4, 6\}$ are all even *natural* (i.e., counting) *numbers* less than or equal to 6, so

$$E = \{x \mid x \text{ is an even natural number } \leq 6\}$$

(c) The sets $A = \{1\}$ and $B = \{o, n, e\}$ do not have the same elements, so they cannot be the same set.

(d) The sets X, V, and Z all have the same elements—i.e., t and o—so $X = V = Z$; and the sets W and Y have the same elements, so $W = Y$. [Notice that sets are considered equal even when their elements appear in different orders and numbers, as long as their elements are the same.]

(e) $Y = \{t, w, o\}$ cannot be a subset of $X = \{t, o\}$ because $w \notin X$, i.e., the element w of set Y is not a member of set X. But $X \subset Y$, i.e., set X is a subset of set Y, because the elements of X are all members of Y.

B. The real number system

- A set has the **closure property** when an arithmetic operation (addition, subtraction, multiplication, division) that is performed on *any* two or more of its members produces *only* members of the set.

Using the idea of closure, we can list certain special sets of numbers that make up the real number system.

N

1. Natural numbers: $N = \{1, 2, 3, \ldots\}$

The set N of natural numbers is made up of the counting numbers 1, 2, 3,..., where the three dots specify continuation of the elements, without end, in the same pattern. If we perform the usual arithmetic operations on any two members of this set, we find that this set is *closed under addition and multiplication*; i.e., the sum of two natural numbers is a natural number and the product of two natural numbers is a natural number.

N ⊂ Z

2. Integers: $Z = \{\ldots, -3, -2, -1, 0, 1, 2, 3, \ldots\}$

The set Z of integers, which includes the natural numbers, their negatives, and zero, is *closed under addition, multiplication, and subtraction*. So set Z has all the properties of set N (i.e., $N \subset Z$) and the additional property that the difference of any two of its members is itself a member of Z.

N ⊂ Z ⊂ Q

3. Rational numbers: $Q = \left\{\dfrac{x}{y} \;\middle|\; x \text{ and } y \text{ are integers and } y \neq 0\right\}$

The set Q of rational numbers is *closed under addition, subtraction, multiplication, and division*—except for division by zero, which is meaningless.

note: Any integer (or natural number) can be written as x/y, where $y = 1$, so $N \subset Z \subset Q$.

Another way of describing the rational numbers Q is as the set of all numbers having a repeating decimal expansion. Thus $\frac{1}{2} = .5000\cdots$ (repeating 0), $\frac{1}{3} = .333\cdots$ (repeating 3), $\frac{13}{99} = .131\,313\cdots$ (repeating 13), $\frac{12}{7} = 1.714\,285\,714\,285\,714\,285\cdots$ (repeating 714 285).

But there are many useful numbers that can't be written as repeating decimals, and therefore cannot be described by the above sets. So we have to define another set:

Q ⊄ I

4. Irrational numbers:

$$I = \left\{x \;\middle|\; x \text{ cannot be expressed as } \frac{p}{q}, \text{ where } p \text{ and } q \text{ are integers}\right\}$$

The set I of irrational numbers includes all nonrepeating, nonterminating decimal numbers, such as $\sqrt{2}$, $\sqrt{5}$, and π (which is $3.1416\cdots$, where the numbers do not repeat).

We finally arrive at the "grandfather" set, which contains all of the others:

5. Real numbers: $R = Q \cup I$

The set R of real numbers is the set of all decimal numbers, both repeating and nonrepeating; i.e., R is the *union* ($\cup$) of the set of all rational and the set of all irrational numbers.

$$R = Q \cup I$$

EXAMPLE 1-2 Given the set $\{-\frac{3}{2}, 5, \sqrt{7}, 6.1, \pi/2, -2.3, 0, \sqrt{2}, -3\}$, find its subset of **(a)** natural numbers, **(b)** integers, **(c)** rational numbers, and **(d)** irrational numbers.

Solution **(a)** $\{5\}$ **(c)** $\{-3, -2.3, -\frac{3}{2}, 0, 5, 6.1\}$
(b) $\{-3, 0, 5\}$ **(d)** $\{\sqrt{2}, \pi/2, \sqrt{7}\}$

1-2. Properties of Real Numbers

A. Axioms

The real numbers R must satisfy the **axioms of equality** as part of the structure of a mathematical system. If x, y, and z are arbitrary real numbers, then the axioms of equality are

AXIOMS OF EQUALITY	**Reflexive law**	$x = x$
	Symmetry law	If $x = y$, then $y = x$
	Transitive law	If $x = y$ and $y = z$, then $x = z$

B. Properties

1. Field properties

The set of real numbers R must also satisfy the **field axioms** or **properties** under the operations of addition and multiplication. If x, y, and z are arbitrary real numbers, then the field properties of addition and multiplication are

	for addition	for multiplication*
Closure law	$x + y$ is a member of R	$x \cdot y$ is a member of R
Commutative law	$x + y = y + x$	$x \cdot y = y \cdot x$
Associative law	$(x + y) + z = x + (y + z)$	$(x \cdot y) \cdot z = x \cdot (y \cdot z)$
Identity law	$x + 0 = x$	$x \cdot 1 = x$
	(0 is the **identity element** for **addition**)	(1 is the **identity element** for **multiplication**)
Inverse law	$x + (-x) = 0$	$x \cdot \dfrac{1}{x} = 1$
[*Note*: 0 doesn't have an inverse]	($-x$ is the **additive inverse** for x)	($1/x$ is the **multiplicative inverse** for x)
Distributive laws [*Note*: These laws link the two operations]	$x \cdot (y + z) = (x \cdot y) + (x \cdot z)$ $(x + y) \cdot z = (x \cdot z) + (y \cdot z)$	

PROPERTIES OF ADDITION AND MULTIPLICATION

*****Note:** Multiplication is indicated here by a center dot, as in $x \cdot y$; but we can also indicate multiplication of x by y as xy. A grouping symbol, such as a pair of parentheses ( ), also indicates multiplication, so $(x \cdot y) \cdot z$ may also be written as $(xy)z$, or—because of the associative law—as $x(yz)$.

Finally, the real numbers must satisfy the zero property of multiplication:

ZERO PROPERTY   If $xy = 0$, then either $x = 0$ or $y = 0$

EXAMPLE 1-3   Identify the axioms of equality or the field properties that verify the correctness of the following:

(a) If $\frac{1}{12}$ of a foot equals 1 inch and 1 inch equals 2.54 centimeters, then $\frac{1}{12}$ of a foot equals 2.54 centimeters.

(b) $3 + 5 = 5 + 3$

(c) $\frac{1}{2} + (-\frac{1}{2}) = 0$

(d) $(3 \cdot 4) \cdot 2 = 3 \cdot (4 \cdot 2)$

(e) $7 \cdot 1 = 7$

(f) $2 + 5$ is a real number

(g) $6 \cdot (2 + 5) = 6 \cdot (7) = 42$
$$= (6 \cdot 2) + (6 \cdot 5)$$
$$= 12 + 30 = 42$$

Solution

(a) Transitive law
(b) Commutativity of addition
(c) Inverse for addition
(d) Associativity for multiplication

(e) Identity for multiplication
(f) Closure for addition
(g) Distributive property

> \+ times + is +
> − times − is +
> \+ times − is −
> − times + is −
>
> Subtraction is just a special case of addition:
> $$a - b = a + (-b)$$
>
> Division is just a special case of multiplication:
> $$\frac{a}{b} \div \frac{c}{d} = \frac{a}{b} \cdot \frac{d}{c}$$

2. Properties under subtraction and division

Real numbers also obey the following properties under subtraction (involving negatives) and division (involving fractions). If a, b, c, and d are real numbers, then

> **PROPERTIES UNDER SUBTRACTION AND DIVISION**

$$a - b = a + (-b)$$
$$-(-a) = a$$
$$-a = (-1)a$$
$$-(ab) = (-a)b = a(-b)$$
$$(-a)(-b) = ab$$
$$a(b - c) = ab - ac$$
$$\frac{a}{b} = a\left(\frac{1}{b}\right)$$
$$\frac{a}{-b} = \frac{-a}{b} = -\frac{a}{b}$$

$$\frac{ac}{bc} = \frac{a}{b} \qquad \text{Reducing or canceling}$$
$$\frac{a}{b} \cdot \frac{c}{d} = \frac{ac}{bd} \qquad \text{Combining}$$
$$\frac{a}{b} + \frac{c}{b} = \frac{a + c}{b} \qquad \begin{array}{l}\text{Same} \\ \text{denominator}\end{array} \Big\} \text{Addition}$$
$$\frac{a}{b} + \frac{c}{d} = \frac{ad + bc}{bd} \qquad \begin{array}{l}\text{Different} \\ \text{denominator}\end{array}$$
$$\frac{a}{b} \div \frac{c}{d} = \frac{a}{b} \cdot \frac{d}{c} = \frac{ad}{bc} \qquad \text{Division}$$

Notice that our list of field properties is preceded by "If x, y, and z are ..." and our list of the properties of subtraction and division is preceded by "If a, b, c, and d are ...." But in both of these lists the different sets of letters are defined the same—as *real numbers*. This practice is typical of algebra (and mathematics in general): Things are what we *say* they are, and we're very careful about what we say they are. The letters themselves don't mean anything until we define them. But the symbols of operation and signs always have a consistent meaning.

EXAMPLE 1-4   Use the properties of division to check the correctness of the following:

(a) $\dfrac{1}{2} + \dfrac{3}{4} = \dfrac{5}{4}$

(b) $\dfrac{28}{49} = \dfrac{4}{7}$

(c) $\dfrac{3}{8} \div \dfrac{9}{4} = \dfrac{1}{6}$

(d) $\dfrac{3}{4} \cdot \dfrac{6}{11} \cdot \dfrac{2}{3} = \dfrac{3}{11}$

Solution

(a) $\overset{\text{addition with a}}{\overset{\text{different denominator}}{\dfrac{1}{2} + \dfrac{3}{4}}} = \dfrac{1 \cdot 4 + 3 \cdot 2}{2 \cdot 4} = \dfrac{4 + 6}{8} = \dfrac{10}{8} = \overset{\text{reducing}}{\dfrac{5 \cdot 2}{4 \cdot 2}} = \dfrac{5}{4}$

(b) $\dfrac{28}{49} = \dfrac{4 \cdot \cancel{7}}{7 \cdot 7} = \dfrac{4}{7}$ *reducing*

(c) $\dfrac{3}{8} \div \dfrac{9}{4} = \dfrac{3}{8} \cdot \dfrac{4}{9} = \dfrac{3 \cdot 4}{8 \cdot 9} = \dfrac{\cancel{3} \cdot \cancel{4}}{2 \cdot \cancel{4} \cdot \cancel{3} \cdot 3} = \dfrac{1}{2 \cdot 3} = \dfrac{1}{6}$ *division combining reducing*

(d) $\dfrac{3}{4} \cdot \dfrac{6}{11} \cdot \dfrac{2}{3} = \dfrac{3 \cdot 6 \cdot 2}{4 \cdot 11 \cdot 3} = \dfrac{36}{132} = \dfrac{3 \cdot \cancel{12}}{11 \cdot \cancel{12}} = \dfrac{3}{11}$ *combining reducing*

1-3. Mixed Operations on Real Numbers

A. Order of operations

When performing operations on mixed arithmetic expressions, we follow a prescribed **order of operations**:

(1) Evaluate within **P**arentheses
(2) Evaluate the **E**xponents
(3) Perform the **M**ultiplication and **D**ivision
(4) Perform the **A**ddition and **S**ubtraction

note: Think of the silly expression: "*P*lease *E*xcuse *M*y *D*ear *A*unt *S*ally," or *PEMDAS*.

When multiplication and division appear in the same expression, we perform them *left-to-right*. We also perform addition and subtraction left-to-right when they appear in the same expression. Finally, if there are two or more sets of *grouping symbols* (e.g., parentheses, brackets), so that one set is inside another, we evaluate the inner set first and work outward.

note: We often use square brackets [ ] to specify an *outer* set of grouping symbols.

EXAMPLE 1-5 Evaluate

(a) $4 + 4 \cdot 5 - 8 \div 2$

(b) $4 + 4(5 - 8) \div 2$

(c) $(4 + 4)5 - 8 \div 2$

(d) $4 + (4 \cdot 5 - 8) \div 2$

(e) $4 + 4[(5 - 8) \div 2]$

(f) $4^2 + 4 \cdot 5^2 - 27 \div (2^2 - 1)$

Solution

(a) Multiply and divide before adding and subtracting.

$$4 + 4 \cdot 5 - 8 \div 2 = 4 + 20 - 4 = 24 - 4 = 20$$

(b) Evaluate within parentheses first; then follow the other rules.

$$4 + 4(5 - 8) \div 2 = 4 + 4(-3) \div 2$$
$$= 4 + (-12) \div 2$$
$$= 4 + (-6) = -2$$

(c) $(4 + 4)5 - 8 \div 2 = 8 \cdot 5 - 8 \div 2 = 40 - 4 = 36$

(d) $4 + (4 \cdot 5 - 8) \div 2 = 4 + (20 - 8) \div 2$
$$= 4 + 12 \div 2$$
$$= 4 + 6 = 10$$

(e) $4 + 4[(5 - 8) \div 2] = 4 + 4(-3 \div 2)$

$$= 4 + 4(-1.5)$$
$$= 4 + (-6) = -2$$

(f) $4^2 + 4 \cdot 5^2 - 27 \div (2^2 - 1) = 4^2 + 4 \cdot 5^2 - 27 \div (4 - 1)$

$$= 16 + 4 \cdot 25 - 27 \div 3$$
$$= 16 + 100 - 9$$
$$= 107$$

B. Simplifying expressions

Taken together, the axioms of equality, the field properties of addition and multiplication, and the properties of subtraction and division make up the rules for the manipulation of all real numbers. We combine the operations, in the correct PEMDAS order, with the rules to *simplify* arithmetic expressions.

EXAMPLE 1-6 Simplify the following expressions:

(a) $\dfrac{11}{4} \div \left(\dfrac{5}{2} + 2 \right)$

(b) $\dfrac{2}{3}\left(\dfrac{1}{4} - \dfrac{3}{8} \right)$

(c) $\dfrac{4}{-2} \div \left(\dfrac{-1}{3} - \dfrac{5}{-2} \right)$

(d) $\left(\dfrac{1}{3} + \dfrac{2}{5} + \dfrac{3}{7} \right)\dfrac{15}{61}$

(e) $\left(1 + \dfrac{1}{10} \right)\left(1 - \dfrac{1}{11} \right)$

Solution

(a) $\dfrac{11}{4} \div \left(\dfrac{5}{2} + 2 \right) = \dfrac{11}{4} \div \left(\dfrac{5}{2} + \dfrac{4}{2} \right) = \dfrac{11}{4} \div \dfrac{9}{2} = \dfrac{11}{4} \cdot \dfrac{2}{9} = \dfrac{22}{36} = \dfrac{11}{18}$

(b) $\dfrac{2}{3}\left(\dfrac{1}{4} - \dfrac{3}{8} \right) = \dfrac{2}{3}\left(\dfrac{2}{8} - \dfrac{3}{8} \right) = \dfrac{2}{3}\left(-\dfrac{1}{8} \right) = -\dfrac{2}{24} = -\dfrac{1}{12}$

(c) $\dfrac{4}{-2} \div \left(\dfrac{-1}{3} - \dfrac{5}{-2} \right) = -\dfrac{4}{2} \div \left[-\dfrac{1}{3} - \left(-\dfrac{5}{2} \right) \right]$

$$= -2 \div \left(-\dfrac{1}{3} + \dfrac{5}{2} \right) = -2 \div \left(-\dfrac{2}{6} + \dfrac{15}{6} \right)$$

$$= -2 \div \left(\dfrac{13}{6} \right) = -2 \cdot \dfrac{6}{13} = -\dfrac{12}{13}$$

(d) $\left(\dfrac{1}{3} + \dfrac{2}{5} + \dfrac{3}{7} \right)\dfrac{15}{61} = \left(\dfrac{5 \cdot 7}{3 \cdot 5 \cdot 7} + \dfrac{2 \cdot 3 \cdot 7}{3 \cdot 5 \cdot 7} + \dfrac{3 \cdot 3 \cdot 5}{3 \cdot 5 \cdot 7} \right)\dfrac{15}{61}$

$$= \left(\dfrac{35 + 42 + 45}{3 \cdot 5 \cdot 7} \right)\dfrac{15}{61} = \dfrac{122}{3 \cdot 5 \cdot 7} \cdot \dfrac{3 \cdot 5}{61}$$

$$= \dfrac{61 \cdot 2}{3 \cdot 5 \cdot 7} \cdot \dfrac{3 \cdot 5}{61} = \dfrac{2}{7}$$

(e) $\left(1 + \dfrac{1}{10} \right)\left(1 - \dfrac{1}{11} \right) = \left(\dfrac{10}{10} + \dfrac{1}{10} \right)\left(\dfrac{11}{11} - \dfrac{1}{11} \right) = \dfrac{11}{10} \cdot \dfrac{10}{11} = 1$

What works for arithmetic expressions also works for algebraic expressions: We combine the properties of real numbers with the operations in the correct PEMDAS order to simplify algebraic expressions. But there's a hitch—we have to account for letters as well as familiar arithmetic numbers. In multiplication (and division), we can apply the field properties directly, so $a \cdot b = ab$, $a \div b = a/b$, and so on. Then, by the associative law we can write $2 \cdot (a \cdot b) = (2 \cdot a) \cdot b = 2ab$, $2 \cdot (a \cdot 3 \cdot b) = (2 \cdot 3) \cdot (a \cdot b) = 6ab$, and so on. But to add (and subtract), we must use the distributive law to group *like terms*, i.e., terms with the same letter and the same exponent. The coefficient (numerical portion) of a term can be added to the coefficient of a like term. Thus, $6x + 4x = (6 + 4)x = 10x$ and $7x^2 - 5x^2 = (7 - 5)x^2 = 2x^2$. But we can't add the coefficients of unlike terms, such as $3x^2$ and $2x$.

EXAMPLE 1-7 Simplify the following expressions:

(a) $4x + 7x$

(b) $\left(\dfrac{1}{3}\right)x + \left(\dfrac{1}{2}\right)x - \left(\dfrac{1}{6}\right)x$

(c) $(5z - 2) + (4z + 3)$

(d) $2 + 4(3 + y) - 2y$

(e) $3 + (2 + x)(2 + x) - x^2 - (x + 1)$

Solution   Add the coefficients of the like terms.

(a) $4x + 7x = (4 + 7)x = 11x$

(b) $\left(\dfrac{1}{3}\right)x + \left(\dfrac{1}{2}\right)x - \left(\dfrac{1}{6}\right)x = \left(\dfrac{1}{3} + \dfrac{1}{2} - \dfrac{1}{6}\right)x = \left(\dfrac{2}{6} + \dfrac{3}{6} - \dfrac{1}{6}\right)x$

$$= \left(\dfrac{2 + 3 - 1}{6}\right)x = \left(\dfrac{4}{6}\right)x = \left(\dfrac{2 \cdot 2}{3 \cdot 2}\right)x = \left(\dfrac{2}{3}\right)x$$

(c) $(5z - 2) + (4z + 3) = (5z + 4z) + (-2 + 3) = 9z + 1$

Expand first; then add the coefficients of the like terms:

(d) $2 + 4(3 + y) - 2y = 2 + (12 + 4y) - 2y = (2 + 12) + (4y - 2y) = 14 + 2y$

(e) $3 + (2 + x)(2 + x) - x^2 - (x + 1) = 3 + (2 + x)2 + (2 + x)x - x^2 - x - 1$

$$= 3 + (4 + 2x) + (2x + x^2) - x^2 - x - 1$$
$$= (3 + 4) + (2x + 2x) + (x^2 - x^2) - x - 1$$
$$= 7 + 4x + 0 - x - 1 = (7 - 1) + (4x - x)$$
$$= 6 + 3x$$

1-4. Polynomials

A. Terminology

We'll need to define some terms in order to discuss polynomials.

- In algebraic expressions, some of the members can have *arbitrary* (or unknown) values while other members have *fixed* (or known) values:

  A letter (such as x) used to denote an *arbitrary* member of a set is called a **variable**.

  A letter or number (such as k or 5 or $\frac{16}{3}$ or π) that has a *fixed*, or known, value is called a **constant**.

- If n is a natural number and x is any real number, then we can write an expression x^n for a **positive power** of x (where x is the *base* and n is the *exponent*), such that

$$x^n = \underbrace{x \cdot x \cdot x \cdots x}_{n \text{ times}}$$

For example, $4^3 = 4 \cdot 4 \cdot 4$ and $(-2)^4 = (-2)(-2)(-2)(-2)$, while $5^1 = 5$. Also, $x^0 = 1$ for any nonzero value of x; that is, *the value of any real number raised to the power of 0 is always 1*.

- The basic building block of algebraic expressions is the **monomial cx^n**, where

    c is a constant   (c is any real number)

    x is a variable

    n is a nonnegative integer   ($n = 0, 1, 2, 3, \ldots$)

    We refer to the constant in the monomial as the **coefficient**. The expressions $3x^4$ and $\frac{16}{3}y^5$ are monomials, whose coefficients are 3 and $\frac{16}{3}$ and whose variables are x and y, respectively. But $3/x$ and $3\sqrt{x}$ are not monomials, because the variable in a monomial must not have a negative or rational (fractional) exponent.

    note: A negative exponent on any number, such as x^{-1}, indicates the *reciprocal* of the number; i.e., $x^{-1} = 1/x$.
    A rational exponent, such as $x^{1/2}$, indicates a root, so $x^{1/2} = \sqrt{x}$. [See also Chapter 3.]

- The sum of a finite number (one or more) of monomials is called a **polynomial**. The monomials in a polynomial are called *terms* of the polynomial. So a monomial cx^n is a polynomial with one term. And if we add two monomials ax^m and bx^n, we get a **binomial** $ax^m + bx^n$, which is a polynomial with two terms.

    note: It's understood here that a, b, and c are constants (coefficients) and that the exponents m and n are nonnegative integers.

EXAMPLE 1-8 Explain why the following are polynomials:

(a) 1

(b) $2^2 y$

(c) $5x - 3$

(d) $\pi t^2 + 8t + 11$

(e) $\sqrt{5}x^4 - \frac{\pi}{2}x^2 + 4x$

(f) $5z^2 + 4^{-2}z + 3$

Solution Each of these expressions is made up of one or more monomials with the form cx^n:

(a) The algebraic expression 1 is a monomial with a coefficient 1 and a variable with exponent zero: $1 = 1 \cdot x^0 = 1 \cdot 1$.

(b) The expression $2^2 y$ is a monomial with coefficient $c = 2^2 = 4$ and exponent 1 on the variable y.

(c) The expression $5x - 3$ is the difference of two monomials, i.e., the sum of $5x$ and -3, and hence is a polynomial with two terms, or a binomial.

(d) The coefficients of a monomial may be *any* real numbers; and since π, 8, and 11 are real, and the variable t is raised to a positive power, this expression is the sum of three monomials.

(e) The coefficients $\sqrt{5}$, $-\pi/2$, and 4 are real and the exponents on the variable x are nonnegative integers, so this expression has three distinct monomials added to form a polynomial.

(f) Each term is a monomial since the exponent on the variable x is a nonnegative integer and the coefficients 5, 4^{-2}, and 3 are real.

There is no restriction that a term in a polynomial must have just one variable. An algebraic expression such as $3x^2 y^2$ is a monomial in two variables (x and y) and $-6xy^2 z^4$ is a monomial in three variables (x, y, and z). These multivariable monomials may also be used to build polynomials.

EXAMPLE 1-9 Explain why the following are NOT polynomials:

(a) $\dfrac{2}{x} - 5$ **(b)** $6xy^{-2} + 3x^2y$ **(c)** $\dfrac{x}{y} + \dfrac{5z}{7} + 5^{-2}$

(d) $s^2 + 2t^{1/2}$ **(e)** $s^2 + 2\sqrt{t}$

Solution

(a) No variable is allowed in the denominator: $2/x - 5 = 2x^{-1} - 5$.
(b) Negative exponents are not allowed on the variable.
(c) No variable is allowed in the denominator.
(d) No fractional exponents are allowed on a variable.
(e) A variable under a radical sign is not allowed: $s^2 + 2\sqrt{t} = s^2 + 2t^{1/2}$.

B. Classifying polynomials

We begin classifying polynomials by the number of variables they have. A polynomial may have many variables, or it may have just one. After we determine how many variables there are, we can classify polynomials by **degree**:

- The **degree of a polynomial in one variable** is the highest exponent that is found on the (single) variable.
  Thus the polynomial $x^2 + 6x + 9$ has one variable, x, and a degree of 2, because the highest exponent on the single variable x is 2.
- The **degree of a polynomial in more than one variable** is the highest *sum* of exponents on the variables in any one term.
  Thus the polynomial $x^4y^2 + 4x^3y^2 + 2xy$ has two variables, x and y, and a degree of 6, because the exponents of the variables x and y in the first term add up to 6—a sum that exceeds the sum of the exponents of the variables in the second term (5) and the third term (2).

EXAMPLE 1-10 Determine the type and degree of each of the following polynomials:

(a) $x^2 - 5x + 3$ **(e)** $2xy^2 + 3$
(b) $y - 2y^2 + \sqrt{6}y + 7$ **(f)** $x^3 + 5x^2y + 8$
(c) $3z^4 - 9z^2 + 4$ **(g)** $12x^2y^3z + 27xy^5z^4 + 19x^4y^2z^2$
(d) 5^2

Solution   Polynomials **(a)**, **(b)**, and **(c)** are all polynomials in one variable.

(a) Degree 2: The term x^2 has the highest exponent on the single variable x.
(b) Degree 2: The term $-2y^2$ has the highest exponent on the single variable y.
(c) Degree 4: The term $3z^4$ has the highest exponent on the single variable z.

Polynomial **(d)** has no obvious variable, but 5^2 is considered to be a polynomial in one variable, where the variable, say x, is raised to the power of zero and equals 1.

(d) Degree 0: The monomial is actually 5^2x^0, so the highest exponent on the single (implied) variable x is 0.

Polynomials **(e)**, **(f)**, and **(g)** have more than one variable.

(e) Degree 3: The term $2xy^2$ has total exponent $1 + 2 = 3$, which exceeds 0 on the term $3 = 3x^0$
(f) Degree 3: Both terms x^3 and $5x^2y$ have total exponents of 3, which exceeds 0.
(g) Degree 10: The middle term $27xy^5z^4$ has total exponent $1 + 5 + 4 = 10$, which exceeds 6 and 8 on the first and third terms, respectively.

This notation looks worse than it is.

- $P(x)$ means, simply, any polynomial whose variable is x.
- The superscript n is the highest exponent on the variable x. Thus n-1 is just what it says it is—the highest exponent minus 1, and so on.
- The integer (i) subscripts, where n, n-1, ..., 2, 1, 0 = i, serve only to tell us the maximum number of terms—excluding the x^0 term, called a **constant term**—that a single-variable polynomial of nth degree *may* have. Thus a third-degree polynomial may have as many as three terms plus a constant term, or

$$\underbrace{a_3x^3 + a_2x^2 + a_1x^1}_{3\ terms} + \underbrace{a_0x^0}_{\substack{constant \\ term}}$$

The general form of a polynomial in one variable x is

$$P(x) = a_nx^n + a_{n-1}x^{n-1} + a_{n-2}x^{n-2} + \cdots + a_2x^2 + a_1x + a_0$$

where each coefficient a_i is a real number. If $a_n \neq 0$, then this polynomial has degree n.

1-5. Operations on Polynomials

A. Addition and subtraction

Since polynomials are made up of terms representing real numbers, all of the rules for the operations on real numbers will apply to them. Thus, if we have two polynomials, $P(x) = a_nx^n + a_{n-1}x^{n-1} + \cdots + a_1x + a_0$ and $Q(x) = b_nx^n + b_{n-1}x^{n-1} + \cdots + b_1x + b_0$, we can find their sum $S(x)$ and difference $T(x)$ by grouping their like terms and adding or subtracting the coefficients of the like terms:

TWO POLYNOMIALS:
$$P(x) = a_nx^n + a_{n-1}x^{n-1} + \cdots + a_1x + a_0$$
$$Q(x) = b_nx^n + b_{n-1}x^{n-1} + \cdots + b_1x + b_0$$

SUM $S(x) = (a_n + b_n)x^n + (a_{n-1} + b_{n-1})x^{n-1} + \cdots + (a_1 + b_1)x + (a_0 + b_0)$

DIFFERENCE $T(x) = (a_n - b_n)x^n + (a_{n-1} - b_{n-1})x^{n-1} + \cdots + (a_1 - b_1)x + (a_0 - b_0)$

EXAMPLE 1-11   Find the sum of the polynomials $P(x) = 3x^2 + 4x - 7$ and $Q(x) = 7x^2 - 9x + 4$.

Solution   Set up the addition by aligning the like terms:

$$3x^2 + 4x - 7$$
$$7x^2 - 9x + 4$$

Add the coefficients: $\overline{10x^2 - 5x - 3}$

Or, simply rearrange the terms—i.e., group like terms—and use the properties for real numbers:

$$
\begin{aligned}
P(x) + Q(x) &= (3x^2 + 4x - 7) + (7x^2 - 9x + 4) \\
&= (3x^2 + 7x^2) + (4x - 9x) + (-7 + 4) &&\text{[Group like terms]} \\
&= (3 + 7)x^2 + (4 - 9)x - 7 + 4 &&\text{[Add coefficients]} \\
&= 10x^2 - 5x - 3
\end{aligned}
$$

EXAMPLE 1-12   Find the sum and difference of the following polynomials:

(a) $P(x) = x + 7; Q(x) = 2x^2 - x + 5$

(b) $P(x) = x^3 + 2; Q(x) = 2x^2 + x - 2$

(c) $P(x) = \frac{2}{3}x^2y - 7x + \frac{2}{5}; Q(x) = x^2y + \frac{3}{5}x - 4$

Solution

(a)
$$
\begin{aligned}
P(x) + Q(x) &= (x + 7) + (2x^2 - x + 5) \\
&= x + 7 + 2x^2 - x + 5 \\
&= 2x^2 + (x - x) + (7 + 5) \\
&= 2x^2 + 12
\end{aligned}
$$

$$P(x) - Q(x) = (x + 7) - (2x^2 - x + 5)$$
$$= x + 7 - 2x^2 + x - 5$$
$$= -2x^2 + (x + x) + (7 - 5)$$
$$= -2x^2 + 2x + 2$$

(b) $P(x) + Q(x) = (x^3 + 2) + (2x^2 + x - 2)$
$$= x^3 + 2x^2 + x + (2 - 2)$$
$$= x^3 + 2x^2 + x$$

$P(x) - Q(x) = (x^3 + 2) - (2x^2 + x - 2)$
$$= x^3 - 2x^2 - x + (2 + 2)$$
$$= x^3 - 2x^2 - x + 4$$

(c) $P(x) + Q(x) = (\frac{2}{3}x^2y - 7x + \frac{2}{5}) + (x^2y + \frac{3}{5}x - 4)$
$$= (\frac{2}{3} + 1)x^2y + (-7 + \frac{3}{5})x + (\frac{2}{5} - 4)$$
$$= \frac{5}{3}x^2y - \frac{32}{5}x - \frac{18}{5}$$

$P(x) - Q(x) = (\frac{2}{3}x^2y - 7x + \frac{2}{5}) - (x^2y + \frac{3}{5}x - 4)$
$$= (\frac{2}{3} - 1)x^2y + (-7 - \frac{3}{5})x + (\frac{2}{5} + 4)$$
$$= -\frac{1}{3}x^2y - \frac{38}{5}x + \frac{22}{5}$$

B. Multiplication

In order to multiply polynomials, we need to extend the definition of an exponent so we can multiply numbers with the same base raised to a power. To do this, we use the rule of exponents $x^n \cdot x^m = x^{n+m}$; that is,

$$x^n \cdot x^m = \underbrace{x \cdot x \cdot x \cdots x}_{n \text{ times}} \cdot \underbrace{x \cdot x \cdot x \cdots x}_{m \text{ times}} = x^{n+m}$$
$$\underbrace{\qquad\qquad\qquad\qquad}_{n + m \text{ times}}$$

note: This rule can be extended to $x^n \div x^m = x^{n-m}$ and $(x^m)^n = x^{m \cdot n} = x^{mn}$. Thus, $x^6 \div x^3 = x^{6-3} = x^3$ and $(x^2)^3 = x^{2 \cdot 3} = x^6$. [See also Chapter 3.]

Now we can find the product of two polynomials $P(x) \cdot Q(x)$ using the distributive laws and rules of exponents.

EXAMPLE 1-13   Find the product of $P(x) = x^2 + 3$ and $Q(x) = x + 4$.

Solution   Set up the multiplication the old "arithmetic" way:

$$
\begin{array}{r}
P(x) = x^2 + 3 \\
\times\ Q(x) = \times\ \ x + 4 \\
\hline
+ 4x^2 + 12 \\
x^3 + 3x \\
\hline
P(x) \cdot Q(x) = x^3 + 4x^2 + 3x\ + 12
\end{array}
$$

[Multiply each member of the top expression by 4]

[Multiply each member of the top expression by x]

[Add]

Or, use the distributive law directly:

$$P(x) \cdot Q(x) = (x^2 + 3) \cdot (x + 4)$$
$$= x(x^2 + 3) + 4(x^2 + 3) \quad \text{[Distributive law]}$$
$$= (x^3 + 3x) + (4x^2 + 12)$$
$$= x^3 + 4x^2 + 3x + 12$$

This result may also be written "backward":
$$x^3 - y^3 = (x - y)(x^2 + xy + y^2)$$
You will see this form again.

EXAMPLE 1-14 Find the product of $P(x) = x - y$ and $Q(x) = x^2 + xy + y^2$.

Solution

$$
\begin{aligned}
P(x) \cdot Q(x) &= (x - y)(x^2 + xy + y^2) \\
&= (x - y)x^2 + (x - y)(xy) + (x - y)y^2 \quad \text{[Distributive law]} \\
&= (x^3 - x^2y) + (x^2y - xy^2) + (xy^2 - y^3) \\
&= x^3 - \cancel{x^2y} + \cancel{x^2y} - \cancel{xy^2} + \cancel{xy^2} - y^3 \quad \text{[Group like terms and cancel]} \\
&= x^3 - y^3
\end{aligned}
$$

The product of three polynomials is obtained in a two-step process: We find the product of two of the polynomials first; then we multiply this "intermediate" product by the third polynomial to get the final product.

$$
\begin{aligned}
P(x) \cdot Q(x) \cdot R(x) &= [P(x) \cdot Q(x)] \cdot R(x) \\
&= P(x) \cdot [Q(x) \cdot R(x)]
\end{aligned}
$$

EXAMPLE 1-15 Find the product $(x - 3)(2x + 5)(x + 4)$.

Solution

$$
\begin{aligned}
(x - 3)(2x + 5)(x + 4) &= [(x - 3)(2x + 5)] \cdot (x + 4) \\
&= [(x - 3)2x + (x - 3)5] \cdot (x + 4) \\
&= (2x^2 - 6x + 5x - 15) \cdot (x + 4) \\
&= (2x^2 - x - 15)(x + 4) \quad \text{[Intermediate product]} \\
&= (2x^2 - x - 15)x + (2x^2 - x - 15)4 \\
&= 2x^3 - x^2 - 15x + 8x^2 - 4x - 60 \\
&= 2x^3 + 7x^2 - 19x - 60 \quad \text{[Final product]}
\end{aligned}
$$

EXAMPLE 1-16 Find the product $(x - 2)^2(x + 3)^2$.

Solution Treat this product as a pair of pairs:

$$
\begin{aligned}
(x - 2)^2(x + 3)^2 &= [(x - 2)(x - 2)] \cdot [(x + 3)(x + 3)] \\
&= [x(x - 2) - 2(x - 2)] \cdot [x(x + 3) + 3(x + 3)] \\
&= (x^2 - 2x - 2x + 4)(x^2 + 3x + 3x + 9) \\
&= (x^2 - 4x + 4)(x^2 + 6x + 9) \quad \text{[Intermediate product]} \\
&= x^2(x^2 + 6x + 9) - 4x(x^2 + 6x + 9) + 4(x^2 + 6x + 9) \\
&= x^4 + 6x^3 + 9x^2 - 4x^3 - 24x^2 - 36x + 4x^2 + 24x + 36 \\
&= x^4 + 2x^3 - 11x^2 - 12x + 36 \quad \text{[Final product]}
\end{aligned}
$$

FOIL

There is a memory device we can use to form the product of two binomials. As a model, consider the product $(2x + 3)(x - 5)$. Using the distributive laws, we obtain

$$
\begin{aligned}
(2x + 3)(x - 5) &= 2x(x - 5) + 3(x - 5) \\
&= 2x \cdot x - 2x \cdot 5 + 3 \cdot x - 3 \cdot 5 \\
&= 2x^2 - 10x + 3x - 15 \\
&= 2x^2 - 7x - 15
\end{aligned}
$$

Examining the steps in this multiplication closely, we see that

$2x^2$ comes from multiplying the *First* terms in each binomial, $2x$ and x

$-10x$ comes from multiplying the *Outside* terms, $2x$ and -5

$+3x$ comes from multiplying the *Inside* terms, 3 and x

-15 comes from multiplying the *Last* terms in each binomial, 3 and -5

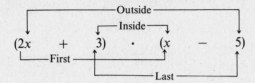

The method of multiplying First, Outside, Inside, and Last terms of two binomials is known by the mnemonic acronym **FOIL**.

EXAMPLE 1-17 Use the FOIL method to find the products of
(a) $(x + 4)(2x - 3)$, (b) $(3x - 1)(5x + 2)$, and (c) $(3a - 7b)(2a + 5b)$.

Solution

(a) $(x + 4)(2x - 3) = 2x^2 - 3x + 8x - 12 = 2x^2 + 5x - 12$
$$\qquad\qquad\qquad\quad\ \underset{F}{\uparrow}\quad \underset{O}{\uparrow}\quad \underset{I}{\uparrow}\quad \underset{L}{\uparrow}$$

(b) $(3x - 1)(5x + 2) = 15x^2 + 6x - 5x - 2 = 15x^2 + x - 2$
$$\qquad\qquad\qquad\quad\ \underset{F}{\uparrow}\quad\ \underset{O}{\uparrow}\quad\ \underset{I}{\uparrow}\quad\ \underset{L}{\uparrow}$$

(c) $(3a - 7b)(2a + 5b) = 6a^2 + 15ab - 14ab - 35b^2 = 6a^2 + ab - 35b^2$
$$\qquad\qquad\qquad\qquad\quad\ \underset{F}{\uparrow}\qquad \underset{O}{\uparrow}\qquad \underset{I}{\uparrow}\qquad \underset{L}{\uparrow}$$

Remember that the FOIL method applies *only* to the product of two binomials. But a combination of this method with the distributive law will solve almost any product problem.

C. Factoring

Factoring is the process of writing a given polynomial as the product of two or more lower-degree polynomials; i.e., factoring is the reverse of multiplication.

1. Special products

There are certain special products that occur often in algebraic manipulation. Some of these special products are multiplied out, or *expanded*, in Example 1-18.

EXAMPLE 1-18 Find the following special products:

(a) $(x - y)(x + y)$

(b) $(x + y)^2$

(c) $(x - y)^2$

(d) $(x + a)(x + b)$

(e) $(cx + d)(ex + f)$

(f) $(x + y)^3$

(g) $(x - y)^3$

(h) $(x + a)(x^2 - ax + a^2)$

(i) $(x - a)(x^2 + ax + a^2)$

Check each answer, using the FOIL method when possible.

Solution

(a) $(x - y)(x + y) = x(x + y) - y(x + y)$

    FOIL $\longrightarrow = x^2 + xy - xy - y^2$

    $= x^2 - y^2$

(b) $(x + y)^2 = (x + y)(x + y) = x(x + y) + y(x + y)$

    FOIL $\longrightarrow = x^2 + xy + xy + y^2$

    $= x^2 + 2xy + y^2$

(c) $(x - y)^2 = (x - y)(x - y) = x(x - y) - y(x - y)$

    FOIL $\longrightarrow = x^2 - xy - xy + y^2$

    $= x^2 - 2xy + y^2$

(d) $(x + a)(x + b) = x(x + b) + a(x + b)$

    FOIL $\longrightarrow = x^2 + bx + ax + ab$

    $= x^2 + (a + b)x + ab$

(e) $(cx + d)(ex + f) = cx(ex + f) + d(ex + f)$

    FOIL $\longrightarrow = cex^2 + cfx + dex + df$

    $= cex^2 + (cf + de)x + df$

(f) $(x + y)^3 = \quad (x + y)(x + y)(x + y)$

    FOIL $\longrightarrow = (x^2 + 2xy + y^2) \cdot (x + y)$ [Intermediate product]

    $= (x^2 + 2xy + y^2)x + (x^2 + 2xy + y^2)y$

    $= x^3 + 2x^2y + xy^2 + x^2y + 2xy^2 + y^3$

    $= x^3 + 3x^2y + 3xy^2 + y^3$

(g) $(x - y)^3 = \quad (x - y)(x - y)(x - y)$

    FOIL $\longrightarrow = (x^2 - 2xy + y^2) \cdot (x - y)$ [Intermediate product]

    $= (x^2 - 2xy + y^2)x - (x^2 - 2xy + y^2)y$

    $= x^3 - 2x^2y + xy^2 - x^2y + 2xy^2 - y^3$

    $= x^3 - 3x^2y + 3xy^2 - y^3$

(h) $(x + a)(x^2 - ax + a^2) = x(x^2 - ax + a^2) + a(x^2 - ax + a^2)$

    $= x^3 - ax^2 + a^2x + ax^2 - a^2x + a^3$

    $= x^3 + a^3$

(i) $(x - a)(x^2 + ax + a^2) = x(x^2 + ax + a^2) - a(x^2 + ax + a^2)$

    $= x^3 + ax^2 + a^2x - ax^2 - a^2x - a^3$

    $= x^3 - a^3$

---

2. Factorization formulas

If a polynomial is written as the product of two or more polynomials of lower degree, then each product polynomial is called a **factor** of the original polynomial. The process of factoring is an important operation in algebra because we often need simpler expressions to analyze. Using the special products found in Example 1-18, we obtain the following standard factorization formulas:

• **Factoring the difference of two squares:** $x^2 - y^2 = (x - y)(x + y)$

EXAMPLE 1-19   Factor (a) $4a^2 - 25$, (b) $64u^4 - 9v^2$, and (c) $(m + n)^2 - 16$.

Solution

(a) $4a^2 - 25 = (2a)^2 - (5)^2 = (2a - 5)(2a + 5)$

(b) $64u^4 - 9v^2 = (8u^2)^2 - (3v)^2 = (8u^2 - 3v)(8u^2 + 3v)$

(c) $(m + n)^2 - 16 = (m + n)^2 - 4^2 = [(m + n) - 4][(m + n) + 4]$

- **Factoring a perfect square:**
$$x^2 + 2xy + y^2 = (x + y)(x + y)$$
$$x^2 - 2xy + y^2 = (x - y)(x - y)$$

EXAMPLE 1-20   Factor (a) $x^2 + 6x + 9$, (b) $x^2 + 4xy + 4y^2$, and (c) $9x^2 - 30xy + 25y^2$.

Solution

(a) Identify 9 as the square of 3, notice that the multiplier of the middle term is twice 3, and apply the factorization formula:
$$x^2 + 6x + 9 = (x + 3)(x + 3)$$

(b) Similarly, identify $4y^2$ as the square of $2y$ and write
$$x^2 + 4xy + 4y^2 = (x)^2 + 2x(2y) + (2y)^2 = (x + 2y)(x + 2y)$$

(c) Again, identify the perfect squares first:
$$9x^2 - 30xy + 25y^2 = (3x)^2 - 30xy + (5y)^2$$
$$= (3x)^2 - 2(3x)(5y) + (5y)^2 = (3x - 5y)(3x - 5y)$$

- **Factoring by trial and error:**
$$x^2 + (a + b)x + ab = (x + a)(x + b)$$
$$x^2 - (a + b)x + ab = (x - a)(x - b)$$

In this pair, we identify the factors of the constant term as a and b and attempt to match their sum to the coefficient of the x term.

EXAMPLE 1-21   Factor (a) $x^2 - 7x + 6$, (b) $x^2 + 8x + 15$, and (c) $x^2 - x - 12$.

Solution

(a) In $x^2 - 7x + 6$, you identify $ab = 6$. So your choices for the pair a and b are -1 and -6, -2 and -3, 2 and 3, 1 and 6. Since the x term shows $a + b = -7$, you have only one choice: $a = -1$ and $b = -6$. Thus
$$x^2 - 7x + 6 = (x - 1)(x - 6)$$

(b) Here $ab = 15$ and $a + b = 8$. A quick check of possibilities shows that $3 + 5 = 8$ and $3 \cdot 5 = 15$, so $a = 3$ and $b = 5$. Thus
$$x^2 + 8x + 15 = (x + 3)(x + 5)$$

(c) Now $ab = -12$, so one of the values, a or b, must be negative; and their sum $a + b$ is -1. The pair you need is $a = -4$ and $b = 3$. Thus
$$x^2 - x - 12 = (x - 4)(x + 3)$$

- **Factoring the sum and difference of two cubes:**
$$x^3 + a^3 = (x + a)(x^2 - ax + a^2)$$
$$x^3 - a^3 = (x - a)(x^2 + ax + a^2)$$

EXAMPLE 1-22 Factor (a) $x^3 - 1$, (b) $x^3 + 8$, and (c) $x^6 - 3^6$.

Solution

(a) Use the formula with $a^3 = 1$, so that $a = 1$:

$$x^3 - 1 = (x - 1)(x^2 + x + 1)$$

(b) In this case $a^3 = 8$, so $a = 2$:

$$x^3 + 8 = (x + 2)(x^2 - 2x + 4)$$

(c) First rewrite the terms as cubes:

$$x^6 - 3^6 = (x^2)^3 - (3^2)^3$$

Then $a = 3^2$ and $x = x^2$, so

$$(x^2)^3 - (3^2)^3 = (x^2 - 3^2)[(x^2)^2 + 3^2(x^2) + (3^2)^2]$$
$$= (x^2 - 9)(x^4 + 9x^2 + 81)$$

And since $x^2 - 9$ is the difference of two squares,

$$(x^2 - 9)(x^4 + 9x^2 + 81) = (x - 3)(x + 3)(x^4 + 9x^2 + 81)$$

Although there are other factorization formulas that could be written, most are special cases of those listed above. (Table 1-2 provides a summary of the important special products and their factors.)

TABLE 1-2: Special Products	
Factors	Product
$(x - y)(x + y)$	$= x^2 - y^2$
$(x + y)^2$	$= x^2 + 2xy + y^2$
$(x - y)^2$	$= x^2 - 2xy + y^2$
$(x + a)(x + b)$	$= x^2 + (a + b)x + ab$
$(x - a)(x - b)$	$= x^2 - (a + b)x + ab$
$(x + y)^3$	$= x^3 + 3x^2y + 3xy^2 + y^3$
$(x - y)^3$	$= x^3 - 3x^2y + 3xy^2 - y^3$
$(x + a)(x^2 - ax + a^2) = x^3 + a^3$	
$(x - a)(x^2 + ax + a^2) = x^3 - a^3$	

3. Factoring guidelines

• If every term of the polynomial contains a **common factor**, pull that term out first and then attempt to reduce the remaining part.

EXAMPLE 1-23 Factor (a) $x^3 - 4x^2 + 3x$ and (b) $3x^3 + 5x^2$.

Solution

(a) The common factor is x, so

$$x^3 - 4x^2 + 3x = x(x^2 - 4x + 3)$$

But the trial-and-error technique shown in Example 1-21 provides further reduction since $x^2 - 4x + 3 = (x - 3)(x - 1)$. Thus

$$x^3 - 4x^2 + 3x = x(x - 3)(x - 1)$$

(b) Here every term contains x^2, so we factor that part out:

$$3x^3 + 5x^2 = x^2(3x + 5)$$

• If a common factor appears in each of a group of terms in the polynomial, pull this common factor out and rearrange the remaining part. This technique is called **factoring by grouping**.

EXAMPLE 1-24 Factor (a) $(2x + 1)x^2 - 4(2x + 1)$, (b) $(x^2 - 1) + 5(x - 1)$, and (c) $x^3 + x^2 - x - 1$ by the grouping method.

Solution

(a) The common factor in each group is $2x + 1$. Thus

$$(2x + 1)x^2 - 4(2x + 1) = (2x + 1)(x^2 - 4)$$

And since $(x^2 - 4)$ is factorable,

$$(2x + 1)(x^2 - 4) = (2x + 1)[(x - 2)(x + 2)]$$

(b) Note first that the term $x^2 - 1$ can be factored:

$$(x^2 - 1) + 5(x - 1) = (x - 1)(x + 1) + 5(x - 1)$$

Now the term $x - 1$ is common, so factor it out:

$$(x - 1)[(x + 1) + 5] = (x - 1)(x + 6)$$

(c) In this case, group the terms first:

$$x^3 + x^2 - x - 1 = (x^3 + x^2) - (x + 1)$$

Then remove the common x^2 from the first group:

$$x^2(x + 1) - (x + 1)$$

Now the term $(x + 1)$ is common to each group:

$$(x + 1)(x^2 - 1) = (x + 1)[(x - 1)(x + 1)]$$

SUMMARY

1. There are five major sets of numbers that make up the real number system:

Natural numbers: $N = \{1, 2, 3, \ldots\}$

Integers: $Z = \{\ldots, -2, -1, 0, 1, 2, \ldots\}$

Rational numbers: $Q = \{\frac{x}{y} \mid x \text{ and } y \text{ are integers and } y \neq 0\}$

$\left. \right\} N \subset Z \subset Q$

Irrational numbers: $I = \{x \mid x \text{ cannot be expressed as } \frac{p}{q}, \text{ where } p \text{ and } q \text{ are integers}\}$

Real numbers: $R = Q \cup I$

2. The real numbers R satisfy the axioms of equality, the field properties of addition and multiplication, and the properties of subtraction and division. These properties are the rules that allow the manipulation of numbers.

3. The order in which arithmetic operations are done is the PEMDAS order: (1) Terms within parentheses are evaluated; (2) Exponents are evaluated; (3) Multiplication and Division are performed (left-to-right); (4) Addition and Subtraction are performed (left-to-right).

4. A monomial cx^n consists of a constant coefficient c multiplying a variable x raised to a nonnegative integer power n.

5. A polynomial is the sum of a finite number of monomials. A polynomial may have several variables, or just one variable with the general form

$$a_n x^n + a_{n-1} x^{n-1} + a_{n-2} x^{n-2} + \cdots + a_2 x^2 + a_1 x + a_0$$

6. The degree of a polynomial in one variable is the highest value of the exponent that appears on the variable. The degree of a polynomial in more than one variable is the highest sum of the exponents on the variables in any one term.

7. The coefficients in polynomials are real numbers and are subject to all the rules of real number manipulation.

8. One polynomial $P(x)$ may be added to (or subtracted from) another polynomial $Q(x)$ by grouping like terms and adding (or subtracting) their coefficients:

$$
\begin{aligned}
P(x) &= a_n x^n && + a_{n-1} x^{n-1} && + \cdots + a_1 x && + a_0 \\
\pm \; Q(x) &= b_n x^n && + b_{n-1} x^{n-1} && + \cdots + b_1 x && + b_0
\end{aligned}
$$

Sum: $S(x) = (a_n + b_n)x^n + (a_{n-1} + b_{n-1})x^{n-1} + \cdots + (a_1 + b_1)x + (a_0 + b_0)$

Difference: $T(x) = (a_n - b_n)x^n + (a_{n-1} - b_{n-1})x^{n-1} + \cdots + (a_1 - b_1)x + (a_0 - b_0)$

9. A polynomial may be multiplied by another polynomial by using the distributive law and the laws of exponents.

10. Factoring is the reverse of multiplication.

RAISE YOUR GRADES

Can you . . . ?

- ☑ explain two different set notations
- ☑ explain the difference between a rational and an irrational number
- ☑ write all of the axioms and properties of real numbers
- ☑ carry out arithmetic operations in the right order
- ☑ define a monomial and a polynomial
- ☑ find the degree of any polynomial
- ☑ add (subtract) one polynomial to (from) another by grouping like terms
- ☑ multiply any polynomial by another and explain the FOIL method of multiplication
- ☑ factor the difference of two squares and a perfect square
- ☑ factor by trial-and-error
- ☑ write the formulas for factoring the sum and difference of two cubes
- ☑ explain factoring by a common factor and by grouping
- ☑ write the formulas for factoring a perfect cube

SOLVED PROBLEMS

Sets and the Real Number System

PROBLEM 1-1   Describe the set S whose elements are odd positive integers with one digit. Write the set in list notation and set-builder notation.

Solution   There are five elements (1, 3, 5, 7, and 9) in the set. Write it as

$$\text{List:} \quad S = \{1, 3, 5, 7, 9\}$$

$$\text{Set-builder:} \quad S = \{x \mid x \text{ is an odd positive integer} < 10\}$$

PROBLEM 1-2   List all the elements described by the set-builder notation $\{x \mid x$ is a natural number $< 6\}$. Write the set in list notation.

Solution   The elements satisfying the property "a natural number less than 6" are 1, 2, 3, 4, and 5, so the set may be written as $S = \{1, 2, 3, 4, 5\}$.

PROBLEM 1-3   Write the set that details the outcome of **(a)** tossing a fair coin; and **(b)** rolling an honest die.

Solution

(a) There are exactly two possible outcomes of the toss, so $T = \{\text{Head, Tail}\}$.
(b) There are six possible outcomes of the roll, so $R = \{1, 2, 3, 4, 5, 6\}$.

PROBLEM 1-4   List the subsets of natural numbers, integers, rational numbers, irrational numbers, and real numbers in the set

$$U = \{-\tfrac{1}{2}, 4\sqrt{2}, 3.2, 3\pi, -2, \tfrac{5}{4}, 8, 117.38, 0, .333, 1, \sqrt{5}\}$$

Solution The subset of natural numbers is $\{1, 8\}$. The subset of integers is $\{-2, 0, 1, 8\}$. The subset of rational numbers is $\{-2, -\frac{1}{2}, 0, .333, 1, \frac{5}{4}, 3.2, 8, 117.38\}$. The subset of irrational numbers is $\{\sqrt{5}, 4\sqrt{2}, 3\pi\}$. All of the elements of this set are real numbers.

Properties of Real Numbers

PROBLEM 1-5 Identify the field property illustrated in each of the following statements:

(a) $4 + 5 = 5 + 4$

(b) $7(3 + 8) = 7 \cdot 3 + 7 \cdot 8$

(c) $(-2) + (2) = 0$

(d) $1 \cdot 3 = 3$

(e) $5 \cdot \frac{1}{5} = 1$

(f) $(6 + 2) + 3 = 6 + (2 + 3)$

Solution

(a) Commutative property for addition
(b) Distributive property
(c) Inverse property for addition

(d) Identity property for multiplication
(e) Inverse property for multiplication
(f) Associative property for addition

Mixed Operations on Real Numbers

PROBLEM 1-6 Perform the following operations, showing how the rules for performing operations on real numbers are used.

(a) $\frac{16}{36} \cdot \frac{85}{20}$

(b) $(\frac{1}{2} + \frac{5}{4}) \div (\frac{2}{3} - \frac{1}{6})$

(c) $(\frac{5}{4} - \frac{1}{10}) \cdot (\frac{1}{3} + \frac{1}{6}) \div (\frac{7}{8} - \frac{1}{4})$

Solution

(a) Reduce each fraction first, canceling wherever possible:

$$\frac{16}{36} \cdot \frac{85}{20} = \frac{4 \cdot \cancel{4}}{9 \cdot \cancel{4}} \cdot \frac{17 \cdot \cancel{5}}{4 \cdot \cancel{5}} = \frac{4}{9} \cdot \frac{17}{4} = \frac{17 \cdot \cancel{4}}{9 \cdot \cancel{4}} = \frac{17}{9}$$

(b) Work inside each set of parentheses first, adding with a different denominator:

$$\left(\frac{1}{2} + \frac{5}{4}\right) \div \left(\frac{2}{3} - \frac{1}{6}\right) = \left(\frac{1 \cdot 4 + 2 \cdot 5}{2 \cdot 4}\right) \div \left(\frac{2 \cdot 6 - 3 \cdot 1}{3 \cdot 6}\right)$$

$$= \frac{4 + 10}{8} \div \frac{12 - 3}{18} = \frac{14}{8} \div \frac{9}{18}$$

Now, since $a/b \div c/d = a/b \cdot d/c$,

$$\frac{14}{8} \div \frac{9}{18} = \frac{14}{8} \cdot \frac{18}{9} = \frac{7 \cdot \cancel{2}}{4 \cdot \cancel{2}} \cdot \frac{2 \cdot \cancel{9}}{1 \cdot \cancel{9}} = \frac{7}{4} \cdot \frac{2}{1} = \frac{7 \cdot \cancel{2}}{2 \cdot \cancel{2}} = \frac{7}{2}$$

(c) Again, perform the operations inside the parentheses first. So, for each group

$$\left(\frac{5}{4} - \frac{1}{10}\right) = \frac{5 \cdot 10 - 4 \cdot 1}{4 \cdot 10} = \frac{50 - 4}{40} = \frac{46}{40} = \frac{23 \cdot \cancel{2}}{20 \cdot \cancel{2}} = \frac{23}{20}$$

$$\left(\frac{1}{3} + \frac{1}{6}\right) = \frac{2 \cdot 1}{2 \cdot 3} + \frac{1}{6} = \frac{2}{6} + \frac{1}{6} = \frac{3}{6} = \frac{1 \cdot \cancel{3}}{2 \cdot \cancel{3}} = \frac{1}{2}$$

$$\left(\frac{7}{8} - \frac{1}{4}\right) = \frac{7}{8} - \frac{2 \cdot 1}{2 \cdot 4} = \frac{7}{8} - \frac{2}{8} = \frac{5}{8}$$

Thus,

$$\left(\frac{5}{4} - \frac{1}{10}\right) \cdot \left(\frac{1}{3} + \frac{1}{6}\right) \div \left(\frac{7}{8} - \frac{1}{4}\right) = \frac{23}{20} \cdot \frac{1}{2} \div \frac{5}{8}$$

$$= \frac{23 \cdot 1}{20 \cdot 2} \div \frac{5}{8} = \frac{23}{40} \cdot \frac{8}{5} = \frac{23}{5 \cdot \cancel{8}} \cdot \frac{\cancel{8}}{5}$$

$$= \frac{23}{25}$$

PROBLEM 1-7 Evaluate

(a) $3 - 4 \div 2 + 5 \cdot 8 - 3$

(b) $3 - 4 \div (2 + 5) \cdot (8 - 3)$

(c) $[(3 - 4) \div 2] + 5 \cdot 8 - 3$

(d) $3 - 4 \div [2 + (5 \cdot 8)] - 3$

(e) $3 - (4^2 \div 2) + (5^2 \cdot 8 - 3)$

Solution Perform the operations left-to-right, using the PEMDAS order rules:

(a) $3 - \underset{(1)}{4 \div 2} + \underset{(2)}{5 \cdot 8} - 3 = 3 - \underset{(3)}{2 + 40} - 3$

(1) Divide and (2) multiply, left-to-right; (3) subtract; (4) add and subtract, left-to-right.

$$= \underset{(4)}{1 + 40 - 3}$$

$$= 38$$

(b) $3 - 4 \div \underset{(1)}{(2 + 5)} \cdot \underset{(1)}{(8 - 3)} = 3 - \underset{(2)}{4 \div 7} \cdot 5$

(1) Evaluate parentheses first; (2) divide and (3) multiply; (4) subtract.

$$= 3 - \underset{(3)}{\tfrac{4}{7} \cdot 5} = 3 - \tfrac{20}{7}$$

$$= \underset{(4)}{\tfrac{21}{7} - \tfrac{20}{7}} = \tfrac{1}{7}$$

(c) $[\underset{(1)}{(3 - 4)} \div 2] + 5 \cdot 8 - 3 = \underset{(2)}{(-1 \div 2)} + \underset{(3)}{5 \cdot 8} - 3$

(1) Evaluate inner parentheses; (2) evaluate outer grouping symbols; (3) multiply; (4) add and subtract.

$$= \underset{(4)}{-\tfrac{1}{2} + 40 - 3}$$

$$= 36\tfrac{1}{2}$$

(d) $3 - 4 \div [2 + (5 \cdot 8)] - 3 = 3 - 4 \div (2 + 40) - 3$

$$= 3 - 4 \div 42 - 3$$

$$= 3 - \tfrac{4}{42} - 3 = -\tfrac{4}{42} = -\tfrac{2}{21}$$

(e) $3 - (4^2 \div 2) + (5^2 \cdot 8 - 3) = 3 - (16 \div 2) + (25 \cdot 8 - 3)$

$$= 3 - 8 + (200 - 3)$$

$$= 3 - 8 + 197$$

$$= 192$$

PROBLEM 1-8 Simplify these algebraic expressions:

(a) $5(y + 3) - 7y + 2(-1 + 2y)$ (b) $-4 + 2(5 - 2x) + 6(x - 2)$

(c) $[3 - (z - 2)] - [-4z - 2(3 + z)]$

Solution

(a) Multiply out the parentheses and combine:

$$5(y + 3) - 7y + 2(-1 + 2y) = 5y + 15 - 7y - 2 + 4y$$

$$= (5y - 7y + 4y) + 15 - 2$$

$$= 2y + 13$$

(b)
$$-4 + 2(5 - 2x) + 6(x - 2) = -4 + 10 - 4x + 6x - 12$$
$$= (-4 + 10 - 12) + (-4x + 6x)$$
$$= -6 + 2x$$

(c) Work inside the grouping symbols (parentheses and square brackets) first; then combine:

$$[3 - (z - 2)] - [-4z - 2(3 + z)] = (3 - z + 2) - (-4z - 6 - 2z)$$
$$= (5 - z) - (-6z - 6)$$
$$= 5 - z + 6z + 6$$
$$= (5 + 6) + (-z + 6z)$$
$$= 11 + 5z$$

Polynomials

PROBLEM 1-9   Which of the following are *not* polynomials? Why not?

(a) $\sqrt{2}x^2 + \frac{1}{2}x$

(b) $y^{-2} + 3y^{-1} + 4$

(c) $\frac{1}{2}x^2y - \frac{1}{2}xy^2$

(d) $(5x^2y - 4y) + (3z - 2yz^3)$

(e) $2\sqrt{x} - 3\sqrt{y} + 8\sqrt{z}$

(f) $4x^3 + 9x - \dfrac{3}{x^2}$

Solution   Expressions **(b)**, **(e)**, and **(f)** are not polynomials: in **(b)** the exponents on the variable y are negative; in **(e)** the variables x, y, and z are all radicals (i.e., their exponents are not integers); and in **(f)** the variable x is in the denominator (i.e., its exponent is negative).

PROBLEM 1-10   What is the degree of each of the following polynomials?

(a) $5 + 3z + 6z^2 - 8z^3$

(b) $(4 - 3x^2)^2 + 6x - 11$

(c) $\sqrt{5}y^2 - \sqrt{7}y^3 + \sqrt{2}y - \sqrt{17}$

(d) $8x^2y^3 - 2x^4y^2$

(e) $\sqrt{\pi}z^4 - \frac{19}{2}z^2 - 92$

(f) $(2xy^2)^3 + 5x^4y - 82$

Solution

(a) The highest power of the single variable z is 3. The degree is 3.
(b) Square the quantity, so that the highest power on the single variable x is 4. The degree is 4.
(c) Don't inspect the coefficients—only the exponents on the single variable y. The degree is 3.
(d) The highest sum of the exponents on the *two variables* x and y is 6. The degree is 6.
(e) The degree is 4.
(f) Expanding the term $(2xy^2)^3$, you get $8x^3y^6$, the sum of whose exponents is 9. The degree is 9.

Operations on Polynomials

PROBLEM 1-11   Find the sum and difference of $P(x) = 2 + x - 6x^2$ and $Q(x) = 7x^2 - 2x - 3$.

Solution

$$P(x) + Q(x) = (2 + x - 6x^2) + (7x^2 - 2x - 3)$$
$$= 2 + x - 6x^2 + 7x^2 - 2x - 3$$
$$= (-6x^2 + 7x^2) + (x - 2x) + (2 - 3) \qquad \text{[Combine like terms]}$$
$$= x^2 - x - 1 \qquad \text{[Add]}$$

$$P(x) - Q(x) = (2 + x - 6x^2) - (7x^2 - 2x - 3)$$
$$= 2 + x - 6x^2 - 7x^2 + 2x + 3$$
$$= (-6x^2 - 7x^2) + (x + 2x) + (2 + 3) \qquad \text{[Combine like terms]}$$
$$= -13x^2 + 3x + 5 \qquad \text{[Subtract]}$$

PROBLEM 1-12 Perform the indicated operations:

(a) $(\frac{2}{3}x^3 - \frac{1}{2}x^2 + 3x - \frac{1}{4}) - (\frac{3}{4}x^2 + 4x + \frac{3}{4})$

(b) $(4x^2 - 1) + 2(x^3 - x + 3) - (2x^3 - x + 3)$

(c) $(3z + 1)^2 - 9z^2 - 6z$

(d) $(\frac{1}{2}a^2b - \frac{3}{4}ab^2 + \frac{1}{8}a^2b^2) - (\frac{1}{4}a^2b + ab^2 + \frac{3}{8}a^2b^2)$

Solution

(a) Remove the parentheses, reorganize the terms, and combine:

$$\frac{2}{3}x^3 - \frac{1}{2}x^2 + 3x - \frac{1}{4} - \frac{3}{4}x^2 - 4x - \frac{3}{4} = \frac{2}{3}x^3 - \frac{1}{2}x^2 - \frac{3}{4}x^2 + 3x - 4x - \frac{1}{4} - \frac{3}{4}$$
$$= \frac{2}{3}x^3 - \frac{5}{4}x^2 - x - 1$$

(b) Multiply the coefficient, remove the parentheses, and combine:

$$4x^2 - 1 + 2x^3 - 2x + 6 - 2x^3 + x - 3 = 2x^3 - 2x^3 + 4x^2 - 2x + x - 1 + 6 - 3$$
$$= 4x^2 - x + 2$$

(c) Square the first term (use the FOIL method or the special product) and then combine:

$$9z^2 + 6z + 1 - 9z^2 - 6z = 9z^2 - 9z^2 + 6z - 6z + 1 = 1$$

(d) Remove the parentheses, reorganize the terms, and combine:

$$\frac{1}{2}a^2b - \frac{1}{4}a^2b - \frac{3}{4}ab^2 - ab^2 + \frac{1}{8}a^2b^2 - \frac{3}{8}a^2b^2$$
$$= (\frac{1}{2} - \frac{1}{4})a^2b + (-\frac{3}{4} - 1)ab^2 + (\frac{1}{8} - \frac{3}{8})a^2b^2$$
$$= \frac{1}{4}a^2b - \frac{7}{4}ab^2 - \frac{1}{4}a^2b^2$$

PROBLEM 1-13 Find the product of $S(x) = x^2 + 1$ and $T(x) = x - 6$.

Solution $S(x) \cdot T(x) = (x^2 + 1)(x - 6)$
$$= x^2(x - 6) + 1(x - 6)$$
$$= x^3 - 6x^2 + x - 6 \qquad \text{(The FOIL method could also be used here.)}$$

PROBLEM 1-14 Find the product of $S(y) = y^2 + 3y$ and $T(y) = y^3 - 5y + 5$.

Solution $S(y) \cdot T(y) = (y^2 + 3y)(y^3 - 5y + 5)$
$$= y^2(y^3 - 5y + 5) + 3y(y^3 - 5y + 5)$$
$$= (y^5 - 5y^3 + 5y^2) + (3y^4 - 15y^2 + 15y)$$
$$= y^5 + 3y^4 - 5y^3 - 10y^2 + 15y$$

PROBLEM 1-15 Find the product $(x^2 - 3y^2)(x + 5y^3)$.

Solution $(x^2 - 3y^2)(x + 5y^3) = x^2(x + 5y^3) - 3y^2(x + 5y^3)$
$$= (x^3 + 5x^2y^3) - (3xy^2 + 15y^5)$$
$$= x^3 + 5x^2y^3 - 3xy^2 - 15y^5 \qquad \text{(Also use FOIL.)}$$

PROBLEM 1-16 Find the product of $P(x) = x^2 - 7x + 12$ and $Q(x) = 2x^2 + 5x - 8$.

Solution Use the same process (the distributive law) you used in the preceding problems, but write the terms in the vertical "number multiplication form":

$$
\begin{array}{r}
x^2 - 7x + 12 \\
\times \quad 2x^2 + 5x - 8 \\
\hline
-8x^2 + 56x - 96 \\
5x^3 - 35x^2 + 60x \\
2x^4 - 14x^3 + 24x^2 \\
\hline
2x^4 - 9x^3 - 19x^2 + 116x - 96
\end{array}
$$

(-8 times each top term)
($5x$ times each top term)
($2x^2$ times each top term)
(Add to get the product polynomial.)

PROBLEM 1-17 Find the product $(3u - 5v)(3u + 5v)$.

Solution
$$(3u - 5v)(3u + 5v) = 3u(3u + 5v) - 5v(3u + 5v)$$
$$= 9u^2 + 15uv - 15uv - 25v^2$$
$$= 9u^2 - 25v^2$$

Notice that this product is the difference of two squares, $x^2 - y^2 = (x - y)(x + y)$, where $x = 3u$ and $y = 5v$.

PROBLEM 1-18 Find the expanded form of $(3s^2 + 4t^2)^2$.

Solution
$$(3s^2 + 4t^2)(3s^2 + 4t^2) = 3s^2(3s^2 + 4t^2) + 4t^2(3s^2 + 4t^2)$$
$$= 9s^4 + 12s^2t^2 + 12s^2t^2 + 16t^4$$
$$= 9s^4 + 24s^2t^2 + 16t^4$$

Notice that this product is a perfect square, $x^2 + 2xy + y^2 = (x + y)^2$, where $x = 3s^2$ and $y = 4t^2$.

PROBLEM 1-19 Find the product $(a - 2b + 3c)(3a + 2b - c)$.

Solution Use the "number multiplication form":

$$
\begin{array}{r}
a - 2b + 3c \\
\times \quad 3a + 2b - c \\
\hline
- ac + 2bc - 3c^2 \\
+ 2ab - 4b^2 \qquad + 6bc \\
3a^2 - 6ab \qquad + 9ac \\
\hline
3a^2 - 4ab - 4b^2 + 8ac + 8bc - 3c^2
\end{array}
$$

PROBLEM 1-20 Find the expanded form of $(g + 3h)^3$.

Solution
$$(g + 3h)^3 = (g + 3h)^2 \cdot (g + 3h)$$
$$= (g^2 + 6gh + 9h^2)(g + 3h)$$
$$= (g^2 + 6gh + 9h^2)g + (g^2 + 6gh + 9h^2)3h$$
$$= g^3 + 6g^2h + 9gh^2 + 3g^2h + 18gh^2 + 27h^3$$
$$= g^3 + 9g^2h + 27gh^2 + 27h^3$$

Notice that this product is a perfect cube $(x + y)^3 = x^3 + 3x^2y + 3xy^2 + y^3$, where $x = g$ and $y = 3h$.

PROBLEM 1-21 Use the FOIL method to find the following products:

(a) $(2x + 1)(4x - 3)$ (b) $(s + 2t)(5s + 8t)$ (c) $(8a^2b - ab^2)^2$

Solution

(a) $(2x + 1)(4x - 3) = 8x^2 - 6x + 4x - 3$

    F   O   I   L

$$= 8x^2 - 2x - 3$$

(b) $(s + 2t)(5s + 8t) = 5s^2 + 8st + 10st + 16t^2$

    F   O   I   L

$$= 5s^2 + 18st + 16t^2$$

(c) $(8a^2b - ab^2)^2 = (8a^2b - ab^2)(8a^2b - ab^2)$

$$= \underbrace{(8a^2b)(8a^2b)}_{F} - \underbrace{(8a^2b)(ab^2)}_{O} - \underbrace{(ab^2)(8a^2b)}_{I} + \underbrace{(ab^2)(ab^2)}_{L}$$

$$= 64a^4b^2 - 8a^3b^3 - 8a^3b^3 + a^2b^4$$

$$= 64a^4b^2 - 16a^3b^3 + a^2b^4$$

PROBLEM 1-22   Find the factors of $x^2 - 25$.

Solution   This is the difference of two squares. Thus $x^2 - 25 = x^2 - 5^2 = (x - 5)(x + 5)$.

PROBLEM 1-23   Find the factors of $x^2 + 12x + 36$.

Solution   The first term x^2 is a square, the trailing number 36 is the perfect square of 6, and the coefficient of x is twice 6, so you know that this is a case of the perfect square: $x^2 + 12x + 36 = (x + 6)^2 = (x + 6)(x + 6)$.

PROBLEM 1-24   Factor $x^2 - 12x + 2$.

Solution   The factors of $x^2 - 12x + 2$ would have to take the form $(x - a)(x - b) = x^2 - (a + b)x + ab$, so that $ab = 2$ and $a + b = -12$. But this is not possible with rational numbers, so the polynomial is *not factorable*. A nonfactorable polynomial is called a **prime polynomial**.

note: Actually, we're lying just a little here. What we SHOULD say is "This polynomial is not factorable *over the field of rational numbers*." Then we have to say that "a nonfactorable polynomial over a given field of numbers is called a *prime polynomial over that field*." In other words, only if we confine our choices [**field**] of factors to rational numbers can we say that $x^2 - 12x + 2$ is a prime polynomial.

PROBLEM 1-25   Factor $r^4 - 10r^3 - 24r^2$.

Solution   Each term contains r^2 so factor that out first:

$$r^4 - 10r^3 - 24r^2 = r^2(r^2 - 10r - 24)$$

To factor further, you need to find the values of a and b so that $ab = -24$ and $a + b = -10$. After sorting through the possible values, you find $a = -12$ and $b = 2$. Thus,

$$r^4 - 10r^3 - 24r^2 = r^2(r^2 - 10r - 24) = r^2(r - 12)(r + 2)$$

PROBLEM 1-26   Factor $x^2(x + 3) - 8x(x + 3) + 6(x + 3)$.

Solution   There is an obvious $(x + 3)$ group in each term, so factor out that group, following the grouping technique:

$$x^2(x + 3) - 8x(x + 3) + 6(x + 3) = (x + 3)(x^2 - 8x + 6)$$

But to factor further, you need a pair a and b so that $ab = 6$ and $a + b = -8$. There are no integers or rational numbers for which this is true, so the factorization is complete.

PROBLEM 1-27   Factor $2x^3 - x^2 + 2x - 1$.

Solution   Notice the pattern of the coefficients: $2, -1, 2, -1$. This pattern is a clue to try for a factoring by grouping—where the group is $(2x - 1)$:

$$2x^3 - x^2 + 2x - 1 = x^2(2x - 1) + 1(2x - 1) = (2x - 1)(x^2 + 1)$$

Since $x^2 + 1$ is not factorable (there is no real number whose square is -1), the factorization is complete.

PROBLEM 1-28   Use the factorization formulas for the sum and difference of two cubes to factor

(a) $8x^3 + 1$ **(b)** $x^3 - 8y^3$ **(c)** $27a^3 + 8b^3c^6$

Solution   The formulas for the sum and difference of two cubes are $x^3 + a^3 = (x + a)(x^2 - ax + a^2)$ and $x^3 - a^3 = (x - a)(x^2 + ax + a^2)$, respectively.

(a)   In $8x^3 + 1$ the first term is a perfect cube of $2x$, so

$$8x^3 + 1 = (2x)^3 + 1^3 = (2x + 1)[(2x)^2 - 1(2x) + 1^2]$$
$$= (2x + 1)(4x^2 - 2x + 1)$$

(b) In $x^3 - 8y^3$, both terms are perfect cubes, so

$$x^3 - 8y^3 = (x)^3 - (2y)^3 = (x - 2y)[x^2 + x(2y) + (2y)^2]$$
$$= (x - 2y)(x^2 + 2xy + 4y^2)$$

(c) Again, rewrite each term as a perfect cube and follow the formula for factoring the sum of two cubes:

$$27a^3 + 8b^3c^6 = (3a)^3 + (2bc^2)^3$$
$$= (3a + 2bc^2)[(3a)^2 - (3a)(2bc^2) + (2bc^2)^2]$$
$$= (3a + 2bc^2)(9a^2 - 6abc^2 + 4b^2c^4)$$

PROBLEM 1-29 Use the factorization formulas for a perfect cube to factor **(a)** $x^9 - y^9$, **(b)** $a^3b^3 - 1000$, **(c)** $x^4 + 8x$, and **(d)** $27x^6 - 1$.

Solution Rewrite each term as a cube; then use the factorization formula.

(a)
$$x^9 - y^9 = (x^3)^3 - (y^3)^3 = (x^3 - y^3)[(x^3)^2 + (x^3)(y^3) + (y^3)^2]$$
$$= (x^3 - y^3)(x^6 + x^3y^3 + y^6)$$
$$= (x - y)(x^2 + xy + y^2)(x^6 + x^3y^3 + y^6)$$

(b)
$$a^3b^3 - 1000 = (ab)^3 - (10)^3$$
$$= (ab - 10)[(ab)^2 + (ab)(10) + (10)^2]$$
$$= (ab - 10)(a^2b^2 + 10ab + 100)$$

(c) First factor out the common x, then use the "cubes" formula:

$$x^4 + 8x = x(x^3 + 8)$$
$$= x[(x)^3 + (2)^3]$$
$$= x(x + 2)(x^2 - 2x + 4)$$

(d)
$$27x^6 - 1 = (3x^2)^3 - (1)^3$$
$$= (3x^2 - 1)[(3x^2)^2 + (3x^2)(1) + (1)^2]$$
$$= (3x^2 - 1)(9x^4 + 3x^2 + 1)$$

PROBLEM 1-30 Factor the following expressions:

(a) $x^2 - x - 132$

(b) $x^3 + 3x^2 - x - 3$

(c) $x^2 + 2x - 35$

(d) $25 - 9x^4$

(e) $8 - 27t^6$

(f) $4x^2 + 12x + 9$

(g) $a^2 - 4ab + 4b^2$

(h) $u^3 - 3u^2v + 3uv^2 - v^3$

(i) $x^3 + 6x + 7 = (x^3 + 1) + 6(x + 1)$

Solution

(a) Use the "trial and error" process. You need $a + b = -1$ and $ab = -132$. These values are $a = -12$ and $b = 11$, so the factors are

$$x^2 - x - 132 = (x - 12)(x + 11)$$

(b) Notice the coefficient pattern: $1, 3, -1, -3$. This suggests grouping:

$$x^3 + 3x^2 - x - 3 = x^2(x + 3) - 1(x + 3)$$
$$= (x + 3)(x^2 - 1)$$
$$= (x + 3)(x - 1)(x + 1)$$

(c) You have $a + b = 2$ and $ab = -35$. The numbers are $a = 7$ and $b = -5$, so the factors are

$$x^2 + 2x - 35 = (x + 7)(x - 5)$$

(d) Treat this as the difference of two squares:

$$25 - 9x^4 = (5)^2 - (3x^2)^2 = (5 - 3x^2)(5 + 3x^2)$$
$$= [(\sqrt{5})^2 - (\sqrt{3}x)^2](5 + 3x^2)$$
$$= (\sqrt{5} - \sqrt{3}x)(\sqrt{5} + \sqrt{3}x)(5 + 3x^2)$$

note: Note that the factorization of $5 - 3x^2$ is done over the field of real numbers. Treat 5 as $(\sqrt{5})^2$ and $3x^2$ as $(\sqrt{3}x)^2$. You now have the difference of two squares. Similarly,

$$2 - 3t^2 = (\sqrt{2})^2 - (\sqrt{3}t)^2$$
$$= (\sqrt{2} - \sqrt{3}t)(\sqrt{2} + \sqrt{3}t)$$

is factored over the field of real numbers. If you were restricted to the field of rational numbers these terms would not be factorable.

(e) Treat this as the difference of two cubes:

$$8 - 27t^6 = (2)^3 - (3t^2)^3$$
$$= (2 - 3t^2)[(2)^2 + (2)(3t^2) + (3t^2)^2]$$
$$= (2 - 3t^2)(4 + 6t^2 + 9t^4)$$
$$= (\sqrt{2} - \sqrt{3}t)(\sqrt{2} + \sqrt{3}t)(4 + 6t^2 + 9t^4)$$

(f) In this expression, the first term is the square of $2x$ and the last term is the square of 3. In addition, the middle term is twice the product of these two. Thus you have

$$4x^2 + 12x + 9 = (2x + 3)(2x + 3) = (2x + 3)^2$$

(g) In this expression, the first term is the square of a and the last term is the square of $2b$. In addition, the middle term is the negative of twice the product of these two. Thus you have

$$a^2 - 4ab + 4b^2 = (a - 2b)(a - 2b) = (a - 2b)^2$$

(h) Maybe you immediately recognize this as the expansion of $(u - v)^3$. But suppose not. Then factor by grouping,

$$u^3 - 3u^2v + 3uv^2 - v^3 = (u^3 - v^3) + (-3u^2v + 3uv^2)$$
$$= (u - v)(u^2 + uv + v^2) - 3uv(u - v)$$
$$= (u - v)(u^2 + uv + v^2 - 3uv)$$
$$= (u - v)(u^2 - 2uv + v^2)$$
$$= (u - v)[(u - v)^2] = (u - v)^3$$

(i) Until you observe the breakdown of the cubic polynomial $x^3 + 6x + 7$ into the form $(x^3 + 1) + 6(x + 1)$ this is a hard problem. But go with the given fact: The first grouping is the sum of two cubes and may be factored:

$$(x^3 + 1) + 6(x + 1) = (x + 1)(x^2 + x + 1) + 6(x + 1)$$
$$= (x + 1)[(x^2 + x + 1) + 6]$$
$$= (x + 1)(x^2 + x + 7)$$

Supplementary Exercises

PROBLEM 1-31 Which of the following sets are equal?

$$A = \{2, 4\}, \quad B = \{t, w, o, f, u, r\}, \quad C = \{4, 2, 0\}, \quad D = \{f, o, r, t, u, w\}$$
$$E = \{2, 0, 4, 0, 2\}, \quad F = \{two, four\}, \quad G = \{r, o, w, t, u, f\}$$

PROBLEM 1-32 Describe each of the following sets in set-builder notation and in list form:

(a) C = colors on a traffic light

(b) V = number of continents you could visit on a world tour. [Assume you could visit *all* the continents.]

(c) A = number of aces you could have in a bridge hand

(d) L = letters between K and T in the alphabet

(e) E = odd natural numbers smaller than 100

PROBLEM 1-33 Simplify the following arithmetic expressions:

(a) $\dfrac{8}{20} + \dfrac{15}{5} - \dfrac{4}{10}$

(b) $\dfrac{6}{15} \div \left(\dfrac{2}{3} - \dfrac{1}{6}\right)$

(c) $\left(\dfrac{-1}{2} + \dfrac{1}{-2}\right) + \left[\dfrac{-2}{3} + 4\left(\dfrac{-2}{-3}\right)\right]$

(d) $\left(\dfrac{5}{12} + \dfrac{1}{16}\right) \div \left(\dfrac{3}{8} - \dfrac{1}{4}\right)$

PROBLEM 1-34 Write $\dfrac{a}{a+b} + \dfrac{b}{a-b}$ as a single fraction.

PROBLEM 1-35 Which of the following is (are) true?

(a) $\dfrac{1}{a} + \dfrac{1}{b} \overset{?}{=} \dfrac{1}{a+b}$

(b) $(a+b)^2 \overset{?}{=} a^2 + b^2$

(c) $\dfrac{1}{(a/b)} \overset{?}{=} \dfrac{b}{a}$

PROBLEM 1-36 Which of the following is (are) true?

(a) $\dfrac{a}{b+c} \overset{?}{=} \dfrac{a}{b} + \dfrac{a}{c}$

(b) $\sqrt{a+b} \overset{?}{=} \sqrt{a} + \sqrt{b}$

(c) $a \overset{?}{=} -a$ if and only if $a = 0$

PROBLEM 1-37 Use the PEMDAS order of operations to evaluate

(a) $5 + 2 \div 5 + 3 - 4 \cdot 6 \div 3 - 1$

(b) $(5 + 2) \div 5 + (3 - 4) \cdot 6 \div 3 - 1$

(c) $5 + (2 \div 5) + 3 - 4 \cdot (6 \div 3) - 1$

(d) $5 + 2 \div [5 + (3 - 4)] \cdot 6 \div (3 - 1)$

(e) $5 + [2 \div (5 + 3)] - 4 \cdot 6 \div 3 - 1$

PROBLEM 1-38 Use as many rules and operations as you need to simplify this "monster."

$$\left(\dfrac{1}{8} \div \dfrac{1}{4}\right) + (\sqrt{2})^{-2} - \left[\left(-\dfrac{3}{8} \div \dfrac{\sqrt{36}}{(-2)^3}\right) + \left(1 + \dfrac{1}{10}\right) \cdot \dfrac{21}{55} \cdot \left(\dfrac{1}{7} + 2\right) \cdot \left(5 \div \left(\dfrac{1}{3}\right)^{-1}\right)\right]$$

PROBLEM 1-39 Which field property is illustrated in these statements?

(a) $3 \cdot 4 = 4 \cdot 3$

(b) $\frac{1}{2} + \frac{3}{8}$ is a real number

(c) $6 \cdot 1 = 1 \cdot 6 = 6$

(d) $(-1 + 4) \cdot 3 = -1 \cdot 3 + 4 \cdot 3$

(e) $-37 + 37 = 0$

PROBLEM 1-40 Simplify the following algebraic expressions:

(a) $(\frac{2}{3}x + \frac{1}{5}x) + (\frac{1}{8}y - \frac{3}{4}y) + x - 6y$

(b) $\frac{1}{3}z - \frac{1}{2}(z + 2) + 4(\frac{1}{2}z - 2) + 9$

(c) $(1 + x)^2 - (1 - x)^2 + 2(1 + x) - 2(1 - x)$

(d) $\frac{1}{2}a^2 + \frac{2}{3}b^2 - \frac{3}{4}a^2 - \frac{4}{5}b^2 + \frac{5}{6}a^2 - \frac{7}{8}b^2$

PROBLEM 1-41 What is the value of each polynomial if $x = 1$?

(a) $x^3 - 6x^2 + 7x + 2$

(b) $\frac{1}{2}x^4 + \frac{3}{8}x^3 - \frac{1}{4}x^2 - \frac{3}{8}x + 1$

(c) $(2x^2 + x - 5) - (x^2 + 5x - 8)$

PROBLEM 1-42   Find the sum and difference of each polynomial pair:

(a) $P(x) = x^2 - 4x + 2$, $Q(x) = 3x^2 + 5x - 1$

(b) $S(y) = \frac{1}{2}y^2 + 8y - 3$, $T(y) = \frac{1}{8}y^2 + \frac{3}{4}y - 7$

(c) $F(x) = 7x^4 - 9x^3 + 2x - 8$, $G(x) = 3x^4 + x^3 - 8x^2 + 1$

(d) $S(t) = 16t^2 + \frac{3}{4}t - \frac{9}{8}$, $V(t) = \frac{5}{8}t + 5$

PROBLEM 1-43   Find the product of each polynomial pair:

(a) $F(x) = x^2 + 4$, $G(x) = x - 6$

(b) $P(x) = x^2 + 3x - 5$, $Q(x) = x^2 + x + 1$

(c) $M = x^2 + xy$, $N = xy + y^2$

PROBLEM 1-44   Use the FOIL method to find the products:

(a) $(2x + 1)(x - 3)$

(b) $(x + 4)(5x - 1)$

(c) $(x + 3y)(2x - y)$

(d) $(7x - 4y)(2x + 5y)$

PROBLEM 1-45   Find each special product:

(a) $(x + 2y)^3$

(b) $(x + 3)(2x + 4)(3x + 5)$

(c) $(2a - 3b)^3$

(d) $(r^2 + 3s)^3$

(e) $\left(\frac{1}{u} - \frac{1}{v}\right)^2$

PROBLEM 1-46   Factor the following polynomials:

(a) $4x^2 - 9$

(b) $x^2 - 7x + 12$

(c) $x^2 + 8x + 16$

(d) $(2a + b)^2 - 25$

PROBLEM 1-47   Factor the following polynomials:

(a) $y^2 + 2y - 15$

(b) $t^4 + t^3 - 20t^2$

(c) $r^5 - 9r^3$

(d) $x^2 + 2xy - 8y^2$

PROBLEM 1-48   Factor the following polynomials:

(a) $x^3 - 3x^2 - 2x + 6$

(b) $x^2(x + 3) - 4x - 12$

(c) $(x^2 - 4) + 3(x + 2)$

(d) $9x^2 + 6x + 1$

PROBLEM 1-49   Factor the following polynomials:

(a) $x^3 - 27$

(b) $8y^3 + 1$

(c) $x^2 - 4x + 4 - 9y^2$

(d) $t^9 + 1$

Answers to Supplementary Exercises

1-31 $B = D = G$ and $C = E$

1-32   (a) $C = \{\text{color} \mid \text{the color is on a traffic light}\}$
       $C = \{\text{red, yellow, green}\}$

(b) $V = \{n \mid n$ is an integer describing the number of continents you could visit on a world tour$\}$
$V = \{1, 2, 3, 4, 5, 6, 7\}$

(c) $A = \{n \mid n$ is an integer denoting the number of aces you could have in a bridge hand$\}$
$A = \{0, 1, 2, 3, 4\}$

(d) $L = \{$letter $\mid$ the letter occurs between K and T in the alphabet$\}$
$L = \{L, M, N, O, P, Q, R, S\}$

(e) $E = \{$natural number $\mid$ the number is odd and less than 100$\}$
$E = \{1, 3, 5, 7, 9, \ldots, 91, 93, 95, 97, 99\}$

1-33 (a) 3   (c) 1

(b) $\dfrac{4}{5}$ (d) $\dfrac{23}{6}$

1-34 $\dfrac{a^2 + b^2}{a^2 - b^2}$

1-35 (a) False

(b) False

(c) True

1-36 (a) False

(b) False, unless $a = 0$ or $b = 0$

(c) True

1-37 (a) $-\dfrac{3}{5}$ (d) $\dfrac{13}{2}$

(b) $-\dfrac{8}{5}$ (e) $-\dfrac{15}{4}$

(c) $-\dfrac{3}{5}$

1-38 -1

1-39 (a) Commutativity of multiplication

(b) Closure under addition

(c) Identity and commutativity for multiplication

(d) Distributivity

(e) Inverse for addition

1-40 (a) $\frac{28}{15}x - \frac{53}{8}y$

(b) $\frac{17}{10}z$

(c) $8x$

(d) $\frac{7}{12}a^2 - \frac{121}{120}b^2$

1-41 (a) 4

(b) $\frac{5}{4}$

(c) 0

1-42 (a) Sum $= 4x^2 + x + 1$
Diff. $= -2x^2 - 9x + 3$

(b) Sum $= \frac{5}{8}y^2 + \frac{35}{4}y - 10$
Diff. $= \frac{3}{8}y^2 + \frac{29}{4}y + 4$

(c) Sum $= 10x^4 - 8x^3 - 8x^2 + 2x - 7$
Diff. $= 4x^4 - 10x^3 + 8x^2 + 2x - 9$

(d) Sum $= 16t^2 + \frac{11}{8}t + \frac{31}{8}$
Diff. $= 16t^2 + \frac{1}{8}t - \frac{49}{88}$

1-43 (a) $x^3 - 6x^2 + 4x - 24$

(b) $x^4 + 4x^3 - x^2 - 2x - 5$

(c) $xy(x + y)^2$

1-44 (a) $2x^2 - 5x - 3$

(b) $5x^2 + 19x - 4$

(c) $2x^2 + 5xy - 3y^2$

(d) $14x^2 + 27xy - 20y^2$

1-45 (a) $x^3 + 6x^2y + 12xy^2 + 8y^3$

(b) $6x^3 + 40x^2 + 86x + 60$

(c) $8a^3 - 36a^2b + 54ab^2 - 27b^3$

(d) $r^6 + 9r^4s + 27r^2s^2 + 27s^3$

(e) $\dfrac{1}{u^2} - \dfrac{2}{uv} + \dfrac{1}{v^2}$

1-46 (a) $(2x - 3)(2x + 3)$

(b) $(x - 4)(x - 3)$

(c) $(x + 4)^2$

(d) $(2a + b - 5)(2a + b + 5)$

1-47 (a) $(y + 5)(y - 3)$

(b) $t^2(t + 5)(t - 4)$

(c) $r^3(r - 3)(r + 3)$

(d) $(x + 4y)(x - 2y)$

1-48 (a) $(x - 3)(x^2 - 2) = (x - 3)(x - \sqrt{2})(x + \sqrt{2})$

(b) $(x + 3)(x - 2)(x + 2)$

(c) $(x + 2)(x + 1)$

(d) $(3x + 1)^2$

1-49 (a) $(x - 3)(x^2 + 3x + 9)$

(b) $(2y + 1)(4y^2 - 2y + 1)$

(c) $(x - 2 - 3y)(x - 2 + 3y)$

(d) $(t^3 + 1)(t^6 - t^3 + 1)$

2 RATIONAL EXPRESSIONS

THIS CHAPTER IS ABOUT

☑ **Recognizing and Reducing Rational Expressions**
☑ **Operations on Rational Expressions**

2-1. Recognizing and Reducing Rational Expressions

A. Definition of rational expressions

Polynomial addition, subtraction, and multiplication all produce another polynomial. But when we *divide* one polynomial by another, we do not necessarily—or even generally—get another polynomial. The quotient of two polynomials is a *rational expression*. Remember that a rational *number* is defined as the quotient x/y, where x and y are integers and $y \neq 0$. So

- A **rational expression** is the quotient $P(x)/Q(x)$ of two polynomials $P(x)$ and $Q(x)$ for which $Q(x) \neq 0$.

note: When we say that $Q(x) \neq 0$, we mean that the **domain** (i.e., the value or set of values that can be substituted for the variable) must consist of numbers that do not give a denominator of zero. In other words, a rational expression is *well-defined* by the condition $Q(x) \neq 0$.

The set of rational expressions is to the set of polynomials what the set of rational numbers is to the set of integers. So any polynomial (integer) is a rational expression (number) that has a denominator of 1.

EXAMPLE 2-1 Under what conditions would the following rational expressions *not* be well-defined?

(a) $\dfrac{x^2 + 1}{x - 3}$ (b) $\dfrac{2x}{6x + 12}$ (c) $\dfrac{1}{12yz}$ (d) $\dfrac{6x^2y + 3x}{5xy^2 + 9}$

Solution   (a) $x = 3$ (b) $x = -2$ (c) $yz = 0$ (d) $5xy^2 = -9$

B. Reducing rational expressions to their lowest terms

We often combine rational expressions and simplify the result. To do this, we use the real-number rules for division (see Section 1-2B). Of particular importance is the rule $\dfrac{ac}{bc} = \dfrac{a}{b}$ for simplifying or reducing an expression to its simplest form, which can be written for rational expressions as

$$\begin{array}{c} \text{REDUCING A} \\ \text{RATIONAL EXPRESSION} \end{array} \qquad \frac{P(x) \cdot R(x)}{Q(x) \cdot R(x)} = \frac{P(x)}{Q(x)} \qquad (2.1)$$

Using this rule involves factoring the numerator and denominator into irreducible factors called **prime factors** and canceling the common factors.

EXAMPLE 2-2 Reduce the following rational expressions to their lowest terms:

$$\text{(a) } \frac{x^2 - 9}{x + 3} \qquad \text{(b) } \frac{x^2 + x - 6}{x^4 + 2x^3 - 3x^2} \qquad \text{(c) } \frac{x + y}{x^2 + 4xy + 3y^2}$$

Solution

(a)
$$\frac{x^2 - 9}{x + 3} = \frac{(x - 3)\cancel{(x + 3)}}{\cancel{x + 3}} = \frac{x - 3}{1} = x - 3$$

(b)
$$\frac{x^2 + x - 6}{x^4 + 2x^3 - 3x^2} = \frac{(x + 3)(x - 2)}{x^2(x^2 + 2x - 3)} = \frac{\cancel{(x + 3)}(x - 2)}{x^2\cancel{(x + 3)}(x - 1)} = \frac{x - 2}{x^2(x - 1)}$$

(c)
$$\frac{x + y}{x^2 + 4xy + 3y^2} = \frac{\cancel{x + y}}{\cancel{(x + y)}(x + 3y)} = \frac{1}{x + 3y}$$

2-2. Operations on Rational Expressions

A. Multiplication and division

The rules for multiplication and division of rational expressions are the logical extensions of the corresponding rules for real numbers.

Since $\dfrac{a}{b} \cdot \dfrac{c}{d} = \dfrac{ac}{bd}$,

| PRODUCT OF RATIONAL EXPRESSIONS | $\dfrac{P(x)}{Q(x)} \cdot \dfrac{R(x)}{T(x)} = \dfrac{P(x) \cdot R(x)}{Q(x) \cdot T(x)}$ | (2.2) |

and since $\dfrac{a}{b} \div \dfrac{c}{d} = \dfrac{ad}{bc}$,

| QUOTIENT OF RATIONAL EXPRESSIONS | $\dfrac{P(x)}{Q(x)} \div \dfrac{R(x)}{T(x)} = \dfrac{P(x)}{Q(x)} \cdot \dfrac{T(x)}{R(x)} = \dfrac{P(x) \cdot T(x)}{Q(x) \cdot R(x)}$ | (2.3) |

EXAMPLE 2-3 Perform the indicated operations in (a) $\dfrac{x^2 - 4}{x + 3} \cdot \dfrac{x^2 + 7x + 12}{x^2 + 2x - 8}$ and (b) $\dfrac{x - 6}{x + 5} \div \dfrac{x^2 - 7x + 6}{x^2 + 6x + 5}$ and simplify the results.

Solution

(a) First factor each numerator and denominator as far as possible; then look for cancelations:

$$\frac{x^2 - 4}{x + 3} \cdot \frac{x^2 + 7x + 12}{x^2 + 2x - 8} = \frac{(x - 2)(x + 2)}{x + 3} \cdot \frac{(x + 3)(x + 4)}{(x + 4)(x - 2)}$$

$$= \frac{\cancel{(x - 2)}(x + 2)\cancel{(x + 3)}\cancel{(x + 4)}}{\cancel{(x + 3)}\cancel{(x + 4)}\cancel{(x - 2)}} = x + 2$$

(b) Factor as usual—but watch the division sign!

$$\frac{x - 6}{x + 5} \div \frac{x^2 - 7x + 6}{x^2 + 6x + 5} = \frac{x - 6}{x + 5} \div \frac{(x - 6)(x - 1)}{(x + 5)(x + 1)}$$

$$= \frac{x - 6}{x + 5} \cdot \frac{(x + 5)(x + 1)}{(x - 6)(x - 1)} \qquad \text{[Flip to get multiplication]}$$

Now cancel:

$$\frac{\cancel{(x-6)}\cancel{(x+5)}(x+1)}{\cancel{(x+5)}\cancel{(x-6)}(x-1)} = \frac{x+1}{x-1}$$

EXAMPLE 2-4 Perform the indicated operations in
(a) $\dfrac{x^3 - x}{x^2 + 5x + 4} \cdot \dfrac{x^2 + 2x - 8}{x^3 + x^2 - 2x}$ and (b) $\dfrac{2x - 1}{x - 3} \div \dfrac{2x^2 - x}{x^2 - 9}$
and simplify the results.

Solution

(a)
$$\frac{x(x-1)(x+1)}{(x+4)(x+1)} \cdot \frac{(x+4)(x-2)}{x(x+2)(x-1)} = \frac{x-2}{x+2}$$

(b)
$$\frac{2x-1}{x-3} \cdot \frac{(x-3)(x+3)}{x(2x-1)} = \frac{x+3}{x}$$

Division also gives us a means of simplifying rational expressions. There are two special cases worth considering

(1) If the denominator of the rational expression consists of a *single term*, we actually perform the indicated division, dividing each term in the numerator by the denominator.

EXAMPLE 2-5 Simplify $\dfrac{x^3 + 5x^2 + 6x + 1}{2x}$ by dividing.

Solution Since the denominator has only one term, we divide each term of the numerator individually by that single term:

$$\frac{x^3 + 5x^2 + 6x + 1}{2x} = \frac{x^3}{2x} + \frac{5x^2}{2x} + \frac{6x}{2x} + \frac{1}{2x}$$

Now we can reduce each resulting term individually:

Remember that $x^n \cdot x^m = x^{n+m}$, so $x^n \div x^m = x^{n-m}$.

$$\frac{x^3}{2x} + \frac{5x^2}{2x} + \frac{6x}{2x} + \frac{1}{2x} = \frac{1}{2}x^2 + \frac{5}{2}x + 3 + \frac{1}{2x}$$

Note that, in general, the result is not a polynomial.

---

(2) If the denominator of a rational expression consists of *two terms*, we perform the indicated division by the arithmetic procedure of long division.

EXAMPLE 2-6 Simplify $\dfrac{x^3 + 2x^2 + 3x + 4}{x + 1}$ by dividing.

Solution The denominator has two terms, so we can use long division:

$$\begin{array}{r} x^2 \\ x+1{\overline{\smash{\big)}\,x^3 + 2x^2 + 3x + 4}} \\ \underline{-(x^3 + x^2)} \\ x^2 \end{array}$$

Step 1: Divide the first term of the numerator (dividend) by the first term of the denominator (divisor) to obtain the first term of the quotient. Multiply the partial quotient by the divisor and subtract the result from the first two terms of the dividend.

$$\begin{array}{r} x^2 + x \\ x + 1 \overline{\smash{\big)}\, x^3 + 2x^2 + 3x + 4} \\ -(x^3 + x^2) \\ \hline x^2 + 3x \\ -(x^2 + x) \\ \hline 2x \end{array}$$

Step 2: Carry down the third term of the original dividend to form a new dividend. Divide the first term in the new dividend by the first term in the divisor and put the result on top. Multiply again and subtract.

$$\begin{array}{r} x^2 + x + 2 \\ x + 1 \overline{\smash{\big)}\, x^3 + 2x^2 + 3x + 4} \\ -(x^3 + x^2) \\ \hline x^2 + 3x \\ -(x^2 + x) \\ \hline 2x + 4 \\ -(2x + 2) \\ \hline \text{remainder} \quad 2 \end{array}$$

Step 3: Repeat the process in Step 2, carrying down the next term, forming a new dividend, and dividing until you run out of divisible dividend.

The long-division process ends when the degree of the remainder polynomial (here, the degree of the remainder, 2, is zero) is less than the degree of the divisor (here, that degree is one).

Now you can write $\quad \dfrac{x^3 + 2x^2 + 3x + 4}{x + 1} = x^2 + x + 2 + \dfrac{2}{x + 1}$

$$\frac{\text{Dividend}}{\text{Divisor}} = \text{Quotient} + \frac{\text{Remainder}}{\text{Divisor}}$$

B. Addition and subtraction

1. Common denominator

When we add or subtract rational expressions whose denominators have *no factors in common*, we follow the real-number rules and apply them to the polynomials. We use the rules $\dfrac{a}{b} + \dfrac{c}{d} = \dfrac{ad + bc}{bd}$ and $\dfrac{a}{b} - \dfrac{c}{d} = \dfrac{ad - bc}{bd}$ to write

SUM OF RATIONAL EXPRESSIONS $\qquad \dfrac{P(x)}{Q(x)} + \dfrac{R(x)}{T(x)} = \dfrac{P(x) \cdot T(x) + Q(x) \cdot R(x)}{Q(x) \cdot T(x)}$ **(2.4)**

DIFFERENCE OF RATIONAL EXPRESSIONS $\qquad \dfrac{P(x)}{Q(x)} - \dfrac{R(x)}{T(x)} = \dfrac{P(x) \cdot T(x) - Q(x) \cdot R(x)}{Q(x) \cdot T(x)}$ **(2.5)**

In other words, we have to find a *common denominator*, $Q(x) \cdot T(x)$.

EXAMPLE 2-7 Find the sum or difference of (a) $\dfrac{x + 2}{x - 1} + \dfrac{x - 2}{x + 3}$ and

(b) $\dfrac{x^2}{x + 4} - \dfrac{x^2 + 2}{x - 3}$ and simplify the results.

Solution

(a) $\qquad \dfrac{x + 2}{x - 1} + \dfrac{x - 2}{x + 3} = \dfrac{(x + 2)(x + 3) + (x - 1)(x - 2)}{(x - 1)(x + 3)}$

$\qquad\qquad = \dfrac{x^2 + 5x + 6 + x^2 - 3x + 2}{(x - 1)(x + 3)}$

$\qquad\qquad = \dfrac{2x^2 + 2x + 8}{(x - 1)(x + 3)}$

(b)
$$\frac{x^2}{x+4} - \frac{x^2+2}{x-3} = \frac{x^2(x-3) - (x+4)(x^2+2)}{(x+4)(x-3)}$$

$$= \frac{x^3 - 3x^2 - (x^3 + 4x^2 + 2x + 8)}{(x+4)(x-3)}$$

$$= \frac{-7x^2 - 2x - 8}{(x+4)(x-3)}$$

2. Least common denominator

When we add or subtract rational expressions that have *common factors in their denominators*, we follow the above rules—but we simplify the computation by using the **least common denominator (lcd)**.

- The **lcd** is obtained by inspecting the *prime factorization* for each denominator and forming the product of the different prime [irreducible] factors, using the highest exponent that appears on each of these.

Easier done than said—a numerical example will show the details.

EXAMPLE 2-8 Find the value of $\frac{7}{30} + \frac{5}{24}$.

Solution The denominators are factored into prime factors:

$$30 = 2 \cdot 3 \cdot 5 \quad \text{and} \quad 24 = 2 \cdot 2 \cdot 2 \cdot 3$$

The lcd is formed from the prime factors 2, 3, and 5, using the highest exponent in each case. Thus the lcd is

$$2^3 \cdot 3^1 \cdot 5^1 = 120$$

Next, knowing that the lcd $= 120$, you determine what each denominator needs to be multiplied by to convert it to the lcd: $\frac{120}{30} = 4$, and $\frac{120}{24} = 5$. So you write

$$\frac{7}{30} \cdot \frac{4}{4} + \frac{5}{24} \cdot \frac{5}{5} = \frac{28}{120} + \frac{25}{120} = \frac{53}{120}$$

Notice that creating the lcd involves multiplying each term by 1 in the general form $\frac{P(x)}{P(x)} = 1$. Thus you can find the lcd by multiplying each term by $\frac{x}{x} = 1$ or $\frac{x^2}{x^2} = 1$ or $\frac{x+a}{x+a} = 1$ or any other variation, just as long as the denominator is the same as the numerator.

EXAMPLE 2-9 Find the equivalent of the rational expression $\frac{-3x^2 + 5}{x(2x+1)} - \frac{3}{x^2} + \frac{4x}{2x+1}$ in simplest form.

Solution Since the denominators are already factored, the lcd can be formed from x and $(2x + 1)$. It is

$$x^2(2x + 1)$$

Determine what each denominator needs to make it into $x^2(2x + 1)$. Multiply into each term the appropriate $\frac{P(x)}{P(x)}$:

$$\frac{-3x^2 + 5}{x(2x+1)} \cdot \frac{x}{x} - \frac{3}{x^2} \cdot \frac{2x+1}{2x+1} + \frac{4x}{2x+1} \cdot \frac{x^2}{x^2}$$

$$= \frac{-3x^3 + 5x}{x^2(2x+1)} - \frac{6x+3}{x^2(2x+1)} + \frac{4x^3}{x^2(2x+1)}$$

$$= \frac{-3x^3 + 5x - 6x - 3 + 4x^3}{x^2(2x+1)} = \frac{x^3 - x - 3}{x^2(2x+1)}$$

EXAMPLE 2-10   Simplify the expression $\dfrac{1}{x+4} - \dfrac{2x}{x^2-16} + \dfrac{x+3}{x^2+8x+16}$.

Solution   First, factor each denominator completely:

$$\frac{1}{x+4} - \frac{2x}{(x+4)(x-4)} + \frac{x+3}{(x+4)^2}$$

Now the factors for the lcd become clear. The lcd is $(x+4)^2(x-4)$. Then multiply appropriately to make all denominators match the lcd:

$$\frac{1}{x+4} \cdot \frac{(x+4)(x-4)}{(x+4)(x-4)} - \frac{2x}{(x+4)(x-4)} \cdot \frac{x+4}{x+4} + \frac{x+3}{(x+4)^2} \cdot \frac{x-4}{x-4}$$

$$= \frac{x^2-16}{(x+4)^2(x-4)} - \frac{2x^2+8x}{(x+4)^2(x-4)} + \frac{x^2-x-12}{(x+4)^2(x-4)}$$

$$= \frac{x^2-16-2x^2-8x+x^2-x-12}{(x+4)^2(x-4)}$$

$$= \frac{-9x-28}{(x+4)^2(x-4)}$$

C. Combinations of rational expressions

Sometimes a problem turns into a manipulation of rational expressions.

EXAMPLE 2-11   Simplify the expression

$$\frac{\left(\dfrac{9}{x+2}\right) - 3}{1 - \left(\dfrac{1}{x}\right)}$$

Solution   Begin by working on the numerator and denominator separately. Simplify the numerator first:

$$\frac{9}{x+2} - 3 = \frac{9}{x+2} - \frac{3}{1} \cdot \left(\frac{x+2}{x+2}\right)$$

$$= \frac{9-3x-6}{x+2} = \frac{3-3x}{x+2} = \frac{-3(x-1)}{x+2}$$

Then simplify the denominator:

$$1 - \frac{1}{x} = \frac{1}{1} \cdot \left(\frac{x}{x}\right) - \frac{1}{x} = \frac{x-1}{x}$$

So the original expression can be rewritten as a problem in division:

$$\frac{-3(x-1)}{x+2} \div \frac{x-1}{x} = \frac{-3\cancel{(x-1)}}{x+2} \cdot \frac{x}{\cancel{x-1}} = \frac{-3x}{x+2}$$

EXAMPLE 2-12   Simplify

$$\frac{\dfrac{x-1}{x+1} - \dfrac{x+2}{x-3}}{\dfrac{x+7}{x+2} - \dfrac{x+3}{x+1}}$$

Solution Simplify the numerator and denominator independently.

$$\frac{x-1}{x+1} - \frac{x+2}{x-3} = \frac{(x-1)(x-3) - (x+1)(x+2)}{(x+1)(x-3)}$$

$$= \frac{x^2 - 4x + 3 - x^2 - 3x - 2}{(x+1)(x-3)} = \frac{-7x+1}{(x+1)(x-3)}$$

$$\frac{x+7}{x+2} - \frac{x+3}{x+1} = \frac{(x+7)(x+1) - (x+2)(x+3)}{(x+2)(x+1)}$$

$$= \frac{x^2 + 8x + 7 - x^2 - 5x - 6}{(x+2)(x+1)} = \frac{3x+1}{(x+2)(x+1)}$$

So the original expression is

$$\frac{-7x+1}{(x+1)(x-3)} \div \frac{3x+1}{(x+2)(x+1)} = \frac{-7x+1}{(x+1)(x-3)} \cdot \frac{(x+2)(x+1)}{3x+1}$$

$$= \frac{(-7x+1)(x+2)}{(x-3)(3x+1)}$$

EXAMPLE 2-13 Simplify $\frac{x}{y} + \frac{z}{x} + \frac{y}{z}$.

Solution Since the denominators are distinct, the lcd is the product xyz. So we multiply each term appropriately:

$$\frac{x}{y} \cdot \frac{xz}{xz} + \frac{z}{x} \cdot \frac{yz}{yz} + \frac{y}{z} \cdot \frac{xy}{xy} = \frac{x^2z + yz^2 + xy^2}{xyz}$$

note: We still use the PEMDAS order of operations detailed in Chapter 1. [Remember "*Please Excuse My Dear Aunt Sally*"?]

EXAMPLE 2-14 Simplify

(a) $\frac{2}{x^2} - \frac{1}{xy} \div \frac{x}{y} + \frac{x+1}{x^2}$ (b) $\left(\frac{2}{x^2} - \frac{1}{xy}\right) \div \left(\frac{x}{y} + \frac{x+1}{x^2}\right)$

Solution

(a) There are no parentheses and no exponents to expand, so we multiply and divide, left-to-right, then add (also left-to-right):

$$\frac{2}{x^2} - \frac{1}{xy} \div \frac{x}{y} + \frac{x+1}{x^2} = \frac{2}{x^2} + \left(-\frac{1}{xy} \cdot \frac{y}{x}\right) + \frac{x+1}{x^2}$$

$$= \frac{2}{x^2} + \left(\frac{-1}{x^2}\right) + \frac{x+1}{x^2}$$

$$= \frac{2 - 1 + x + 1}{x^2} = \frac{2+x}{x^2}$$

(b) This is the same expression, but it has parentheses—which changes the result. First, simplify inside each set of parentheses:

$$\left(\frac{2}{x^2} \cdot \frac{y}{y} - \frac{1}{xy} \cdot \frac{x}{x}\right) \div \left(\frac{x}{y} \cdot \frac{x^2}{x^2} + \frac{x+1}{x^2} \cdot \frac{y}{y}\right) = \frac{2y - x}{x^2y} \div \frac{x^3 + xy + y}{x^2y}$$

Then divide:

$$\frac{2y-x}{x^2y} \div \frac{x^3+xy+y}{x^2y} = \frac{2y-x}{x^2y} \cdot \frac{x^2y}{x^3+xy+y} = \frac{2y-x}{x^3+xy+y}$$

Remember that $\frac{A}{B} \div \frac{C}{D} = \frac{A}{B} \cdot \frac{D}{C}$

SUMMARY

1. A rational expression is the quotient of two polynomials. It is well-defined for the value(s) of the variable that give a nonzero denominator.
2. We operate on rational expressions using all the rules of real-number operation:

$$\frac{P(x) \cdot R(x)}{Q(x) \cdot R(x)} = \frac{P(x)}{Q(x)} \qquad \left(\text{cf.} \quad \frac{a \cdot c}{b \cdot c} = \frac{a}{b} \right)$$

$$\frac{P(x)}{Q(x)} \cdot \frac{R(x)}{T(x)} = \frac{P(x) \cdot R(x)}{Q(x) \cdot T(x)} \qquad \left(\text{cf.} \quad \frac{a}{b} \cdot \frac{c}{d} = \frac{ac}{bd} \right)$$

$$\frac{P(x)}{Q(x)} \div \frac{R(x)}{T(x)} = \frac{P(x) \cdot T(x)}{Q(x) \cdot R(x)} \qquad \left(\text{cf.} \quad \frac{a}{b} \div \frac{c}{d} = \frac{ad}{bc} \right)$$

$$\frac{P(x)}{Q(x)} + \frac{R(x)}{T(x)} = \frac{P(x) \cdot T(x) + Q(x) \cdot R(x)}{Q(x) \cdot T(x)} \qquad \left(\text{cf.} \quad \frac{a}{b} + \frac{c}{d} = \frac{ad + bc}{bd} \right)$$

$$\frac{P(x)}{Q(x)} - \frac{R(x)}{T(x)} = \frac{P(x) \cdot T(x) - Q(x) \cdot R(x)}{Q(x) \cdot T(x)} \qquad \left(\text{cf.} \quad \frac{a}{b} - \frac{c}{d} = \frac{ad - bc}{bd} \right)$$

3. We use rules of division to simplify rational expressions.
4. To simplify combinations of rational expressions whose denominators have common factors, we use the least common denominator (lcd).
5. The PEMDAS order-of-operation rules also apply to manipulation of rational expressions.

RAISE YOUR GRADES

Can you...?

☑ simplify rational expressions by factoring and canceling
☑ explain how to divide a polynomial by a denominator containing a single term
☑ explain how to divide a polynomial by a denominator containing two terms
☑ use the long-division process on a rational number expression
☑ define the least common denominator
☑ remember and use all the special products and their factors

SOLVED PROBLEMS

Recognizing and Reducing Rational Expressions

You can't work efficiently with rational expressions without being very good at factoring. Make sure you can recognize and use all the special products and their factors listed in Table 1-2.

PROBLEM 2-1 Reduce $\dfrac{x^2 - 36}{6 - x}$ to its lowest terms.

Solution Set up the problem so you can use the canceling rule (2.1) $\dfrac{P(x) \cdot R(x)}{Q(x) \cdot R(x)} = \dfrac{P(x)}{Q(x)}$. First factor the numerator:

$$\frac{x^2 - 36}{6 - x} = \frac{(x - 6)(x + 6)}{6 - x}$$

You can't cancel yet, but you can change the denominator by factoring out the quantity -1, which gives you $6 - x = -1(x - 6)$. Thus,

$$\frac{x^2 - 36}{6 - x} = \frac{\cancel{(x - 6)}(x + 6)}{-1\cancel{(x - 6)}} = \frac{x + 6}{-1} = -(x + 6)$$

PROBLEM 2-2   Reduce $\dfrac{3x + 9}{x^2 - 9}$ to its lowest terms.

Solution   Factor both numerator and denominator and check for cancelations:

$$\frac{3x + 9}{x^2 - 9} = \frac{3\cancel{(x + 3)}}{(x - 3)\cancel{(x + 3)}} = \frac{3}{x - 3}$$

PROBLEM 2-3   Reduce $\dfrac{a^2 - b^2}{a^3 - b^3}$.

Solution   Both the numerator and the denominator are factorable—the numerator as the difference of two squares, and the denominator as the difference of two cubes. Thus

$$\frac{a^2 - b^2}{a^3 - b^3} = \frac{(a - b)(a + b)}{(a - b)(a^2 + ab + b^2)} = \frac{a + b}{a^2 + ab + b^2}$$

PROBLEM 2-4   Reduce $\dfrac{x^3 - 3x^2 - 4x + 12}{x^2 - x - 6}$.

Solution   Factor the numerator twice—once by grouping (note the coefficient pattern), and once by the difference of two squares. Then factor the denominator and cancel:

$$\frac{x^3 - 3x^2 - 4x + 12}{x^2 - x - 6} = \frac{x^2(x - 3) - 4(x - 3)}{x^2 - x - 6}$$

$$= \frac{(x - 3)(x^2 - 4)}{x^2 - x - 6}$$

$$= \frac{(x - 3)(x - 2)(x + 2)}{(x - 3)(x + 2)} = x - 2$$

PROBLEM 2-5   Reduce $\dfrac{x^2 - 9y^2}{x^2 + xy - 6y^2}$ to its lowest terms.

Solution

$$\frac{x^2 - 9y^2}{x^2 + xy - 6y^2} = \frac{(x - 3y)(x + 3y)}{(x + 3y)(x - 2y)} = \frac{x - 3y}{x - 2y}$$

Operations on Rational Expressions

PROBLEM 2-6   Perform the indicated operation in $\dfrac{12x^5 - 28x^4 + 20x^3}{4x^2}$ and simplify the result.

Solution   The denominator has only one term. So you can divide each numerator term individually by the denominator and simplify each resulting term individually:

$$\frac{12x^5 - 28x^4 + 20x^3}{4x^2} = \frac{12x^5}{4x^2} - \frac{28x^4}{4x^2} + \frac{20x^3}{4x^2}$$

$$= 3x^3 - 7x^2 + 5x$$

PROBLEM 2-7 Perform the indicated operations in (a) $\dfrac{x-3}{x^2+x-12} \cdot \dfrac{x^2+8x+16}{x+4}$ and

(b) $\dfrac{x^3-x^2}{x^2+3x} \div \dfrac{x^2+x}{x^2+4x+3}$ and simplify the results.

Solution The rules of multiplication and division for rational expressions are the same as those for real numbers. In each case, factor first; then look for cancelations.

(a) $\qquad \dfrac{x-3}{x^2+x-12} \cdot \dfrac{x^2+8x+16}{x+4} = \dfrac{x-3}{(x+4)(x-3)} \cdot \dfrac{(x+4)(x+4)}{x+4} = 1$

$\qquad\qquad\qquad\qquad\qquad$ [Everything cancels? Yes!]

(b) $\qquad \dfrac{x^3-x^2}{x^2+3x} \div \dfrac{x^2+x}{x^2+4x+3} = \dfrac{x^2(x-1)}{x(x+3)} \div \dfrac{x(x+1)}{(x+3)(x+1)}$

$\qquad$ [You can cancel immediately $\;=\dfrac{x(x-1)}{x+3} \div \dfrac{x}{x+3}$
$\qquad$ in each term.]

$\qquad\qquad\qquad\qquad\qquad =\dfrac{x(x-1)}{x+3} \cdot \dfrac{x+3}{x} \qquad$ [Flip for multiplication and cancel.]

$\qquad\qquad\qquad\qquad\qquad = x-1$

PROBLEM 2-8 Find the simplified form of (a) $\dfrac{(x+h)^2-x^2}{h}$ and (b) $\dfrac{(x+h)^3-x^3}{h}$.

Solution Expand the numerator first and collect the terms; then divide by the monomial and simplify the resulting terms.

(a) $\qquad\qquad \dfrac{(x+h)^2-x^2}{h} = \dfrac{(x^2+2xh+h^2)-x^2}{h} = \dfrac{2xh+h^2}{h}$

$\qquad\qquad\qquad\qquad\quad =\dfrac{2xh}{h} + \dfrac{h^2}{h}$

$\qquad\qquad\qquad\qquad\quad = 2x+h$

(b) $\qquad\qquad \dfrac{(x+h)^3-x^3}{h} = \dfrac{(x^3+3x^2h+3xh^2+h^3)-x^3}{h}$

$\qquad\qquad\qquad\qquad\quad =\dfrac{3x^2h+3xh^2+h^3}{h}$

$\qquad\qquad\qquad\qquad\quad =\dfrac{3x^2h}{h} + \dfrac{3xh^2}{h} + \dfrac{h^3}{h}$

$\qquad\qquad\qquad\qquad\quad = 3x^2+3xh+h^2$

PROBLEM 2-9 Perform the indicated operations in (a) $\dfrac{x^3+8}{x^2-4} \cdot \dfrac{x-2}{x+2}$ and

(b) $\dfrac{y-2}{y^3-8} \div \dfrac{y^2+4y+4}{y^4-16}$ and simplify the results.

Solution You need to use the formulas for factoring the sum and difference of two cubes, then cancel where possible and simplify.

(a) $\dfrac{x^3+8}{x^2-4} \cdot \dfrac{x-2}{x+2} = \dfrac{(x+2)(x^2-2x+4)}{(x-2)(x+2)} \cdot \dfrac{x-2}{x+2} = \dfrac{x^2-2x+4}{x+2}$

(b) $\dfrac{y-2}{y^3-8} \div \dfrac{y^2+4y+4}{y^4-16} = \dfrac{y-2}{(y-2)(y^2+2y+4)} \div \dfrac{(y+2)(y+2)}{(y^2-4)(y^2+4)}$

$\qquad\qquad\qquad\qquad = \dfrac{1}{y^2+2y+4} \cdot \dfrac{(y-2)(y+2)(y^2+4)}{(y+2)(y+2)} = \dfrac{(y-2)(y^2+4)}{(y+2)(y^2+2y+4)}$

PROBLEM 2-10 Use the long-division process to reduce (a) $\dfrac{x^2 - 4x + 8}{x - 3}$ and

(b) $\dfrac{x^3 + 4x^2 - 3x - 2}{x - 1}$.

Solution

(a)
$$
\begin{array}{r}
x \\
x - 3 \,\overline{\big)\, x^2 - 4x + 8} \\
\underline{-(x^2 - 3x)} \\
-x
\end{array}
$$

Step 1: Divide the first term of the divisor, x, into the first term of the dividend, x^2, putting the result x into the quotient. Then multiply x by $x - 3$ and subtract the result from $x^2 - 4x$.

$$
\begin{array}{r}
x \; - 1 \\
x - 3 \,\overline{\big)\, x^2 - 4x + 8} \\
\underline{-(x^2 - 3x)} \\
-x + 8 \\
\underline{-(-x + 3)} \\
\text{remainder} \;\; 5
\end{array}
$$

Step 2: Carry down the next term of the original dividend, 8, to form the new dividend $-x + 8$. Divide x into $-x$ and put the result on top. Then multiply and subtract as in Step 1.

The process ends here since the remainder 5 has degree 0 while the divisor $x - 3$ has degree 1.

Now you can write

$$
\frac{x^2 - 4x + 8}{x - 3} = x - 1 + \frac{5}{x - 3}
$$

$$
\frac{\text{Dividend}}{\text{Divisor}} = \text{Quotient} + \frac{\text{Remainder}}{\text{Divisor}}
$$

(b)
$$
\begin{array}{r}
x^2 + 5x \; + 2 \\
x - 1 \,\overline{\big)\, x^3 + 4x^2 - 3x - 2} \\
\underline{-(x^3 - \; x^2)} \\
5x^2 - 3x \\
\underline{-(5x^2 - 5x)} \\
2x - 2 \\
\underline{-(2x - 2)} \\
0
\end{array}
$$

Now you have

$$
\frac{x^3 + 4x^2 - 3x - 2}{x - 1} = x^2 + 5x + 2
$$

PROBLEM 2-11 Perform the indicated operations in (a) $\dfrac{2x - 3}{x + 1} + \dfrac{x - 4}{3x + 5}$ and

(b) $\dfrac{x^2 + 1}{2x - 1} - \dfrac{x}{x - \frac{1}{2}}$ and simplify the results.

Solution Since all terms are already in lowest factored form, you need only apply the rules for addition and subtraction:

$$
\frac{P(x)}{Q(x)} + \frac{R(x)}{T(x)} = \frac{P(x) \cdot T(x) + Q(x) \cdot R(x)}{Q(x) \cdot T(x)} \quad \text{and} \quad \frac{P(x)}{Q(x)} - \frac{R(x)}{T(x)} = \frac{P(x) \cdot T(x) - Q(x) \cdot R(x)}{Q(x) \cdot T(x)}
$$

(a)
$$
\frac{2x - 3}{x + 1} + \frac{x - 4}{3x + 5} = \frac{(2x - 3)(3x + 5) + (x + 1)(x - 4)}{(x + 1)(3x + 5)}
$$

$$
= \frac{(6x^2 + 10x - 9x - 15) + (x^2 - 4x + x - 4)}{(x + 1)(3x + 5)}
$$

$$
= \frac{7x^2 - 2x - 19}{(x + 1)(3x + 5)}
$$

(b)
$$\frac{x^2+1}{2x-1} - \frac{x}{x-\frac{1}{2}} = \frac{(x^2+1)(x-\frac{1}{2}) - (2x-1)x}{(2x-1)(x-\frac{1}{2})}$$

$$= \frac{x^3 - \frac{1}{2}x^2 + x - \frac{1}{2} - 2x^2 + x}{(2x-1)(x-\frac{1}{2})}$$

$$= \frac{x^3 - \frac{5}{2}x^2 + 2x - \frac{1}{2}}{(2x-1)(x-\frac{1}{2})}$$

This result appears to be in lowest factored form—but it isn't! Look back at the original problem and note that the denominators are related: i.e., $(2x-1) = 2(x-\frac{1}{2})$. So there's an lcd here of $2(x-\frac{1}{2})$—which means that the approach to the problem has to change:

$$\frac{x^2+1}{2x-1} - \frac{x}{x-\frac{1}{2}} = \frac{x^2+1}{2(x-\frac{1}{2})} - \frac{x}{x-\frac{1}{2}} = \frac{x^2+1}{2(x-\frac{1}{2})} - \left(\frac{x}{x-\frac{1}{2}} \cdot \frac{2}{2}\right)$$

$$= \frac{x^2+1}{2(x-\frac{1}{2})} - \frac{2x}{2(x-\frac{1}{2})} = \frac{x^2-2x+1}{2(x-\frac{1}{2})} = \frac{(x-1)^2}{2(x-\frac{1}{2})}$$

$$= \frac{(x-1)^2}{2x-1}$$

By comparing the results, you can see that there should be a cancelation of the term $(x-\frac{1}{2})$ with an identical factor in the numerator. That factor of $(x-\frac{1}{2})$ in the numerator is *not* readily apparent.

Check:
$$
\begin{array}{r}
x^2 - 2x + 1 \\
x-\frac{1}{2}\overline{)x^3 - \frac{5}{2}x^2 + 2x - \frac{1}{2}} \\
-(x^3 - \frac{1}{2}x^2) \\
\hline
-2x^2 + 2x \\
-(-2x^2 + x) \\
\hline
x - \frac{1}{2} \\
-(x - \frac{1}{2}) \\
\hline
0
\end{array}
$$

[The remainder is zero; the division is perfect and the quotient is $x^2 - 2x + 1 = (x-1)^2$.]

[Exactly as expected.]

PROBLEM 2-12 Find **(a)** the least common denominator for the following fractions and **(b)** their sum:

$$\frac{7}{15} + \frac{3}{40} + \frac{9}{10}$$

Solution

(a) Find the prime factors of each denominator:

$$15 = 3 \cdot 5$$
$$40 = 2 \cdot 2 \cdot 2 \cdot 5$$
$$10 = 2 \cdot 5$$

So the lcd is $2^3 \cdot 3 \cdot 5 = 8 \cdot 3 \cdot 5 = 120$.

(b) Multiply each fraction by the appropriate numbers (i.e., the appropriate form of $n/n = 1$) to form the lcd in each case.

$$\frac{7}{15} + \frac{3}{40} + \frac{9}{10} = \frac{7}{15} \cdot \frac{8}{8} + \frac{3}{40} \cdot \frac{3}{3} + \frac{9}{10} \cdot \frac{12}{12}$$

$$= \frac{56}{120} + \frac{9}{120} + \frac{108}{120} = \frac{56+9+108}{120} = \frac{173}{120}$$

PROBLEM 2-13 Given $\dfrac{4}{x} - \dfrac{5}{x+1} + \dfrac{x-3}{x^2+x}$, find the equivalent rational expression in simplest form.

Solution   Factor each denominator first to get the prime factors:

$$\left.\begin{array}{l} x = x \\ x + 1 = (x + 1) \\ x^2 + x = x(x + 1) \end{array}\right\} \text{lcd is } x(x + 1)$$

Now multiply each term by the appropriate factors $(P(x)/P(x) = 1)$ to create the same lcd in each denominator:

$$\frac{4}{x} - \frac{5}{x + 1} + \frac{x - 3}{x^2 + x} = \frac{4}{x}\left(\frac{x + 1}{x + 1}\right) - \frac{5}{x + 1}\left(\frac{x}{x}\right) + \frac{x - 3}{x(x + 1)}$$

$$= \frac{4(x + 1)}{x(x + 1)} - \frac{5x}{x(x + 1)} + \frac{x - 3}{x(x + 1)}$$

$$= \frac{4x + 4 - 5x + x - 3}{x(x + 1)} = \frac{1}{x(x + 1)}$$

PROBLEM 2-14   Find the least common denominator for the rational expression $\frac{3}{x + 3} + \frac{1 - x}{2x^2 + 6x} - \frac{2x + 1}{x^2}$ and perform the indicated operations.

Solution   First factor each denominator completely:

$$\left.\begin{array}{l} x + 3 = (x + 3) \\ 2x^2 + 6x = 2x(x + 3) \\ x^2 = x^2 \end{array}\right\} \text{lcd is } 2x^2(x + 3)$$

Now multiply each term by an appropriate factor to create the lcd in each denominator:

$$\frac{3}{x + 3} + \frac{1 - x}{2x^2 + 6x} - \frac{2x + 1}{x^2} = \frac{3}{x + 3}\left(\frac{2x^2}{2x^2}\right) + \frac{1 - x}{2x^2 + 6x}\left(\frac{x}{x}\right) - \frac{2x + 1}{x^2}\left(\frac{2(x + 3)}{2(x + 3)}\right)$$

$$= \frac{6x^2}{2x^2(x + 3)} + \frac{(1 - x)x}{2x^2(x + 3)} - \frac{(2x + 1)(2x + 6)}{2x^2(x + 3)}$$

$$= \frac{6x^2 + (x - x^2) - (4x^2 + 14x + 6)}{2x^2(x + 3)}$$

$$= \frac{x^2 - 13x - 6}{2x^2(x + 3)}$$

PROBLEM 2-15   Simplify the expression $\dfrac{2x - \dfrac{8}{x}}{\dfrac{3x - 6}{x^2 + x}}$.

Solution   Work first in the numerator to create a single rational expression:

$$2x - \frac{8}{x} = 2x\left(\frac{x}{x}\right) - \frac{8}{x} = \frac{2x^2}{x} - \frac{8}{x} = \frac{2x^2 - 8}{x} = \frac{2(x^2 - 4)}{x} = \frac{2(x - 2)(x + 2)}{x}$$

Then work on the complete factorization of the denominator:

$$\frac{3x - 6}{x^2 + x} = \frac{3(x - 2)}{x(x + 1)}$$

Now rewrite the original expression and cancel the common terms:

$$\frac{2(x-2)(x+2)}{x} \div \frac{3(x-2)}{x(x+1)} = \frac{2(x-2)(x+2)}{x} \cdot \frac{x(x+1)}{3(x-2)}$$

$$= \frac{2}{3}(x+2)(x+1)$$

PROBLEM 2-16 Simplify $\dfrac{1}{h}\left[\dfrac{1}{(x+h)^2} - \dfrac{1}{x^2}\right]$.

Solution   First, combine the terms inside the brackets into a single expression:

$$\frac{1}{(x+h)^2} - \frac{1}{x^2} = \frac{1}{(x+h)^2}\left(\frac{x^2}{x^2}\right) - \frac{1}{x^2}\left(\frac{(x+h)^2}{(x+h)^2}\right)$$

$$= \frac{x^2}{x^2(x+h)^2} - \frac{(x+h)^2}{x^2(x+h)^2} = \frac{x^2 - (x+h)^2}{x^2(x+h)^2}$$

$$= \frac{x^2 - (x^2 + 2xh + h^2)}{x^2(x+h)^2} = \frac{x^2 - x^2 - 2xh - h^2}{x^2(x+h)^2}$$

$$= \frac{-2xh - h^2}{x^2(x+h)^2}$$

Now combine this result with the multiplier $1/h$ from the original problem to obtain

$$\frac{1}{h}\left[\frac{1}{(x+h)^2} - \frac{1}{x^2}\right] = \frac{1}{h}\left[\frac{-2xh - h^2}{x^2(x+h)^2}\right] = \frac{1}{h}\left[\frac{h(-2x-h)}{x^2(x+h)^2}\right] = \frac{-2x-h}{x^2(x+h)^2}$$

PROBLEM 2-17 Simplify $\left(\dfrac{2x}{y} + \dfrac{2y}{x}\right) \div \left(\dfrac{x^2 + y^2}{(x+y)^2 - (x-y)^2}\right)$.

Solution   Work first inside the parentheses. The dividend is

$$\frac{2x}{y} + \frac{2y}{x} = \frac{2x}{y}\left(\frac{x}{x}\right) + \frac{2y}{x}\left(\frac{y}{y}\right) = \frac{2x^2 + 2y^2}{xy}$$

and the divisor is

$$\frac{x^2 + y^2}{(x+y)^2 - (x-y)^2} = \frac{x^2 + y^2}{(x^2 + 2xy + y^2) - (x^2 - 2xy + y^2)} = \frac{x^2 + y^2}{4xy}$$

So the original expression is now

$$\frac{2x^2 + 2y^2}{xy} \div \frac{x^2 + y^2}{4xy} = \frac{2(x^2 + y^2)}{xy} \cdot \frac{4xy}{x^2 + y^2} = 8$$

PROBLEM 2-18 Simplify $\dfrac{\dfrac{x+3}{x} - 2}{4 - \dfrac{x-1}{x}}$.

Solution   If

$$\frac{x+3}{x} - 2 = \frac{x+3}{x} - 2\left(\frac{x}{x}\right) = \frac{x+3}{x} - \frac{2x}{x} = \frac{x+3-2x}{x} = \frac{-x+3}{x}$$

and

$$4 - \frac{x-1}{x} = 4\left(\frac{x}{x}\right) - \frac{x-1}{x} = \frac{4x}{x} - \frac{x-1}{x} = \frac{4x - x + 1}{x} = \frac{3x+1}{x}$$

then
$$\frac{-x+3}{x} \div \frac{3x+1}{x} = \frac{3-x}{x} \cdot \frac{x}{3x+1} = \frac{3-x}{3x+1}$$

PROBLEM 2-19   Use the PEMDAS order of operations to simplify the following:

(a) $\dfrac{y}{x} - \dfrac{xy+1}{x^2} \div \dfrac{2}{x-1} + \left(\dfrac{1}{xy}\right)^2 \cdot \dfrac{1-xy}{2}$ (b) $\left(\dfrac{y}{x} - \dfrac{xy+1}{x^2}\right) \div \left(\dfrac{2}{x-1} + \dfrac{1}{xy}\right)^2 \cdot \dfrac{1-xy}{2}$

(c) $\left(\dfrac{y}{x} - \dfrac{xy+1}{x^2}\right) \div \left[\dfrac{2}{x-1} + \left(\dfrac{1}{xy}\right)^2 \cdot \dfrac{1-xy}{2}\right]$

Solution

(a) $\dfrac{y}{x} - \dfrac{xy+1}{x^2} \div \dfrac{2}{x-1} + \left(\dfrac{1}{xy}\right)^2 \cdot \dfrac{1-xy}{2}$

$= \dfrac{y}{x} - \dfrac{xy+1}{x^2} \div \dfrac{2}{x-1} + \dfrac{1}{x^2y^2} \cdot \dfrac{1-xy}{2}$ Expand the exponent expression.

$= \dfrac{y}{x} - \dfrac{xy+1}{x^2} \cdot \dfrac{x-1}{2} + \dfrac{1-xy}{2x^2y^2}$ Divide (invert and multiply).

$= \dfrac{y}{x} - \dfrac{(xy+1)(x-1)}{2x^2} + \dfrac{1-xy}{2x^2y^2}$ Multiply.
(lcd is $2x^2y^2$)

$= \dfrac{y}{x}\left(\dfrac{2xy^2}{2xy^2}\right) - \dfrac{x^2y+x-xy-1}{2x^2}\left(\dfrac{y^2}{y^2}\right) + \dfrac{1-xy}{2x^2y^2}$

$= \dfrac{2xy^3}{2x^2y^2} - \dfrac{(x^2y^3+xy^2-xy^3-y^2)}{2x^2y^2} + \dfrac{1-xy}{2x^2y^2}$ Form correct denominators.

$= \dfrac{2xy^3 - x^2y^3 - xy^2 + xy^3 + y^2 + 1 - xy}{2x^2y^2}$ Collect like terms.

$= \dfrac{3xy^3 - x^2y^3 - xy^2 + y^2 + 1 - xy}{2x^2y^2}$ Voila!

(b) $\left(\dfrac{y}{x} - \dfrac{xy+1}{x^2}\right) \div \left(\dfrac{2}{x-1} + \dfrac{1}{xy}\right)^2 \cdot \dfrac{1-xy}{2}$ Work inside the parentheses.

$= \left(\dfrac{xy}{x^2} - \dfrac{xy+1}{x^2}\right) \div \left(\dfrac{2xy+x-1}{(x-1)xy}\right)^2 \left(\dfrac{1-xy}{2}\right)$

$= \dfrac{xy-xy-1}{x^2} \div \left(\dfrac{2xy+x-1}{(x-1)xy}\right)^2 \left(\dfrac{1-xy}{2}\right)$ Note: The placement (or nonplacement) of parentheses is very important.

$= \left(\dfrac{-1}{x^2}\right)\left(\dfrac{(x-1)^2x^2y^2}{(2xy+x-1)^2}\right)\left(\dfrac{1-xy}{2}\right)$ $\dfrac{A}{B} \div \dfrac{C}{D} \cdot \dfrac{E}{F} = \dfrac{A}{B} \cdot \dfrac{D}{C} \cdot \dfrac{E}{F}$

$= \dfrac{-y^2(x-1)^2(1-xy)}{2(2xy+x-1)^2}$ $\dfrac{A}{B} \div \left(\dfrac{C}{D} \cdot \dfrac{E}{F}\right) = \dfrac{A}{B} \div \dfrac{CE}{DF}$

$= \dfrac{y^2(x-1)^2(xy-1)}{2(2xy+x-1)^2}$ $= \dfrac{A}{B} \cdot \dfrac{DF}{CE} = \dfrac{A}{B} \cdot \dfrac{D}{C} \cdot \dfrac{F}{E}$

(c) $\left(\dfrac{y}{x} - \dfrac{xy+1}{x^2}\right) \div \left[\dfrac{2}{x-1} + \left(\dfrac{1}{xy}\right)^2 \cdot \dfrac{1-xy}{2}\right]$

$= \left(\dfrac{y}{x} - \dfrac{xy+1}{x^2}\right) \div \left(\dfrac{2}{x-1} + \dfrac{1}{x^2y^2} \cdot \dfrac{1-xy}{2}\right)$ Evaluate inside the inner grouping symbols.

$= \left(\dfrac{xy}{x^2} - \dfrac{xy+1}{x^2}\right) \div \left(\dfrac{2}{x-1} + \dfrac{1-xy}{2x^2y^2}\right)$ Evaluate inside the outer grouping symbols.

$$= \left(\frac{xy - xy - 1}{x^2}\right) \div \left(\frac{4x^2y^2 + (x-1)(1-xy)}{(x-1)2x^2y^2}\right)$$

$$= \left(\frac{-1}{x^2}\right)\left(\frac{2x^2y^2(x-1)}{4x^2y^2 + x - 1 - x^2y + xy}\right)$$

$$= \frac{-2y^2(x-1)}{4x^2y^2 - x^2y + xy + x - 1}$$

PROBLEM 2-20   Evaluate the following expressions for the given variable value:

(a) $\dfrac{x^2 + 4x}{x^3 - 2x^2 + 5x - 3}$ at $x = 1$ (b) $\dfrac{(x + 5)(x - 5)}{x^2 + 9}$ at $x = \sqrt{3}$

Solution

(a) Insert the given x value, calculate the numbers, and simplify:

$$\frac{x^2 + 4x}{x^3 - 2x^2 + 5x - 3} \text{ at } x = 1 \Rightarrow \frac{(1)^2 + 4(1)}{(1)^3 - 2(1)^2 + 5(1) - 3} = \frac{1 + 4}{1 - 2 + 5 - 3} = \frac{5}{1} = 5$$

(b) Multiply the numerator out and *then* evaluate:

$$\frac{(x + 5)(x - 5)}{x^2 + 9} = \frac{x^2 - 25}{x^2 + 9} \text{ at } x = \sqrt{3} \Rightarrow \frac{(\sqrt{3})^2 - 25}{(\sqrt{3})^2 + 9} = \frac{3 - 25}{3 + 9} = \frac{-22}{12} = -\frac{11}{6}$$

Supplementary Exercises

Reduce each of the following expressions to lowest terms.

PROBLEM 2-21

(a) $\dfrac{x + 1}{x^2 - 8x - 9}$

(d) $\dfrac{9t^2 - 1}{3t^2 + 4t + 1}$

(b) $\dfrac{1}{x} + \dfrac{2}{x - 1}$

(e) $\dfrac{y - 2}{y^2 - 4y} + \dfrac{-1}{y - 4}$

(c) $\dfrac{x^2 + y^2}{x^4 - y^4}$

(f) $\dfrac{1}{R_1} + \dfrac{1}{R_2}$

PROBLEM 2-22   (a) $\dfrac{x^2(2x + 1)}{x - 3} + \dfrac{9(2x + 1)}{3 - x}$ (b) $\dfrac{\dfrac{1}{xy}}{\dfrac{x^2y^2}{x - y}}$ (c) $\left(\dfrac{\frac{5}{24} + \frac{7}{18} + \frac{11}{6}}{\frac{1}{5} + \frac{7}{15}}\right)$

PROBLEM 2-23

(a) $\dfrac{(x + h)^2 - 3(x + h) - (x^2 - 3x)}{h}$

(b) $\dfrac{4(x + h)^2 - 3(x + h) - 4x^2 + 3x}{h}$

PROBLEM 2-24   (a) $\dfrac{4a^2 - 9b^2}{a - \frac{3}{2}b}$ (b) $\dfrac{16u^2 - 25v^2}{4u^2 - uv - 5v^2}$ (c) $\dfrac{x^2 + 2x + 1}{x^3 + 3x^2 + 3x + 1}$

Perform the indicated operations in each of the following and simplify.

PROBLEM 2-25

(a) $\dfrac{4x^2 - 1}{6x^2 + x - 1} \cdot \dfrac{x^2 - 5x + 6}{x^2 + x - 12}$
(b) $\dfrac{3x^2y - 6x^2 - 3xy}{3xy}$
(c) $\dfrac{x - 3}{x + 2} \div \dfrac{x - 3}{x - 2}$

PROBLEM 2-26 (a) $\dfrac{x^2 + 3x + 2}{x^2 + 2x + 1} \div \dfrac{4x + 8}{3x + 3}$
(b) $(a^2 \div b^2) \div \left(\dfrac{a^2 + ab}{ab + b^2} \div \dfrac{ab - a^2}{ab - b^2} \right)$

PROBLEM 2-27

(a) $\left(\dfrac{y^2}{y^2 - 1} + \dfrac{2y}{y - 1} \right) \div \dfrac{y^3}{y - 1}$
(b) $\left(\dfrac{x}{x - 2} + \dfrac{1}{(x - 2)^2} \right) \cdot \left(\dfrac{1}{x + 2} - \dfrac{x}{(x + 2)^2} \right)$

PROBLEM 2-28 (a) $\dfrac{x^3 + y^3}{x^2 - y^2} \cdot \dfrac{x - y}{xy}$
(b) $\left(\dfrac{1}{a^2} - \dfrac{1}{a^3} \right)\left(\dfrac{1}{a} + \dfrac{1}{a^2} \right)$
(c) $\dfrac{y}{y + x} - \dfrac{x}{x - y}$

PROBLEM 2-29

(a) $\dfrac{(-10x^3y^2)(-4x^3y^4)}{8x^2y^3}$
(c) $\dfrac{5t - 15}{t^2 + 2t - 15} - \dfrac{t^2 + 5t}{t^2 + 8t + 15}$

(b) $\dfrac{3x^2 + 7x - 6}{x + 3} - \dfrac{3x^2 - 5x + 2}{x - 1}$
(d) $\dfrac{x + 5}{x^3 + 125} - \dfrac{1}{x + 5} + \dfrac{x - 5}{x^2 - 25}$

PROBLEM 2-30

(a) $\dfrac{x}{(y - x)(z - x)} + \dfrac{y}{(z - y)(x - y)} + \dfrac{z}{(x - z)(y - z)}$

(b) $\dfrac{2 - x}{x + 1} \div \dfrac{x^2 - 4}{x^2 + 3x + 2}$

(c) $\dfrac{9u^2 - 4v^2}{2u^2 + 5u - 3} \cdot \dfrac{u + 3}{2v - 3u}$

(d) $\dfrac{x^3 + y^3}{x^3 - y^3} \div \dfrac{x^2 - xy + y^2}{x^2 + xy + y^2}$

PROBLEM 2-31   Use the long-division process to reduce

(a) $\dfrac{5x^2 + 7x - 5}{x + 2}$
(c) $\dfrac{x^3 + 8x^2 + 9x + 3}{x^2 + 3}$

(b) $\dfrac{y^2 - 14y + 22}{y - 2}$
(d) $\dfrac{y^3 - 7y^2 + 3y - 8}{y - 4}$

PROBLEM 2-32   Simplify each of the following:

(a) $\dfrac{y}{y - \frac{1}{2}} + \dfrac{3 - 2y}{2y - 1}$
(c) $\dfrac{\frac{1}{20}}{(x - \frac{1}{4})(x - \frac{1}{5})} - \dfrac{20x^2 - 9x}{(4x - 1)^2(5x - 1)^2}$

(b) $\left(\dfrac{u}{v} - \dfrac{v}{u} \right) \div \left(\dfrac{u}{v} + \dfrac{v}{u} \right) \cdot \left(\dfrac{u^2 + v^2}{u^2 - v^2} \right)$
(d) $1 - \dfrac{1}{1 - \dfrac{1}{1 - x}}$

PROBLEM 2-33 Find the lcd and the sum for each of the following expressions:

(a) $\dfrac{1}{18} + \dfrac{1}{12} + \dfrac{1}{24}$ (b) $\dfrac{35 - 8\frac{3}{4}}{3\frac{1}{4} + \frac{1}{2}}$ (c) $\dfrac{11}{84} + \dfrac{4}{49} - \dfrac{5}{28}$

PROBLEM 2-34 Find the equivalent rational expressions in simplest form:

(a) $\dfrac{2}{x-3} - \dfrac{x}{x+1} + \dfrac{x^2 - 2x}{x^2 - 2x - 3}$

(c) $\dfrac{1}{y^3 + y^2 - 2y} - \dfrac{\frac{3}{2}}{y+2} + \dfrac{1}{2 - y - y^2}$

(b) $\dfrac{2-x}{x^2 + 4x} + \dfrac{5}{x} - \dfrac{4-x}{x+4}$

(d) $\dfrac{2x + 12}{x^2 - 4} - \dfrac{2}{\frac{1}{2}x - 1} + \dfrac{x-1}{x+2}$

PROBLEM 2-35 Simplify

(a) $\dfrac{\dfrac{1}{x+1} - \dfrac{1}{x-1}}{\dfrac{1}{x-3} - \dfrac{1}{x-1}}$ (b) $\dfrac{\dfrac{1}{2} - \dfrac{4}{x^3}}{\dfrac{1}{x^2} + \dfrac{1}{4} + \dfrac{1}{2x}}$ (c) $\dfrac{-\dfrac{1}{u} - \dfrac{1}{v}}{\dfrac{1}{v^2} - \dfrac{1}{u^2}}$

PROBLEM 2-36 Simplify

(a) $\dfrac{1}{x} \div \dfrac{y}{x^2} + \dfrac{x}{y^2} \cdot \dfrac{x-y}{xy} - x$

(c) $\left(\dfrac{1}{x} \div \dfrac{y}{x^2}\right) + \left(\dfrac{x}{y^2} \cdot \dfrac{x-y}{xy} - x\right)$

(b) $\dfrac{1}{x} \div \left(\dfrac{y}{x^2} + \dfrac{x}{y^2}\right) \cdot \left(\dfrac{x-y}{xy} - x\right)$

(d) $\dfrac{1}{x} \div \left[\dfrac{y}{x^2} + \left(\dfrac{x}{y^2} \cdot \dfrac{x-y}{xy}\right)\right] - x$

PROBLEM 2-37 Test your skill on this "potpourri" expression:

$$\left[\dfrac{2x - 2y}{x^2 y^2} \cdot \left(1 + \dfrac{y-1}{x+1}\right) \div \dfrac{x^2 - y^2}{x+1}\right] - \left[\left(\dfrac{x}{y} - \dfrac{y}{x}\right)^2 \cdot \dfrac{1}{(x-y)^2} \div \dfrac{(x+y)^2}{2}\right]$$

Answers to Supplementary Exercises

2-21 (a) $\dfrac{1}{x-9}$

(b) $\dfrac{3x-1}{x(x-1)}$

(c) $\dfrac{1}{(x-y)(x+y)}$

(d) $\dfrac{3t-1}{t+1}$

(e) $\dfrac{-2}{y(y-4)}$

(f) $\dfrac{R_1 + R_2}{R_1 R_2}$

2-22 (a) $(2x+1)(x+3)$

(b) $(x-y)/(x^3 y^3)$

(c) $175/48$

2-23 (a) $(2x-3) + h$

(b) $(8x-3) + 4h$

2-24 (a) $2(2a + 3b)$

(b) $\dfrac{4u + 5v}{u+v}$

(c) $\dfrac{1}{x+1}$

2-25 (a) $\dfrac{(2x-1)(x-2)}{(3x-1)(x+4)}$

(b) $\dfrac{1}{y}(xy - 2x - y)$

(c) $\dfrac{x-2}{x+2}$

2-26 (a) $\dfrac{3}{4}$

(b) $-\dfrac{a^2}{b^2}$

2-27 (a) $\dfrac{3y+2}{y^2(y+1)}$

(b) $\dfrac{2(x-1)^2}{(x^2-4)^2}$

2-28 (a) $\dfrac{x^2 - xy + y^2}{xy}$

(b) $\dfrac{a^2 - 1}{a^5}$

(c) $\dfrac{-x^2 - y^2}{x^2 - y^2}$

2-29 (a) $5x^4y^3$

(b) 0

(c) $\dfrac{15 - t^2}{(t + 5)(t + 3)}$

(d) $\dfrac{1}{x^2 - 5x + 25}$

2-30 (a) 0

(b) -1

(c) $\dfrac{3u + 2v}{1 - 2u}$

(d) $\dfrac{x + y}{x - y}$

2-31 (a) $5x - 3 + \dfrac{1}{x + 2}$

(b) $y - 12 - \dfrac{2}{y - 2}$

(c) $x + 8 + \dfrac{6x - 21}{x^2 + 3}$

(d) $y^2 - 3y - 9 - \dfrac{44}{y - 4}$

2-32 (a) $\dfrac{3}{2y - 1}$

(b) 1

(c) $\dfrac{1}{(4x - 1)^2(5x - 1)^2}$

(d) $\dfrac{1}{x}$

2-33 (a) $\dfrac{13}{72}$

(b) 7

(c) $\dfrac{5}{147}$

2-34 (a) $\dfrac{3x + 2}{(x - 3)(x + 1)}$

(b) $\dfrac{x^2 + 22}{x^2 + 4x}$

(c) $\dfrac{-\dfrac{3}{2}y^2 + \dfrac{1}{2}y + 1}{y(y + 2)(y - 1)}$

(d) $\dfrac{x - 3}{x + 2}$

2-35 (a) $\dfrac{3 - x}{x + 1}$

(b) $\dfrac{2x - 4}{x}$

(c) $\dfrac{uv}{v - u}$

2-36 (a) $\dfrac{xy^2 + x - y - xy^3}{y^3}$

(b) $\dfrac{y(x - y - x^2y)}{y^3 + x^3}$

(c) $\dfrac{xy^2 + x - y - xy^3}{y^3}$

(d) $\dfrac{xy^3 - xy^4 - x^4 + x^3y}{y^4 + x^3 - x^2y}$

2-37 0

3
EXPONENTS, RADICALS, AND COMPLEX NUMBERS

THIS CHAPTER IS ABOUT

- ☑ **Integral Exponents**
- ☑ **Radicals**
- ☑ **Rational Exponents**
- ☑ **Complex Numbers**

3-1. Integral Exponents

A. Integral exponents—a reminder

The expression a^n (read "a to the nth power") signifies the product of n values of a, where a is a real number and n is an integer. In the expression a^n, a is the **base** and the power n is the **exponent** [see also Section 1-4A]:

$$\text{base} \xrightarrow{\text{exponent}} a^n = \underbrace{a \cdot a \cdots a}_{n \text{ times}}$$

We obtain the product $a^m \cdot a^n$, where the base is the same and n and m are both integers, by adding exponents:

$$a^m \cdot a^n = \underbrace{(a \cdot a \cdots a)}_{m \text{ times}} \cdot \underbrace{(a \cdot a \cdots a)}_{n \text{ times}} = a^{m+n}$$

And we raise a^m to the nth power by multiplying exponents:

$$(a^n)^m = \underbrace{(a^n \cdot a^n \cdots a^n)}_{m \text{ times}} = \underbrace{\underbrace{(a \cdot a \cdots a)}_{n \text{ times}} \cdot \underbrace{(a \cdot a \cdots a)}_{n \text{ times}} \cdots \underbrace{(a \cdot a \cdots a)}_{n \text{ times}}}_{m \cdot n \text{ times}} = a^{nm}$$

[See also Section 1-5B.]

$$(a^2)^4 = \underbrace{a^2 \cdot a^2 \cdot a^2 \cdot a^2}_{4 \text{ times}}$$

$$= \underbrace{(a \cdot a) \cdot (a \cdot a) \cdot (a \cdot a) \cdot (a \cdot a)}_{8 \text{ repetitions of } a}$$

$$= a^8$$

EXAMPLE 3-1 If you deposit A dollars in a savings account earning 8% annual interest, how much money will you have in the account after n years?

Solution After one year, you'll have $100\% + 8\% = 108\%$ of your initial amount, or

$$A + .08A = 1.08A$$

And, after two years, you'll have 108% of what you had at the end of the first year, or A times 1.08 times itself:

$$\underbrace{A(1.08)(1.08)}_{2 \text{ times}} = A(1.08)^2$$

So, after n years, you'll have A times 1.08 times itself n times:

$$\underbrace{A(1.08 \cdot 1.08 \cdot 1.08 \cdots 1.08)}_{n \text{ times}} = A(1.08)^n$$

If $A = \$100.00$, after 5 years you would have $\$(100)(1.08)(1.08) \times (1.08)(1.08)(1.08) = \$146.93$

EXAMPLE 3-2   A colony consisting of 1000 bacteria has the property that it doubles its size every day. How many bacteria will there be at the end of n days?

Solution

$$1000\underbrace{(2 \cdot 2 \cdot 2 \cdots 2)}_{n \text{ times}} = 2^n(1000)$$

EXAMPLE 3-3   If a radioactive substance loses $\frac{1}{5}$ of its mass each week in the process of decay, how much of the substance will be left after n weeks?

Solution   Call the initial mass M. Then, after one week, the mass will be

$$M - \tfrac{1}{5}M = M(1 - \tfrac{1}{5}) = M(.8)$$

And, after two weeks,

$$M(.8)(.8) = M(.8)^2$$

So, after n weeks,

$$M\underbrace{[(.8)(.8)\cdots(.8)]}_{n \text{ times}} = M(.8)^n$$

B.   General rules of integral exponents

If a and b are real numbers and m and n are integers, then

RULES OF INTEGRAL EXPONENTS

1. $a^m a^n = a^{m+n}$ **4.** $\left(\dfrac{a}{b}\right)^n = \dfrac{a^n}{b^n}$ for $b \neq 0$

2. $(a^m)^n = a^{mn}$ **5.** $a^{-n} = \dfrac{1}{a^n}$ for $a \neq 0$

3. $(ab)^n = a^n b^n$ **6.** $\dfrac{a^m}{a^n} = a^{m-n}$ for $a \neq 0$

note: (1) From Rule 6 it follows that if $m = n$, $\dfrac{a^m}{a^n} = \dfrac{a^n}{a^n} = a^{n-n} = a^0 = 1$.

   (2) The restrictions on Rules 4, 5, and 6 are important: Division by zero is *meaningless*.

   (3) Rule 4 can also be written as $\left(\dfrac{a}{b}\right)^n = a^n b^{-n}$.

EXAMPLE 3-4   Use the rules of integral exponents to simplify the following expressions:

(a) $3^7 \cdot 3^{-5}$ (b) $[(-4)^3]^2$ (c) $(2 \cdot 3)^3$ (d) $\dfrac{4^{12}}{4^7}$ (e) $\dfrac{2^3}{2^3}$ (f) $\left(\dfrac{\pi}{3}\right)^2$

Solution

(a) $3^7 \cdot 3^{-5} = \dfrac{3^7}{3^5}$ $\left(Rule\ 5:\ a^{-n} = \dfrac{1}{a^n}\right)$

   then $\dfrac{3^7}{3^5} = 3^{7-5} = 3^2 = 9$ $\left(Rule\ 6:\ \dfrac{a^m}{a^n} = a^{m-n}\right)$

   or

   $3^7 \cdot 3^{-5} = 3^{7+(-5)} = 3^{7-5} = 3^2 = 9$ (*Rule 1:* $a^m a^n = a^{m+n}$)

(b) $[(-4)^3]^2 = (-4)^6 = 4096$ (*Rule 2:* $(a^m)^n = a^{mn}$)

note: $(-a)^n$ is nonnegative if n is even.
$(-a)^n$ is negative if n is odd.

(c) $\quad (2\cdot 3)^3 = 2^3\cdot 3^3 = 8\cdot 27 = 216 \qquad$ (*Rule 3:* $(ab)^n = a^n b^n$)

(d) $\quad \dfrac{4^{12}}{4^7} = 4^{12-7} = 4^5 = 1024 \qquad \left(\text{*Rule 6:* } \dfrac{a^m}{a^n} = a^{m-n}\right)$

(e) $\qquad\qquad \dfrac{2^3}{2^3} = 2^{3-3} = 2^0 \qquad$ (*Rule 6*)

but $\qquad\qquad \dfrac{2^3}{2^3} = \dfrac{8}{8} = 1 \qquad$ Hence $2^0 = 1$

(f) $\qquad\qquad \left(\dfrac{\pi}{3}\right)^2 = \dfrac{\pi^2}{3^2} = \dfrac{\pi^2}{9} \qquad \left(\text{*Rule 4:* } \left(\dfrac{a}{b}\right)^n = \dfrac{a^n}{b^n}\right)$

EXAMPLE 3-5 Simplify

(a) $(xy^2)^3$ $\quad$ **(b)** $\left(\dfrac{a^2 b}{ab^3}\right)^3$ $\quad$ **(c)** $\dfrac{(a^{-2})^3}{(a^{-3})^2}$ $\quad$ **(d)** $\dfrac{8}{2^{-4}} \div \dfrac{4^3}{2^{-2}}$

Solution

(a) $\qquad (xy^2)^3 = x^3(y^2)^3 = x^3 y^6 \qquad$ (*Rules 2 and 3*)

(b) $\qquad \left(\dfrac{a^2 b}{ab^3}\right)^3 = \left(\dfrac{a^2}{a}\cdot\dfrac{b}{b^3}\right)^3 = \left(\dfrac{a}{b^2}\right)^3 \qquad$ (Cancel first; *Rules 5 and 6*)

$\qquad \dfrac{a^3}{(b^2)^3} = \dfrac{a^3}{b^6} \qquad$ (*Rules 2 and 4*)

or

$\qquad \left(\dfrac{a^2 b}{ab^3}\right)^3 = \left(\dfrac{a^6 b^3}{a^3 b^9}\right) = \dfrac{a^3}{b^6}$

(c) $\qquad \dfrac{(a^{-2})^3}{(a^{-3})^2} = \dfrac{a^{-6}}{a^{-6}} = a^{-6-(-6)}$

$\qquad\qquad = a^{-6+6} = a^0 = 1 \qquad$ (*Rules 2 and 6*)

(d) $\qquad \dfrac{8}{2^{-4}} \div \dfrac{4^3}{2^{-2}} = 8\cdot 2^4 \div 4^3 \cdot 2^2 \qquad$ (*Rule 5*)

$\qquad\qquad = (8\cdot 16) \div (64\cdot 4) = \dfrac{1}{2}$

note: If $a^{-n} = \dfrac{1}{a^n}$, then $a^n = \dfrac{1}{a^{-n}}$.

EXAMPLE 3-6 Simplify

(a) $\dfrac{(-3x)^2(-x^2)^3}{(x^{-4})^2}$ $\quad$ **(b)** $\dfrac{2^{-1}+2^0}{2^{-2}+2^{-4}}$ $\quad$ **(c)** $\dfrac{x^{-1}-y^{-1}}{x^{-1}+y^{-1}}$

Solution

(a) Expand the powers first (including the coefficients!):

$$\dfrac{(-3x)^2(-x^2)^3}{(x^{-4})^2} = \dfrac{[(-3)^2 x^2](-1)^3 x^6}{x^{-8}} = \dfrac{(9x^2)(-x^6)}{x^{-8}}$$

$$= -\dfrac{9x^8}{x^{-8}} = -9x^8\cdot x^8 = -9x^{16}$$

(b) Remove the negative exponents via Rule 5:

$$\frac{2^{-1} + 2^0}{2^{-2} + 2^{-4}} = \frac{\dfrac{1}{2^1} + 1}{\dfrac{1}{2^2} + \dfrac{1}{2^4}} = \frac{\dfrac{1}{2} + 1}{\dfrac{1}{4} + \dfrac{1}{16}} = \frac{\dfrac{3}{2}}{\dfrac{1}{4} + \dfrac{1}{16}}$$

The lcd is 16, so revise each fraction:

$$\frac{\left(\dfrac{3}{2} \cdot \dfrac{8}{8}\right)}{\left(\dfrac{1}{4} \cdot \dfrac{4}{4}\right) + \dfrac{1}{16}} = \frac{\dfrac{24}{16}}{\dfrac{4}{16} + \dfrac{1}{16}} = \frac{\dfrac{24}{16}}{\dfrac{5}{16}} = \frac{24}{16} \cdot \frac{16}{5} = \frac{24}{5}$$

(c) Remove the negative exponents first, find the lcd, and divide:

$$\frac{x^{-1} - y^{-1}}{x^{-1} + y^{-1}} = \frac{\dfrac{1}{x} - \dfrac{1}{y}}{\dfrac{1}{x} + \dfrac{1}{y}} = \frac{\dfrac{1}{x}\left(\dfrac{y}{y}\right) - \dfrac{1}{y}\left(\dfrac{x}{x}\right)}{\dfrac{1}{x}\left(\dfrac{y}{y}\right) + \dfrac{1}{y}\left(\dfrac{x}{x}\right)}$$

$$= \frac{\dfrac{y}{xy} - \dfrac{x}{xy}}{\dfrac{y}{xy} + \dfrac{x}{xy}} = \frac{\dfrac{y-x}{xy}}{\dfrac{y+x}{xy}} = \frac{y-x}{xy} \cdot \frac{xy}{y+x} = \frac{y-x}{y+x}$$

C. Extended rules of integral exponents

The exponential rules are not limited to the case of two product numbers: They can be extended. For example,

$$a^m a^n a^p = a^{m+n+p}$$

$$(abc)^n = a^n b^n c^n$$

EXAMPLE 3-7 Simplify

(a) $(2x^2)^3(xy)^4(2^2x^3y^2)^{-2}$ and   **(b)** $\dfrac{(4u^{-2}v^2)^2(2uv^3)^{-2}}{(3^0u^{-3}v^{-1})^{-2}}$.

Solution

(a)

$$(2x^2)^3(xy)^4(2^2x^3y^2)^{-2} = (2^3x^6)(x^4y^4)(2^{-4}x^{-6}y^{-4})$$

$$= (8x^6)(x^4y^4)\left(\frac{1}{16} \cdot \frac{1}{x^6} \cdot \frac{1}{y^4}\right)$$

$$= (8x^{6+4}y^4)\left(\frac{1}{16} \cdot \frac{1}{x^6} \cdot \frac{1}{y^4}\right)$$

$$= \frac{8x^{10}y^4}{16x^6y^4} = \frac{x^4}{2}$$

or

$$(2x^2)^3(xy)^4(2^2x^3y^2)^{-2} = (2^3x^6)(x^4y^4)(2^{-4}x^{-6}y^{-4})$$

[Remember that $a^0 = 1$.]

$$= (2^{3-4})(x^{6+4-6}y^{4-4})$$

$$= \frac{x^4y^0}{2} = \frac{x^4}{2}$$

(b)

$$\frac{(4u^{-2}v^2)^2(2uv^3)^{-2}}{(3^0u^{-3}v^{-1})^{-2}} = \frac{(4^2u^{-4}v^4)(2^{-2}u^{-2}v^{-6})}{1^{-2}u^6v^2}$$

$$= \frac{16u^{-4}v^4 \cdot \frac{1}{4}u^{-2}v^{-6}}{\left(\frac{1}{1^2}\right)u^6v^2} = \frac{\left(\frac{16}{4}\right)u^{-4+(-2)}v^{4+(-6)}}{u^6v^2}$$

$$= \frac{4u^{-6}v^{-2}}{u^6v^2} = 4u^{-12}v^{-4} = \frac{4}{u^{12}v^4}$$

3-2. Radicals

A. Square roots and radicals

The symbols $\sqrt{2}, \sqrt{3}, \sqrt{7}$, etc., may be used to denote certain real numbers, which are the values of the length of a side of a square having an area, respectively, of 2, 3, 7, etc.

- If a is any real number greater than or equal to zero,

$$\sqrt{a} = b \text{ if and only if } b^2 = a \text{ and } b \geq 0$$

We call b the **(principal) square root** of a. The expression $\sqrt{a}$ is also called a **radical**, the value a the **radicand**, and the symbol $\sqrt{}$ the **radical sign**.

Since $(-\sqrt{a})^2 = (-1)^2(\sqrt{a})^2 = (\sqrt{a})^2 = a$, for $a \geq 0$, the squares of $-a$ and $+a$ are the same.

If $a > 0$ and $b^2 = a$, then $(-b)^2 = a$, too. So there are always *two* real numbers—one positive and one negative—that give the value a when squared. But, by our definition, $\sqrt{a}$ will mean *only* the POSITIVE real number whose square is a. For example, $\sqrt{9} = 3$, but $\sqrt{9} \neq -3$.

EXAMPLE 3-8 Find **(a)** $\sqrt{9}$, **(b)** $\sqrt{25}$, **(c)** $\sqrt{169}$, and **(d)** $\sqrt{(\sqrt{3})^2}$.

Solution

(a) $\sqrt{9} = 3$, since $3^2 = 9$ and $3 \geq 0$.
(b) $\sqrt{25} = 5$, since $5^2 = 25$, and $5 \geq 0$.
(c) $\sqrt{169} = 13$, since $13^2 = 169$ and $13 \geq 0$.
(d) $\sqrt{(\sqrt{3})^2} = \sqrt{3}$, since $(\sqrt{3})^2 = (\sqrt{3})^2$ and $\sqrt{3} \geq 0$.

B. Combining radicals

Two of the basic **rules for combining radicals** are

RADICAL PRODUCT $\qquad \sqrt{ab} = \sqrt{a}\sqrt{b}$

RADICAL QUOTIENT $\qquad \sqrt{\dfrac{a}{b}} = \dfrac{\sqrt{a}}{\sqrt{b}}$

EXAMPLE 3-9 Simplify **(a)** $\sqrt{\dfrac{49}{64}}$, **(b)** $\sqrt{18}\sqrt{8}$, and **(c)** $\dfrac{\sqrt{80}}{\sqrt{125}}$.

Solution Use the rules for combining radicals.

(a)

$$\sqrt{\frac{49}{64}} = \frac{\sqrt{49}}{\sqrt{64}} = \frac{\sqrt{7^2}}{\sqrt{8^2}} = \frac{7}{8}$$

(b)

$$\sqrt{18}\sqrt{8} = \sqrt{18 \cdot 8} = \sqrt{144} = 12$$

(c) Look for a common factor and cancel:

$$\frac{\sqrt{80}}{\sqrt{125}} = \frac{\sqrt{16 \cdot 5}}{\sqrt{25 \cdot 5}} = \frac{\sqrt{16}\sqrt{5}}{\sqrt{25}\sqrt{5}} = \frac{\sqrt{16}}{\sqrt{25}} = \frac{4}{5}$$

EXAMPLE 3-10 Assuming that a, b, and c are positive quantities, simplify the following expressions:

(a) $\sqrt{a^4b^2}$ (b) $\sqrt{9a^6b^3}\sqrt{4a^2b^5}$ (c) $\dfrac{\sqrt{a^3b^4c^{-4}}}{\sqrt{16ab^{-2}c^6}}$

Solution Remember the rules of integral exponents.

(a) $\sqrt{a^4b^2} = \sqrt{a^4}\sqrt{b^2} = \sqrt{(a^2)^2}\sqrt{b^2} = a^2b$

(b) $\sqrt{9a^6b^3}\sqrt{4a^2b^5} = (\sqrt{9}\sqrt{a^6}\sqrt{b^3})(\sqrt{4}\sqrt{a^2}\sqrt{b^5})$

$= (\sqrt{9}\sqrt{(a^3)^2}\sqrt{b^2\cdot b})(\sqrt{4}\sqrt{a^2}\sqrt{b^4\cdot b})$ [Factor to find perfect square. Take roots of the squares.]

$= (3a^3b\sqrt{b})(2ab^2\sqrt{b})$

$= 6a^4b^3\sqrt{b}\sqrt{b} = 6a^4b^3b$

$= 6a^4b^4$

(c) $\dfrac{\sqrt{a^3b^4c^{-4}}}{\sqrt{16ab^{-2}c^6}} = \dfrac{\sqrt{a^2\cdot a}\sqrt{b^4}\sqrt{c^{-4}}}{\sqrt{16}\sqrt{a}\sqrt{b^{-2}}\sqrt{c^6}} = \dfrac{a\sqrt{a}b^2\sqrt{c^{-4}}}{4\sqrt{a}\sqrt{b^{-2}}c^3}$

$= \dfrac{ab^2\sqrt{c^{-4}}}{4c^3\sqrt{b^{-2}}} = \dfrac{ab^2\sqrt{\dfrac{1}{c^4}}}{4c^3\sqrt{\dfrac{1}{b^2}}} = \dfrac{ab^2\dfrac{1}{\sqrt{c^4}}}{4c^3\dfrac{1}{\sqrt{b^2}}}$

$= \dfrac{ab^2\dfrac{1}{c^2}}{4c^3\dfrac{1}{b}} = \dfrac{ab^2}{c^2}\cdot\dfrac{b}{4c^3} = \dfrac{ab^3}{4c^5}$

Alternatively, we can use the rules of exponents directly to write

$$\frac{\sqrt{a^3b^4c^{-4}}}{\sqrt{16ab^{-2}c^6}} = \sqrt{\frac{a^3b^4c^{-4}}{16ab^{-2}c^6}} = \sqrt{\frac{a^2b^6}{16c^{10}}} = \frac{ab^3}{4c^5}$$

C. nth Roots and radicals

The concepts of radicals may be extended to nth roots for positive integer n, so that

- $\sqrt[n]{a} = b$ if and only if $b^n = a$ $\begin{cases} \text{for any real number } a \text{ if } n \text{ is odd} \\ \text{for } a \geq 0 \text{ and } b \geq 0 \text{ if } n \text{ is even} \end{cases}$

EXAMPLE 3-11 Find the value of $\sqrt[3]{8}$, $\sqrt[4]{256}$, and $\sqrt[5]{-32}$.

Solution $\sqrt[3]{8} = 2$, since $2^3 = 8$.

$\sqrt[4]{256} = 4$, since $4^4 = 256$.

$\sqrt[5]{-32} = -2$, since $(-2)^5 = -32$.

D. Rules of nth roots

Since the value for n is an integer, it follows that

$$(\sqrt[n]{a})^n = a$$

and if $a > 0$, or if $a < 0$ and n is an odd positive integer, then

$$\sqrt[n]{a^n} = a$$

The rules for combining nth roots are extensions of the rules for combining radicals and the rules of integral exponents:

- $\sqrt[n]{ab} = \sqrt[n]{a}\sqrt[n]{b}$

- $\sqrt[n]{\dfrac{a}{b}} = \dfrac{\sqrt[n]{a}}{\sqrt[n]{b}}$ (if $b \neq 0$)

- $\sqrt[m]{\sqrt[n]{a}} = \sqrt[mn]{a}$

Notice that $(\sqrt[4]{2})^4 = 2$, $(\sqrt[3]{3})^3 = 3$, $\sqrt[3]{(-2)^3} = -2$ but $\sqrt[4]{(-2)^4} = \sqrt[4]{16} \neq -2$, by definition.

EXAMPLE 3-12   Simplify **(a)** $\sqrt{72}$, **(b)** $\sqrt[3]{-\dfrac{1}{64}}$, **(c)** $\sqrt[4]{128}$, and **(d)** $\sqrt[4]{.0016}$.

Solution   In each case reduce the radicand into factors that are perfect nth powers wherever possible.

(a) Factor out the perfect square:

$$\sqrt{72} = \sqrt{36 \cdot 2} = \sqrt{36}\sqrt{2} = 6\sqrt{2}$$

(b) Factor out the perfect cube (3rd power):

$$\sqrt[3]{-\frac{1}{64}} = \frac{\sqrt[3]{-1}}{\sqrt[3]{64}} = \frac{-1}{\sqrt[3]{4^3}} = \frac{-1}{4}$$

(c) Factor out the perfect 4th power:

$$\sqrt[4]{128} = \sqrt[4]{16 \cdot 8} = \sqrt[4]{16}\sqrt[4]{8} = \sqrt[4]{2^4}\sqrt[4]{8} = 2\sqrt[4]{8}$$

(d)
$$\sqrt[4]{.0016} = \sqrt[4]{\frac{16}{10,000}} = \sqrt[4]{\frac{2^4}{10^4}} = \frac{\sqrt[4]{2^4}}{\sqrt[4]{10^4}} = \frac{2}{10} = .2$$

Check: $(.2)^4 = (.2)(.2)(.2)(.2) = (.04)(.2)(.2) = (.008)(.2) = .0016$.

EXAMPLE 3-13   Simplify **(a)** $\sqrt[3]{3x^2y^2}\sqrt[3]{9xy^4}$, **(b)** $(3\sqrt[3]{x})(4\sqrt[3]{x})(8\sqrt[3]{x})$, and **(c)** $\sqrt[4]{\sqrt{512}}$.

Solution

(a) $\sqrt[3]{3x^2y^2}\sqrt[3]{9xy^4} = \sqrt[3]{(3x^2y^2)(9xy^4)}$

$\qquad\qquad = \sqrt[3]{27x^3y^6}$

$\qquad\qquad = \sqrt[3]{27}\sqrt[3]{x^3}\sqrt[3]{y^6} = \sqrt[3]{3^3}\sqrt[3]{x^3}\sqrt[3]{y^6}$

$\qquad\qquad = 3xy^2$

(b) $(3\sqrt[3]{x})(4\sqrt[3]{x})(8\sqrt[3]{x}) = (3 \cdot 4 \cdot 8)\sqrt[3]{x}\sqrt[3]{x}\sqrt[3]{x}$

$\qquad\qquad\qquad\qquad = (96)(\sqrt[3]{x})^3$

$\qquad\qquad\qquad\qquad = 96x$

(c) $\sqrt[4]{\sqrt{512}} = \sqrt[8]{512} = \sqrt[8]{256 \cdot 2} = \sqrt[8]{256}\sqrt[8]{2} = 2\sqrt[8]{2}$

E. Rationalizing

If we begin with a quotient that contains a radical in the denominator, we can simplify the quotient by multiplying *both* numerator and denominator by an expression that removes the radical from the denominator. This kind of simplification is called **rationalizing the denominator**.

EXAMPLE 3-14   Rationalize the denominators of

(a) $\dfrac{1}{\sqrt{7}}$ (b) $\sqrt{\dfrac{4}{5}}$ (c) $\dfrac{\sqrt{6}}{2 + \sqrt{6}}$ (d) $\dfrac{4 - \sqrt{3}}{4 + \sqrt{3}}$

Solution   Multiply the given expression by 1 in the form of a quotient $r/r = 1$, whose denominator r removes the radical from the given denominator.

(a) $\dfrac{1}{\sqrt{7}} = \dfrac{1}{\sqrt{7}} \cdot \left(\dfrac{\sqrt{7}}{\sqrt{7}}\right) = \dfrac{\sqrt{7}}{(\sqrt{7})^2} = \dfrac{\sqrt{7}}{7}$

(b) $\sqrt{\dfrac{4}{5}} = \dfrac{\sqrt{4}}{\sqrt{5}} = \dfrac{2}{\sqrt{5}} = \dfrac{2}{\sqrt{5}} \cdot \dfrac{\sqrt{5}}{\sqrt{5}} = \dfrac{2\sqrt{5}}{5}$

Remember that if the denominator of a fraction is of the form $a + \sqrt{b}$ (or $a - \sqrt{b}$), then multiplying by $\dfrac{a - \sqrt{b}}{a - \sqrt{b}} \left(\text{or } \dfrac{a + \sqrt{b}}{a + \sqrt{b}}\right)$ will clear the denominator of the irrational expression $\sqrt{b}$.

(c) $\dfrac{\sqrt{6}}{2 + \sqrt{6}} = \dfrac{\sqrt{6}}{2 + \sqrt{6}} \cdot \left(\dfrac{2 - \sqrt{6}}{2 - \sqrt{6}}\right)$

$= \dfrac{\sqrt{6}(2 - \sqrt{6})}{(2 + \sqrt{6})(2 - \sqrt{6})}$ [Use FOIL on the bottom.]

$= \dfrac{2\sqrt{6} - 6}{4 - 2\sqrt{6} + 2\sqrt{6} - 6} = \dfrac{2\sqrt{6} - 6}{4 - 6} = \dfrac{2\sqrt{6} - 6}{-2}$

$= \dfrac{6 - 2\sqrt{6}}{2} = 3 - \sqrt{6}$

(d) $\dfrac{4 - \sqrt{3}}{4 + \sqrt{3}} = \dfrac{4 - \sqrt{3}}{4 + \sqrt{3}} \cdot \left(\dfrac{4 - \sqrt{3}}{4 - \sqrt{3}}\right)$ [Use FOIL on top and bottom.]

$= \dfrac{16 - 4\sqrt{3} - 4\sqrt{3} + 3}{16 - 4\sqrt{3} + 4\sqrt{3} - 3} = \dfrac{19 - 8\sqrt{3}}{13}$

EXAMPLE 3-15   Simplify

(a) $(\sqrt{x} - 2\sqrt{2})^2$ (b) $\dfrac{\sqrt{x} - 2\sqrt{t}}{\sqrt{x} + 2\sqrt{t}}$ (c) $\dfrac{\sqrt{x + h} - \sqrt{x}}{h}$

Solution

(a) $(\sqrt{x} - 2\sqrt{2})^2 = (\sqrt{x} - 2\sqrt{2})(\sqrt{x} - 2\sqrt{2})$

$= (\sqrt{x})^2 - 2\sqrt{2}\sqrt{x} - 2\sqrt{2}\sqrt{x} + (2\sqrt{2})^2$

$= x - 2\sqrt{2x} - 2\sqrt{2x} + 2^2 \cdot 2$

$= x - 4\sqrt{2x} + 8$

(b) $\dfrac{\sqrt{x} - 2\sqrt{t}}{\sqrt{x} + 2\sqrt{t}} = \dfrac{\sqrt{x} - 2\sqrt{t}}{\sqrt{x} + 2\sqrt{t}} \cdot \left(\dfrac{\sqrt{x} - 2\sqrt{t}}{\sqrt{x} - 2\sqrt{t}}\right)$

$= \dfrac{(\sqrt{x})^2 - 2\sqrt{t}\sqrt{x} - 2\sqrt{t}\sqrt{x} + 4(\sqrt{t})^2}{(\sqrt{x})^2 - 2\sqrt{t}\sqrt{x} + 2\sqrt{t}\sqrt{x} - 4(\sqrt{t})^2}$

$= \dfrac{x - 4\sqrt{tx} + 4t}{x - 4t} = \dfrac{x + 4t - 4\sqrt{tx}}{x - 4t}$

(c) This one's tricky: In this case we **rationalize the numerator** to get the simplification:

$\dfrac{\sqrt{x + h} - \sqrt{x}}{h} = \dfrac{\sqrt{x + h} - \sqrt{x}}{h} \cdot \left(\dfrac{\sqrt{x + h} + \sqrt{x}}{\sqrt{x + h} + \sqrt{x}}\right)$

$$\frac{\sqrt{x+h}-\sqrt{x}}{h} = \frac{(\sqrt{x+h})^2 + \sqrt{x}\sqrt{x+h} - \sqrt{x}\sqrt{x+h} - (\sqrt{x})^2}{h(\sqrt{x+h}+\sqrt{x})}$$

$$= \frac{x+h-x}{h(\sqrt{x+h}+\sqrt{x})} = \frac{h}{h(\sqrt{x+h}+\sqrt{x})}$$

$$= \frac{1}{\sqrt{x+h}+\sqrt{x}}$$

note: The result is simpler than the original since an h has been removed.

3-3. Rational Exponents

The definitions and rules of integral exponents and radicals allow us to define a **rational power** of a number. We can see that

$$(a^{1/n})^n = a^{(1/n)n} = a^1 = a = (\sqrt[n]{a})^n$$

So we define an arbitrary rational exponent as a power of an nth root:

RATIONAL EXPONENT $\qquad a^{1/n} = \sqrt[n]{a} \quad \begin{cases} \text{for any } a \text{ if } n \text{ is odd} \\ \text{for } a \geq 0 \text{ if } n \text{ is even} \end{cases}$

EXAMPLE 3-16 Compute the values of (a) $9^{1/2}$, (b) $27^{1/3}$, (c) $(-\frac{1}{8})^{1/3}$, and (d) $1^{1/5}$

Solution

(a) $9^{1/2} = \sqrt{9} = 3$

(b) $27^{1/3} = \sqrt[3]{27} = 3$

(c) $\left(-\frac{1}{8}\right)^{1/3} = \sqrt[3]{\frac{-1}{8}} = \frac{\sqrt[3]{-1}}{\sqrt[3]{8}} = \frac{-1}{2} = -\frac{1}{2}$

(d) $1^{1/5} = \sqrt[5]{1} = 1$

* All the rules for integral exponents remain valid for rational exponents, so we have

RULES OF RATIONAL EXPONENTS
$$(a^{1/n})^m = a^{(1/n)m} = a^{m/n}$$
$$(a^m)^{1/n} = a^{m(1/n)} = a^{m/n}$$

EXAMPLE 3-17 Compute the values of (a) $9^{3/2}$, (b) $4^{5/2}$, (c) $(-27)^{2/3}$, and (d) $(u^6v^3)^{1/2}$.

Solution

(a) $9^{3/2} = (9^{1/2})^3 = (\sqrt{9})^3 = 3^3 = 27$
   or
   $9^{3/2} = (9^3)^{1/2} = (9^2 \cdot 9)^{1/2} = 9(9)^{1/2} = 9\sqrt{9} = 9 \cdot 3 = 27$

(b) $4^{5/2} = (4^{1/2})^5 = (\sqrt{4})^5 = 2^5 = 32$

(c) $(-27)^{2/3} = (-27^{1/3})^2 = (\sqrt[3]{-27})^2 = (-3)^2 = 9$

(d) $(u^6v^3)^{1/2} = (u^6)^{1/2} \cdot (v^3)^{1/2} = u^{6/2} \cdot v^{3/2} = u^3v^{3/2}$

EXAMPLE 3-18 Simplify

(a) $\left(\frac{1}{81}\right)^{1/4}(8)^{2/3}$

(b) $(125)^{-2/3}(25)^{1/2}$

(c) $\left(\frac{4x^{2/3}}{y^{1/8}}\right)^2 \left(\frac{y^{5/4}}{2x^{1/3}}\right)$

Solution

(a) $\left(\dfrac{1}{81}\right)^{1/4}(8)^{2/3} = \dfrac{(1)^{1/4}}{(81)^{1/4}}(8^{1/3})^2 = \dfrac{1}{(9^2)^{1/4}}(8^{1/3})^2 = \dfrac{1}{9^{1/2}}\cdot 2^2 = \dfrac{1}{3}\cdot 4 = \dfrac{4}{3}$

or $\qquad\qquad\qquad = \dfrac{1}{\sqrt[4]{81}}(\sqrt[3]{8})^2 = \dfrac{1}{3}\cdot 2^2 = \dfrac{4}{3}$

(b) $(125)^{-2/3}(25)^{1/2} = \dfrac{1}{(125)^{2/3}}(25)^{1/2} = \dfrac{1}{(5^3)^{2/3}}(25)^{1/2} = \dfrac{1}{5^2}\cdot 5 = \dfrac{1}{5}$

or $\qquad\qquad\qquad = \dfrac{1}{(\sqrt[3]{125})^2}\cdot\sqrt{25} = \dfrac{5}{5^2} = \dfrac{1}{5}$

(c) $\left(\dfrac{4x^{2/3}}{y^{1/8}}\right)^2\left(\dfrac{y^{5/4}}{2x^{1/3}}\right) = \left(\dfrac{4^2(x^{2/3})^2}{(y^{1/8})^2}\right)\left(\dfrac{y^{5/4}}{2x^{1/3}}\right)$

$\qquad\qquad = \left(\dfrac{16x^{4/3}}{y^{1/4}}\right)\left(\dfrac{y^{5/4}}{2x^{1/3}}\right)$

$\qquad\qquad = \left(\dfrac{16}{2}\right)\left(\dfrac{x^{4/3}}{x^{1/3}}\right)\left(\dfrac{y^{5/4}}{y^{1/4}}\right) = 8x^{(4/3-1/3)}y^{(5/4-1/4)} = 8xy$

As in the case of integral exponents, we can extend the rules to operations on more than two products.

EXAMPLE 3-19 Simplify (a) $\left(\dfrac{a^2b^{-3}c^4}{(ab^{-2}c^2)^2}\right)^{3/2}$ and (b) $\left(\dfrac{x^{1/3}\sqrt{y+z}}{x^2(y+z)}\right)^6$.

Solution

(a) $\left(\dfrac{a^2b^{-3}c^4}{(ab^{-2}c^2)^2}\right)^{3/2} = \left(\dfrac{a^2b^{-3}c^4}{a^2b^{-4}c^4}\right)^{3/2} = (b^{-3+4})^{3/2} = b^{3/2}$

(b) $\left(\dfrac{x^{1/3}\sqrt{y+z}}{x^2(y+z)}\right)^6 = \dfrac{[x^{1/3}(y+z)^{1/2}]^6}{[x^2(y+z)]^6} = \dfrac{x^{6/3}(y+z)^{6/2}}{x^{12}(y+z)^6}$

$\qquad\qquad = \dfrac{x^2(y+z)^3}{x^{12}(y+z)^6} = \dfrac{1}{x^{10}(y+z)^3}$

3-4. Complex Numbers

A. Definition of complex numbers

The definitions and rules of exponents and radicals leave a gap for the value of $\sqrt[n]{a}$ when $a < 0$ and n is even; i.e., the even root of a negative number is undefined. For example, $\sqrt{-9}$ and $\sqrt{-16}$ can't be real numbers, because there is no real number that can be multiplied by itself to give -9 or -16. But we can see that

$$\sqrt{-9} = \sqrt{9(-1)} = \sqrt{9}\cdot\sqrt{-1}$$

and

$$\sqrt{-16} = \sqrt{16(-1)} = \sqrt{16}\cdot\sqrt{-1}$$

So we could have a workable quantity if we had some way to denote the $\sqrt{-1}$ part of these numbers. We do this by using the symbol i (for *imaginary*):

$$i \equiv \sqrt{-1}$$

Then we can write

$$\sqrt{-9} = \sqrt{9}\cdot\sqrt{-1} = 3i \qquad \text{and} \qquad \sqrt{-16} = \sqrt{16}\cdot\sqrt{-1} = 4i$$

Check: $(3i)^2 = 9i^2 = 9(\sqrt{-1})^2 = 9(-1) = -9$

$\qquad (4i)^2 = 16i^2 = 16(\sqrt{-1})^2 = 16(-1) = -16$

Now we can define a new set:

- If a and b are real numbers, then the set of all numbers of the form

$$a + bi \qquad \text{where } i = \sqrt{-1}$$

is called the set of **complex numbers**. The number a is called the *real part* and the number b is called the *imaginary part* of the complex number.

> The imaginary part of a complex number is the MULTIPLIER of i—it does NOT INCLUDE i.

EXAMPLE 3-20 Identify the real and imaginary parts of (a) $4 - 3i$, (b) $-7i$, and (c) 5.

Solution

(a) If $a + bi = 4 - 3i$, then the real part is $a = 4$ and the imaginary part is $b = -3$.
(b) If $a + bi = -7i$, then the real part is $a = 0$ and the imaginary part is $b = -7$.
(c) If $a + bi = 5$, then the real part is $a = 5$ and the imaginary part is $b = 0$.

With the definition of i, it is easy to see that

$$i^2 = -1, \qquad i^3 = i^2 \cdot i = -1 \cdot i = -1(i) = -i, \qquad i^4 = i^2 \cdot i^2 = (-1)(-1) = 1$$

EXAMPLE 3-21 Find the reduced form of (a) i^9 and (b) i^{-6}.

Solution

(a) $i^9 = i^8 \cdot i = i^4 \cdot i^4 \cdot i = (1)(1)(i) = i$

(b) $i^{-6} = \dfrac{1}{i^6} = \dfrac{1}{i^4 \cdot i^2} = \dfrac{1}{(1)(-1)} = -1$

or

$$i^{-6} = \frac{1}{i^6} = \frac{1}{i^3 \cdot i^3} = \frac{1}{(-i)(-i)} = \frac{1}{i^2} = \frac{1}{-1} = -1$$

B. Operations on complex numbers

1. The sum, difference, and product of complex numbers

- If $a, b, c,$ and d are real numbers, then

SUM OF COMPLEX NUMBERS
$$(a + bi) + (c + di) = (a + c) + (bi + di)$$
$$= (a + c) + (b + d)i$$

DIFFERENCE OF COMPLEX NUMBERS
$$(a + bi) - (c + di) = (a - c) + (bi - di)$$
$$= (a - c) + (b - d)i$$

PRODUCT OF COMPLEX NUMBERS
$$(a + bi)(c + di) = ac + adi + bci + bdi^2$$
$$= ac + (ad + bc)i + bd(-1)$$
$$= (ac - bd) + (ad + bc)i$$

note: The rules of manipulation for real and complex numbers are very nearly identical: The axioms of equality, the field properties, and all the rules under the four arithmetic operations all apply to complex numbers.

EXAMPLE 3-22   Find the simplified value of

(a) $\dfrac{1 + 2i}{3} + (1 - i)^2$ 　　　　　　　**(c)** $(3 + 5i)(2 - 3i)$

(b) $(2 - 3i)^2 - (4 + i) + 3$ 　　　　　　**(d)** $(a + bi)(a - bi)$

Solution   Begin by treating i as if it were an ordinary (real) number. Then look for powers of i that produce a real number (e.g., $i^2 = -1$), and substitute that real number into the expression.

(a) $\dfrac{1 + 2i}{3} + (1 - i)^2 = \left(\dfrac{1}{3} + \dfrac{2}{3}i\right) + (1 - 2i + i^2)$ 　　　$[i^2 = -1]$

$$= \left(\dfrac{1}{3} + \dfrac{2}{3}i\right) + (1 - 2i - 1) \qquad \text{[Substitute and cancel]}$$

$$= \left(\dfrac{1}{3} + \dfrac{2}{3}i\right) + (-2i)$$

$$= \dfrac{1}{3} + \dfrac{2}{3}i - \dfrac{6}{3}i = \dfrac{1}{3} + \left(\dfrac{2}{3} - \dfrac{6}{3}\right)i$$

$$= \dfrac{1}{3} - \dfrac{4}{3}i$$

(b) $(2 - 3i)^2 - (4 + i) + 3 = (4 - 12i + 9i^2) - (4 + i) + 3$

$$= 4 - 12i + 9(-1) - 4 - i + 3$$

$$= 4 - 9 - 4 + 3 - 12i - i$$

$$= -6 - 13i$$

(c) $(3 + 5i)(2 - 3i) = 6 - 9i + 10i - 15i^2$

$$= 6 + i - 15(-1) = 6 + i + 15$$

$$= 21 + i$$

(d) $(a + bi)(a - bi) = a^2 - abi + abi - b^2i^2$

$$= a^2 - b^2i^2$$

$$= a^2 - b^2(-1)$$

$$= a^2 + b^2$$

If $a + bi$ is a complex number, then the complex number $a - bi$ is called the **complex conjugate**. *Multiplying a complex number by its complex conjugate produces a real number*, as Example 3-22d illustrates. The complex conjugate is often useful in simplifying complex fractions.

2. The quotient of two complex numbers

- If a, b, c, and d are real numbers, and c and d are not both zero, then

QUOTIENT OF COMPLEX NUMBERS $\qquad \dfrac{a + bi}{c + di} = \dfrac{ac + bd}{c^2 + d^2} + \left(\dfrac{bc - ad}{c^2 + d^2}\right)i$

- This result follows from complex conjugate multiplication:

$$\dfrac{a + bi}{c + di} = \left(\dfrac{a + bi}{c + di}\right)\left(\dfrac{c - di}{c - di}\right) = \dfrac{ac - adi + bci - bdi^2}{c^2 - cdi + cdi - d^2i^2}$$

$$= \dfrac{ac - bd(-1) + (bc - ad)i}{c^2 - d^2(-1)}$$

$$\frac{a+bi}{c+di} = \left(\frac{a+bi}{c+di}\right)\left(\frac{c-di}{c-di}\right) = \frac{ac+bd+(bc-ad)i}{c^2+d^2}$$

$$= \frac{ac+bd}{c^2+d^2} + \left(\frac{bc-ad}{c^2+d^2}\right)i$$

note: This result really IS simpler—there's only one i (and it's in the general form, $a + bi$, for complex numbers).

EXAMPLE 3-23   Simplify each of the following:

(a) $\dfrac{4}{i}$ **(b)** $\dfrac{1-2i}{5+2i}$ **(c)** $\dfrac{4+3i}{7-i}$ **(d)** $\dfrac{(\sqrt{2}-3i)^2}{i^3}$

Solution   In each case, find the complex conjugate of the denominator and multiply.

(a) The complex conjugate of i is $-i$:

$$\frac{4}{i} = \frac{4}{i}\left(\frac{-i}{-i}\right) = \frac{-4i}{-i^2} = \frac{-4i}{-(-1)} = -4i \qquad \left[\text{Multiply by } \frac{-i}{-i}\right]$$

(b) The complex conjugate of $5 + 2i$ is $5 - 2i$:

$$\frac{1-2i}{5+2i} = \left(\frac{1-2i}{5+2i}\right)\left(\frac{5-2i}{5-2i}\right) = \frac{(1-2i)(5-2i)}{(5+2i)(5-2i)} \qquad \left[\text{Multiply by } \frac{5-2i}{5-2i}\right]$$

$$= \frac{5-2i-10i+4i^2}{25-10i+10i-4i^2} = \frac{5-2i-10i+4(-1)}{25-10i+10i-4(-1)}$$

$$= \frac{5-4-12i}{25+4} = \frac{1-12i}{29}$$

$$= \frac{1}{29} - \frac{12}{29}i$$

(c) The complex conjugate of $7 - i$ is $7 + i$:

$$\frac{4+3i}{7-i} = \frac{4+3i}{7-i}\left(\frac{7+i}{7+i}\right) = \frac{(4+3i)(7+i)}{(7-i)(7+i)}$$

$$= \frac{28+4i+21i+3i^2}{49+7i-7i-i^2} = \frac{28+3(-1)+25i}{49+(-i^2)}$$

$$= \frac{25+25i}{50} = \frac{25(1+i)}{50} = \frac{1+i}{2}$$

$$= \frac{1}{2} + \frac{1}{2}i$$

(d) The complex conjugate of i^3 is i:

$$\frac{(\sqrt{2}-3i)^2}{i^3} = \frac{(\sqrt{2}-3i)(\sqrt{2}-3i)}{i^2 \cdot i} = \frac{2-3\sqrt{2}i-3\sqrt{2}i+9i^2}{(-1)i}$$

$$= \frac{2+9(-1)-6\sqrt{2}i}{(-1)i} = \frac{-7-6\sqrt{2}i}{-i}$$

$$= \frac{-7-6\sqrt{2}i}{-i}\left(\frac{i}{i}\right)$$

$$= \frac{-7i-6\sqrt{2}i^2}{-i^2} = \frac{-7i-6\sqrt{2}(-1)}{-(-1)}$$

$$= 6\sqrt{2} - 7i$$

SUMMARY

1. The rules of integral exponents, where a and b are real numbers and m and n are integers, are

$$a^m a^n = a^{m+n} \qquad \left(\frac{a}{b}\right)^n = \frac{a^n}{b^n} \qquad (b \neq 0)$$

$$(a^m)^n = a^{mn} \qquad a^{-n} = \frac{1}{a^n} \qquad (a \neq 0)$$

$$(ab)^n = a^n b^n \qquad \frac{a^m}{a^n} = a^{m-n} \qquad (a \neq 0)$$

and $a^0 = 1$. These rules also work with rational exponents and radicals.

2. The principal square root b of a radical $\sqrt{a}$ is defined as

$$\sqrt{a} = b \quad \text{if and only if} \quad b^2 = a \quad \text{and} \quad b \geq 0$$

3. Radicals are combined by two basic rules:

$$\sqrt{ab} = \sqrt{a}\sqrt{b} \quad \text{and} \quad \sqrt{a/b} = \sqrt{a}/\sqrt{b}$$

4. The nth root of a number is defined by $(\sqrt[n]{a})^n = a$, and if $a > 0$, or if $a < 0$ and n is odd, $\sqrt[n]{a^n} = a$.

5. An arbitrary rational exponent is a power of an nth root: The rational exponent $\frac{1}{2}$ is directly related to the radical; other rational exponents relate to nth roots.

$$a^{1/2} \equiv \sqrt{a} \qquad a^{1/n} \equiv \sqrt[n]{a}$$

6. The complex numbers are an extension of the real numbers by the addition of the quantity $i = \sqrt{-1}$.

7. Complex numbers of the form $a + bi$ are manipulated by rules nearly identical with those for real numbers.

RAISE YOUR GRADES

Can you...?

- ☑ write the rules for integral exponents
- ☑ show that $a^0 = 1$ for any nonzero value of a
- ☑ define a radical, a radicand, and an nth root
- ☑ explain how to rationalize a denominator
- ☑ explain the connection between radicals and rational exponents
- ☑ define the set of complex numbers
- ☑ identify the real and imaginary parts of a complex number
- ☑ add, subtract, and multiply complex numbers
- ☑ define the complex conjugate of a complex number
- ☑ explain how to divide a complex number into another complex number

SOLVED PROBLEMS

Integral Exponents

PROBLEM 3-1 If you deposit $100 in a savings account earning 7% interest compounded annually, how large will your account be in five years?

Solution If your initial deposit P is \$100 and the rate of interest is 7%, then at the end of one year you'll have

$$.07P + P = 1.07P$$

and at the end of five years you'll have

$$P(1.07)^5 = 100(1.40255) \cong 140.26 \text{ dollars}$$

PROBLEM 3-2 Using the rules of integral exponents, show the validity of the following rules: If a is a real number and m, n, and p are integers, then

$$a^m \cdot a^n \cdot a^p = a^{m+n+p} \qquad \text{and} \qquad \frac{a^m \cdot a^n}{a^p} = a^{m+n-p}$$

Solution Use the associative law and the appropriate rule of integral exponents.

$$a^m \cdot a^n \cdot a^p = (a^{m+n}) \cdot a^p = a^{(m+n)+p} = a^{m+n+p} \qquad [\textit{Rule 1}]$$

$$\frac{a^m \cdot a^n}{a^p} = \frac{a^{m+n}}{a^p} = a^{(m+n)-p} = a^{m+n-p} \qquad [\textit{Rule 6}]$$

PROBLEM 3-3 Use the rules of integral exponents to simplify the following expressions:

(a) $4^3 \cdot 4^{-5} \div 4^{-4}$ (b) $\left(\dfrac{3}{4}\right)^2 \left(\dfrac{4}{5}\right)^2$ (c) $[(-3)^{-2}]^{-1}$ (d) $\left(\dfrac{6^{-12}}{5^{18}}\right)^0$

Solution

(a)
$$4^3 \cdot 4^{-5} \div 4^{-4} = 4^{3+(-5)} \div 4^{-4} \qquad [a^m \cdot a^n = a^{mn}]$$

$$= 4^{-2} \div 4^{-4} = 4^{-2-(-4)} \qquad [a^m/a^n = a^{m-n}]$$

$$= 4^2 = 16$$

or

$$4^3 \cdot 4^{-5} \div 4^{-4} = 4^{-2} \div 4^{-4} = \frac{1}{4^2} \div \frac{1}{4^4} \qquad \left[a^{-n} = \frac{1}{a^n}\right]$$

$$= \frac{1}{4^2} \cdot \frac{4^4}{1} = \frac{4^4}{4^2} = 4^2 = 16$$

(b)
$$\left(\frac{3}{4}\right)^2 \left(\frac{4}{5}\right)^2 = \frac{3^2}{4^2} \cdot \frac{4^2}{5^2} = \frac{3^2}{5^2} = \frac{9}{25} \qquad \left[\left(\frac{a}{b}\right)^n = \frac{a^n}{b^n}\right]$$

(c)
$$[(-3)^{-2}]^{-1} = (-3)^{-2(-1)} = -3^2 = 9 \qquad [(a^m)^n = a^{mn}]$$

or
$$[(-3)^{-2}]^{-1} = \left(\frac{1}{(-3)^2}\right)^{-1} = \left(\frac{1}{9}\right)^{-1} = 9 \qquad \left[a^{-n} = \frac{1}{a^n}\right]$$

(d)
$$\left(\frac{6^{-12}}{5^{18}}\right)^0 = 1 \qquad [a^0 = 1 \text{ for any nonzero value } a]$$

PROBLEM 3-4 Use the rules of exponents to simplify

(a) $\dfrac{(x^3 y^0)^2}{(x^3 - x^2)^2}$ (b) $\left(\dfrac{u^{-2}}{v^{-3}}\right)^{-2}$ (c) $\dfrac{xy^3}{4^3} \div \dfrac{x^2 y^2}{2^5}$ (d) $\dfrac{3(abc)^3}{a^{-1}b^{-1}c^{-1}}$

Solution

(a) First expand the numerator and factor the denominator:

$$\frac{(x^3 y^0)^2}{(x^3 - x^2)^2} = \frac{(x^3)^2(1)^2}{[x^2(x-1)]^2} = \frac{x^6}{x^4(x-1)^2} = \frac{x^2}{(x-1)^2} = \left(\frac{x}{x-1}\right)^2$$

(b)
$$\left(\frac{u^{-2}}{v^{-3}}\right)^{-2} = \frac{u^{-2(-2)}}{v^{-3(-2)}} = \frac{u^4}{v^6}$$

or

$$\left(\frac{u^{-2}}{v^{-3}}\right)^{-2} = \left(\frac{\frac{1}{u^2}}{\frac{1}{v^3}}\right)^{-2} = \left(\frac{1}{u^2} \cdot \frac{v^3}{1}\right)^{-2} = \left(\frac{v^3}{u^2}\right)^{-2} = \frac{v^{-6}}{u^{-4}} = \frac{u^4}{v^6}$$

(c)

$$\frac{xy^3}{4^3} \div \frac{x^2y^2}{2^5} = \frac{xy^3}{(2^2)^3} \cdot \frac{2^5}{x^2y^2} = \frac{2^5}{2^6} \cdot \frac{xy^3}{x^2y^2} = \frac{1}{2} \cdot \frac{y}{x} = \frac{y}{2x}$$

[Note that $4 = 2^2$, so $4^3 = (2^2)^3 = 2^6$. This kind of base changing can often eliminate a lot of tedious arithmetic!]

(d)

$$\frac{3(abc)^3}{a^{-1}b^{-1}c^{-1}} = 3a^{3-(-1)}b^{3-(-1)}c^{3-(-1)} = 3a^4b^4c^4$$

or

$$\frac{3(abc)^3}{a^{-1}b^{-1}c^{-1}} = \frac{3a^3b^3c^3}{\frac{1}{a} \cdot \frac{1}{b} \cdot \frac{1}{c}} = \frac{3a^3b^3c^3}{\frac{1}{abc}} = 3a^3b^3c^3 \cdot \frac{abc}{1} = 3a^4b^4c^4$$

PROBLEM 3-5   Use the rules of exponents to simplify

(a) $\dfrac{y^{-2} + y^{-3}}{y^{-1}}$ (b) $\dfrac{a + b^{-1}}{a - b^{-1}}$ (c) $\left(\dfrac{3a}{b^{-1}}\right)^2 \left(\dfrac{b^{-2}}{a^{-4}}\right) \left(\dfrac{a^2}{b^2}\right)^{-1}$

Solution   Remember that you can change the negative exponents to positive by inverting.

(a) Factor first:

$$\frac{y^{-2} + y^{-3}}{y^{-1}} = \frac{y^{-1}(y^{-1} + y^{-2})}{y^{-1}} = y^{-1} + y^{-2}$$

then invert and add:

$$= \frac{1}{y} + \frac{1}{y^2} = \frac{1}{y}\left(\frac{y}{y}\right) + \frac{1}{y^2}$$

$$= \frac{y}{y^2} + \frac{1}{y^2} = \frac{y+1}{y^2}$$

or

$$\frac{y^{-2} + y^{-3}}{y^{-1}} = \frac{\frac{1}{y^2} + \frac{1}{y^3}}{\frac{1}{y}} = \frac{\frac{1}{y^2}\left(\frac{y}{y}\right) + \frac{1}{y^3}}{\frac{1}{y}} = \frac{\frac{y+1}{y^3}}{\frac{1}{y}} = \frac{y+1}{y^3} \cdot \frac{y}{1} = \frac{y+1}{y^2}$$

(b)

$$\frac{a + b^{-1}}{a - b^{-1}} = \frac{a + \frac{1}{b}}{a - \frac{1}{b}} = \frac{a\left(\frac{b}{b}\right) + \frac{1}{b}}{a\left(\frac{b}{b}\right) - \frac{1}{b}} = \frac{\frac{ab}{b} + \frac{1}{b}}{\frac{ab}{b} - \frac{1}{b}} = \frac{\frac{ab+1}{b}}{\frac{ab-1}{b}}$$

$$= \frac{ab+1}{b} \cdot \frac{b}{ab-1} = \frac{ab+1}{ab-1}$$

(c) Expand the exponents on the outside first:

$$\left(\frac{3a}{b^{-1}}\right)^2 \left(\frac{b^{-2}}{a^{-4}}\right) \left(\frac{a^2}{b^2}\right)^{-1} = \left(\frac{9a^2}{b^{-2}}\right) \left(\frac{b^{-2}}{a^{-4}}\right) \left(\frac{b^2}{a^2}\right) \qquad \left[\text{Note that } \left(\frac{M}{N}\right)^{-1} = \frac{N}{M}.\right]$$

$$= \frac{9a^2 b^{-2} b^2}{b^{-2} a^{-4} a^2} = \frac{9b^2}{a^{-4}} = 9a^4b^2$$

PROBLEM 3-6 Simplify (a) $\dfrac{(2p^2q)^3}{5p^4}\left(\dfrac{p^{-1}q^3}{3p^{-2}q}\right)^{-2}$ and (b) $\dfrac{(-2a^2b^{-1}cd^{-2})^3}{(4a^{-3}bc^2d^{-3})^{-2}}$.

Solution

(a) Cancel everything that cancels:

$$\frac{(2p^2q)^3}{5p^4}\left(\frac{p^{-1}q^3}{3p^{-2}q}\right)^{-2} = \frac{2^3p^6q^3}{5p^4}\left(\frac{p^2q^{-6}}{3^{-2}p^4q^{-2}}\right)$$

$$= \frac{2^3p^8q^{-3}}{5\cdot3^{-2}p^8q^{-2}} = \frac{2^3\cdot3^2}{5q} = \frac{72}{5q}$$

or, do it the hard way:

$$= \frac{8p^6q^3}{5p^4}\left(\frac{p^2\cdot\dfrac{1}{q^6}}{\dfrac{1}{9}\cdot p^4\cdot\dfrac{1}{q^2}}\right) = \frac{8}{5}p^2q^3\left(\frac{\dfrac{p^2}{q^6}}{\dfrac{p^4}{9q^2}}\right)$$

$$= \frac{8}{5}p^2q^3\left(\frac{p^2}{q^6}\cdot\frac{9q^2}{p^4}\right) = \frac{8}{5}p^2q^3\left(\frac{9}{q^4p^2}\right) = \frac{72}{5q}$$

note: The easiest way of doing an algebra problem is usually the best way. In short, cancel everything you can, then work with what's left.

(b)

$$\frac{(-2a^2b^{-1}cd^{-2})^3}{(4a^{-3}bc^2d^{-3})^{-2}} = \frac{(-2)^3a^6b^{-3}c^3d^{-6}}{4^{-2}a^6b^{-2}c^{-4}d^6} = (-8)(16)(1)(b^{-1})(c^7)(d^{-12})$$

$$= \frac{-128c^7}{bd^{12}}$$

Radicals

PROBLEM 3-7 Simplify.

(a) $\sqrt{a^6b^4}$ (b) $\sqrt{\dfrac{81}{225}}$ (c) $\sqrt{90r^3s^4}\sqrt{30r^7s^6}$

Solution Use the rules for combining radicals: $\sqrt{ab} = \sqrt{a}\sqrt{b}$ and $\sqrt{a/b} = \sqrt{a}/\sqrt{b}$.

(a) $\sqrt{a^6b^4} = \sqrt{a^6}\sqrt{b^4} = \sqrt{(a^3)^2}\sqrt{(b^2)^2} = a^3b^2$

(b) $\sqrt{\dfrac{81}{225}} = \dfrac{\sqrt{81}}{\sqrt{225}} = \dfrac{\sqrt{9^2}}{\sqrt{15^2}} = \dfrac{9}{15} = \dfrac{3}{5}$

(c) $\sqrt{90r^3s^4}\sqrt{30r^7s^6} = \sqrt{2700r^{10}s^{10}}$

$$= \sqrt{27}\sqrt{100}\sqrt{r^{10}}\sqrt{s^{10}}$$

$$= \sqrt{9\cdot3}\cdot10\cdot r^5\cdot s^5$$

$$= 3\sqrt{3}\cdot10r^5s^5$$

$$= 30\sqrt{3}r^5s^5$$

PROBLEM 3-8 Simplify

(a) $\sqrt{5xy^9}\sqrt{125x^5y^3}$ (b) $\dfrac{\sqrt{8a^3b^2c}}{\sqrt{32ab^4c^5}}$ (c) $\dfrac{\sqrt{9u^{-2}v^{-4}w^{-6}}}{\sqrt{72u^{-6}v^{-4}w^{-2}}}$

Solution

(a) $\sqrt{5xy^9}\sqrt{125x^5y^3} = \sqrt{5\cdot125x^6y^{12}} = \sqrt{5\cdot5^3x^6y^{12}} = \sqrt{5^4x^6y^{12}}$

$$= \sqrt{5^4}\sqrt{x^6}\sqrt{y^{12}} = 5^2x^3y^6$$

$$= 25x^3y^6$$

(b) $\dfrac{\sqrt{8a^3b^2c}}{\sqrt{32ab^4c^5}} = \sqrt{\dfrac{2^3a^3b^2c}{2^5ab^4c^5}} = \sqrt{\dfrac{a^2}{2^2b^2c^4}} = \dfrac{\sqrt{a^2}}{\sqrt{2^2b^2c^4}} = \dfrac{a}{2bc^2}$

(c) $\dfrac{\sqrt{9u^{-2}v^{-4}w^{-6}}}{\sqrt{72u^{-6}v^{-4}w^{-2}}} = \sqrt{\dfrac{9u^{-2}v^{-4}w^{-6}}{9 \cdot 8u^{-6}v^{-4}w^{-2}}} = \sqrt{\dfrac{u^{-2-(-6)}v^{-4-(-4)}w^{-6-(-2)}}{2^3}} = \sqrt{\dfrac{u^4}{2^2 \cdot 2w^4}}$

$$= \dfrac{u^2}{2w^2\sqrt{2}}\left(\dfrac{\sqrt{2}}{\sqrt{2}}\right) = \dfrac{\sqrt{2}u^2}{4w^2} \qquad \text{[Rationalize the denominator.]}$$

PROBLEM 3-9   Find the values of **(a)** $\sqrt[3]{-1}$, **(b)** $\sqrt[5]{2^5}$, **(c)** $\sqrt[4]{80}$, and **(d)** $\sqrt[10]{1024}$.

Solution

(a) $\sqrt[3]{-1} = -1$, since $(-1)^3 = -1$

(b) $\sqrt[5]{2^5} = 2$

(c) $\sqrt[4]{80} = \sqrt[4]{16 \cdot 5} = \sqrt[4]{16} \cdot \sqrt[4]{5} = 2\sqrt[4]{5}$

(d) $\sqrt[10]{1024} = \sqrt[10]{2^{10}} = 2$, since $1024 = 2 \cdot 512 = 2^2 \cdot 256 = 2^3 \cdot 128 = 2^4 \cdot 64 = 2^5 \cdot 32$
$$= 2^6 \cdot 16 = 2^7 \cdot 8 = 2^8 \cdot 4 = 2^9 \cdot 2 = 2^{10}$$

PROBLEM 3-10   Simplify

(a) $\sqrt[3]{\dfrac{1}{8x^6}}$ **(b)** $\sqrt[5]{\dfrac{4a^4}{b^7}}\,\sqrt[5]{\dfrac{8a^6}{b^{-2}}}$ **(c)** $\dfrac{\sqrt[3]{(p-4q)^3}}{\sqrt[4]{(p+4q)^4}}$ **(d)** $\sqrt[3]{(4ab^2)^0}$

Solution

(a) $\sqrt[3]{\dfrac{1}{8x^6}} = \dfrac{\sqrt[3]{1}}{\sqrt[3]{8x^6}} = \dfrac{1}{\sqrt[3]{8}\sqrt[3]{x^6}} = \dfrac{1}{2x^2}$

(b) $\sqrt[5]{\dfrac{4a^4}{b^7}}\,\sqrt[5]{\dfrac{8a^6}{b^{-2}}} = \sqrt[5]{\dfrac{4a^4}{b^7} \cdot \dfrac{8a^6}{b^{-2}}} = \sqrt[5]{\dfrac{32a^{10}}{b^5}} = \dfrac{\sqrt[5]{2^5}\sqrt[5]{a^{10}}}{\sqrt[5]{b^5}} = \dfrac{2a^2}{b}$

(c) $\dfrac{\sqrt[3]{(p-4q)^3}}{\sqrt[4]{(p+4q)^4}} = \dfrac{p-4q}{p+4q}$ [Assuming $p + 4q$ is positive.]

(d) Since $a^0 = 1$ if $a \neq 0$, then

$$\sqrt[3]{(4ab^2)^0} = \sqrt[3]{1} = 1$$

PROBLEM 3-11   Rationalize the denominators of the following and simplify the results:

(a) $\sqrt{\dfrac{5}{12}}$ **(b)** $\dfrac{1+\sqrt{2}}{1-\sqrt{2}}$ **(c)** $\dfrac{2}{2-\sqrt{2}}$ **(d)** $\dfrac{1}{1+\sqrt{10}}$

Solution

(a) $\sqrt{\dfrac{5}{12}} = \dfrac{\sqrt{5}}{\sqrt{12}}\left(\dfrac{\sqrt{12}}{\sqrt{12}}\right) = \dfrac{\sqrt{60}}{12} = \dfrac{\sqrt{4 \cdot 15}}{12} = \dfrac{\sqrt{4}\sqrt{15}}{12} = \dfrac{2\sqrt{15}}{12} = \dfrac{\sqrt{15}}{6}$

(b) $\dfrac{1+\sqrt{2}}{1-\sqrt{2}} = \left(\dfrac{1+\sqrt{2}}{1-\sqrt{2}}\right)\left(\dfrac{1+\sqrt{2}}{1+\sqrt{2}}\right) = \dfrac{1+\sqrt{2}+\sqrt{2}+2}{1+\sqrt{2}-\sqrt{2}-2}$

$$= \dfrac{3+2\sqrt{2}}{-1} = -3 - 2\sqrt{2}$$

(c) $\dfrac{2}{2-\sqrt{2}} = \left(\dfrac{2}{2-\sqrt{2}}\right)\left(\dfrac{2+\sqrt{2}}{2+\sqrt{2}}\right) = \dfrac{4+2\sqrt{2}}{4+2\sqrt{2}-2\sqrt{2}-2} = \dfrac{4+2\sqrt{2}}{2} = 2+\sqrt{2}$

(d) $\dfrac{1}{1+\sqrt{10}} = \left(\dfrac{1}{1+\sqrt{10}}\right)\left(\dfrac{1-\sqrt{10}}{1-\sqrt{10}}\right) = \dfrac{1-\sqrt{10}}{1-\sqrt{10}+\sqrt{10}-10} = \dfrac{1-\sqrt{10}}{-9} = -\dfrac{1}{9}+\dfrac{1}{9}\sqrt{10}$

PROBLEM 3-12 Simplify

(a) $\sqrt[3]{27x^3 - 54x^5}$ (b) $\dfrac{(\sqrt{x} - \sqrt{y})(\sqrt{x} + \sqrt{y})}{x - y}$ (c) $\dfrac{\sqrt{u}}{\sqrt{u} + \sqrt{v}}$

Solution

(a) Factor out the perfect cube:

$$\sqrt[3]{27x^3 - 54x^5} = \sqrt[3]{27x^3(1 - 2x^2)} = \sqrt[3]{27x^3}\sqrt[3]{1 - 2x^2} = 3x\sqrt[3]{1 - 2x^2}$$

(b) Expand the numerator:

$$\frac{(\sqrt{x} - \sqrt{y})(\sqrt{x} + \sqrt{y})}{x - y} = \frac{x - \sqrt{x}\sqrt{y} + \sqrt{x}\sqrt{y} - y}{x - y} = \frac{x - y}{x - y} = 1$$

(c) Rationalize the denominator:

$$\frac{\sqrt{u}}{\sqrt{u} + \sqrt{v}} = \left(\frac{\sqrt{u}}{\sqrt{u} + \sqrt{v}}\right)\left(\frac{\sqrt{u} - \sqrt{v}}{\sqrt{u} - \sqrt{v}}\right) = \frac{u - \sqrt{u}\sqrt{v}}{u - \sqrt{u}\sqrt{v} + \sqrt{u}\sqrt{v} - v} = \frac{u - \sqrt{uv}}{u - v}$$

PROBLEM 3-13 Simplify

(a) $(a\sqrt{b} + b\sqrt{a})^2$ (b) $(4\sqrt{u} + \sqrt{uv})^2$ (c) $\dfrac{y - \sqrt{y^2 + 4y + 4}}{2}$

Solution

(a) $(a\sqrt{b} + b\sqrt{a})^2 = (a\sqrt{b})^2 + 2(a\sqrt{b})(b\sqrt{a}) + (b\sqrt{a})^2$

$\qquad\qquad\qquad = a^2b + 2ab\sqrt{ab} + b^2a$ [Expand]

$\qquad\qquad\qquad = ab(a + 2\sqrt{ab} + b)$ [Factor: Common factor]

$\qquad\qquad\qquad = ab(\sqrt{a} + \sqrt{b})^2$ [Factor: Perfect square]

(b) $(4\sqrt{u} + \sqrt{uv})^2 = (4\sqrt{u})^2 + 2(4\sqrt{u})(\sqrt{uv}) + (\sqrt{uv})^2$

$\qquad\qquad\qquad = 16u + 8\sqrt{u^2v} + uv$

$\qquad\qquad\qquad = 16u + 8u\sqrt{v} + uv$

$\qquad\qquad\qquad = u(16 + 8\sqrt{v} + v)$

$\qquad\qquad\qquad = u(4 + \sqrt{v})^2$

(c) $\dfrac{y - \sqrt{y^2 + 4y + 4}}{2} = \dfrac{y - \sqrt{(y + 2)^2}}{2} = \dfrac{y - (y + 2)}{2} = \dfrac{y - y - 2}{2} = -1$

PROBLEM 3-14 Evaluate

(a) $(\sqrt{50}) + \sqrt{27})(\sqrt{18} + \sqrt{75})$ (b) $(\sqrt{20} - \sqrt{27})(\sqrt{45} - \sqrt{108})$

Solution Before you multiply, simplify each radical.

(a) $(\sqrt{50} + \sqrt{27})(\sqrt{18} + \sqrt{75}) = (\sqrt{25 \cdot 2} + \sqrt{9 \cdot 3})(\sqrt{9 \cdot 2} + \sqrt{25 \cdot 3})$

$\qquad\qquad\qquad\qquad\qquad = (5\sqrt{2} + 3\sqrt{3})(3\sqrt{2} + 5\sqrt{3})$

Now use FOIL: $\qquad\qquad = (15 \cdot 2) + 25\sqrt{2}\sqrt{3} + 9\sqrt{2}\sqrt{3} + (15 \cdot 3)$

$\qquad\qquad\qquad\qquad\qquad = 30 + 25\sqrt{6} + 9\sqrt{6} + 45$

$\qquad\qquad\qquad\qquad\qquad = 75 + 34\sqrt{6}$

(b) $(\sqrt{20} - \sqrt{27})(\sqrt{45} - \sqrt{108}) = (\sqrt{4 \cdot 5} - \sqrt{9 \cdot 3})(\sqrt{9 \cdot 5} - \sqrt{36 \cdot 3})$

$\qquad\qquad\qquad\qquad\qquad = (2\sqrt{5} - 3\sqrt{3})(3\sqrt{5} - 6\sqrt{3})$

Use FOIL: $\qquad\qquad\quad = (6 \cdot 5) - 12\sqrt{3}\sqrt{5} - 9\sqrt{3}\sqrt{5} + (18 \cdot 3)$

$\qquad\qquad\qquad\qquad\qquad = 30 - 12\sqrt{15} - 9\sqrt{15} + 54$

$\qquad\qquad\qquad\qquad\qquad = 84 - 21\sqrt{15}$

[*moral:* Simplify as much as you can before you do anything drastic!]

Rational Exponents

PROBLEM 3-15   Compute the values of

(a) $8^{3/2}$ (b) $4^{2/3} \cdot 4^{4/3}$ (c) $(8x^{2/3})(xy)^{-1/3}$ (d) $(a^3 b^9 c^2)^{2/3}$

Solution

(a) $8^{3/2} = (8^{1/2})^3 = (\sqrt{2^3})^3 = (\sqrt{2^2 \cdot 2})^3 = (2\sqrt{2})^3$
$= 2^3(\sqrt{2})^3 = 8(\sqrt{2})^2(\sqrt{2}) = 8 \cdot 2\sqrt{2} = 16\sqrt{2}$

(b) $4^{2/3} \cdot 4^{4/3} = 4^{2/3+4/3} = 4^{6/3} = 4^2 = 16$

(c) $(8x^{2/3})(xy)^{-1/3} = \dfrac{8x^{2/3}}{(xy)^{1/3}} = \dfrac{8x^{2/3}}{x^{1/3}y^{1/3}} = \dfrac{8x^{1/3}}{y^{1/3}}$

(d) $(a^3 b^9 c^2)^{2/3} = (a^3)^{2/3}(b^9)^{2/3}(c^2)^{2/3} = a^{6/3}b^{18/3}c^{4/3} = a^2 b^6 c^{4/3}$

PROBLEM 3-16   Simplify

(a) $\dfrac{\sqrt[3]{4}}{\sqrt{2}} - \sqrt[6]{2}$ (b) $x^{3/5}y^{-1/3}(x^2 y^4)^{1/5}$ (c) $\dfrac{(5xy^2)^{1/3}(27x^{1/3}y^{2/3})^2}{\sqrt[3]{5x^3 y^3}}$

Solution

(a) $\dfrac{\sqrt[3]{4}}{\sqrt{2}} - \sqrt[6]{2} = \dfrac{4^{1/3}}{2^{1/2}} - 2^{1/6} = \dfrac{(2^2)^{1/3}}{2^{1/2}} - 2^{1/6}$

$= \dfrac{2^{2/3}}{2^{1/2}} - 2^{1/6} = 2^{(2/3-1/2)} - 2^{1/6}$

$= 2^{1/6} - 2^{1/6} = 0$

(b) $x^{3/5}y^{-1/3}(x^2 y^4)^{1/5} = x^{3/5}y^{-1/3}(x^{2/5}y^{4/5})$
$= x^{3/5+2/5}y^{-1/3+4/5}$
$= x^1 y^{7/15} = xy^{7/15}$

(c) $\dfrac{(5xy^2)^{1/3}(27x^{1/3}y^{2/3})^2}{\sqrt[3]{5x^3 y^3}} = \dfrac{(5xy^2)^{1/3}(27x^{1/3}y^{2/3})^2}{(5x^3 y^3)^{1/3}} = \dfrac{(5^{1/3}x^{1/3}y^{2/3})[(27)^2 x^{2/3}y^{4/3}]}{5^{1/3}xy}$

$= \dfrac{(27)^2 x^{3/3}y^{6/3}}{xy} = \dfrac{27^2 xy^2}{xy} = 729y$

PROBLEM 3-17   (a) Simplify $\dfrac{\dfrac{1}{\sqrt{x+h}} - \dfrac{1}{\sqrt{x}}}{h}$. (b) Insert the value $h = 0$ and simplify again.

Solution

(a) $\dfrac{\dfrac{1}{\sqrt{x+h}} - \dfrac{1}{\sqrt{x}}}{h} = \dfrac{\dfrac{1}{\sqrt{x+h}}\left(\dfrac{\sqrt{x}}{\sqrt{x}}\right) - \dfrac{1}{\sqrt{x}}\left(\dfrac{\sqrt{x+h}}{\sqrt{x+h}}\right)}{h}$

$= \dfrac{\dfrac{\sqrt{x} - \sqrt{x+h}}{\sqrt{x}\sqrt{x+h}}}{h} = \dfrac{\sqrt{x} - \sqrt{x+h}}{h\sqrt{x}\sqrt{x+h}}$

(b) You can't set $h = 0$ yet, because division by zero is meaningless. So rationalize the numerator:

$$\dfrac{\sqrt{x} - \sqrt{x+h}}{h\sqrt{x}\sqrt{x+h}} = \left(\dfrac{\sqrt{x} - \sqrt{x+h}}{h\sqrt{x}\sqrt{x+h}}\right)\left(\dfrac{\sqrt{x} + \sqrt{x+h}}{\sqrt{x} + \sqrt{x+h}}\right)$$

$$= \frac{x - (x + h)}{h\sqrt{x}\sqrt{x+h}(\sqrt{x} + \sqrt{x+h})} = \frac{-h}{h\sqrt{x}\sqrt{x+h}(\sqrt{x} + \sqrt{x+h})}$$

and cancel the h's:

$$= \frac{-1}{\sqrt{x}\sqrt{x+h}(\sqrt{x} + \sqrt{x+h})}$$

Now set $h = 0$ in the resulting expression:

$$\left. \frac{\frac{1}{\sqrt{x+h}} - \frac{1}{\sqrt{x}}}{h} \right|_{\text{at } h=0} = \frac{-1}{\sqrt{x}\sqrt{x}(\sqrt{x} + \sqrt{x})} = \frac{-1}{x(2\sqrt{x})} = -\frac{1}{2} \cdot \frac{1}{x^{3/2}} = -\frac{1}{2}x^{-3/2}$$

Complex Numbers

PROBLEM 3-18 Find the simplified form of (a) $i^9 - i^5$, (b) $(2 + 3i) - (4 - 5i)$, and (c) $(1 + i)^2 - (1 - i)^2$.

Solution Remember that $i = \sqrt{-1}$ and $i^2 = (\sqrt{-1})^2 = -1$.

(a) Since $i^2 = -1$, then $i^4 = 1$ and $i^8 = 1$. So you can write

$$i^9 - i^5 = (i^8 \cdot i) - (i^4 \cdot i)$$
$$= (1 \cdot i) - (1 \cdot i)$$
$$= i - i = 0$$

(b) Treat i as if it were any variable or constant:

$$(2 + 3i) - (4 - 5i) = 2 + 3i - 4 + 5i$$
$$= (2 - 4) + (3i + 5i)$$
$$= -2 + 8i$$

(c) Expand; then look for powers of i that are real numbers:

$$(1 + i)^2 - (1 - i)^2 = 1 + 2i + i^2 - (1 - 2i + i^2)$$
$$= 1 + 2i - 1 - (1 - 2i - 1)$$
$$= 2i + 2i = 4i$$

PROBLEM 3-19 Simplify (a) $\sqrt{(5 - 4i)(5 + 4i)} - \left(\frac{9^2 + 1}{2}\right)^{1/2}$ and (b) $(3 + i)(2 - 5i)(i - 3)$.

Solution

(a) $\sqrt{(5 - 4i)(5 + 4i)} - \left(\frac{9^2 + 1}{2}\right)^{1/2} = \sqrt{25 - 16i^2} - \left(\frac{82}{2}\right)^{1/2}$

$$= \sqrt{25 - 16(-1)} - (41)^{1/2}$$
$$= (41)^{1/2} - (41)^{1/2} = 0$$

(b) $(3 + i)(2 - 5i)(i - 3) = (i + 3)(i - 3)(2 - 5i)$
$$= (i^2 - 9)(2 - 5i) = (-1 - 9)(2 - 5i)$$
$$= -10(2 - 5i) = -20 + 50i$$

PROBLEM 3-20 Use the complex conjugate to simplify

(a) $\dfrac{3 + \sqrt{-7}}{2 - \sqrt{-3}}$ (b) $\dfrac{\sqrt{2} - \sqrt{3}i}{\sqrt{2} + \sqrt{3}i}$ (c) $\dfrac{4}{i} \cdot \dfrac{3 - i}{i^3} \cdot \dfrac{5 + i}{i^5} \cdot \dfrac{i^9}{8}$

Solution

(a) $\dfrac{3 + \sqrt{-7}}{2 - \sqrt{-3}} = \dfrac{3 + \sqrt{7}\sqrt{-1}}{2 - \sqrt{3}\sqrt{-1}} = \dfrac{3 + \sqrt{7}i}{2 - \sqrt{3}i}$

$$\frac{3 + \sqrt{-7}}{2 - \sqrt{-3}} = \left(\frac{3 + \sqrt{7}i}{2 - \sqrt{3}i}\right)\left(\frac{2 + \sqrt{3}i}{2 + \sqrt{3}i}\right) = \frac{6 + 3\sqrt{3}i + 2\sqrt{7}i + \sqrt{21}i^2}{4 + 2\sqrt{3}i - 2\sqrt{3}i - 3i^2}$$

$$= \frac{6 + \sqrt{21}(-1) + 3\sqrt{3}i + 2\sqrt{7}i}{4 - 3(-1) + 2\sqrt{3}i - 2\sqrt{3}i} = \frac{6 - \sqrt{21} + (3\sqrt{3} + 2\sqrt{7})i}{7}$$

(b) $\dfrac{\sqrt{2} - \sqrt{3}i}{\sqrt{2} + \sqrt{3}i} = \left(\dfrac{\sqrt{2} - \sqrt{3}i}{\sqrt{2} + \sqrt{3}i}\right)\left(\dfrac{\sqrt{2} - \sqrt{3}i}{\sqrt{2} - \sqrt{3}i}\right) = \dfrac{2 - \sqrt{6}i - \sqrt{6}i + 3i^2}{2 - \sqrt{6}i + \sqrt{6}i - 3i^2}$

$$= \frac{2 + 3(-1) - 2\sqrt{6}i}{2 - 3(-1) - \sqrt{6}i + \sqrt{6}i} = \frac{-1 - 2\sqrt{6}i}{5} = -\frac{1}{5} - \frac{2\sqrt{6}}{5}i$$

(c) $\dfrac{4}{i} \cdot \dfrac{3 - i}{i^3} \cdot \dfrac{5 + i}{i^5} \cdot \dfrac{i^9}{8} = \left(\dfrac{4}{i} \cdot \dfrac{3 - i}{i^3}\right)\left(\dfrac{5 + i}{i^4 \cdot i}\right)\left(\dfrac{i^8 \cdot i}{8}\right) = \left(\dfrac{12 - 4i}{i^4}\right)\left(\dfrac{5 + i}{1 \cdot i}\right)\left(\dfrac{1 \cdot i}{8}\right)$

$$= \frac{(12 - 4i)(5 + i)}{8} = \frac{60 + 12i - 20i - 4i^2}{8} = \frac{64 - 8i}{8} = 8 - i$$

PROBLEM 3-21   Use everything you know to simplify this "monster":

$$\left[\frac{(1 + i)^3}{2} - (-1 + i)\right]^5 + \left[\frac{i^{20} + i^{17}}{(1 - i) \cdot 10^0} - \frac{(1 + i)^2}{16^{1/4}}\right]^7$$

Solution   Work on selected pieces first:

(1) $(1 + i)^3 = 1 + 3i + 3i^2 + i^3$

$\qquad\qquad = 1 + 3i + 3(-1) - i = -2 + 2i$ [Special product for a cube]

(2) $i^{20} + i^{17} = (i^4)^5 + (i^{16} \cdot i) = 1^5 + (1 \cdot i) = 1 + i$

(3) $10^0 = 1$

(4) $16^{1/4} = 2$

Then put the pieces back together:

$$\left[\frac{(1 + i)^3}{2} - (-1 + i)\right]^5 + \left[\frac{i^{20} + i^{17}}{(1 - i) \cdot 10^0} - \frac{(1 + i)^2}{16^{1/4}}\right]^7$$

$$= \left[\frac{-2 + 2i}{2} - (-1 + i)\right]^5 + \left[\frac{1 + i}{1(1 - i)} - \frac{(1 + i)^2}{2}\right]^7$$

$$= [-1 + i - (-1 + i)]^5 + \left[\left(\frac{1 + i}{1 - i}\right)\left(\frac{1 + i}{1 + i}\right) - \frac{(1 + i)^2}{2}\right]^7$$

$$= (-1 + i + 1 - i)^5 + \left[\frac{(1 + i)^2}{1 - i^2} - \frac{(1 + i)^2}{2}\right]^7$$

$$= 0^5 + \left[\frac{(1 + i)^2}{2} - \frac{(1 + i)^2}{2}\right]^7$$

$$= 0 + 0^7$$

$$= 0$$

Supplementary Exercises

Simplify the following expressions.

PROBLEM 3-22   **(a)** $\dfrac{5^3 \cdot 5^{-2}}{5^4}$ **(b)** $\dfrac{[(-2)^3]^2}{[4^2]^3}$ **(c)** $\dfrac{16^{-2}}{4^{-4}}$ **(d)** $\dfrac{3^0}{4^2 - 2^4 + 2^0}$

PROBLEM 3-23   (a) $\left(\dfrac{x^2}{y}\right)^3$ (b) $\left(\dfrac{5uv^{-2}}{7u^{-2}v}\right)^2$ (c) $\left(\dfrac{p^{-4}}{p^{-5}}\right)^{-3}$ (d) $\dfrac{2x^2y^3z^{-2}}{8^{-1}x^5y^{-1}z^4}$

PROBLEM 3-24   (a) $\dfrac{(2x^{-3})^2(8x^{-3}y)^{-2}}{(x^{-4}y^2)^{-2}}$ (b) $\dfrac{(xy)^{-1}-x^{-1}}{y^{-1}+(xy)^{-1}}$ (c) $\dfrac{a^{-2}+b^{-2}}{a^2b^2(a^{-1}-b^{-1})}$

PROBLEM 3-25   (a) $\dfrac{x^{-1}+y^{-1}+z^{-1}}{(xyz)^{-1}}$ (b) $\dfrac{a^{-3}}{a^{-2}+a^{-1}}\cdot\dfrac{a^{-2}}{a^{-2}-a^{-1}}$

PROBLEM 3-26   (a) $\sqrt{2x}\sqrt{4x}\sqrt{8x}\sqrt{16x}$ (b) $\dfrac{\sqrt{128}\sqrt{625}}{\sqrt{169}\sqrt{98}}$ (c) $\sqrt{\left(\dfrac{37a^2b^5}{93a^{17}}\right)^0}$

PROBLEM 3-27   Simplify the following expressions, assuming that a, b, and c are positive quantities.

(a) $\sqrt{a^6b^{-4}c^8}$ (b) $\dfrac{\sqrt{12a^{-2}b^5c^{-4}}}{\sqrt{18a^{-4}b^{-1}c^2}}$ (c) $\sqrt{a^{-2}b^{-2}}-\dfrac{1}{\sqrt{a^2b^2}}$

PROBLEM 3-28   Find the values of

(a) $\sqrt[3]{216}$ (b) $\sqrt[4]{256}$ (c) $\sqrt[3]{-8}\,\sqrt[5]{-32}$ (d) $\sqrt[3]{\sqrt{10^0}}$

PROBLEM 3-29   Find the values of

(a) $\sqrt[3]{-\dfrac{1}{343}}$ (b) $\sqrt[5]{.00032}$ (c) $\sqrt[3]{-.512}$ (d) $\sqrt[4]{\sqrt{10\,000}}$

PROBLEM 3-30   Rationalize the denominator to simplify

(a) $\dfrac{\sqrt{3}}{\sqrt{5}-\sqrt{3}}$ (b) $\dfrac{\sqrt{.04}}{\sqrt{.01}+\sqrt{.04}}$ (c) $\dfrac{\sqrt{7}+\sqrt{5}}{\sqrt{7}-\sqrt{5}}$

PROBLEM 3-31   Simplify

(a) $\dfrac{\sqrt{u}-4\sqrt{v}}{\sqrt{u}+\sqrt{v}}$ (b) $\dfrac{\sqrt{2x+h}-\sqrt{2x}}{h}$ (c) $\dfrac{(\sqrt{a}-\sqrt{b})^2}{a+b}$

PROBLEM 3-32   Find the values of

(a) $\dfrac{(27)^{1/3}}{(49)^{1/2}}$ (b) $(-1)^{5/3}$ (c) $(2^{1/2}+3^{1/2})^2$

PROBLEM 3-33   Simplify the rational exponent expressions (a) $x^{1/3}\cdot x^{1/6}$, (b) $x^{1/3}\div x^{1/6}$, and (c) $(x^{-3})^{-2/3}$.

PROBLEM 3-34   Combine similar terms to simplify

(a) $\left(\dfrac{x-y}{x+y}\right)^{1/2}-\left(\dfrac{x+y}{x-y}\right)^{1/2}+\left(\dfrac{x^2}{x^2-y^2}\right)^{1/2}$ and (b) $\dfrac{3}{2}(108)^{1/2}-9^{1/4}+4(18)^{1/2}$.

PROBLEM 3-35   Simplify

(a) $\dfrac{1}{\sqrt{3}+\sqrt{5}+\sqrt{7}}$ (b) $\dfrac{1}{\sqrt[3]{32}+\sqrt[3]{4}}$ (c) $\left(\dfrac{2x^{3/2}y^{1/3}}{z^{1/2}}\right)^2\div\left(\dfrac{x^6y^2}{z}\right)^{1/2}$

PROBLEM 3-36   Find the simplified forms of (a) $(5\sqrt{3}+2\sqrt{5})(4\sqrt{3}-\sqrt{5})$ and (b) $(7\sqrt{3}-9\sqrt{5})(9\sqrt{3}+7\sqrt{5})$.

PROBLEM 3-37   Simplify

(a) $\left(\dfrac{a^{-1}b^4}{a^{-5}b^0}\right)^{-1}\left(\dfrac{a^4b^{-1}}{ab^{-3}}\right)$ (b) $\dfrac{x^{n-5}}{x^{n-8}}$ (c) $\left(\dfrac{1}{R_1}+\dfrac{1}{R_2}+\dfrac{1}{R^3}\right)^{-1}$

PROBLEM 3-38   Find the reduced forms of

(a) $i^{10}-i^5$ (b) $\dfrac{i^3-i^{11}}{a}$ (c) i^{5228}

PROBLEM 3-39   Identify the real and imaginary parts of

(a) $\dfrac{17}{2}+\dfrac{\sqrt{87}}{5}i$ (b) $(4-3i)(2+i)$ (c) $5\dfrac{1}{2}$

PROBLEM 3-40   Find the values of (a) $\dfrac{(4-i)^2}{3}-\dfrac{8i^3}{3}$ and (b) $(4i+1)(2-3i)(1-4i)(3i+2)$.

PROBLEM 3-41   Simplify

(a) $\dfrac{2-i}{i^3}$ (b) $\dfrac{4-7i}{2+5i}$ (c) $\dfrac{(2i+1)^2}{-3i^5}$ (d) $\dfrac{(\sqrt{2}+\sqrt{3}i)^2}{\sqrt{2}-\sqrt{3}i}$

PROBLEM 3-42   Simplify the "monster" expression

$$\frac{i^5-7i^4+3i^3-9i^2+4i+3}{(i^2-4)^2}\div\left(\frac{-i^{-2}+i^{-4}}{\sqrt{5}i^{-1}}\right)^2$$

Answers to Supplementary Exercises

3-22 (a) $5^{-3}=\frac{1}{125}$

(b) $2^{-6}=\frac{1}{64}$

(c) 1

(d) 1

3-23 (a) $\dfrac{x^6}{y^3}$

(b) $\dfrac{25u^6}{49v^6}$

(c) p^{-3}

(d) $\dfrac{16y^4}{x^3z^6}$

3-24 (a) $\dfrac{y^2}{16x^8}$

(b) $\dfrac{1-y}{x+1}$

(c) $\dfrac{a^2+b^2}{a^3b^3(b-a)}$

3-25 (a) $yz+xz+xy$

(b) $\dfrac{1}{1-a^2}$

3-26 (a) $32x^2$

(b) $\dfrac{200}{91}$

(c) 1

3-27 (a) $\dfrac{a^3c^4}{b^2}$

(b) $\sqrt{\frac{2}{3}}ab^3c^{-3}$

(c) 0

3-28 (a) 6

(b) 4

(c) 4

(d) 1

3-29 (a) $-\frac{1}{7}$

(b) $\frac{1}{5}$

(c) $-\frac{4}{5}$

(d) $\sqrt{10}$

3-30 (a) $\dfrac{\sqrt{15}+3}{2}$

(b) $\frac{2}{3}$

(c) $6+\sqrt{35}$

3-31 (a) $\dfrac{u - 5\sqrt{uv} + 4v}{u - v}$

(b) $\dfrac{1}{\sqrt{2x + h} + \sqrt{2x}}$

(c) $1 - \dfrac{2\sqrt{ab}}{a + b}$

3-32 (a) $\frac{3}{7}$

(b) -1

(c) $5 + 2\sqrt{6}$

3-33 (a) $x^{1/2}$

(b) $x^{1/6}$

(c) x^2

3-34 (a) $\dfrac{x - 2y}{\sqrt{x^2 - y^2}}$

(b) $8\sqrt{3} + 12\sqrt{2}$

3-35 (a) $\dfrac{9\sqrt{3} + 5\sqrt{5} + \sqrt{7} - 2\sqrt{105}}{59}$

(b) $\frac{1}{3} \cdot 4^{-1/3}$

(c) $\dfrac{4}{z^{1/2} y^{1/3}}$

3-36 (a) $50 + 3\sqrt{15}$

(b) $-126 - 32\sqrt{15}$

3-37 (a) $\dfrac{1}{ab^2}$

(b) x^3

(c) $\dfrac{R_1 R_2 R_3}{R_2 R_3 + R_1 R_3 + R_1 R_2}$

3-38 (a) $-1 - i$

(b) 0

(c) 1

3-39 (a) Real part $= \dfrac{17}{2}$

Imaginary part $= \dfrac{\sqrt{87}}{5}$

(b) Real part $= 11$

Imaginary part $= -2$

(c) Real part $= \dfrac{11}{2}$

Imaginary part $= 0$

3-40 (a) 5

(b) 221

3-41 (a) $1 + 2i$

(b) $\dfrac{-27 - 34i}{29}$

(c) $\dfrac{-4 - 3i}{3}$

(d) $\dfrac{-7\sqrt{2} + 3\sqrt{3}\,i}{5}$

3-42 $\dfrac{-5 - 2i}{20}$

EXAM 1 (Chapters 1–3)

1. Find the values of the following expressions:

   (a) $(\frac{1}{3} + \frac{3}{5}) \div (\frac{1}{4} - \frac{1}{6})$

   (b) $[4 + (2 \div 3)] \cdot 9 + 5 \div 6$

   (c) $8^{1/3} + 4^{1/2} - 16^{1/4} - 2 \cdot 5^0$

   (d) $\frac{3}{5} + \frac{7}{20} - \frac{1}{2} + \frac{4}{35}$

2. Simplify the following expressions:

   (a) $[-34 + (x - 5)^2] - [(2x + 1)^2 - 3(\frac{4}{3}x + 5)]$

   (b) $(x^2 - 3)^2 + (\sqrt{6}x + 1)^2 - (x^4 + 10)$

   (c) $(x + 3)(3x + 5)(x - 7)$

3. Find the products (a) $(a + 4b - 7c)(2a - b - c)$ and (b) $(2x^2y^3 - 3x^3y)^2$.

4. Find the product $(2x - 3)(x + 1)^2(3x - 1)$.

5. Factor (a) $x^5 - 7x^4 + 6x^3$, (b) $x^2(x + 2) - 2x(x + 2) - 15x - 30$, and (c) $64a^3 - b^3$.

6. Factor (a) $2x^2 - 11x - 21$, (b) $25x^2 - 10x + 1$, and (c) $r^3t^6 + 1$.

7. Reduce the following to lowest terms:

   (a) $\dfrac{x^3 + 4x^2 - 12x}{x^2 - 2x}$

   (b) $\dfrac{y - 1}{y^2 + 5y + 6} \div \dfrac{y^2 - 1}{y^2 + 4y + 3}$

   (c) $\dfrac{\dfrac{2x - 7}{x} - \dfrac{4}{x^2}}{3 - \dfrac{12}{x}}$

8. Use the long-division process to reduce $\dfrac{x^4 + 2x^2 + 5x - 7}{x + 4}$.

9. Simplify

   (a) $\dfrac{(x^2y^3)^2}{(xy)^{-2}} \div \dfrac{x^6y^0}{(y^{-4})^2}$

   (b) $\dfrac{\sqrt{45a^4b^{-2}c^3}\sqrt{2a^{-2}bc^{-5}}}{\sqrt{10a^2b^{-1}c^2}}$

   (c) $(4\sqrt{10} - 3\sqrt{5})(\sqrt{2} - \sqrt{4})$

   (d) $\dfrac{(81u^6v^2)^{1/4}(2^0u^{1/3}v^{1/2})^4}{(u^9v^{15/2})^{1/3}}$

10. Simplify (a) $(i - 1)^2 + (2 - i)^2 + (2i + 1)^2$, (b) $(2 + 5i)(3 - 6i)(14 - 2i)$, and (c) $i^{15} - 7i^9 + 4i^3$.

Answers to Exam 1

1. (a) $\frac{56}{5}$ (c) 0

   (b) $42\frac{5}{6}$ (d) $\frac{79}{140}$

2. (a) $-3x^2 - 10x + 5$

   (b) $2\sqrt{6}x$

   (c) $3x^3 - 7x^2 - 83x - 105$

3. (a) $2a^2 - 4b^2 + 7c^2 + 7ab + 3bc - 15ac$

   (b) $x^4 y^2 (4y^4 - 12xy^2 + 9x^2)$

4. $6x^4 + x^3 - 13x^2 - 5x + 3$

5. (a) $x^3(x - 6)(x - 1)$

   (b) $(x + 2)(x - 5)(x + 3)$

   (c) $(4a - b)(16a^2 + 4ab + b^2)$

6. (a) $(2x + 3)(x - 7)$

   (b) $(5x - 1)^2$

   (c) $(rt^2 + 1)(r^2 t^4 - rt^2 + 1)$

7. (a) $x + 6$

   (b) $\dfrac{1}{y + 2}$

   (c) $\dfrac{2x + 1}{3x}$

8. $x^3 - 4x^2 + 18x - 67 + \dfrac{261}{x + 4}$

9. (a) 1   (c) $\sqrt{10}(7\sqrt{2} - 19)$

   (b) $\dfrac{3}{c^2}$ (d) $3u^{-1/6}$

10. (a) $-2i$

   (b) $30(17 - i)$

   (c) $-12i$

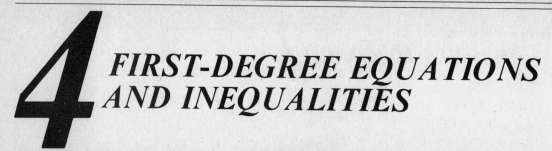

4 FIRST-DEGREE EQUATIONS AND INEQUALITIES

4-1. First-Degree Equations in One Variable

A. Definitions and terminology

An **equation** is a statement of equality between two algebraic expressions.

- An **identity** is an equation that is always true.
- A **conditional equation** is an equation that is true when the variable used meets the conditions of the equation.

EXAMPLE 4-1 Which of the following is an identity and which is a conditional equation? (a) $\frac{6}{3} + 4 = 6$ (b) $x + 5 = 8$

Solution

(a) $\frac{6}{3} + 4 = 6$ is an identity because it is always true.

(b) $x + 5 = 8$ is a conditional equation because its truth depends on the value(s) of the variable x.

Conditional equations that involve only one variable raised to the first power are called **linear** or **first-degree equations** in one variable. If we let x be a variable and a and b be constants, we can write the general form of a linear equation as

LINEAR EQUATION $ax + b = 0,$ a and b are real numbers, $a \neq 0$

If we replace the variable in an equation by a value or an expression that makes the equation an identity, then that value or expression is a **solution** to the equation.

EXAMPLE 4-2 Change the conditional equations (a) $3x - 6 = 0$ and (b) $\frac{1}{2}x = 12$ to identities.

Solution

(a) $3x - 6 = 0$

$3(2) - 6 = 0$ identity

The linear equation has the value 2 as its solution.

(b) $\frac{1}{2}x = 12$

$\frac{1}{2}(24) = 12$ identity

The linear equation has the value 24 as its solution.

B. Rules of algebraic operation

While the equations in Example 4-2 are simple enough to solve by inspection, it's not usually that easy. So we need a set of rules or operations that we can use to solve first-degree equations.

- If $A(x)$, $B(x)$, and $C(x)$ are algebraic expressions, then the algebraic equation $A(x) = B(x)$ may be expressed as

Rule 1.	$A(x) + C(x) = B(x) + C(x)$	Equals added to equals are equal.
Rule 2.	$A(x) - C(x) = B(x) - C(x)$	Equals subtracted from equals are equal.
Rule 3.	$A(x) \cdot C(x) = B(x) \cdot C(x)$ where $C(x) \neq 0$	Equals multiplied by equal, nonzero expressions are equal.
Rule 4.	$\dfrac{A(x)}{C(x)} = \dfrac{B(x)}{C(x)}$ where $C(x) \neq 0$	Equals divided by equal, nonzero expressions are equal.

RULES OF ALGEBRAIC OPERATION

Two equations that have identical solutions are called **equivalent equations**. In using the rules of algebraic operation, we are *creating successive equations that are equivalent*. This process continues until a solution is reached.

EXAMPLE 4-3   Solve $4x + 3 = x - 7$.

Solution   Produce equivalent equations via the rules:

$$4x + 3 = x - 7$$

$$(4x + 3) - 3 = (x - 7) - 3 \qquad \text{[\textit{Rule 2:} Subtract equals from equals]}$$

$$4x + \cancel{3} - \cancel{3} = x - 10 \qquad \text{[Cancel and combine]}$$

$$4x - x = (x - 10) - x \qquad \text{[\textit{Rule 2:} Subtract equals from equals]}$$

$$3x = \cancel{x} - 10 - \cancel{x} \qquad \text{[Cancel and combine]}$$

$$\frac{3x}{3} = \frac{-10}{3} \qquad \text{[\textit{Rule 4:} Divide equals by equal, nonzero expressions]}$$

$$x = \frac{-10}{3}$$

Check: $4\left(\dfrac{-10}{3}\right) + 3 \overset{?}{=} \dfrac{-10}{3} - 7$

$$\frac{-40}{3} + \frac{9}{3} \overset{?}{=} \frac{-10}{3} - \frac{21}{3}$$

$$\frac{-31}{3} \overset{\checkmark}{=} \frac{-31}{3}$$

C. Procedure for solving linear equations

Example 4-3 illustrates the procedure for solving linear equations. This procedure is detailed in the following steps:

(1) Add or subtract the same constant terms on both sides to bring all the constant numbers to one side. Then simplify the equation.

(2) Add or subtract the same variable terms on both sides to bring all the variable terms to the other side of the equation. Simplify the result.

(3) Multiply or divide both sides by the number that will make the variable stand alone. Simplify to obtain the solution.

(4) Substitute the solution into the original equation to check its validity.

EXAMPLE 4-4   Solve $x - 3 + 3(x + 5) = 4(2x - 7)$.

Solution   First simplify the given equation as much as possible:

$$x - 3 + 3(x + 5) = 4(2x - 7)$$

$$x - 3 + 3x + 15 = 8x - 28 \qquad \text{[Multiply out]}$$

$$4x + 12 = 8x - 28 \qquad \text{[Combine]}$$

(1)   Bring all the constant numbers to one side:

$$4x + \cancel{12} - \cancel{12} = 8x - 28 - 12 \qquad [\textit{Rule 2:} \text{ Subtract 12 from each side}]$$

(2)   Bring all the variable terms to the other side:

$$4x - 8x = \cancel{8x} - 40 - \cancel{8x} \qquad [\textit{Rule 2:} \text{ Subtract } 8x \text{ from each side}]$$

$$-4x = -40 \qquad \text{[Combine]}$$

(3)   Divide both sides by the same number

$$\frac{-4x}{-4} = \frac{-40}{-4} \qquad [\textit{Rule 4:} \text{ Divide both sides by } -4]$$

to make the variable stand alone:

$$x = 10$$

(4)   *Check:* Substitute the solution into the original equation:

$$10 - 3 + 3(10 + 5) \overset{?}{=} 4(2 \cdot 10 - 7)$$

$$7 + 3(15) \overset{?}{=} 4(20 - 7)$$

$$7 + 45 \overset{?}{=} 4(13)$$

$$52 \overset{\checkmark}{=} 52$$

note: It doesn't really matter which side you bring the variable term to—but the convention is to put variables on the left-hand side (lhs) and constants on the right-hand side (rhs).

If an equation involves fractions, we simplify the problem by multiplying by the least common denominator (lcd). Multiplying by the lcd gets rid of, or *clears*, the fractions.

EXAMPLE 4-5   Solve $\dfrac{x - 3}{2} - \dfrac{4x + 2}{3} = 2$.

Solution
$$\frac{x - 3}{2} - \frac{4x + 2}{3} = 2$$

$$6\left(\frac{x - 3}{2}\right) - 6\left(\frac{4x + 2}{3}\right) = 6(2) \qquad [\text{Multiply by lcd} = 6 \ (\textit{Rule 3})]$$

$$3(x - 3) - 2(4x + 2) = 12 \qquad \text{[Simplify]}$$

$$3x - 9 - 8x - 4 = 12 \qquad \text{[Expand]}$$

$$-5x - 13 = 12 \qquad \text{[Combine]}$$

$$-5x - \cancel{13} + \cancel{13} = 12 + 13 \qquad [\textit{Rule 1}]$$

$$-5x = 25 \qquad \text{[Simplify]}$$

$$\frac{-5x}{-5} = \frac{25}{-5} \qquad [Rule\ 4]$$

$$x = -5$$

Check: $\dfrac{-5-3}{2} - \dfrac{4(-5)+2}{3} \overset{?}{=} 2$

$$\dfrac{-8}{2} - \dfrac{-20+2}{3} \overset{?}{=} 2$$

$$-4 - (-6) \overset{?}{=} 2$$

$$-4 + 6 \overset{\checkmark}{=} 2$$

The process used to solve for the unknown x can also be used to solve **literal equations**, i.e., equations involving one or more letters besides the one whose value is being found.

EXAMPLE 4-6   Solve for x in $mx - 3a = 7b$.

Solution $mx - 3a = 7b$

$$mx - 3a + 3a = 7b + 3a \qquad [Rule\ 1]$$

$$mx = 7b + 3a \qquad [\text{Simplify}]$$

$$\frac{mx}{m} = \frac{7b + 3a}{m} \qquad [Rule\ 4\ (\text{assuming } m \neq 0)]$$

$$x = \frac{7b + 3a}{m}$$

Check: $\dfrac{m(7b + 3a)}{m} - 3a \overset{?}{=} 7b$

$$(7b + 3a) - 3a \overset{?}{=} 7b$$

$$7b + 3a - 3a \overset{?}{=} 7b$$

$$7b \overset{\checkmark}{=} 7b$$

EXAMPLE 4-7   The formula for converting degrees Fahrenheit ($°F$) into degrees Celsius ($°C$) is $°C = \frac{5}{9}(°F - 32)$. Solve this formula for $°F$.

Solution $°C = \frac{5}{9}(°F - 32)$

$$\tfrac{9}{5}°C = \tfrac{9}{5} \cdot \tfrac{5}{9}(°F - 32) \qquad [Rule\ 3]$$

$$\tfrac{9}{5}°C = °F - 32 \qquad [\text{Cancel}]$$

$$\tfrac{9}{5}°C + 32 = °F - 32 + 32 \qquad [Rule\ 1]$$

$$°F = \tfrac{9}{5}°C + 32 \qquad [\text{Symmetry of equality}]$$

Not all equations have a solution. Often the only way to tell is to work the silly thing and see.

EXAMPLE 4-8   Solve the first-degree equation $3(2x + 4) - 7 = 2(3x - 1)$.

Solution   Multiply out the given quantities and follow the rules:

$$3(2x + 4) - 7 = 2(3x - 1)$$

$$6x + 12 - 7 = 6x - 2$$

$$6x + 5 = 6x - 2$$

Subtract $6x$ from each side and obtain $5 = -2$. Impossible! Hence there is NO value of x that satisfies the given equation.

4-2. First-Degree Inequalities in One Variable

A. Definitions and notation

If we are concerned with the *ordering of expressions*, we work with **inequalities**. The four **relationals** that we use to connect unequal expressions are

> $>$ greater than $\geq$ greater than or equal to

> $<$ less than $\leq$ less than or equal to

The related expressions $x + 5 > 10$, $\dfrac{8 + y}{4} \leq 28$, and $.09 < (2r - .5) < .13$

are examples of inequalities.

In working with first-degree inequalities, we again will be looking for *solutions*—the values of the variable that make the relational expressions true. But the solution of an inequality is a *set* of values: This set may have a single value, a finite number of values, or a continuous string (interval) of values.

EXAMPLE 4-9 Express the value(s) that will make these inequalities true:
(a) $4x + 3 > 15$ **(b)** $4x + 3 \geq 15$

Solution

(a) Picking arbitrary numbers, we see that the values $x = 6$ and $x = 4$ are solutions to the inequality $4x + 3 > 15$ because

$$\begin{Bmatrix} 4(6) + 3 > 15 \\ 27 > 15 \end{Bmatrix} \quad \text{and} \quad \begin{Bmatrix} 4(4) + 3 > 15 \\ 19 > 15 \end{Bmatrix}$$

But we're not through yet—ANY real number greater than 3 will work here. In other words, there is a *continuous string*, or *interval*, of numbers greater than but not equal to 3 that will solve this inequality. We can graph this solution set on the number line as

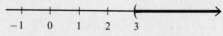

where the single parenthesis "(" at 3 indicates that 3 is not included in the set. We can also express this solution set in two other ways: We can use

(1) SET-BUILDER NOTATION $\{x \mid x > 3\}$, read as "the set of x values such that x is greater than 3," or

(2) INTERVAL NOTATION $(3, \infty)$, where "(3" indicates a lower bound down to but not including 3 and "∞)" indicates an extension of the interval without bound in the positive direction.

(b) Everything that's true of $4x + 3 > 15$ is also true of $4x + 3 \geq 15$, except that the set of values in the solution includes 3; i.e., $4(3) + 3 = 15$. So we can graph this solution as

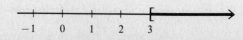

where the single bracket at 3 indicates that 3 is included. Or, we can write **(1)** $\{x \mid x \geq 3\}$ or **(2)** $[3, \infty)$.

note: In interval notation, the symbols $\pm \infty$ *are not real numbers:* Each should be read as "without bound," meaning that the interval goes on forever in the direction specified by the plus or minus sign. A single parenthesis,) or (, at either end of the interval indicates that the given endpoint is not included in the set; and a single bracket,] or [, indicates that the given endpoint is included in the set.

- A **bounded interval** has two finite endpoints, a and b.
- An **unbounded interval** may have two endpoints, $-\infty$ and ∞, neither of which is finite; or it may have one endpoint, a or b, that is finite and one endpoint, $-\infty$ or ∞, that is not finite.
- An **open interval** does not include the endpoints in the set.
- A **closed interval** includes the endpoints in the set.
- A **half-open interval** has two finite endpoints, only one of which is included in the set.

The types of intervals are listed in Tables 4-1 and 4-2.

TABLE 4-1: Unbounded Intervals*		
Open	(a, ∞)	$\equiv \{x \mid x > a\}$
	$(-\infty, a)$	$\equiv \{x \mid x < a\}$
	$(-\infty, \infty)$	$\equiv \{x \mid x \in R\}$
		$\equiv$ all real numbers
Closed	$[a, \infty)$	$\equiv \{x \mid x \geq a\}$
	$(-\infty, a]$	$\equiv \{x \mid x \leq a\}$

*a is real; $\infty \equiv$ without positive bound; $-\infty \equiv$ without negative bound; $R =$ set of real numbers.

TABLE 4-2: Bounded Intervals*		
Open	(a, b)	$\equiv \{x \mid a < x < b\}$
Closed	$[a, b]$	$\equiv \{x \mid a \leq x \leq b\}$
Half-open	$(a, b]$	$\equiv \{x \mid a < x \leq b\}$
	$[a, b)$	$\equiv \{x \mid a \leq x < b\}$

*a and *b* are real.

B. Rules for solving inequalities

The rules for solving inequalities are the same as the rules for solving equations—except that the multiplication (*Rule 3*) and division (*Rule 4*) rules must be modified:

Rule 3′. If $A < B$
then $\quad AC < BC \quad$ when $C > 0$ — Multiplying both sides of an inequality by the same *positive number* PRESERVES the direction of the inequality.

or $\quad AC > BC \quad$ when $C < 0$ — Multiplying both sides by the same *negative number* REVERSES the direction of the inequality.

Rule 4′. If $A < B$
then $\quad \dfrac{A}{D} < \dfrac{B}{D} \quad$ when $D > 0$ — Dividing both sides of an inequality by the same *positive number* PRESERVES the direction of the inequality.

or $\quad \dfrac{A}{D} > \dfrac{B}{D} \quad$ when $D < 0$ — Dividing both sides by the same *negative number* REVERSES the direction of the inequality.

EXAMPLE 4-10 Solve **(a)** $-4x < 8$ and **(b)** $\dfrac{2 - x}{3} \geq \dfrac{1}{4}$.

Solution

(a) To solve $-4x < 8$, divide by -4 and reverse the inequality to obtain

$$\frac{-4x}{-4} > \frac{8}{-4}$$

$$x > -2$$

The values of x that satisfy the original inequality are all those values greater than -2, or $\{x \mid x > -2\}$ or $(-2, \infty)$:

(b) To solve $\dfrac{2-x}{3} \geq \dfrac{1}{4}$, first multiply by 12 to clear the expression of fractions and obtain

$$\frac{12(2-x)}{3} \geq \frac{12(1)}{4}$$

$$4(2-x) \geq 3 \qquad \text{[Simplify]}$$

$$8 - 4x \geq 3$$

$$(8 - 4x) - 8 \geq 3 - 8 \qquad \text{[\textit{Rule 2:} Subtract equals]}$$

$$-4x \geq -5$$

$$x \leq \tfrac{5}{4} \qquad \text{[\textit{Rule 4':} Divide by } -4 \text{ and reverse the inequality.]}$$

So $\{x \mid x \leq \tfrac{5}{4}\}$ or $(-\infty, \tfrac{5}{4}]$:

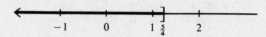

EXAMPLE 4-11 A salesperson receives a salary of \$200 per week, but she also gets a commission of \$20 for each refrigerator she sells. How many refrigerators must she sell to earn \$400 or more per week?

Solution Call the unknown number of refrigerators x, so that total income is expressed by $200 + 20x$. Then solve the inequality $200 + 20x \geq 400$:

$$200 + 20x \geq 400$$

$$20x \geq 200$$

$$x \geq 10 \qquad \text{so } \{x \mid x \geq 10\} \quad \text{or} \quad [10, \infty)$$

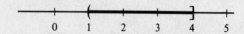

C. Techniques for solving compound inequalities

Inequalities may have more than one relational. In that case we obtain a **compound inequality**, which has its own solution technique.

EXAMPLE 4-12 Solve the continued inequality $2 < 6x - 4 \leq 20$.

Solution We treat this inequality as if we were working on two inequalities at the same time. Then we work to isolate x in the middle.

$$2 < \quad 6x - 4 \quad \leq 20$$

$$2 + 4 < (6x - 4) + 4 \leq 20 + 4 \qquad \text{[Add 4 to each member]}$$

$$6 < \quad 6x \quad \leq 24 \qquad \text{[Simplify]}$$

$$1 < \quad x \quad \leq 4 \qquad \text{[Divide by (positive) 6]}$$

The solution is a set $\{x \mid 1 < x \leq 4\}$, which is an interval $(1, 4]$ of x values greater than 1 and yet less than or equal to 4:

note: You can use any of the various set notations available for expressing solutions of compound inequalities. The notation $\{x \mid 1 < x \leq 4\}$ is equivalent to the notation for the **interval** $(1, 4]$ and the graphical (number line) pictorial is just another way of expressing the same set.

EXAMPLE 4-13   Solve the inequality $2 < \dfrac{-4x + 2}{3} < 6$.

Solution $3(2) < \dfrac{3(-4x + 2)}{3} < 3(6)$ [Multiply each member by $+3$]

$\qquad 6 < \quad -4x + 2 \quad < 18$

$\qquad 6 - 2 < -4x + 2 - 2 < 18 - 2$ [Add -2 to each member]

$\qquad\quad 4 < \qquad -4x \qquad < 16$

$\qquad \dfrac{4}{-4} > \qquad \dfrac{-4x}{-4} \qquad > \dfrac{16}{-4}$ [Divide each member by -4, reversing each inequality]

$\qquad -1 > \qquad x \qquad > -4$

$\qquad -4 < \qquad x \qquad < -1$ [Rewrite the inequalities]

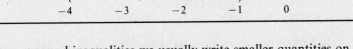

note: In compound inequalities we usually write smaller quantities on the left and larger quantities on the right. And we *never* have opposing directions of inequality signs in a given inequality! (*Right:* $1 < 2 < 3$; *Wrong:* $3 > 2 < 4$.)

4-3. Absolute-Value Equations and Inequalities

A. Absolute value

In mathematics problems we often are interested in the undirected distance between the origin and any other point x on the number line. This undirected distance is called the **absolute value** of x and is defined by

ABSOLUTE VALUE $|x| = \begin{cases} x \text{ if } x \geq 0 \\ -x \text{ if } x < 0 \end{cases}$ **(4.1)**

For example, using the definition, we have $|5| = 5$, $|-5| = 5$, $|-3| = 3$, $|-\frac{1}{2}| = \frac{1}{2}$, and so on.

B. Absolute-value equations
- In general, if $|x| = d$, then $x = d$ and $x = -d$, or $x = \pm d$.

EXAMPLE 4-14   Solve $|x| = 3$.

Solution   Since the left side of the equation is the undirected distance from the origin to a point x on the number line and this distance is 3 units, you can conclude that x has the value $+3$ if the point is to the right of the origin and the value -3 if the point is to the left. Thus the solutions of $|x| = 3$ are $x = \pm 3$.

- For an algebraic expression $A(x)$, if $|A(x)| = k$, then $A(x) = k$ and $A(x) = -k$, or $A(x) = \pm k$.

EXAMPLE 4-15   Solve $|x - 4| = 3$.

Solution   Write this equation as a pair of equations:

$\qquad x - 4 = 3 \qquad$ and $\quad x - 4 = -3 \qquad [\pm k = \pm 3]$

$\qquad\quad x = 4 + 3 \qquad$ and $\qquad\quad x = 4 - 3$

So $\qquad x = 7 \qquad$ and $\qquad x = 1$

and these two values are the solutions.

Check: $|7 - 4| \overset{?}{=} 3 \qquad |1 - 4| \overset{?}{=} 3$

$\qquad\qquad |3| \overset{\checkmark}{=} 3 \qquad |-3| \overset{\checkmark}{=} 3$

C. Absolute-value inequalities

From the definition of absolute value (4.1), the expression

$$|ax + b| < c \qquad \text{where } c > 0$$

is equivalent to the pair of inequalities

$$ax + b < c \qquad \text{where } ax + b \geq 0$$
$$-(ax + b) < c \qquad \text{where } ax + b < 0$$

Then, if we multiply both members of $-(ax + b) < c$ by -1, we reverse the direction of the inequality and obtain the equivalent expression $ax + b > -c$. Now we can draw a conclusion:

- The expression $|ax + b| < c$ is equivalent to the pair

$$ax + b < c \text{ and } ax + b > -c$$

which may be written in one continuous string as $-c < ax + b < c$, or

$$|ax + b| < c \Leftrightarrow -c < ax + b < c \qquad \textbf{(4.2)}$$

The pair or continuous string may then be algebraically changed to obtain the solution(s).

EXAMPLE 4-16 Solve $|x| < 5$.

Solution The solutions of $|x| < 5$ are the values in the interval

$$-5 < x < 5$$

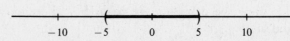

The geometric interpretation of $|x| < 5$ shows the interval to be all points whose undirected distance from the origin is less than 5 units.

EXAMPLE 4-17 Solve $|2x - 3| < 7$.

Solution Write the inequality as the equivalent pair

$$|2x - 3| < 7 \Leftrightarrow \begin{cases} 2x - 3 < 7 \\ 2x - 3 > -7 \end{cases} \qquad \left[|ax + b| < c \Leftrightarrow \begin{cases} ax + b < c \\ ax + b > -c \end{cases} \right]$$

which can be written as the continuous string

$$-7 < 2x - 3 < 7 \qquad [-c < ax + b < c]$$

Now follow algebraic procedure:

$$-7 + 3 < 2x - \cancel{3} + \cancel{3} < 7 + 3 \qquad \text{[Add 3 to each member]}$$

$$-4 < \qquad 2x \qquad < 10 \qquad \text{[Simplify]}$$

$$-\frac{4}{2} < \qquad \frac{2x}{2} \qquad < \frac{10}{2} \qquad \text{[Divide by (positive) 2]}$$

$$-2 < \qquad x \qquad < 5$$

The solution values are all x in the open interval graphed as

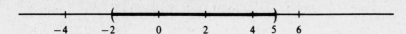

Any x value in this solution interval may be checked into the original problem to see that the original inequality is satisfied.

Solve the equivalent pair separately when the relational is $>$ or $\geq$.

EXAMPLE 4-18   Solve and graph $\left|\dfrac{x-3}{4}\right| \geq 2.$

Solution   This inequality is equivalent to the pair of inequalities

$$\frac{x-3}{4} \geq 2 \quad \text{if} \quad \frac{x-3}{4} > 0 \qquad \text{(is nonnegative)}$$

and

$$-\left(\frac{x-3}{4}\right) \geq 2 \quad \text{if} \quad \frac{x-3}{4} < 0 \qquad \text{(is negative)}$$

Now we solve each one separately to get two solutions:

$$\frac{x-3}{4} \geq 2 \qquad\qquad -\left(\frac{x-3}{4}\right) \geq 2$$

$$x - 3 \geq 8 \qquad\qquad \frac{x-3}{4} \leq -2$$

$$x \geq 11 \qquad\qquad x - 3 \leq -8$$

$$x \leq -5$$

So the solution set is $\{x \mid x \leq -5 \text{ and } x \geq 11\}$:

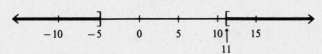

SUMMARY

1. An identity is an equation that is always true; a conditional equation is an equation that is true only when certain isolated values of the variable are substituted for the variable.
2. There are four basic rules for operating on an algebraic equation $A(x) = B(x)$:

$$A(x) + C(x) = B(x) + C(x)$$

$$A(x) - C(x) = B(x) - C(x)$$

$$A(x) \cdot C(x) = B(x) \cdot C(x) \qquad \text{where } C(x) \neq 0$$

$$\frac{A(x)}{C(x)} = \frac{B(x)}{C(x)} \qquad \text{where } C(x) \neq 0$$

3. Algebraic expressions may also be ordered by four relationals ($<$, $\leq$, $>$, $\geq$), which connect nonequal expressions, or inequalities.
4. The rules for operating on algebraic inequalities are the same as those for algebraic equations— except that, when both sides of an inequality are multiplied (or divided) by a negative number, the direction of the inequality is reversed.

5. If the value of x is greater than or equal to zero, the absolute value of x is x itself. If the value of x is less than zero, the absolute value of x is the negative of x; i.e.,

$$|x| = \begin{cases} x \text{ if } x \geq 0 \\ -x \text{ if } x < 0 \end{cases}$$

6. For an algebraic expression $A(x)$, if $|A(x)| = k$, then $A(x) = \pm k$.
7. The expression $|ax + b| < c$ is equivalent to the pair $ax + b < c$ and $ax + b > -c$, which can be written as $-c < ax + b < c$; i.e.,

$$|ax + b| < c \Leftrightarrow -c < ax + b < c$$

RAISE YOUR GRADES

Can you . . . ?

☑ explain the difference between an identity and a conditional equation
☑ define a first-degree equation
☑ use the four rules of algebraic operation
☑ define equivalent equations
☑ outline the procedure for solving linear equations
☑ give examples of the four inequality relations
☑ define absolute value
☑ explain the steps in solving absolute-value equations

SOLVED PROBLEMS

First-Degree Equations in One Variable

PROBLEM 4-1 Which of the following are conditional equations and which are identities?

(a) $3(x - 4) = (x + 7) - (19 - 2x)$
(b) $5x - 7 = 3 - (x + 4)$

(c) $5(2x - 1) = (7x - 10) + (3x + 5)$
(d) $(7x - 5) - (x + 4) = (3x - 2) + 9$

Solution Combine terms and simplify each side of the given equation. If the left- and right-hand sides are the same, then the equation is an identity, so that the equation is always true. If the two sides are different, the equation is conditional, so that only one value of x will make it true.

(a) $3x - 12 = x + 7 - 19 + 2x$

$3x - 12 = 3x - 12$

This is an identity.

(b) $5x - 7 = 3 - x - 4$

$5x - 7 = -x - 1$

This is a conditional equation.

(c) $5(2x - 1) = (7x - 10) + (3x + 5)$

$10x - 5 = 10x - 5$

This is an identity.

(d) $7x - 5 - x - 4 = 3x - 2 + 9$

$\qquad 6x - 9 = 3x + 7$

This is a conditional equation.

PROBLEM 4-2   Solve the conditional equations in Problem 4-1.

Solution

(b)
$$5x - 7 = -x - 1$$

$\qquad (5x - 7) + x = (-x - 1) + x \qquad$ [Add equals]

$\qquad 6x - 7 = -1 \qquad$ [Simplify]

$\qquad (6x - 7) + 7 = (-1) + 7 \qquad$ [Add equals]

$\qquad 6x = 6 \qquad$ [Simplify]

$\qquad x = 1 \qquad$ [Solution]

(d)
$$6x - 9 = 3x + 7$$

$\qquad (6x - 9) - 3x = (3x + 7) - 3x \qquad$ [Subtract equals]

$\qquad 3x - 9 = 7 \qquad$ [Simplify]

$\qquad (3x - 9) + 9 = (7) + 9 \qquad$ [Add equals]

$\qquad 3x = 16 \qquad$ [Simplify]

$\qquad \dfrac{3x}{3} = \dfrac{16}{3} \qquad$ [Divide by equals]

$\qquad x = \frac{16}{3} \qquad$ [Solution]

Check: Going back to the original equations,
$$5x - 7 = 3 - (x + 4) \qquad (7x - 5) - (x + 4) = (3x - 2) + 9$$

insert the solution at each spot where x appears:

$5 \cdot 1 - 7 \overset{?}{=} 3 - (1 + 4) \qquad\qquad (7 \cdot \frac{16}{3} - 5) - (\frac{16}{3} + 4) \overset{?}{=} (3 \cdot \frac{16}{3} - 2) + 9$

$5 - 7 \overset{?}{=} 3 - 5 \qquad\qquad\qquad (\frac{112}{3} - \frac{15}{3}) - (\frac{16}{3} + \frac{12}{3}) \overset{?}{=} (\frac{48}{3} - \frac{6}{3}) + \frac{27}{3}$

$-2 \overset{\checkmark}{=} -2 \qquad\qquad\qquad\qquad \frac{97}{3} - \frac{28}{3} \overset{?}{=} \frac{42}{3} + \frac{27}{3}$

$\qquad\qquad\qquad\qquad\qquad\qquad\qquad\qquad \frac{69}{3} \overset{\checkmark}{=} \frac{69}{3}$

PROBLEM 4-3   Solve the first-degree conditional equation $2x + 9 = -4x + 24$.

Solution   Use the rules of algebraic operation:

$\qquad (2x + 9) - 9 = (-4x + 24) - 9 \qquad$ [Subtract equals]

$\qquad 2x = -4x + 15 \qquad$ [Simplify]

$\qquad (2x) + 4x = (-4x + 15) + 4x \qquad$ [Add equals] $\qquad$ *Check:* $2 \cdot \frac{5}{2} + 9 \overset{?}{=} -4 \cdot \frac{5}{2} + 24$

$\qquad 6x = 15 \qquad$ [Simplify] $\qquad\qquad\qquad\qquad\quad \frac{10}{2} + \frac{18}{2} \overset{?}{=} -\frac{20}{2} + \frac{48}{2}$

$\qquad x = \frac{15}{6} = \frac{5}{2} \qquad$ [Solution] $\qquad\qquad\qquad\qquad\quad 14 \overset{\checkmark}{=} 14$

PROBLEM 4-4   Solve $\dfrac{x - 5}{2} - \dfrac{1}{3} = \dfrac{2x + 7}{3}$.

Solution

$$6\left(\dfrac{x - 5}{2}\right) - 6\left(\dfrac{1}{3}\right) = 6\left(\dfrac{2x + 7}{3}\right) \qquad \text{[Clear fractions (lcd = 6)]}$$

$$3(x - 5) - 2 = 2(2x + 7)$$

$$3x - 15 - 2 = 4x + 14$$

$$3x - 17 = 4x + 14$$

$$(3x - 17) + 17 = (4x + 14) + 17$$

$$3x = 4x + 31$$

$$(3x) - 4x = (4x + 31) - 4x$$

$$-x = 31$$

$$x = -31$$

Check: $\frac{-31-5}{2} - \frac{1}{3} \overset{?}{=} \frac{2(-31)+7}{3}$

$\qquad\quad \frac{-36}{2} - \frac{1}{3} \overset{?}{=} \frac{-62+7}{3}$

$\qquad\quad -\frac{54}{3} - \frac{1}{3} \overset{?}{=} \frac{-55}{3}$

$\qquad\quad \frac{-55}{3} \overset{\checkmark}{=} \frac{-55}{3}$

PROBLEM 4-5 Solve $\frac{1}{2}x + \frac{2}{3} = 3 + \frac{1}{5}x$.

Solution

$$\left(\frac{1}{2}x + \frac{2}{3}\right) - \frac{1}{5}x - \frac{2}{3} = \left(3 + \frac{1}{5}x\right) - \frac{1}{5}x - \frac{2}{3} \quad \text{or}$$

$$\frac{1}{2}x - \frac{1}{5}x = 3 - \frac{2}{3}$$

$$\frac{5}{10}x - \frac{2}{10}x = \frac{9}{3} - \frac{2}{3}$$

$$\frac{3}{10}x = \frac{7}{3}$$

$$\frac{10}{3}\left(\frac{3}{10}x\right) = \frac{10}{3}\left(\frac{7}{3}\right)$$

$$x = \frac{70}{9}$$

$$\frac{x}{2} + \frac{2}{3} = 3 + \frac{x}{5}$$

$$30\left(\frac{x}{2}\right) + 30\left(\frac{2}{3}\right) = 30 \cdot 3 + 30\left(\frac{x}{5}\right)$$

$$15x + 20 = 90 + 6x$$

$$9x = 70$$

$$x = \frac{70}{9}$$

Check: $\frac{1}{2}(\frac{70}{9}) + \frac{2}{3} \overset{?}{=} 3 + \frac{1}{5}(\frac{70}{9})$

$\qquad\quad \frac{35}{9} + \frac{6}{9} \overset{?}{=} \frac{27}{9} + \frac{14}{9}$

$\qquad\quad \frac{41}{9} \overset{\checkmark}{=} \frac{41}{9}$

PROBLEM 4-6 Find the solution of $2x - 5(x - 3) + 4 = 19 - 3x$.

Solution Remove the parentheses and combine terms:

$$2x - 5x + 15 + 4 = 19 - 3x$$

$$-3x + 19 = 19 - 3x$$

You can't find a single-value solution because this is an identity—EVERY value of x on the real number line is a solution.

PROBLEM 4-7 Find the solution of $(x - 4) + (2x + 7) = 3(x - 9)$.

Solution

$$x - 4 + 2x + 7 = 3x - 27$$

$$3x + 3 = 3x - 27$$

$$(3x + 3) - 3x = (3x - 27) - 3x$$

$$3 = -27$$

When you subtract $3x$ from both sides, you end up with a nonsense statement! This is neither an identity nor a conditional equation. It could be classified as an incorrect or ill-posed equation. In any case, the "solution" here is that there is no solution: There is no value of x that satisfies this equation.

PROBLEM 4-8 Solve $\dfrac{x-3}{x+7} = \dfrac{x+2}{x-5}$.

Solution   First multiply each side by $(x - 5)(x + 7)$ to clear:

$$(x-5)(x+7)\left(\frac{x-3}{x+7}\right) = (x-5)(x+7)\left(\frac{x+2}{x-5}\right)$$

Expand:
$$(x-5)(x-3) = (x+7)(x+2)$$

$$x^2 - 8x + 15 = x^2 + 9x + 14$$

Subtract equals
to get rid of x^2:
$$(x^2 - 8x + 15) - x^2 = (x^2 + 9x + 14) - x^2$$

And carry on as usual:
$$-8x + 15 = 9x + 14$$

$$(-8x + 15) - 9x - 15 = (9x + 14) - 9x - 15$$

$$-17x = -1$$

$$x = \frac{-1}{-17} = \frac{1}{17}$$

Check:
$$\frac{\frac{1}{17} - 3}{\frac{1}{17} + 7} \overset{?}{=} \frac{\frac{1}{17} + 2}{\frac{1}{17} - 5}$$

$$\frac{\frac{1}{17} - \frac{51}{17}}{\frac{1}{17} + \frac{119}{17}} \overset{?}{=} \frac{\frac{1}{17} + \frac{34}{17}}{\frac{1}{17} - \frac{85}{17}}$$

$$\frac{-\frac{50}{17}}{\frac{120}{17}} \overset{?}{=} \frac{\frac{35}{17}}{-\frac{84}{17}}$$

$$-\frac{50}{120} \overset{?}{=} -\frac{35}{84}$$

$$-\frac{5}{12} \overset{\checkmark}{=} -\frac{5}{12}$$

PROBLEM 4-9 Solve $\dfrac{1}{t-1} - \dfrac{1}{t+3} = -\dfrac{1}{t+4} + \dfrac{1}{t}$.

Solution   Combine terms on each side first. On the left, multiply the first term by $\dfrac{t+3}{t+3}$, and the second by $\dfrac{t-1}{t-1}$. This gives

$$\left(\frac{t+3}{t+3}\right)\left(\frac{1}{t-1}\right) - \left(\frac{t-1}{t-1}\right)\left(\frac{1}{t+3}\right) =$$

On the right, multiply the first term by $\dfrac{t}{t}$ and the second by $\dfrac{t+4}{t+4}$. This gives

$$= \left(\frac{t}{t}\right)\left(-\frac{1}{t+4}\right) + \left(\frac{t+4}{t+4}\right)\left(\frac{1}{t}\right)$$

You now have
$$\frac{(t+3) - (t-1)}{(t+3)(t-1)} = \frac{-t + (t+4)}{t(t+4)}$$

or
$$\frac{4}{(t+3)(t-1)} = \frac{4}{t(t+4)}$$

Finally, clear the fractions by multiplying both sides by $(t+3)(t-1)(t)(t+4)$ and then follow the usual rules:

$$(t+3)(t-1)(t)(t+4)\left(\frac{4}{(t+3)(t-1)}\right) = (t+3)(t-1)(t)(t+4)\left(\frac{4}{t(t+4)}\right)$$

$$4(t)(t+4) = 4(t+3)(t-1)$$

$$(4t)(t+4) = (4t+12)(t-1)$$

$$4t^2 + 16t = 4t^2 + 8t - 12$$

$$16t = 8t - 12$$

$$8t = -12$$

$$t = -\tfrac{12}{8} = -\tfrac{3}{2}$$

Check: $\dfrac{1}{-\frac{3}{2}-1} - \dfrac{1}{-\frac{3}{2}+3} \overset{?}{=} \dfrac{-1}{-\frac{3}{2}+4} + \dfrac{1}{-\frac{3}{2}}$

$$\dfrac{1}{-\frac{5}{2}} - \dfrac{1}{\frac{3}{2}} \overset{?}{=} \dfrac{-1}{\frac{5}{2}} + \dfrac{1}{-\frac{3}{2}}$$

$$-\tfrac{2}{5} - \tfrac{2}{3} \overset{\checkmark}{=} -\tfrac{2}{5} - \tfrac{2}{3}$$

PROBLEM 4-10   The formula for the surface area S of a closed cylindrical can is $S = 2\pi r^2 + 2\pi rh$, where r is the radius of the cylinder and h is its height. Given the radius and the surface area, how would you find the height?

Solution   Solve the given formula for h:

$$S = 2\pi r^2 + 2\pi rh$$

$$S - 2\pi r^2 = 2\pi rh$$

$$\dfrac{S - 2\pi r^2}{2\pi r} = h$$

PROBLEM 4-11   Given the formula $d = vt - \tfrac{1}{2}gt^2$, solve for g in terms of d, v, and t.

Solution $d = vt - \tfrac{1}{2}gt^2$

$$d - vt = -\tfrac{1}{2}gt^2$$

$$g = \dfrac{d - vt}{-\frac{1}{2}t^2} = \left(\dfrac{-2}{-2}\right)\left(\dfrac{d - vt}{-\frac{1}{2}t^2}\right) = \dfrac{-2d + 2vt}{t^2}$$

PROBLEM 4-12   The sum of two consecutive even integers is 16 more than the larger minus the smaller. What are the integers?

Solution   *Begin by restating the quantities given in the problem in algebraic terms.* Let x be the smaller integer; then the next even integer to follow x is $x + 2$. The sum is $x + (x + 2)$ and the larger minus the smaller is $(x + 2) - x$. *Now restate the conditions of the problem algebraically:* "The sum $[x + (x + 2)]\ldots$ is 16 more $[+16]$ than the larger minus the smaller $[(x + 2) - x]$." Then you can write a statement of equality that can be solved for x:

$$x + (x + 2) = (x + 2) - x + 16$$

$$2x + 2 = 2 + 16$$

$$2x = 16$$

$$x = 8$$

So the numbers are $x = 8$ and $x + 2 = 10$. The sum of $10 + 8$ is 18, which is 16 more than the difference, $10 - 8 = 2$.

PROBLEM 4-13   Old MacDonald has a rectangular flower garden on her farm which is 20 feet longer than it is wide. She uses 160 feet of fence to completely enclose her garden. What are its dimensions?

Solution   Let x denote the width, so that $x + 20$ denotes the length. The perimeter is then $x + (x + 20) + x + (x + 20)$, and the perimeter is covered by 160 feet of fence. Hence

$$x + (20 + x) + x + (20 + x) = 160$$

$$4x + 40 = 160$$

$$4x = 120$$

$$x = 30$$

So the bloomin' garden is 30 feet by 50 feet.

PROBLEM 4-14 McElroy, Connelly, and Lendlease are discussing their shaving habits. In one week they use 23 Tric Razors. Connelly uses three more than Lendlease in a week while McElroy uses twice as many as Lendlease. How many razors does each player use in a week?

Solution Let x be the number of razors that Lendlease uses, $x + 3$ be the number that Connelly uses, and $2x$ be the number that McElroy uses. Then

$$x + (x + 3) + 2x = 23$$

$$4x + 3 = 23$$

$$4x = 20$$

$$x = 5$$

So Lendlease uses 5, Connelly 8, and McElroy 10 razors in one week. (It would appear that McElroy has the cutting edge.)

PROBLEM 4-15 It takes Ma two hours to mow the lawn and it takes Sonny three hours to mow it. How long (in minutes) would it take for the two of them to mow the lawn with two lawnmowers? (Assume that each mows at a steady pace.)

Solution In one hour, Ma can mow one-half of the lawn. In one hour, Sonny can mow one-third of the lawn. This establishes their rate (r) of mowing: Ma $= r_M = \frac{1}{2}$ and Sonny $= r_S = \frac{1}{3}$. The amount of time it takes for the job to get done is t hours, so Ma's portion is $\frac{1}{2}t$ and Sonny's portion is $\frac{1}{3}t$. Now indicate the whole job as 1, which is the sum of these portions:

$$1 = \tfrac{1}{2}t + \tfrac{1}{3}t$$

$$= \tfrac{5}{6}t$$

$$t = \tfrac{6}{5} \text{ hours} = (\tfrac{6}{5}\cancel{h}) \times (60 \text{ min}/\cancel{h}) = 72 \text{ minutes}$$

First-Degree Inequalities in One Variable

PROBLEM 4-16 Solve the inequality $(4x - 2) - (5 - 3x) \le 3 + 2x$ and graph the solution.

Solution

$4x - 2 - 5 + 3x \le 3 + 2x$	[Combine terms]
$7x - 7 \le 3 + 2x$	
$(7x - 7) + 7 - 2x \le (3 + 2x) + 7 - 2x$	[Add 7 to both sides and subtract $2x$ from both sides]
$7x - 2x \le 3 + 7$	
$5x \le 10$	
$x \le 2$	

PROBLEM 4-17   Solve $5(x + 4) - 3(x + 7) \geq 5$ and graph the solution.

Solution $5x + 20 - 3x - 21 \geq 5$ [Multiply out]

$$2x - 1 \geq 5 \quad \text{[Combine]}$$

$$2x \geq 6$$

$$x \geq 3$$

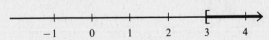

PROBLEM 4-18   Solve the continued inequality $x + 3 < 4x - 6 < x + 15$ and graph the solution.

Solution $(x + 3) - x < (4x - 6) - x < (x + 15) - x$ [Subtract x from every term]

$$3 < 3x - 6 < 15$$

$$9 < 3x < 21$$

$$3 < x < 7$$

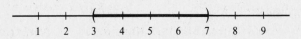

PROBLEM 4-19   Solve for values of x that satisfy $\dfrac{1}{x} < \dfrac{5}{x} + \dfrac{2}{x^2} < \dfrac{8}{x^2} + \dfrac{1}{x}$ and graph the solution.

Solution   To preserve the direction of the inequalities, you have to multiply by a positive value. So to clear the fractions, multiply by x^2 (which is *never* negative):

$$x^2\left(\frac{1}{x}\right) < x^2\left(\frac{5}{x} + \frac{2}{x^2}\right) < x^2\left(\frac{8}{x^2} + \frac{1}{x}\right)$$

$$x < \quad 5x + 2 \quad < \quad 8 + x$$

$$x - x < 5x - x + 2 \; < \; 8 + x - x \qquad \text{[Subtract x from each term]}$$

$$0 < \quad 4x + 2 \quad < \quad 8$$

$$-2 < \quad 4x \quad < \quad 6$$

$$-\frac{2}{4} < \quad \frac{4x}{4} \quad < \quad \frac{6}{4}$$

$$-\frac{1}{2} < \quad x \quad < \quad \frac{3}{2}$$

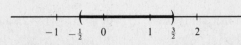

PROBLEM 4-20   Prove that if a is positive and $a < b$, then $a^2 < b^2$.

Solution   Subtract a from both sides of the given inequality $a < b$:

$$a - a < b - a$$

$$0 < b - a \qquad \text{(i.e., } b - a \text{ is positive)}$$

Since a is positive and $a < b$, then $b + a$ is positive; so you can multiply on both sides by $b + a$ without changing the direction of the inequality:

$$0(b + a) < (b - a)(b + a)$$

$$0 < b^2 - a^2 \qquad \qquad \text{(i.e., } b^2 - a^2 \text{ is positive)}$$

Add a^2 to both sides: $a^2 + 0 < b^2 - a^2 + a^2$

so $a^2 < b^2$ *QED**

note: A reversal of the steps will prove the companion statement: If $a^2 < b^2$ then $a < b$.

PROBLEM 4-21 Show that $\sqrt{2} + \sqrt{7} < \sqrt{3} + \sqrt{6}$—*without* finding the values of the roots.

Solution *Assume* the truth of the inequality and apply the result of Problem 4-20 (which you *know* is true because you just proved it):

$$(\sqrt{2} + \sqrt{7})^2 < (\sqrt{3} + \sqrt{6})^2$$

or

$$2 + 2\sqrt{2}\sqrt{7} + 7 < 3 + 2\sqrt{3}\sqrt{6} + 6$$

$$9 + 2\sqrt{14} < 9 + 2\sqrt{18}$$

$$9 - 9 + 2\sqrt{14} < 9 - 9 + 2\sqrt{18}$$

$$\frac{2\sqrt{14}}{2} < \frac{2\sqrt{18}}{2}$$

$$\sqrt{14} < \sqrt{18}$$

Since this result is clearly true, the reversal of the steps proves the original statement. *QED*

note: This is one of the ways to prove a mathematical statement: Assume it's true and use what you *know* to see if it leads to an obvious truth. If it does, it is; if it doesn't, it isn't.

PROBLEM 4-22 Emily scored 89, 79, and 84 on her first three algebra tests. In order to get an A— or better for the course, her average must be more than 87 for the semester. How high must she score on the fourth and last test to earn at least an A—?

Solution Let x denote the final score; then the score average $\dfrac{89 + 79 + 84 + x}{4}$ must be greater than 87, which is a statement of inequality:

$$\frac{89 + 79 + 84 + x}{4} > 87$$

$$252 + x > 348$$

$$x > 348 - 252$$

$$x > 96 \qquad \text{(Emily is going to switch her major to Art History!)}$$

PROBLEM 4-23 Find the values of x satisfying the compound inequality

$$\tfrac{3}{2}t - 1 < \tfrac{5}{3}(t + 3) < (7 - t)$$

Solution To clear of fractions, multiply all terms by 6:

$$6(\tfrac{3}{2}t - 1) < 6(\tfrac{5}{3})(t + 3) < 6(7 - t)$$

$$\tfrac{18}{2}t - 6 < \tfrac{30}{3}(t + 3) < 6(7 - t)$$

$$9t - 6 < 10(t + 3) < 6(7 - t)$$

$$9t - 6 < 10t + 30 < 42 - 6t$$

Subtract 30 from all terms:

$$9t - 36 < 10t < 12 - 6t$$

Subtract $9t$ from all terms:

$$-36 < t < 12 - 15t$$

---

* *QED = Quod Erat Demonstrandum =* [Thus was it proved] $\cong$ Awesome! We *did* it!

Now split the compound inequality into two parts:

$$-36 < t \qquad \text{and} \qquad t < 12 - 15t$$
$$\Downarrow \qquad\qquad\qquad\qquad \Downarrow$$
$$-36 < t \qquad\qquad\qquad 16t < 12$$
$$t < \tfrac{12}{16}$$
$$t < \tfrac{3}{4}$$

Recombine the results:

$$-36 < t < \tfrac{3}{4}$$

Absolute-Value Equations and Inequalities

PROBLEM 4-24   Solve $|3m - 1| = 8$.

Solution   The equation is interpreted as (and is equivalent to)

$$3m - 1 = 8 \quad \text{or} \quad 3m - 1 = -8 \qquad \textit{Check:}$$

Thus $\qquad 3m = 9 \quad \text{or} \qquad 3m = -7 \qquad |3 \cdot 3 - 1| \overset{?}{=} 8 \ \text{ or } \ |3(-\tfrac{7}{3}) - 1| \overset{?}{=} 8$

$$m = 3 \quad \text{or} \qquad m = -\tfrac{7}{3} \qquad |9 - 1| \overset{?}{=} 8 \qquad\qquad |-7 - 1| \overset{?}{=} 8$$
$$8 \overset{\checkmark}{=} 8 \qquad\qquad\qquad 8 \overset{\checkmark}{=} 8$$

PROBLEM 4-25   Solve $|7x - 4| + 3 = 13$.

Solution

$$|7x - 4| + 3 - 3 = 13 - 3 \qquad \text{[Subtract 3 from both sides]}$$
$$|7x - 4| = 10$$

Now write the equation in equivalent form:

Check:

$$7x - 4 = 10 \quad \text{or} \quad 7x - 4 = -10 \qquad |7 \cdot 2 - 4| + 3 \overset{?}{=} 13 \ \text{ or } \ |7(-\tfrac{6}{7}) - 4| \overset{?}{=} 13$$
$$7x = 14 \quad \text{or} \qquad 7x = -6 \qquad |14 - 4| + 3 \overset{?}{=} 13 \qquad |-6 - 4| + 3 \overset{?}{=} 13$$
$$x = 2 \quad \text{or} \qquad x = -\tfrac{6}{7} \qquad |10| + 3 \overset{?}{=} 13 \qquad\quad |-10| + 3 \overset{?}{=} 13$$
$$13 \overset{\checkmark}{=} 13 \qquad\qquad\quad 13 \overset{\checkmark}{=} 13$$

PROBLEM 4-26   Solve $|2x - 3| = |x - 7|$.

Solution   This may be written in its equivalent form

$$2x - 3 = x - 7 \quad \text{or} \quad 2x - 3 = -(x - 7)$$
$$x = -4 \quad \text{or} \quad 2x - 3 = -x + 7 \qquad \textit{Check:}$$
$$3x = 10 \qquad\qquad |2(-4) - 3| \overset{?}{=} |-4 - 7| \ \text{ or } \ |2(\tfrac{10}{3}) - 3| \overset{?}{=} |\tfrac{10}{3} - 7|$$
$$x = \tfrac{10}{3} \qquad\qquad |-8 - 3| \overset{?}{=} |-11| \qquad\quad |\tfrac{20}{3} - \tfrac{9}{3}| \overset{?}{=} |\tfrac{10}{3} - \tfrac{21}{3}|$$
$$11 \overset{\checkmark}{=} 11 \qquad\qquad\qquad \tfrac{11}{3} \overset{\checkmark}{=} \tfrac{11}{3}$$

PROBLEM 4-27   Solve the inequality $|4x - 9| < 3$ and graph the result.

Solution   The inequality is the same as

$$-3 < \quad 4x - 9 \quad < \quad 3$$
$$-3 + 9 < (4x - 9) + 9 < 3 + 9 \qquad \text{[Add 9 to all the terms]}$$
$$6 < \qquad 4x \qquad < \quad 12$$
$$\tfrac{3}{2} < \qquad x \qquad < \quad 3$$

PROBLEM 4-28   Solve $|6a - 7| \geq 3$ and graph the result.

Solution   The inequality is the same as

$$6a - 7 \geq 3 \qquad \text{or} \qquad -(6a - 7) \geq 3$$

$$6a \geq 10 \qquad \text{or} \qquad 6a - 7 \leq -3 \qquad \text{[Multiplying by } -1 \text{ reverses the inequality]}$$

$$a \geq \tfrac{5}{3} \qquad \text{or} \qquad 6a \leq 4$$

$$a \leq \tfrac{2}{3}$$

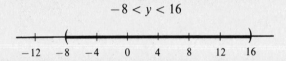

PROBLEM 4-29   Solve $|2 - \tfrac{1}{2}y| < 6$ and graph the result.

Solution   Write $|2 - \tfrac{1}{2}y| < 6$ as

$$2 - \tfrac{1}{2}y < 6 \qquad \text{or} \qquad -(2 - \tfrac{1}{2}y) < 6$$

$$-2(2) - (-2)(\tfrac{1}{2})y > (-2)6 \qquad \text{or} \qquad -2 + \tfrac{1}{2}y < 6 \qquad \text{[Multiplying by } -2 \text{ reverses the inequality]}$$

$$-4 + y > -12 \qquad \text{or} \qquad \tfrac{1}{2}y < 8$$

$$y > -8 \qquad \text{or} \qquad y < 16$$

So

$$-8 < y < 16$$

PROBLEM 4-30   Charles, who lives in City A, is going to bicycle to meet his friend Chuck, who lives in City B. City B is a distance of 40 miles from City A. Each cyclist begins pedaling at noon, with Charles traveling at 10 miles per hour and Chuck at 15 miles per hour. **(a)** At what time will they meet? **(b)** How far will each travel?

Solution   Let $|d_1|$ denote the distance Charles travels, at rate 10, for time period t; so $|d_1| = 10t$. Let $|d_2|$ denote the distance Chuck travels, at rate 15, for the same time period t; so $|d_2| = 15t$. And the combined distances they travel, regardless of direction, must equal the distance between cities: $|d_1| + |d_2| = 40$. Therefore

(a)
$$|d_1| = 10t$$
$$+ |d_2| = 15t$$
$$\overline{|d_1| + |d_2| = 10t + 15t} \qquad \text{[Add equals to equals]}$$

Then, since $|d_1| + |d_2| = 40$,

$$40 = 25t$$

$$t = \frac{40}{25} \text{ hours} = 1 \text{ hour, 36 minutes}$$

So they will meet at 1:36 P.M.

(b) $|d_1| = 10(\tfrac{40}{25}) = 16$ miles $|d_2| = 15(\tfrac{40}{25}) = 24$ miles

PROBLEM 4-31   The expression for the distance between points p and q on the number line is $|p - q|$.

(a) What is the formula for the phrase "a point x is more than 5 units from the point 3"?
(b) What is the interpretation for the expression $|y + 2| \leq 3$?

Solution

(a) One of the points is x and the other is 3. The distance between x and 3 is denoted by $|x - 3|$.

Since this is to be more than 5 units, you get

$$|x - 3| > 5$$

(b) The left side $|y + 2|$ is the distance between the point y and the point -2. Since this distance is ≤ 3, the statement is "the distance between y and -2 is less than 3 or at most equal to 3." An easier, but equivalent expression is "the distance between y and -2 is not more than 3."

Supplementary Exercises

Find the solution of each equation in Problems 4-32 through 4-36:

PROBLEM 4-32

(a) $3x - 12 = 0$

(b) $\frac{1}{3}x + 8 = 0$

(c) $2x + 9 = x - 13$

(d) $\frac{1}{4}x + 5 = 3x - \frac{5}{2}$

(e) $\dfrac{x - 5}{2} - 7 = \dfrac{2 - 3x}{5} + \dfrac{3}{10}$

PROBLEM 4-33

(a) $7y + 8 = 3y - 16$

(b) $\dfrac{3}{t - 3} - \dfrac{t}{3 - t} = 4$

(c) $\sqrt[3]{x + 4} = -2$

(d) $x - (5 - x) = 3$

(e) $v - (v + 4)(v - 1) = 6 - v^2$

PROBLEM 4-34

(a) $6 + \dfrac{z}{5} = -\dfrac{2}{3}$

(b) $\dfrac{1}{3} = \dfrac{u}{u + 1} - \dfrac{4}{3}$

(c) $\sqrt{t + 3} - \frac{1}{2} = 3$

(d) $.5m - 1.3 = .7(m - .6)$

(e) $\dfrac{1}{2r + 2} - 1 = \dfrac{2}{r + 1}$

PROBLEM 4-35

(a) $(w - 8)(w + 3) = (w + 5)(w + 2)$

(b) $-7x + \frac{2}{5} = 5(x - \frac{7}{2})$

(c) $\dfrac{7}{x^2 - 1} = \dfrac{3}{x + 1}$

(d) $\dfrac{4k + 1}{5} - \dfrac{k + 2}{7} = -\dfrac{3}{35}$

(e) $(6z + 1)(4z - 3) + 8 = (8z - 13)(3z + 5) - 16z + 59$

PROBLEM 4-36　Solve for the indicated variable:

(a) $V = \pi r^2 h$　for h

(b) $A = \dfrac{h}{2}(m + n)$　for m

(c) $S = \dfrac{a}{1 - r}$　for r

(d) $R = \dfrac{1}{R_1} + \dfrac{1}{R_2}$　for R_2

(e) $F = k\left(\dfrac{m_1 m_2}{r^2}\right)$　for r

PROBLEM 4-37 Four more than a number is the same as three times the sum of the number and 2. Find the number.

PROBLEM 4-38 The second of two numbers is 4 more than 5 times the first. Seven times the first is equal to the second. Find the two numbers.

PROBLEM 4-39 Solve the following inequalities:

(a) $6a + 5 > -7$

(b) $-2 - 3x \leq 10$

(c) $2d - 8 \geq -d + 3(d - 5)$

(d) $4(y - 1) + 6 < 2(3y - 1) + (3 - 5y)$

(e) $-2(x - 1) + 4(x + 6) \leq 0$

PROBLEM 4-40 Solve the following inequalities:

(a) $-5 < 3a + 4 < 16$ **(b)** $1 \leq -2n + 1 \leq 3$ **(c)** $3 + x < 5x - 2 < x + 13$

PROBLEM 4-41 Your telephone company offers two options: You may make an unlimited number of local calls at a fixed monthly charge of \$17, or you may pay a base rate of \$8 per month plus 10¢ per message unit. Above what usage level does it cost less to have the unlimited service?

PROBLEM 4-42 A manufacturing company spends \$18 000 on developmental expenses and \$5.50 per unit on production, advertising, and sales in the production of a new calculator. If each calculator sells for \$10.00, how many calculators must be sold to show a profit?

PROBLEM 4-43 Solve the following absolute-value equations:

(a) $|3x - 4| = 8$ **(b)** $|-2x + 3| + 5 = 11$ **(c)** $|x + \frac{5}{2}| = |2x + 6|$

PROBLEM 4-44 Solve the following absolute-value inequalities:

(a) $|4y + 3| < 9$ **(b)** $|z - 4| > 17$ **(c)** $|x - 3| + 5 < 8$

PROBLEM 4-45 Which values of x on the number line are less than 5 units away from the point $x = \frac{9}{2}$?

PROBLEM 4-46 The algebraic statement $|x - 2| < 5$ is equivalent to the geometric statement "x is less than 5 units from 2." Give the corresponding statement for the following:

Algebraic Statement	Geometric Statement
(a) $|x + 4| > 1$	_____
(b) $|y| < 3$	_____
(c) _____	u is closer to 0 than v is.
(d) $|z - 2| \leq 4$	_____
(e) _____	y is at least 2 units from 7.

Answers to Supplementary Exercises

4-32 **(a)** $x = 4$
(b) $x = -24$
(c) $x = -22$
(d) $x = \frac{30}{11}$
(e) $x = \frac{102}{11}$

4-33 **(a)** $y = -6$
(b) $t = 5$
(c) $x = -12$
(d) $x = 4$
(e) $v = -1$

4-34 (a) $z = -\frac{100}{3}$ (d) $m = 4.4$
     (b) $u = -\frac{5}{2}$ (e) $r = -\frac{5}{2}$
     (c) $t = \frac{37}{4}$

4-35 (a) $w = -\frac{17}{6}$ (d) $k = 0$
     (b) $x = \frac{179}{120}$ (e) $z = -11$
     (c) $x = \frac{10}{3}$

4-36 (a) $h = \dfrac{V}{\pi r^2}$ (d) $R_2 = \dfrac{R_1}{RR_1 - 1}$

     (b) $m = \dfrac{2A}{h} - n$ (e) $r = \sqrt{\dfrac{km_1m_2}{F}}$

     (c) $r = 1 - \dfrac{a}{S}$

4-37 -1

4-38 2 and 14

4-39 (a) $a > -2$ (d) $y < -\frac{1}{3}$
     (b) $x \geq -4$ (e) $x \leq -13$
     (c) True for all d

4-40 (a) $-3 < a < 4$
     (b) $-1 \leq n \leq 0$
     (c) $\frac{5}{4} < x < \frac{15}{4}$

4-41 $x < 90$

4-42 $x > 4000$

4-43 (a) $4, \frac{-4}{3}$
     (b) $\frac{-3}{2}, \frac{9}{2}$
     (c) $\frac{-7}{2}, \frac{-17}{6}$

4-44 (a) $-3 < y < \frac{3}{2}$
     (b) $z > 21$ or $z < -13$
     (c) $0 < x < 6$

4-45 $-\frac{1}{2} < x < \frac{19}{2}$

4-46 (a) x is more than 1 unit from -4
     (b) y is less than 3 units from the origin
     (c) $|v| - |u| > 0$ (or, simply, $|u| < |v|$)
     (d) z is not more than 4 units from 2
     (e) $|y - 7| \geq 2$

5 QUADRATIC EQUATIONS AND INEQUALITIES

THIS CHAPTER IS ABOUT

☑ **Quadratic Equations**
☑ **Quadratic-Type Equations**
☑ **Quadratic Inequalities**

5-1. Quadratic Equations

- An equation that is in—or can be put into—the form $ax^2 + bx + c = 0$, where a, b, and c are constants and $a \neq 0$, is a **second-degree** or **quadratic equation** in x.

Solving a quadratic equation involves finding the value(s) of x that will satisfy the equation. These values are known as **roots** of the equation, and the set of roots is the solution. A second-degree equation always has two roots:

- Both roots may be real and unequal, so there are two real values in the solution set.
- Both roots may be real and equal, so there is only one real value in the solution set.
- Both roots may be complex numbers, so there are no real values in the solution set.

There are three main ways of solving second-degree equations: factoring, completing the square, and using the quadratic formula.

A. Solving quadratic equations by factoring

The factoring method depends on the zero product principle:

> **ZERO PRODUCT PRINCIPLE**   If $A \cdot B = 0$, then either $A = 0$ or $B = 0$ (or both)

If a quadratic expression is factorable, we can apply this principle by setting each factor equal to zero and finding the value of x from the resulting linear equations.

EXAMPLE 5-1   Solve the quadratic equation $x^2 + 5x + 4 = 0$.

Solution   Factor the left-hand side:

$$x^2 + 5x + 4 = 0$$

$$(x + 4)(x + 1) = 0$$

so either $x + 4$ or $x + 1$ must equal zero. Thus

$$\text{if } x + 4 = 0, \quad \text{then } x = -4$$

$$\text{if } x + 1 = 0, \quad \text{then } x = -1$$

Check: $(-4)^2 + 5(-4) + 4 \overset{?}{=} 0$ $(-1)^2 + 5(-1) + 4 \overset{?}{=} 0$

$16 - 20 + 4 \overset{?}{=} 0$ $1 - 5 + 4 \overset{?}{=} 0$

$0 \overset{\checkmark}{=} 0$ $0 \overset{\checkmark}{=} 0$

There are two real values of x that satisfy this equation; so $x = -4$ and $x = -1$ are the roots of the equation and the solution is the set $\{x \mid x = -4, x = -1\}$.

EXAMPLE 5-2   Solve $y^2 - 5y = 0$.

Solution   The left-hand side of this equation is a quadratic expression (where $a = 1$, $b = -5$, and $c = 0$) and is factorable:

$$y^2 - 5y = 0$$

$$y(y - 5) = 0$$

So by the zero product principle,

$$y = 0 \quad \text{or} \quad y - 5 = 0$$

$$y = 5$$

and the solution set is $\{y \mid y = 0, y = 5\}$.

Check: $0^2 - 5 \cdot 0 \overset{?}{=} 0$ $5^2 - 5 \cdot 5 \overset{?}{=} 0$

$0 - 0 \overset{?}{=} 0$ $25 - 25 \overset{?}{=} 0$

$0 \overset{\checkmark}{=} 0$ $0 \overset{\checkmark}{=} 0$

EXAMPLE 5-3   Solve $t^2 - 16 = 0$.

Solution $t^2 - 16 = 0$

$$(t - 4)(t + 4) = 0$$

Thus $t - 4 = 0 \quad \text{or} \quad t + 4 = 0$

$$t = 4 \qquad\qquad t = -4$$

note: We could also have written this as

$$t^2 = 16$$

$$\sqrt{t^2} = \sqrt{16}$$

$t = \pm 4$ which is the same result as before.

When a quadratic equation is in the form $(Ax + B)^2 = C$, where A, B, and C are constants and $A \neq 0$, we can solve without factoring first. To do this, we use the fact that

$$\text{if } A^2 = B^2, \quad \text{then } A = \pm B$$

And, just as we can multiply and divide equals, we can take the square roots of equals.

EXAMPLE 5-4   Solve $(x - 4)^2 = 9$.

Solution   Take the square root of both sides:

$$(x - 4)^2 = 9$$

$$\sqrt{(x - 4)^2} = \sqrt{9}$$

$$x - 4 = \pm 3$$

$$x = 4 \pm 3$$

So
$$x = 4 + 3 \quad \text{or} \quad x = 4 - 3$$
$$= 7 \qquad\qquad\qquad = 1$$

Check: $(7 - 4)^2 \overset{?}{=} 9 \qquad (1 - 4)^2 \overset{?}{=} 9$

$\qquad\qquad 3^2 \overset{\checkmark}{=} 9 \qquad\quad (-3)^2 \overset{\checkmark}{=} 9$

The roots of equations in the form $(Ax + B)^2 = C$ may not be real: The solution set may include *complex numbers*. Recall that $\sqrt{-1} = i$ and that $i^2 = -1$.

EXAMPLE 5-5 Solve $(2x + 5)^2 = -9$.

Solution $(2x + 5)^2 = -9$

$$2x + 5 = \pm\sqrt{-9} \qquad \text{[Take the square root of both sides]}$$
$$= \pm\sqrt{-1}\sqrt{9}$$
$$= \pm 3i$$

Thus
$$2x = -5 \pm 3i$$
$$x = -\tfrac{5}{2} \pm \tfrac{3}{2}i \qquad \text{[Solutions]}$$

Check:

Let $x = -\tfrac{5}{2} + \tfrac{3}{2}i$. Then

$$(2x + 5) = 2(-\tfrac{5}{2} + \tfrac{3}{2}i) + 5$$
$$= -5 + 3i + 5$$
$$= 3i$$

Let $x = -\tfrac{5}{2} - \tfrac{3}{2}i$. Then

$$(2x + 5) = 2(-\tfrac{5}{2} - \tfrac{3}{2}i) + 5$$
$$= -5 - 3i + 5$$
$$= -3i$$

Substitute into the original equation:

$$(3i)^2 \overset{?}{=} -9 \qquad\qquad\qquad (-3i)^2 \overset{?}{=} -9$$
$$9i^2 \overset{?}{=} -9 \qquad\qquad\qquad 9i^2 \overset{?}{=} -9$$
$$-9 \overset{\checkmark}{=} -9 \qquad\qquad\qquad -9 \overset{\checkmark}{=} -9$$

B. Solving quadratic equations by completing the square

The technique used in Examples 5-4 and 5-5 leads us to the second method. Many quadratic equations can be CHANGED from the form $ax^2 + bx + c = 0$ to the form $(Ax + B)^2 = C$, so that the equation then may be solved by taking square roots. This process—called **completing the square**—involves adding the proper term to both sides of the given equation in order to complete the perfect trinomial square on the left. There are seven steps needed to solve an equation by completing the square:

Step 1: Divide all the terms of the equation by the coefficient of x^2.

Step 2: Subtract the constant term (c/a) from both sides.

Step 3: Compute the number equal to one-half the coefficient of x; then square it.

Step 4: Add the value computed in Step 3 to both sides.

Step 5: Rewrite the left-hand side trinomial as a binomial squared.

Step 6: Simplify the right-hand side.

Step 7: Solve by extracting the roots.

EXAMPLE 5-6   Solve $2x^2 + 8x - 3 = 0$ by completing the square.

Solution   Follow the steps outlined above:

$2x^2 + 8x - 3 = 0$

$x^2 + 4x - \frac{3}{2} = 0$ *[Step 1:* Divide each term by the coefficient of x^2]

$x^2 + 4x \quad\quad = \frac{3}{2}$ *[Step 2:* Subtract the constant term from each side]

Find $\frac{1}{2}(4) = 2$ and calculate $2^2 = 4$.   *[Step 3:* Find one-half the coefficient of x and square it]

$x^2 + 4x + 4 = \frac{3}{2} + 4$ *[Step 4:* Add 4 to each side]

$(x + 2)^2 = \frac{3}{2} + 4$ *[Step 5:* Rewrite $x^2 + 4x + 4$ as a binomial squared]

$(x + 2)^2 = \frac{11}{2}$ *[Step 6:* Simplify the rhs]

$x + 2 = \pm\sqrt{\frac{11}{2}}$ *[Step 7:* Extract the roots]

$x = -2 \pm \sqrt{\frac{11}{2}}$

EXAMPLE 5-7   Solve $x^2 - 7x - 5 = 0$ by completing the square.

Solution $x^2 - 7x - 5 = 0$

$x^2 - 7x \quad\quad = 5$ *[Steps 1* and *2]*

$x^2 - 7x + (\frac{7}{2})^2 = 5 + (\frac{7}{2})^2$ *[Steps 3* and *4]*

$x^2 - 7x + (\frac{7}{2})^2 = 5 + \frac{49}{4}$ *[Simplify]*

$(x - \frac{7}{2})^2 = \frac{69}{4}$ *[Steps 5* and *6]*

$x - \frac{7}{2} = \pm\sqrt{\frac{69}{4}}$ *[Step 7]*

$x = \frac{7}{2} \pm \frac{\sqrt{69}}{2}$ *[Simplify]*

$x = \frac{7 \pm \sqrt{69}}{2}$

EXAMPLE 5-8   Solve the general quadratic equation $ax^2 + bx + c = 0$ by completing the square.

Solution

$ax^2 + bx + c = 0$

$x^2 + \left(\dfrac{b}{a}\right)x + \dfrac{c}{a} = 0$ [Divide by the coefficient of x^2]

$x^2 + \left(\dfrac{b}{a}\right)x \quad\quad = -\dfrac{c}{a}$ [Subtract the constant term]

$\left[\dfrac{1}{2}\left(\dfrac{b}{a}\right)\right]^2 = \left(\dfrac{b}{2a}\right)^2 = \dfrac{b^2}{4a^2}$ [Compute the value to be added and simplify]

$x^2 + \left(\dfrac{b}{a}\right)x + \dfrac{b^2}{4a^2} = -\dfrac{c}{a} + \dfrac{b^2}{4a^2}$ [Add the above value to both sides]

$$\left(x + \frac{b}{2a}\right)^2 = \frac{-4ac}{4a^2} + \frac{b^2}{4a^2} \qquad \text{[Rewrite each side (lcd = } 4a^2\text{)]}$$

$$= \frac{b^2 - 4ac}{4a^2}$$

$$x + \frac{b}{2a} = \pm\sqrt{\frac{b^2 - 4ac}{4a^2}} \qquad \text{[Take square roots of both sides]}$$

$$= \pm\frac{\sqrt{b^2 - 4ac}}{\sqrt{4a^2}}$$

$$= \frac{\pm\sqrt{b^2 - 4ac}}{2a} \qquad \text{[Simplify]}$$

$$x = -\frac{b}{2a} \pm \frac{\sqrt{b^2 - 4ac}}{2a} \qquad \text{[Solve for } x\text{]}$$

$$x = \frac{-b \pm \sqrt{b^2 - 4ac}}{2a} \qquad \text{[Solutions to } ax^2 + bx + c = 0\text{]}$$

C. Solving quadratic equations by using the quadratic formula

Example 5-8 provides the result we need to solve the general quadratic equation $ax^2 + bx + c = 0$, where $a \neq 0$. The solutions are written entirely in terms of the coefficients a, b, c:

QUADRATIC FORMULA $\quad x = \dfrac{-b + \sqrt{b^2 - 4ac}}{2a} \quad$ and $\quad x = \dfrac{-b - \sqrt{b^2 - 4ac}}{2a}$

Memorize these formulas. They will be used *many* times in future math studies!

When we can't use any other method to solve a given quadratic equation, we can always rewrite the given equation in this form and solve for x.

EXAMPLE 5-9 Use the quadratic formula to solve $x^2 - 6x + 9 = 0$.

Solution Here we see that $a = 1$, $b = -6$, and $c = 9$. Then we find the value of $b^2 - 4ac$ to help our computations:

$$b^2 - 4ac = (-6)^2 - 4(1)(9)$$
$$= 36 - 36$$
$$= 0$$

So the values of x are

$$x = \frac{-(-6) + \sqrt{0}}{2} = 3 \quad \text{and} \quad x = \frac{-(-6) - \sqrt{0}}{2} = 3$$

Here the roots are equal, so there is just one value in the solution set, $\{x \mid x = 3\}$.

EXAMPLE 5-10 Use the quadratic formula to solve for x in $3x^2 - 5x - 7 = 0$.

Solution Here $a = 3$, $b = -5$, and $c = -7$; so

$$b^2 - 4ac = 25 - 4(3)(-7)$$
$$= 25 + 84$$
$$= 109$$

The solutions are

$$x = \frac{5 + \sqrt{109}}{6} \quad \text{and} \quad x = \frac{5 - \sqrt{109}}{6}$$

Here the roots are different, so there are two real values in the solution set, $\left\{x \mid x = \dfrac{5 \pm \sqrt{109}}{6}\right\}$.

EXAMPLE 5-11   Solve $\frac{1}{2}x^2 + 3x + 5 = 0$ by the quadratic formula.

Solution   Here $a = \frac{1}{2}$, $b = 3$, and $c = 5$; so

$$b^2 - 4ac = 9 - 4(\tfrac{1}{2})(5)$$
$$= 9 - 10$$
$$= -1$$

Then

$$x = \frac{-3 \pm \sqrt{-1}}{2(\frac{1}{2})} = \frac{-3 \pm i}{1}$$
$$= -3 \pm i$$

Check:

Let $x = -3 - i$. Then

$$x^2 = (-3 - i)(-3 - i)$$
$$= 9 + 6i + i^2$$
$$= 9 + 6i - 1$$
$$= 8 + 6i$$

Let $x = -3 + i$. Then

$$x^2 = (-3 + i)(-3 + i)$$
$$= 9 - 6i + i^2$$
$$= 9 - 6i - 1$$
$$= 8 - 6i$$

Substitute into the original equation:

$$\tfrac{1}{2}(8 + 6i) + 3(-3 - i) + 5 \overset{?}{=} 0$$
$$4 + 3i - 9 - 3i + 5 \overset{?}{=} 0$$
$$9 - 9 \overset{\checkmark}{=} 0$$

$$\tfrac{1}{2}(8 - 6i) + 3(-3 + i) + 5 \overset{?}{=} 0$$
$$4 - 3i - 9 + 3i + 5 \overset{?}{=} 0$$
$$-5 + 5 \overset{\checkmark}{=} 0$$

The nature of the roots of the quadratic equation $ax^2 + bx + c = 0$ depends on the value of the quantity $b^2 - 4ac$, which is called the **discriminant.**

- If $b^2 - 4ac = 0$, the roots are equal real numbers, so there is only one real value in the solution set:

$$x = \frac{-b \pm 0}{2a} = \frac{-b}{2a} \quad \text{or} \quad \left\{x \mid x = \frac{-b}{2a}\right\}$$

- If $b^2 - 4ac$ is positive, the roots are unequal real numbers, so there are two real values in the solution set:

$$x = \frac{-b \pm \sqrt{b^2 - 4ac}}{2a} \quad \text{or} \quad \left\{x \mid x = \frac{-b \pm \sqrt{b^2 - 4ac}}{2a}\right\}$$

- If $b^2 - 4ac$ is negative, the roots are complex numbers, so there are two complex number values in the solution set:

$$x = \frac{-b \pm i\sqrt{4ac - b^2}}{2a} \quad \text{or} \quad \left\{x \mid x = \frac{-b \pm i\sqrt{4ac - b^2}}{2a}\right\}$$

Look again at Examples 5-9, 5-10, and 5-11 to see how the discriminant determines the nature of the solution set.

5-2. Quadratic-Type Equations

Frequently, we encounter equations that can be rewritten in the form of a quadratic equation, so they can be solved by one of the familiar quadratic methods.

- Equations that have the variable under a radical (square root) sign can be rewritten by squaring both sides.

EXAMPLE 5-12 Solve $\sqrt{x + 3} = x + 1$.

Solution Begin by squaring both sides:

$$\sqrt{x + 3} = x + 1$$
$$(\sqrt{x + 3})^2 = (x + 1)^2$$
$$x + 3 = x^2 + 2x + 1$$

Then subtract $(x + 3)$ from both sides and simplify:

$$0 = x^2 + 2x + 1 - x - 3$$
$$0 = x^2 + x - 2$$
$$0 = (x + 2)(x - 1)$$

Thus
$$x = -2 \quad \text{or} \quad x = 1$$

Since we are multiplying by terms involving the unknown x, we must be sure to check each root: We do this by substituting each root into the original equation.

Check:

If $x = -2$:

$$\sqrt{(-2) + 3} \stackrel{?}{=} -2 + 1$$
$$\sqrt{1} \stackrel{?}{=} -1$$
$$1 \neq -1$$

If $x = 1$:

$$\sqrt{1 + 3} \stackrel{?}{=} 1 + 1$$
$$\sqrt{4} \stackrel{\checkmark}{=} 2$$

Thus $x = -2$ is NOT a solution.   Thus $x = 1$ is a solution.

So the given equation has only one member in the solution set, $\{x \mid x = 1\}$.

EXAMPLE 5-13 Solve $\sqrt{t - 3} = \sqrt{t} - 1$.

Solution

$$(\sqrt{t - 3})^2 = (\sqrt{t} - 1)^2 \qquad \text{[Square both sides]}$$
$$(\sqrt{t - 3})^2 = (\sqrt{t} - 1)(\sqrt{t} - 1)$$
$$t - 3 = t - 2\sqrt{t} + 1$$
$$2\sqrt{t} = 4$$
$$\sqrt{t} = 2$$
$$(\sqrt{t})^2 = 2^2 \qquad \text{[Square both sides again]}$$
$$t = 4 \qquad \text{[Solution]}$$

Check: $\sqrt{4 - 3} \stackrel{?}{=} \sqrt{4} - 1$
$$\sqrt{1} \stackrel{?}{=} 2 - 1$$
$$1 \stackrel{\checkmark}{=} 1$$

So the solution set is $\{t \mid t = 4\}$.

EXAMPLE 5-14 Solve $\sqrt{x+5} = \sqrt{x} + \sqrt{5}$.

Solution

$$\sqrt{x+5} = \sqrt{x} + \sqrt{5}$$

$$x + 5 = x + 2\sqrt{x}\sqrt{5} + 5 \qquad \text{[Square both sides]}$$

$$x - x = 2\sqrt{x}\sqrt{5} + 5 - 5 \qquad \text{[Cancel]}$$

$$2\sqrt{x}\sqrt{5} = 0$$

$$\frac{2\sqrt{x}\sqrt{5}}{2\sqrt{5}} = \frac{0}{2\sqrt{5}} \qquad \text{[Divide by } 2\sqrt{5}\text{]}$$

$$\sqrt{x} = 0$$

So
$$x = 0 \qquad \text{[Solution]}$$

Check: $\sqrt{0+5} \overset{?}{=} \sqrt{0} + \sqrt{5}$

$$\sqrt{5} \overset{\checkmark}{=} \sqrt{5}$$

There is a second type of equation that may be solved by the methods of quadratic equations.

- Equations that look like quadratics in form can often be rewritten by expansion or substitution of variables.

EXAMPLE 5-15 Solve for x in $x - \sqrt{x} - 6 = 0$.

Solution To see that this is a disguised quadratic, we replace $\sqrt{x}$ by a new variable, y. This makes x itself equal to y^2 and so we have

$$x - \sqrt{x} - 6 = 0 \qquad (\sqrt{x} = y)$$

$$y^2 - y - 6 = 0$$

Solving this quadratic in the variable y, we get

$$(y - 3)(y + 2) = 0$$

So
$$y = 3 \quad \text{or} \quad y = -2$$

But actually $y = \sqrt{x}$, so we have

$$\sqrt{x} = 3 \quad \text{or} \quad \sqrt{x} = -2$$

Thus

$$x = 9 \quad \text{or} \quad x = 4$$

Check:

If $x = 9$:   If $x = 4$:

$$9 - \sqrt{9} - 6 \overset{?}{=} 0 \qquad\qquad 4 - \sqrt{4} - 6 \overset{?}{=} 0$$

$$9 - 3 - 6 \overset{\checkmark}{=} 0 \qquad\qquad 4 - 2 - 6 \overset{?}{=} 0$$

$$-4 \neq 0$$

Thus $x = 9$ is a solution.

Thus $x = 4$ is not a solution.

So the solution set is $\{x \mid x = 9\}$.

EXAMPLE 5-16 Solve $(x + 2)^2 - 5(x + 2) + 4 = 0$.

Solution   Replace $x + 2$ by another variable t:

$$(x + 2)^2 - 5(x + 2) + 4 = 0$$
$$t^2 - 5t + 4 = 0$$
$$(t - 4)(t - 1) = 0$$

Thus $\qquad\qquad\qquad t = 4 \quad \text{or} \quad t = 1$

Then replacing t by $x + 2$, we have

$$x + 2 = 4 \quad \text{or} \quad x + 2 = 1$$

or $\qquad\qquad\qquad\quad x = 2 \qquad\qquad x = -1$

Check:

If $x = 2$:

$$(2 + 2)^2 - 5(2 + 2) + 4 \overset{?}{=} 0$$
$$4^2 - 5(4) + 4 \overset{?}{=} 0$$
$$0 \overset{\checkmark}{=} 0$$

If $x = -1$:

$$(-1 + 2)^2 - 5(-1 + 2) + 4 \overset{?}{=} 0$$
$$1^2 - 5(1) + 4 \overset{?}{=} 0$$
$$0 \overset{\checkmark}{=} 0$$

So the solution set is $\{x \mid x = 2, -1\}$.

EXAMPLE 5-17   Solve for x in $2x^{2/3} - 3x^{1/3} - 2 = 0$.

Solution   A substitution of variables here again makes the equation quadratic. Let $x^{1/3} = u$, so that $(x^{1/3})^2 = x^{2/3} = u^2$. Then we have

$$2x^{2/3} - 3x^{1/3} - 2 = 0$$
$$2u^2 - 3u - 2 = 0$$
$$(2u + 1)(u - 2) = 0 \qquad \text{[Factor]}$$

Thus $\qquad\qquad\qquad u = -\tfrac{1}{2} \quad \text{or} \quad u = 2$

But, since $u = x^{1/3}$,

$$x^{1/3} = -\tfrac{1}{2} \quad \text{or} \quad x^{1/3} = 2$$

Now cube both sides of each of these to find x:

$$(x^{1/3})^3 = (-\tfrac{1}{2})^3 \quad \text{or} \quad (x^{1/3})^3 = 2^3$$
$$x = -\tfrac{1}{8} \qquad\qquad\qquad x = 8$$

Check:

If $x = -\tfrac{1}{8}$:

$$2(-\tfrac{1}{8})^{2/3} - 3(-\tfrac{1}{8})^{1/3} - 2 \overset{?}{=} 0$$
$$2(\tfrac{1}{4}) - 3(-\tfrac{1}{2}) - 2 \overset{?}{=} 0$$
$$\tfrac{1}{2} + \tfrac{3}{2} - 2 \overset{\checkmark}{=} 0$$

Thus $x = -\tfrac{1}{8}$ is a solution.

If $x = 8$:

$$2(8)^{2/3} - 3(8)^{1/3} - 2 \overset{?}{=} 0$$
$$2(4) - 3(2) - 2 \overset{?}{=} 0$$
$$8 - 6 - 2 \overset{\checkmark}{=} 0$$

Thus $x = 8$ is also a solution.

EXAMPLE 5-18   Solve the fourth-degree equation $x^4 - 13x^2 + 36 = 0$.

Solution   This fourth-degree equation can be made to look quadratic (second-degree) if we replace x^2 by u. Then $x^4 = u^2$, and the equation becomes

$$u^2 - 13u + 36 = 0$$
$$(u - 9)(u - 4) = 0$$

Thus $\qquad\qquad\qquad u = 9 \quad \text{or} \quad u = 4$

But since $u = x^2$,

$$x^2 = 9 \quad \text{or} \quad x^2 = 4$$

and $\qquad\qquad\qquad x = \pm 3 \quad \text{or} \quad x = \pm 2$

So there are four possible values of x in the solution set: $-3, -2, 2,$ and 3.

Check:

If $x = -3$:

$(-3)^4 - 13(-3)^2 + 36 \overset{?}{=} 0$

$9^2 - 13 \cdot 9 + 9 \cdot 4 \overset{?}{=} 0$

$9 - 13 + 4 \overset{?}{=} 0$

$0 \overset{\checkmark}{=} 0$

If $x = 3$:

$3^4 - 13(3)^2 + 36 \overset{?}{=} 0$

$9^2 - 13 \cdot 9 + 9 \cdot 4 \overset{?}{=} 0$

$9 - 13 + 4 \overset{?}{=} 0$

$0 \overset{\checkmark}{=} 0$

If $x = -2$:

$(-2)^4 - 13(-2)^2 + 36 \overset{?}{=} 0$

$(-2)(-2)(-2)(-2) - 13(-2)(-2) + 36 \overset{?}{=} 0$

$16 - 52 + 36 \overset{?}{=} 0$

$0 \overset{\checkmark}{=} 0$

If $x = 2$:

$2^4 - 13(2)^2 + 36 \overset{?}{=} 0$

$(2)(2)(2)(2) - 13(2)(2) + 36 \overset{?}{=} 0$

$16 - 52 + 36 \overset{?}{=} 0$

$0 \overset{\checkmark}{=} 0$

So the solution set is $\{x \mid x = -3, -2, 2, 3\}$.

EXAMPLE 5-19 Find a polynomial whose roots are $2, -3,$ and 1.

Solution If 2 is a root, then $(x - 2)$ is a factor.
If -3 is a root, then $(x + 3)$ is a factor.
If 1 is a root, then $(x - 1)$ is a factor.
And the product of the factors is the desired polynomial:

$$P(x) = (x - 2)(x + 3)(x - 1)$$
$$= x^3 - 7x + 6$$

---

note: Any equation in one variable has a maximum number of roots equal to its degree.

5-3. Quadratic Inequalities

A solution of an inequality in one variable is a value that makes a true statement when that value is substituted back into the given inequality. But an inequality will often have an *infinite solution set*, represented as an interval. Such a set can be graphed on the number line. For example:

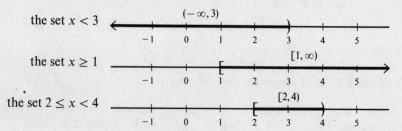

When we solve a quadratic inequality,

$$ax^2 + bx + c > 0$$

the roots of the quadratic equation $ax^2 + bx + c = 0$ play a crucial role.

To obtain the solution set to a quadratic inequality, we determine the sign of the quadratic polynomial in each of the intervals into which the number line is divided by the roots (or zeros) of the polynomial. Then we pick the correct interval(s) as the solution. And in the process of solving quadratic inequalities, we make use of a theorem from number theory:

- Between consecutive values of x for which a polynomial is zero, the polynomial remains constant in sign.

EXAMPLE 5-20   Solve the inequality $(x - 2)(x + 1) < 0$.

Solution   In the given form the zeros of the polynomial are easy to find. By inspection, these values are $x = -1$ and $x = 2$, and serve as points to divide the real number line into three parts: $x < -1$, $-1 < x < 2$, and $x > 2$.

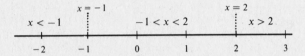

Remembering that the sign of the polynomial does not change in an interval, we choose a representative test value in each of these intervals; for example, $x = -2$, 0, and 3.

Then we find the value of the polynomial $(x - 2)(x + 1) = x^2 - x - 2$ at each chosen point:

$$(-2 - 2)(-2 + 1) = (-4)(-1) = 4 \qquad \text{at } x = -2$$

$$(0 - 2)(0 + 1) = -2 \qquad \text{at } x = 0$$

$$(3 - 2)(3 + 1) = 4 \qquad \text{at } x = 3$$

And now we can see that only in the middle interval is the polynomial negative (i.e., less than zero), as required in the given inequality. Thus the solution to the inequality is the interval $-1 < x < 2$.

EXAMPLE 5-21   Solve $2x^2 - x - 3 > 0$.

Solution   First factor to find the zeros of the quadratic polynomial:

$$2x^2 - x - 3 > 0$$

$$(x + 1)(2x - 3) > 0$$

This shows zero values at $x = -1$ and $x = \frac{3}{2}$, which divide the number line into

Then choose representative test values (just make sure they are somewhere in the interval): Let

$$x = -3 \qquad x = \tfrac{1}{2} \qquad x = 2$$

The polynomial values at these test points are

$$(2(-3) - 3)(-3 + 1) = (-9)(-2) = 18 \qquad \text{at } x = -3$$

$$(2(\tfrac{1}{2}) - 3)(\tfrac{1}{2} + 1) = (-2)(\tfrac{3}{2}) = -3 \qquad \text{at } x = \tfrac{1}{2}$$

$$(2(2) - 3)(2 + 1) = (1)(3) = 3 \qquad \text{at } x = 2$$

In the two outside intervals the polynomial $2x^2 - x - 3$ has positive values, so the given inequality has as its solution those intervals where the polynomial is positive:

$$\{x < -1\} \quad \text{and} \quad \{x > \tfrac{3}{2}\}$$

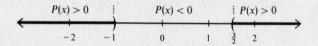

EXAMPLE 5-22 Solve $x^2 + 4x \geq 0$.

Solution

$$x^2 + 4x \geq 0$$

$$x(x + 4) \geq 0$$

The zeros are at 0 and -4, so the intervals are

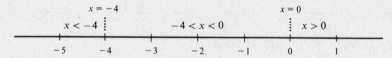

Test values:

$$x = -5 \qquad x = -2 \qquad x = 1$$

The polynomial at the test points has the value

$$(-5)(-5 + 4) = (-5)(-1) = 5 \qquad \text{at } x = -5$$

$$(-2)(-2 + 4) = (-2)(2) = -4 \qquad \text{at } x = -2$$

$$(1)(1 + 4) = 5 \qquad \text{at } x = 1$$

The polynomial is positive on the two outside intervals and is equal at the points $x = 0$ and $x = -4$, so the solution intervals are

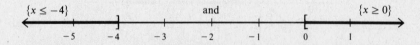

SUMMARY

1. The three main ways of solving a second-degree or quadratic equation $ax^2 + bx + c = 0$ are factoring, completing the square, and using the quadratic formula.
2. The process of completing the square on the general quadratic equation will produce the quadratic formula:

$$x = \frac{-b \pm \sqrt{b^2 - 4ac}}{2a}$$

3. The nature of the solution of a quadratic equation depends on the value of the discriminant, $b^2 - 4ac$:
   (a) If $b^2 - 4ac = 0$, then the roots are equal and there is only one real value in the solution set.

(b) If $b^2 - 4ac > 0$, then the roots are unequal real numbers and there are two real values in the solution set.

(c) If $b^2 - 4ac < 0$, then the roots are complex numbers and there are two complex number values in the solution set.

4. Radical equations are usually solved by a squaring process, and quadratic-like equations may be solved by a substitution process.

5. Between consecutive zeros of a polynomial, the polynomial remains constant in sign.

6. The solution to a quadratic inequality is often an infinite set, or interval, where the zero values of the quadratic polynomial serve as endpoints to define intervals on the number line. The correct solution intervals are those whose signs are the same as the sign designated in the given inequality.

RAISE YOUR GRADES

Can you...?

☑ explain the zero product principle
☑ explain each step in the process for completing the square
☑ derive the quadratic formula
☑ explain the nature of the roots of a quadratic and their relationship to the discriminant
☑ explain solution procedures for radical and other quadratic-type equations
☑ outline the process for solving a quadratic inequality

SOLVED PROBLEMS

Quadratic Equations

PROBLEM 5-1 Solve $x^2 - 11x + 24 = 0$.

Solution Factor the left-hand side:

$$x^2 - 11x + 24 = 0$$
$$(x - 8)(x - 3) = 0$$
$$\text{So} \quad x = 8 \quad \text{or} \quad x = 3$$

Check: $8^2 - 11(8) + 24 \overset{?}{=} 0 \qquad 3^2 - 11(3) + 24 \overset{?}{=} 0$
$64 - 88 + 24 \overset{\checkmark}{=} 0 \qquad 9 - 33 + 24 \overset{\checkmark}{=} 0$

Thus, by the zero product principle, the solution set is $\{x \mid x = 8, 3\}$.

PROBLEM 5-2 Solve $2x^2 - x - 3 = 0$.

Solution Factor the left-hand side and apply the zero product principle:

$$2x^2 - x - 3 = 0$$
$$(2x - 3)(x + 1) = 0$$
$$\text{So} \quad 2x - 3 = 0 \quad \text{or} \quad x + 1 = 0$$
$$x = \tfrac{3}{2} \qquad\qquad x = -1$$

Check: $2(\tfrac{3}{2})^2 - \tfrac{3}{2} - 3 \overset{?}{=} 0 \qquad 2(-1)^2 - (-1) - 3 \overset{?}{=} 0$
$2(\tfrac{9}{4}) - \tfrac{3}{2} - 3 \overset{?}{=} 0 \qquad 2(1) + 1 - 3 \overset{?}{=} 0$
$\tfrac{9}{2} - \tfrac{3}{2} - \tfrac{6}{2} \overset{\checkmark}{=} 0 \qquad 2 + 1 - 3 \overset{\checkmark}{=} 0$

PROBLEM 5-3 Solve $5x^2 + 2x - 16 = 0$.

Solution Since the factors of this one are hard to find by inspection or trial-and-error, you apply the quadratic formula, where $a = 5, b = 2, c = -16$:

$$x = \frac{-b \pm \sqrt{b^2 - 4ac}}{2a}$$

$$x = \frac{-2 \pm \sqrt{2^2 - 4(5)(-16)}}{2(5)}$$

$$= \frac{-2 \pm \sqrt{4 + 320}}{10} = \frac{-2 \pm \sqrt{324}}{10}$$

$$= \frac{-2 \pm 18}{10}$$

Thus $\qquad x = \dfrac{-2 - 18}{10} \quad$ or $\quad x = \dfrac{-2 + 18}{10}$

$$= -2 \qquad\qquad = \tfrac{8}{5}$$

Check: $5(-2)^2 + 2(-2) - 16 \overset{?}{=} 0 \qquad 5(\tfrac{8}{5})^2 + 2(\tfrac{8}{5}) - 16 \overset{?}{=} 0$

$5(4) + 2(-2) - 16 \overset{?}{=} 0 \qquad 5(\tfrac{64}{25}) + \tfrac{16}{5} - \tfrac{80}{5} \overset{?}{=} 0$

$20 - 4 - 16 \overset{\checkmark}{=} 0 \qquad \tfrac{64}{5} + \tfrac{16}{5} - \tfrac{80}{5} \overset{\checkmark}{=} 0$

PROBLEM 5-4 Solve $9x^2 + 54x + 81 = 0$.

Solution Note that each number is a multiple of 9, so first divide by 9:

$$\frac{9x^2}{9} + \frac{54x}{9} + \frac{81}{9} = \frac{0}{9}$$

$$x^2 + 6x + 9 = 0$$

$$(x + 3)(x + 3) = 0 \qquad \text{[Factor]}$$

So $\qquad\qquad x = -3, \quad x = -3 \qquad$ [A repeated root]

Thus the given equation has only one member in the solution set, $\{x \mid x = -3\}$.

PROBLEM 5-5 Solve $x^2 + 2x + 5 = 0$ by using the quadratic formula.

Solution $a = 1, b = 2, c = 5$; so

$$x = \frac{-2 \pm \sqrt{2^2 - 4(1)(5)}}{2(1)}$$

$$= \frac{-2 \pm \sqrt{4 - 20}}{2} = \frac{-2 \pm \sqrt{-16}}{2}$$

$$= \frac{-2 \pm 4i}{2} = \frac{-2}{2} \pm \frac{4i}{2}$$

$$= -1 \pm 2i$$

So the solutions are the complex numbers $x = -1 + 2i$ and $x = -1 - 2i$.

Check: $(-1 + 2i)^2 + 2(-1 + 2i) + 5 \overset{?}{=} 0 \qquad (-1 - 2i)^2 + 2(-1 - 2i) + 5 \overset{?}{=} 0$

$1 - 4i + 4i^2 - 2 + 4i + 5 \overset{?}{=} 0 \qquad 1 + 4i + 4i^2 - 2 - 4i + 5 \overset{?}{=} 0$

$1 + 4(-1) - 2 + 5 \overset{?}{=} 0 \qquad 1 + 4(-1) - 2 + 5 \overset{?}{=} 0$

$1 - 4 - 2 + 5 \overset{\checkmark}{=} 0 \qquad 1 - 4 - 2 + 5 \overset{\checkmark}{=} 0$

PROBLEM 5-6 Solve $3x^2 + 2x - 1 = 0$.

Solution You could try to factor the quadratic expression, or you could apply the quadratic formula. An attempt to factor would produce (if you persist in the trial process)

$$(3x - 1)(x + 1) = 0$$

so $$x = \tfrac{1}{3} \quad \text{or} \quad x = -1$$

Using the quadratic formula,

$$x = \frac{-b \pm \sqrt{b^2 - 4ac}}{2a} = \frac{-2 \pm \sqrt{2^2 - 4(3)(-1)}}{2(3)}$$

$$= \frac{-2 \pm \sqrt{4 + 12}}{6} = \frac{-2 \pm \sqrt{16}}{6}$$

Thus $$x = \frac{-2 + 4}{6} \quad \text{or} \quad x = \frac{-2 - 4}{6}$$

$$= \tfrac{1}{3} \qquad\qquad\qquad = -1$$

PROBLEM 5-7 Solve $7x = -x^2$.

Solution

$$7x = -x^2$$

$$x^2 + 7x = 0 \qquad \text{[Add } x^2 \text{ to both sides]}$$

$$x(x + 7) = 0 \qquad \text{[Factor]}$$

So $$x = 0 \quad \text{or} \quad x = -7$$

Check: $7(0) \overset{?}{=} -(0)^2 \qquad 7(-7) \overset{?}{=} -(-7)^2$

$0 \overset{\checkmark}{=} 0 \qquad\qquad -49 \overset{\checkmark}{=} -49$

PROBLEM 5-8 Solve $(2x - 1)^2 = 16$.

Solution Take the square root of both sides:

$$\sqrt{(2x - 1)^2} = \sqrt{16}$$

$$2x - 1 = \pm 4$$

so that $$2x - 1 = 4 \quad \text{or} \quad 2x - 1 = -4$$

$$2x = 5 \qquad\qquad 2x = -3$$

$$x = \tfrac{5}{2} \qquad\qquad x = -\tfrac{3}{2}$$

Check: $(2 \cdot \tfrac{5}{2} - 1)^2 \overset{?}{=} 16 \qquad [2(-\tfrac{3}{2}) - 1]^2 \overset{?}{=} 16$

$(5 - 1)^2 \overset{?}{=} 16 \qquad\qquad (-3 - 1)^2 \overset{?}{=} 16$

$16 \overset{\checkmark}{=} 16 \qquad\qquad\qquad 16 \overset{\checkmark}{=} 16$

PROBLEM 5-9 Solve $x - \dfrac{1}{x} = \dfrac{3}{2}$.

Solution To clear the fractions, multiply each term by $2x$:

$$2x(x) - 2x\left(\frac{1}{x}\right) = 2x\left(\frac{3}{2}\right)$$

or $$2x^2 - 2 = 3x$$

$$2x^2 - 3x - 2 = 0$$

$$(2x + 1)(x - 2) = 0$$

So $$2x + 1 = 0 \quad \text{or} \quad x - 2 = 0$$

$$x = -\tfrac{1}{2} \qquad\qquad x = 2$$

Check: $-\dfrac{1}{2} - \dfrac{1}{(-\tfrac{1}{2})} \overset{?}{=} \dfrac{3}{2} \qquad 2 - \dfrac{1}{2} \overset{\checkmark}{=} \dfrac{3}{2}$

$-\dfrac{1}{2} + 2 \overset{\checkmark}{=} \dfrac{3}{2}$

PROBLEM 5-10 Solve $(5x - 1)^2 = -20$ by two different methods.

Solution
Method 1: Take the square root of both sides, so that

$$5x - 1 = \pm \sqrt{-20}$$

$$= \pm \sqrt{-1}\sqrt{4}\sqrt{5}$$

$$= \pm i \cdot 2\sqrt{5}$$

or

$$5x = 1 \pm i \cdot 2\sqrt{5}$$

$$x = \tfrac{1}{5} \pm i \cdot \tfrac{2}{5}\sqrt{5} = \tfrac{1}{5} \pm \tfrac{2}{5}\sqrt{5}i$$

So the solution set is a pair of complex numbers:

$$\{\tfrac{1}{5} + \tfrac{2}{5}\sqrt{5}i, \tfrac{1}{5} - \tfrac{2}{5}\sqrt{5}i\}$$

Method 2: Multiply out the left-hand side first and simplify:

$$(5x - 1)(5x - 1) = -20$$

$$25x^2 - 10x + 1 = -20$$

$$25x^2 - 10x + 21 = 0$$

Then use the quadratic formula ($a = 25$, $b = -10$, $c = 21$):

$$x = \frac{-(-10) \pm \sqrt{(-10)^2 - 4(25)(21)}}{2 \cdot 25} = \frac{10 \pm \sqrt{100 - 2100}}{50}$$

$$= \frac{10 \pm \sqrt{-2000}}{50} = \frac{10 \pm \sqrt{(-1)(10^2)(20)}}{50} = \frac{10 \pm i \cdot 10\sqrt{20}}{50}$$

$$= \frac{10}{50} \pm i \cdot \frac{10}{50}\sqrt{20} = \frac{1}{5} \pm i \cdot \frac{1}{5}\sqrt{5 \cdot 2^2} = \frac{1}{5} \pm i \cdot \frac{1}{5} \cdot 2\sqrt{5}$$

$$= \frac{1}{5} \pm \frac{2}{5}\sqrt{5}i$$

PROBLEM 5-11 Solve $b^2 - 2b + 5 = 0$.

Solution This equation can be solved by completing the square.

Rewrite: $\qquad\qquad\qquad\qquad b^2 - 2b = -5$

Compute ($\frac{1}{2} \times$ coefficient of b) $= \frac{1}{2}(-2) = -1$ and square it: $(-1)^2 = 1$. Add 1 to both sides:

$$b^2 - 2b + 1 = -5 + 1$$

Rewrite the left-hand-side trinomial as a binomial squared and simplify:

$$(b - 1)^2 = -4$$

Take the square root of both sides:

$$(b - 1) = \pm\sqrt{-4} = \pm 2\sqrt{-1} = \pm 2i$$

$$b = 1 \pm 2i$$

Thus the solution set is the complex pair $\{1 + 2i, 1 - 2i\}$.

PROBLEM 5-12 Solve $3y^2 - 9y - 12 = 0$ by completing the square.

Solution

$$y^2 - 3y - 4 = 0 \qquad \text{[Divide all terms by 3]}$$

$$y^2 - 3y = 4 \qquad \text{[Transpose the constant term]}$$

Find the square of half the coefficient of y: $(-\tfrac{3}{2})^2 = \tfrac{9}{4}$.

$$y^2 - 3y + \tfrac{9}{4} = 4 + \tfrac{9}{4} \qquad\qquad \text{[Add } \tfrac{9}{4} \text{ to both sides]}$$

$$(y - \tfrac{3}{2})^2 = \tfrac{25}{4}$$

$$(y - \tfrac{3}{2}) = \pm\sqrt{\tfrac{25}{4}} = \pm\tfrac{5}{2} \qquad \text{[Extract the roots]}$$

$$y = \tfrac{3}{2} \pm \tfrac{5}{2}$$

So the solution set is $\{\tfrac{3}{2} + \tfrac{5}{2}, \tfrac{3}{2} - \tfrac{5}{2}\}$ or $\{4, -1\}$.

PROBLEM 5-13 The quadratic formula applied to $ax^2 + bx + c = 0$ gives the roots

$$x = \frac{-b + \sqrt{b^2 - 4ac}}{2a} \quad \text{and} \quad x = \frac{-b - \sqrt{b^2 - 4ac}}{2a}$$

Find the sum and the product of the roots.

Solution The sum is

$$\left(-\frac{b}{2a} + \frac{\sqrt{b^2 - 4ac}}{2a}\right) + \left(-\frac{b}{2a} - \frac{\sqrt{b^2 - 4ac}}{2a}\right) = -\frac{b}{2a} + \left(-\frac{b}{2a}\right) = -\frac{2b}{2a} = -\frac{b}{a}$$

The product is

$$\left(\frac{-b + \sqrt{b^2 - 4ac}}{2a}\right)\left(\frac{-b - \sqrt{b^2 - 4ac}}{2a}\right) = \frac{(-b)^2 - (\sqrt{b^2 - 4ac})^2}{4a^2}$$

$$= \frac{b^2 - (b^2 - 4ac)}{4a^2}$$

$$= \frac{4ac}{4a^2} = \frac{c}{a}$$

PROBLEM 5-14 You can use the results of Problem 5-13 to help you factor a quadratic expression. Suppose you need to factor

$$x^2 + Bx + C = 0$$

You have found that the sum of the roots is $-b/a = -B/1 = -B$ and the product is $c/a = C/1 = C$. (Did you note that $A = 1$ in the problem?) Now use what you know to factor and solve
(a) $x^2 - 5x + 6 = 0$ and **(b)** $2x^2 - 4x - 126$.

Solution

(a) The sum of the roots of $x^2 - 5x + 6 = 0$ must be $-B = -(-5) = 5$ and the product of the roots must be $C = 6$. These, of course, are the values 2 and 3, since $2 + 3 = 5$ and $2 \cdot 3 = 6$. Thus the factors are $(x - 2)(x - 3) = 0$, so that $x - 2 = 0$ or $x - 3 = 0$ and the solution is $\{x = 2, x = 3\}$.

(b) Divide $2x^2 - 4x - 126$ through by the coefficient of x^2 to make $A = 1$:

$$\frac{2x^2}{2} - \frac{4x}{2} - \frac{126}{2} = x^2 - 2x - 63$$

Now the sum of the roots is $-B = 2$, and the product of the roots is $C = -63$, which give the values 9 and -7. The factors are $(x + 7)(x - 9) = 0$. Thus $x + 7 = 0$ or $x - 9 = 0$, so that the solution set is $\{x = -7, x = 9\}$.

PROBLEM 5-15 Use the sum and product factors to solve $x^2 - 13x + 36 = 0$.

Solution You need to find two numbers whose sum is 13 and whose product is 36. A few trials will soon produce the values 9 and 4. Thus $(x - 9)(x - 4) = 0$ and $x = 9, x = 4$ are the solutions.

PROBLEM 5-16 Solve $\dfrac{x^2}{3} - \dfrac{1}{2} = \dfrac{5x}{6}$.

Solution It helps to clear the fractions here. The least common denominator is 6, so multiply by that value:

$$6\left(\frac{x^2}{3}\right) - 6\left(\frac{1}{2}\right) = 6\left(\frac{5x}{6}\right)$$

$$2x^2 - 3 = 5x$$

$$2x^2 - 5x - 3 = 0$$

Then the quadratic formula supplies the roots:

$$x = \frac{-(-5) \pm \sqrt{(-5)^2 - 4(2)(-3)}}{2(2)}$$

$$= \frac{5 \pm \sqrt{25 + 24}}{4}$$

$$= \frac{5 \pm \sqrt{49}}{4}$$

$$= \frac{5 \pm 7}{4}$$

note: You *could* factor here; but fractional factors are awkward. *The quadratic formula always works*—use it when other methods become cumbersome.

So
$$x = \tfrac{12}{4} = 3 \qquad \text{or} \qquad x = -\tfrac{2}{4} = -\tfrac{1}{2}$$

Check: Notice that the sum of the roots is $3 + (-\tfrac{1}{2}) = \tfrac{5}{2}$ and the product is $3(-\tfrac{1}{2}) = -\tfrac{3}{2}$. If you rewrite the original equation so that the coefficient of x^2 is 1, you get

$$3\left(\frac{x^2}{3}\right) - 3\left(\frac{1}{2}\right) = 3\left(\frac{5x}{6}\right)$$

$$x^2 - \frac{3}{2} = \frac{5x}{2}$$

$$x^2 - \tfrac{5}{2}x - \tfrac{3}{2} = 0$$

Now note that $-B = \tfrac{5}{2}$ and $C = -\tfrac{3}{2}$.

PROBLEM 5-17 Use the quadratic formula to solve $\tfrac{1}{3}x^2 - 2x + 3 = 0$.

Solution Since $a = \tfrac{1}{3}$, $b = -2$, and $c = 3$, the roots are

$$x = \frac{-(-2) \pm \sqrt{(-2)^2 - 4(\tfrac{1}{3})(3)}}{2(\tfrac{1}{3})}$$

$$= \frac{2 \pm \sqrt{4 - 4}}{\tfrac{2}{3}}$$

$$= \frac{2 \pm 0}{\tfrac{2}{3}} = 2\left(\frac{3}{2}\right) = 3$$

There is only one solution to this quadratic equation. (Remember that when the discriminant $b^2 - 4ac = 0$, the roots are equal.)

note: This equation is easily factorable if you multiply through by 3 first:

$$3(\tfrac{1}{3})x^2 - 3(2x) + 3(3) = 0$$

$$x^2 - 6x + 9 = 0$$

$$(x - 3)^2 = 0$$

$$x = 3$$

Quadratic-Type Equations

PROBLEM 5-18 Solve $\sqrt{2y - 1} - 2 = \sqrt{y - 4}$.

Solution Clear the radicals first by squaring both sides:

$$(\sqrt{2y - 1} - 2)^2 = (\sqrt{y - 4})^2$$

$$(2y - 1) - 4\sqrt{2y - 1} + 4 = y - 4$$

Isolate the remaining radical on one side of the equation and simplify:

$$2y - 1 + 4 - y + 4 = 4\sqrt{2y - 1}$$
$$y + 7 = 4\sqrt{2y - 1}$$

Square both sides again:

$$(y + 7)^2 = 16(2y - 1)$$

then transpose and combine terms:

$$y^2 + 14y + 49 = 32y - 16$$
$$y^2 - 18y + 65 = 0$$

Now use the sum and product property. This one is easy since the only factors of 65 are 13 and 5 and their sum is 18. Thus

$$(y - 13)(y - 5) = 0$$

And $y = 13$ or $y = 5$.

Check: You *have* to check because the equation was squared.

If $y = 13$:

$$\sqrt{2(13) - 1} - 2 \overset{?}{=} \sqrt{13 - 4}$$
$$\sqrt{25} - 2 \overset{?}{=} \sqrt{9}$$
$$5 - 2 \overset{\checkmark}{=} 3$$

If $y = 5$:

$$\sqrt{2(5) - 1} - 2 \overset{?}{=} \sqrt{5 - 4}$$
$$\sqrt{9} - 2 \overset{?}{=} \sqrt{1}$$
$$3 - 2 \overset{\checkmark}{=} 1$$

So the solution set is $\{y \mid y = 5, 13\}$.

PROBLEM 5-19 Solve $2t^4 + t^2 - 15 = 0$.

Solution This equation can be made quadratic by changing variables. Let $t^2 = v$, so that $t^4 = v^2$:

$$2t^4 + t^2 - 15 = 0$$
$$2v^2 + v - 15 = 0$$
$$(2v - 5)(v + 3) = 0$$

Thus
$$v = \tfrac{5}{2} \quad \text{or} \quad v = -3$$

But $v = t^2$, so you have
$$t^2 = \tfrac{5}{2} \quad \text{or} \quad t^2 = -3$$

Then
$$t = \pm\sqrt{\tfrac{5}{2}} \quad \text{or} \quad t = \pm\sqrt{-3} = \pm\sqrt{3}i$$

So there are two real roots, $t = \sqrt{\tfrac{5}{2}}$ and $t = -\sqrt{\tfrac{5}{2}}$, and two complex roots, $t = \sqrt{3}i$ and $t = -\sqrt{3}i$.

Check:

$$2(\sqrt{\tfrac{5}{2}})^4 + (\sqrt{\tfrac{5}{2}})^2 - 15 \overset{?}{=} 0 \qquad 2(-\sqrt{\tfrac{5}{2}})^4 + (-\sqrt{\tfrac{5}{2}})^2 - 15 \overset{?}{=} 0$$
$$2(\tfrac{5}{2})^2 + \tfrac{5}{2} - 15 \overset{?}{=} 0 \qquad\qquad 2(\tfrac{5}{2})^2 + \tfrac{5}{2} - 15 \overset{?}{=} 0$$
$$2(\tfrac{25}{4}) + \tfrac{5}{2} - 15 \overset{?}{=} 0 \qquad\qquad\qquad 15 - 15 \overset{\checkmark}{=} 0$$
$$\tfrac{25}{2} + \tfrac{5}{2} - 15 \overset{?}{=} 0$$
$$15 - 15 \overset{\checkmark}{=} 0$$

$$2(\sqrt{3}i)^4 + (\sqrt{3}i)^2 - 15 \overset{?}{=} 0 \qquad 2(-\sqrt{3}i)^4 + (-\sqrt{3}i)^2 - 15 \overset{?}{=} 0$$
$$2(1)(9) + (-1)(3) - 15 \overset{?}{=} 0 \qquad 2(i^4)(9) + (i^2)(3) - 15 \overset{?}{=} 0$$
$$18 - \qquad 3 - 15 \overset{?}{=} 0 \qquad\qquad 2(1)(9) + (-1)(3) - 15 \overset{?}{=} 0$$
$$18 - 18 \overset{\checkmark}{=} 0 \qquad\qquad\qquad 18 - 18 \overset{\checkmark}{=} 0$$

All four roots check, so the solution set is $\{t \mid t = \pm\sqrt{\tfrac{5}{2}}, \pm\sqrt{3}i\}$.

PROBLEM 5-20   Solve $16x^4 - 49 = 0$.

Solution   Let $x^2 = v$, so that $x^4 = v^2$:

$$16v^2 - 49 = 0$$

$$(4v - 7(4v + 7) = 0$$

So

$$v = \tfrac{7}{4} \quad \text{or} \quad v = -\tfrac{7}{4}$$

But $v = x^2$, so $x^2 = \tfrac{7}{4}$ or $x^2 = -\tfrac{7}{4}$.
These give you four solutions:

$$x = \sqrt{\frac{7}{4}} = \frac{\sqrt{7}}{2}, \quad x = \frac{-\sqrt{7}}{2}, \quad x = \sqrt{-\frac{7}{4}} = \frac{\sqrt{7}i}{2}, \quad x = \frac{-\sqrt{7}i}{2}$$

Check:

$$16\left(\frac{\sqrt{7}}{2}\right)^4 - 49 \overset{?}{=} 0 \qquad 16\left(-\frac{\sqrt{7}}{2}\right)^4 - 49 \overset{?}{=} 0$$

$$16(\tfrac{49}{16}) - 49 \overset{?}{=} 0 \qquad\qquad 16(\tfrac{7}{4})^2 - 49 \overset{?}{=} 0$$

$$49 - 49 \overset{\checkmark}{=} 0 \qquad\qquad 49 - 49 \overset{\checkmark}{=} 0$$

$$16\left(\frac{\sqrt{7}i}{2}\right)^4 - 49 \overset{?}{=} 0 \qquad 16\left(-\frac{\sqrt{7}i}{2}\right)^4 - 49 \overset{?}{=} 0$$

$$16(i^4)(\tfrac{49}{16}) - 49 \overset{?}{=} 0 \qquad 16(i^2)^2(\tfrac{7}{4})^2 - 49 \overset{?}{=} 0$$

$$16(1)(\tfrac{49}{16}) - 49 \overset{?}{=} 0 \qquad\quad 16(1)(\tfrac{49}{16}) - 49 \overset{?}{=} 0$$

$$49 - 49 \overset{\checkmark}{=} 0 \qquad\qquad\quad 49 - 49 \overset{\checkmark}{=} 0$$

PROBLEM 5-21   Solve $3x + 1 - 7\sqrt{3x + 1} + 12 = 0$.

Solution   You could isolate the radical and square or use the quadratic formula:

$$3x + 1 - 7\sqrt{3x + 1} + 12 = 0$$

$$3x + 13 = 7\sqrt{3x + 1} \qquad \text{[Transpose]}$$

$$9x^2 + 78x + 169 = 49(3x + 1) \qquad \text{[Square both sides]}$$

$$9x^2 - 69x + 120 = 0 \qquad \text{[Transpose again]}$$

$$3x^2 - 23x + 40 = 0 \qquad \text{[Divide each term by 3]}$$

Complete the square:

$$x^2 - \frac{23x}{3} + \frac{40}{3} = 0$$

$$x^2 - \frac{23x}{3} = -\frac{40}{3}$$

$$x^2 - \frac{23x}{3} + \left(\frac{1}{2} \cdot \frac{23}{3}\right)^2 = \left(\frac{1}{2} \cdot \frac{23}{3}\right)^2 - \frac{40}{3}$$

$$= \left(\frac{23}{6}\right)^2 - \frac{40}{3}$$

$$\left(x - \frac{23}{6}\right)^2 = \frac{49}{36}$$

$$x - \frac{23}{6} = \pm \frac{7}{6}$$

Use the quadratic formula:

$$x = \frac{-(-23) \pm \sqrt{(-23)^2 - 4(3)(40)}}{2 \cdot 3}$$

$$= \frac{23 \pm \sqrt{529 - 480}}{6}$$

$$= \frac{23 \pm \sqrt{49}}{6}$$

$$x = \frac{23}{6} \pm \frac{7}{6}$$

so

$$x = 5, \quad x = \frac{8}{3}$$

But you may also make a variable change to turn the given equation into a quadratic.

Let $\sqrt{3x + 1} = z$, so that $3x + 1 = z^2$:

Then
$$z^2 - 7z + 12 = 0$$
$$(z - 4)(z - 3) = 0$$

so
$$z = 4 \quad \text{or} \quad z = 3$$

But $z = \sqrt{3x + 1}$, so you have

$$\sqrt{3x + 1} = 4 \qquad \text{or} \qquad \sqrt{3x + 1} = 3$$
$$3x + 1 = 16 \qquad\qquad 3x + 1 = 9$$
$$3x = 15 \qquad\qquad 3x = 8$$
$$x = 5 \qquad\qquad x = \tfrac{8}{3}$$

Check:

$$3(5) + 1 - 7\sqrt{3(5) + 1} + 12 \overset{?}{=} 0 \qquad 3(\tfrac{8}{3}) + 1 - 7\sqrt{3(\tfrac{8}{3}) + 1} + 12 \overset{?}{=} 0$$
$$15 + 1 - 7\sqrt{15 + 1} + 12 \overset{?}{=} 0 \qquad\quad 8 + 1 - 7\sqrt{8 + 1} + 12 \overset{?}{=} 0$$
$$16 - 7\sqrt{16} + 12 \overset{?}{=} 0 \qquad\qquad 9 - 7\sqrt{9} + 12 \overset{?}{=} 0$$
$$16 - 7(4) + 12 \overset{?}{=} 0 \qquad\qquad 9 - 7(3) + 12 \overset{?}{=} 0$$
$$28 - 28 \overset{\checkmark}{=} 0 \qquad\qquad\qquad 21 - 21 \overset{\checkmark}{=} 0$$

moral: Since the easiest way is usually the best way in algebra, you should probably choose the method of variable change and substitution for a problem like this one. But be sure you know *all* the methods, so you can pick the one that's easiest for *you.*

PROBLEM 5-22 Solve $6x^{2/5} + 11x^{1/5} + 3 = 0$.

Solution This equation can be turned into a quadratic if you set $x^{1/5} = w$, so that $x^{2/5} = w^2$. Then

$$6w^2 + 11w + 3 = 0$$

Now apply the quadratic formula:

$$w = \frac{-11 \pm \sqrt{11^2 - 4(6)(3)}}{2(6)} = \frac{-11 \pm \sqrt{121 - 72}}{12}$$
$$= \frac{-11 \pm \sqrt{49}}{12}$$
$$= \frac{-11 \pm 7}{12}$$

Thus
$$w = \frac{-11 - 7}{12} \qquad \text{or} \qquad w = \frac{-11 + 7}{12}$$
$$= \frac{-18}{12} = -\frac{3}{2} \qquad\qquad = -\frac{4}{12} = -\frac{1}{3}$$

But $w = x^{1/5}$, so
$$x^{1/5} = -\frac{3}{2} \qquad \text{or} \qquad x^{1/5} = -\frac{1}{3}$$

To get x, you have to raise both sides to the fifth power:

$$x = \left(-\frac{3}{2}\right)^5 \qquad \text{or} \qquad x = \left(-\frac{1}{3}\right)^5$$
$$= \frac{-3^5}{2^5} = -\frac{243}{32} \qquad\qquad = \frac{-1^5}{3^5} = \frac{-1}{243}$$

You should check these solutions.

Quadratic Inequalities

PROBLEM 5-23　Find the values of x that satisfy $2x^2 - 9x < 5$.

Solution　Subtract 5 from both sides and factor the resulting trinomial lhs:

$$2x^2 - 9x < 5$$

$$2x^2 - 9x - 5 < 0$$

$$(2x + 1)(x - 5) < 0$$

So the zero values of the polynomial $2x^2 - 9x - 5$ are

$$2x + 1 = 0 \qquad \text{and} \qquad x - 5 = 0$$

$$x = -\frac{1}{2} \qquad\qquad x = 5$$

which divide the real number line into three parts:

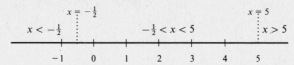

Since the sign of the polynomial does not change inside any of these intervals, you can choose a representative value in each one, such as

$$x = -1 \qquad x = 0 \qquad x = 6$$

Then find the value of the quadratic polynomial at these points:

$$(2(-1) + 1)(-1 - 5) = (-1)(-6) \qquad (2(0) + 1)(0 - 5) = 1(-5) \qquad (2(6) + 1)(6 - 5) = 13(1)$$

$$= 6 \qquad\qquad\qquad = -5 \qquad\qquad\qquad = 13$$

Only in the middle interval is the polynomial negative, as required in the given inequality $2x^2 - 9x - 5 < 0$. This interval is the solution interval $-\frac{1}{2} < x < 5$, or

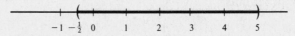

PROBLEM 5-24　Solve the inequality $x^2 - 4 > 0$.

Solution　You need values for x whose squares are more than 4. You might think that this would mean x values more than 2. But using algebraic techniques, you see that

$$x^2 - 4 > 0$$

$$(x - 2)(x + 2) > 0$$

The zeros are at $x = -2$ and $x = 2$, so the intervals of the number line are

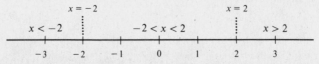

Then, picking representative values, you get

If　　　　　　　　$x = -3$:　　　　　　　$x = 0$:　　　　　　　$x = 3$:

then　　　　　$(-3)^2 - 4 = 9 - 4 = 5$　　　$0^2 - 4 = -4$　　　$3^2 - 4 = 5$

So the polynomial is positive not only for $x > 2$, but also for $x < -2$. Hence the solution set is $\{x < -2 \text{ or } x > 2\}$:

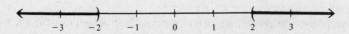

PROBLEM 5-25 Solve the cubic inequality $x^3 - 5x^2 - 6x > 0$.

Solution Proceed as usual, factoring the polynomial:

$$x^3 - 5x^2 - 6x > 0$$

$$x(x^2 - 5x - 6) > 0$$

$$x(x - 6)(x + 1) > 0$$

so the zeros of the cubic polynomial are at $x = 0$, $x = 6$, and $x = -1$. These divide the number line into *four* parts:

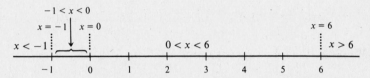

If $x = -2$: $x = -\frac{1}{2}$: $x = 3$: $x = 7$:

Then $(-2)(-2 - 6)(-2 + 1)$ $(-\frac{1}{2})(-\frac{1}{2} - 6)(-\frac{1}{2} + 1)$ $3(3 - 6)(3 + 1)$ $7(7 - 6)(7 + 1)$

$\quad = (-2)(-8)(-1)$ $\quad = (-\frac{1}{2})(-\frac{13}{2})(\frac{1}{2})$ $= 3(-3)(4)$ $= 7(1)(8)$

$\quad = -16$ $\quad = \frac{13}{8}$ $= -36$ $= 56$

Thus the cubic polynomial is positive throughout the intervals $-1 < x < 0$ and $x > 6$, which are the solution intervals:

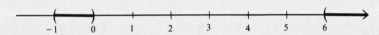

PROBLEM 5-26 Solve $12x^2 - 5x \geq 7$.

Solution Reorganize to $12x^2 - 5x - 7 \geq 0$. Then find the zeros of the polynomial by using the quadratic formula:

$$x = \frac{-(-5) \pm \sqrt{(-5)^2 - 4(12)(-7)}}{2(12)} = \frac{5 \pm \sqrt{25 + 336}}{24} = \frac{5 \pm \sqrt{361}}{24} = \frac{5 \pm 19}{24}$$

so the zeros are $x = \dfrac{5 + 19}{24} = 1$ and $x = \dfrac{5 - 19}{24} = \dfrac{-7}{12}$. Then the intervals are

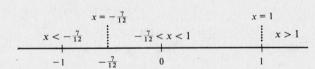

If $x = -1$: $x = 0$: $x = 2$:

Then $12(-1)^2 - 5(-1) - 7$ $12(0) - 5(0) - 7$ $12(2)^2 - 5(2) - 7$

$\quad = 12 + 5 - 7$ $\quad = -7$ $\quad = 48 - 10 - 7$

$\quad = 10$ $\quad = 31$

So the solution intervals, where the polynomial is greater than or equal to zero, are $x \leq -\frac{7}{12}$ and $x \geq 1$.

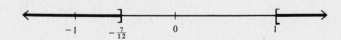

Supplementary Exercises

Solve the following second-degree equations.

PROBLEM 5-27

(a) $x^2 - 7x + 6 = 0$
(b) $y^2 - 25 = 0$
(c) $9t^2 - 1 = 0$
(d) $4t^2 - 4t + 1 = 0$
(e) $3x^2 - x = 4$

PROBLEM 5-28

(a) $9x^2 + 24x = 0$
(b) $8z^2 = 3 - 10z$
(c) $x^2 = 9x + 10$
(d) $(t + 1)(t - 6) + 12 = 0$
(e) $1 - 9x = -20x^2$

PROBLEM 5-29

(a) $(2x + 3)^2 = 16$
(b) $2x^2 + 5x - 13 = 0$
(c) $-x^2 + 7x - 2 = 0$
(d) $(y + 2)^2 = 9y$
(e) $(y + 2)^2 = 12$

PROBLEM 5-30 Solve the following equations:

(a) $3 + \dfrac{2}{x} - \dfrac{4}{x^2} = 0$

(b) $1 - \dfrac{2}{t} = -3t$

(c) $z + \dfrac{1}{3} = \dfrac{4}{3z}$

(d) $\sqrt{6x^2 - 10x} = 2$

(e) $\dfrac{1}{a + 3} + \dfrac{6}{a^2 - 9} = 1$

PROBLEM 5-31 Verify that

$$Ax^2 + Bx + C = A\left[x + \frac{B + \sqrt{B^2 - 4AC}}{2A}\right]\left[x + \frac{B - \sqrt{B^2 - 4AC}}{2A}\right]$$

Then use it to solve $4x^2 - 5x + 1 = 0$.

Solve the following equations.

PROBLEM 5-32

(a) $(2s + 3)^2 = 10$
(b) $\sqrt{t + 4} = 5 - \sqrt{t - 1}$
(c) $\sqrt[3]{v^2 + 3} = 2$
(d) $x - \sqrt{x} = 12$
(e) $\sqrt{y + 4} - \sqrt{y + 20} = -2$

PROBLEM 5-33

(a) $(2x - 1)^2 - 5(2x - 1) - 6 = 0$
(b) $x^{2/5} = -2x^{1/5} - 1$
(c) $3t^6 - 13t^3 - 10 = 0$
(d) $\dfrac{2x}{\sqrt{x^2 - 1}} + \dfrac{2x - 1}{\sqrt{x^3 - x + 3}} = 0$

Solve the following quadratic inequalities.

PROBLEM 5-34

(a) $x^2 + 2x - 8 \le 0$
(b) $x^2 - x - 12 > 0$
(c) $4x^2 - 9 < 0$
(d) $6y^2 - 1 < 5y - 2$
(e) $(x - 1)(x - 2)(x - 3) < 0$

PROBLEM 5-35

(a) $x^2 - 4x + 4 \ge 0$
(b) $2x^2 > 3 - 5x$
(c) $(x + 3)(x - 6) > (x + 1)(x - 8)$
(d) $x^3 - 7x^2 - 8x \le 0$
(e) $\dfrac{(x - 2)(x + 5)}{2} - 4 \le (x + 1)^2 - \dfrac{x^2}{2}$

Answers to Supplementary Exercises

5-27 (a) $x = 1, x = 6$

(b) $y = \pm 5$

(c) $t = \pm \frac{1}{3}$

(d) $t = \frac{1}{2}$

(e) $x = -1, x = \frac{4}{3}$

5-28 (a) $x = 0, x = -\frac{8}{3}$

(b) $z = \frac{1}{4}, z = -\frac{3}{2}$

(c) $x = 10, x = -1$

(d) $t = 3, t = 2$

(e) $x = \frac{1}{5}, x = \frac{1}{4}$

5-29 (a) $x = \frac{1}{2}, x = -\frac{7}{2}$

(b) $x = \dfrac{-5 \pm \sqrt{129}}{4}$

(c) $x = \dfrac{7 \pm \sqrt{41}}{2}$

(d) $y = 1, y = 4$

(e) $y = -2 \pm \sqrt{12}$

5-30 (a) $x = \dfrac{-1 \pm \sqrt{13}}{3}$

(b) $t = \frac{2}{3}, t = -1$

(c) $z = -\frac{4}{3}, z = 1$

(d) $x = -\frac{1}{3}, x = 2$

(e) $a = 4$

5-31 This is the quadratic formula;
$x = \frac{1}{4}, x = 1$

5-32 (a) $s = \dfrac{-3 \pm \sqrt{10}}{2}$

(b) $t = 5$

(c) $v = \pm \sqrt{5}$

(d) $x = 16$

(e) $y = 5$

5-33 (a) $x = \frac{7}{2}, x = 0$

(b) $x = -1$

(c) $t = \sqrt[3]{-2/3}, t = \sqrt[3]{5}$

(d) $x = \dfrac{2 \pm i\sqrt{11}}{15}$

5-34 (a) $-4 \leq x \leq 2$

(b) $x < -3, x > 4$

(c) $-\frac{3}{2} < x < \frac{3}{2}$

(d) $\frac{1}{3} < y < \frac{1}{2}$

(e) $x < 1, 2 < x < 3$

5-35 (a) All x

(b) $x < -3, x > \frac{1}{2}$

(c) $x > \frac{5}{2}$

(d) $x \leq -1, 0 \leq x \leq 8$

(e) $x \geq -20$

6

WORD PROBLEMS— APPLICATIONS

THIS CHAPTER IS ABOUT

- ☑ **How to Solve Word Problems**
- ☑ **Word Problems for First-Order Equations**
- ☑ **Word Problems for First-Order Inequalities**
- ☑ **Word Problems for Quadratic Equations**
- ☑ **Word Problems for Quadratic Inequalities**

6-1. How to Solve Word Problems

Algebra is more than symbol manipulation. You also have to apply the theory and techniques you learn to practical situations. In fact, applications (word) problems are the most important part of algebra. There are some general steps that can help you to solve word problems:

Step 1: Pick a letter to represent some quantity that you're looking for in the problem.

Step 2: Create other expressions, using this same letter, to represent other quantities in the problem.

Step 3: Develop an equation or inequality to connect the quantity you want with quantities you know.

> ***note:*** *Any* equation (or inequality) will do: You do *not* have to find an equation that says $x = $ [something]—that's what algebraic manipulations are for!

Step 4: Solve this equation using the algebraic techniques that are easiest for you.

Step 5: Check your answer(s) to see if your results satisfy all the conditions of the problem; i.e., make sure you've answered the right question(s).

Steps 1 through 3 are the hardest. These require the translation of English words into the language of mathematics. For example:

"The sum of x and y" $\Leftrightarrow x + y$
"Three times the sum of p and q" $\Leftrightarrow 3(p + q)$
"The larger minus the smaller" $\Leftrightarrow L - s$
"The sum of twice m and 6" $\Leftrightarrow 2m + 6$
"The sum of a and b is more than 10" $\Leftrightarrow a + b > 10$
"The sum of a number and 12 is twice the number" $\Leftrightarrow n + 12 = 2n$

6-2. Word Problems for First-Order Equations

EXAMPLE 6-1 A salesperson earned $150 more commission in the second month of employment than in the first month. In the third month the commission was double that of the second month. The three-month total commission was $1650. How much commission did the salesperson earn in the first month? the second month? the third month?

Solution

Step 1: Let x be the first month's commission.

Step 2: Then $x + 150$ was the commission in the second month and $2(x + 150)$ was the commission in the third month.

Step 3:
$$x + (x + 150) + (2x + 300) = 1650$$

Step 4:
$$4x + 450 = 1650$$
$$4x = 1200$$
$$x = 300$$

Step 5: The commissions were

$$
\begin{array}{lll}
\text{Month 1} & x = \$\ 300 \\
\text{Month 2} & x + 150 = \$\ 450 \\
\text{Month 3} & 2(x + 150) = \underline{\$\ 900} \\
& \text{Total} \quad \$1650
\end{array}
$$

Note that carefully following Steps 1–5 detailed in Section 6-1 helps to "organize" the problem and makes word problems seem easier.

EXAMPLE 6-2 Carlos can paint a house in 6 days and Juan can paint the same house in 5 days. How many days would it take to paint the house if they worked together?

Solution

Step 1: Let n be the number of days it takes for both to paint the house.

Step 2: Since Carlos needs 6 days and Juan needs 5 days, Carlos can do 1/6 of the job in one day and Juan can do 1/5 of the job in one day. And since it takes n days for both, they can do $1/n$ of the job in one day by working together. Thus

$$\begin{pmatrix} \text{Carlos' work} \\ \text{for 1 day} \end{pmatrix} + \begin{pmatrix} \text{Juan's work} \\ \text{for 1 day} \end{pmatrix} = \begin{pmatrix} \text{combined work} \\ \text{for 1 day} \end{pmatrix}$$

Step 3:
$$\frac{1}{6} + \frac{1}{5} = \frac{1}{n}$$

Step 4:
$$30\left(\frac{1}{6}\right) + 30\left(\frac{1}{5}\right) = 30\left(\frac{1}{n}\right)$$

$$5 + 6 = \frac{30}{n}$$

$$11 = \frac{30}{n}$$

$$11n = 30$$

Step 5:
$$n = \tfrac{30}{11} \text{ days or } \sim 2.72 \text{ days}$$

Sometimes two letters are needed at the beginning to identify different quantities. But additional information is always available to connect these quantities, and this new equation can be substituted appropriately to reduce the problem to one variable.

EXAMPLE 6-3   A coin-operated vending machine was emptied and found to contain \$30 in quarters and dimes. If there were 150 coins in the machine, how many were quarters and how many were dimes?

Solution

Step 1: Let d = number of dimes and q = number of quarters.
Step 2: Then

$$d + q = 150 \text{ coins} \tag{1}$$

The total value (in cents) of the dimes is $10d$ and the total value of the quarters is $25q$. Thus the total amount of money (in cents) is

$$10d + 25q = 3000 \tag{2}$$

Step 3: From the relation $d + q = 150$, you know that $d = 150 - q$; substitute this value into Eq. (2), so the money total now becomes

$$10(150 - q) + 25q = 3000 \tag{3}$$

Step 4:
$$1500 - 10q + 25q = 3000$$

$$15q = 1500$$

$$q = 100$$

and
$$d = 150 - q = 50$$

Step 5:
$$100 \text{ quarters} = \$25.00$$

$$\underline{50 \text{ dimes} \quad = \$\ 5.00}$$

$$150 \text{ coins} \quad = \$30.00$$

6-3. Word Problems for First-Order Inequalities

EXAMPLE 6-4   To be eligible for a contest, you have to be over a certain age. Twice your age plus 3 years must be more than your age plus 18 years. How old must you be?

Solution

Step 1: Let A denote your age.
Step 2: Then twice your age is $2A$ and twice your age plus 3 years is $2A + 3$.
Step 3: The conditions say that "twice your age plus 3 years must be more than your age plus 18 years," so $2A + 3 > A + 18$.

Step 4:
$$(2A + 3) - A > (A + 18) - A$$

$$A + 3 > 18$$

$$(A + 3) - 3 > (18) - 3$$

$$A > 15$$

Step 5: So you need to be *older than* 15 years to participate.

Note again that setting up Steps 1–5 helps in the solution process.

EXAMPLE 6-5   The scores on your first three tests in algebra were 88, 92, and 83. If it takes an average of 90 or above on the four semester tests to get an A for the course, how high must your score be on the last test to ensure an A?

Solution

Step 1: Let x be the score you need to get.

Step 2: Then you must have an average of $\dfrac{(88 + 92 + 83 + x)}{4}$.

Step 3:
$$\frac{88 + 92 + 83 + x}{4} \geq 90$$

Step 4:
$$88 + 92 + 83 + x \geq 360$$
$$x + 263 \geq 360$$
$$x \geq 97$$

Step 5: So a grade of $x \geq 97$ will get you an A.

6-4. Word Problems for Quadratic Equations

EXAMPLE 6-6 The sum of a number and five times its reciprocal is 6. What is the number?

Solution

Step 1: Let the number be x.
Step 2: Its reciprocal is $1/x$.
Step 3: You are told that the sum of x and five times $1/x$ is six, so
$$x + 5\left(\frac{1}{x}\right) = 6$$

Step 4:
$$(x)(x) + (x)\left(\frac{5}{x}\right) = (x)(6)$$
$$x^2 + 5 = 6x$$
$$x^2 - 6x + 5 = 0$$
$$(x - 5)(x - 1) = 0$$

Step 5: Hence the number is $x = 5$ or $x = 1$: There are *two* solutions.

Check: $5 + 5(\frac{1}{5}) \overset{?}{=} 6 \qquad 1 + 5(\frac{1}{1}) \overset{?}{=} 6$

$\qquad\quad 5 + 1 \overset{\checkmark}{=} 6 \qquad\quad 1 + 5 \overset{\checkmark}{=} 6$

When the problem produces a quadratic equation, we may use any of our available methods to find the solutions.

EXAMPLE 6-7 The length of a rectangle is 8 inches greater than its width and the area is 128 square inches. Find the length and width.

Solution

Step 1: Let x be the width in inches.
Step 2: Then $x + 8$ expresses the length.
Step 3: Since the area is length times width, you have

$$x(x + 8) = 128$$

or $\qquad\qquad\qquad\qquad x^2 + 8x = 128$

Step 4: $\qquad\qquad\qquad x^2 + 8x - 128 = 0$

Using the quadratic formula:

$$x = \frac{-8 \pm \sqrt{64 - 4(1)(-128)}}{2 \cdot 1}$$

$$= \frac{-8 \pm \sqrt{64 + 512}}{2}$$

$$= \frac{-8 \pm \sqrt{576}}{2}$$

$$= \frac{-8 \pm 24}{2}$$

Thus
$$x = \frac{-8 + 24}{2} \quad \text{or} \quad x = \frac{-8 - 24}{2}$$

$$= \frac{16}{2} \qquad\qquad\qquad = \frac{-32}{2}$$

$$= 8 \qquad\qquad\qquad\quad = -16$$

Step 5: The second value, $x = -16$, is rejected because the width cannot be negative. So

$$\text{width} = x = 8$$

$$\text{length} = x + 8 = 16$$

Check: area = (width)(length)

$$= (8)(16)$$

$$= 128$$

Even a problem from physics becomes workable with our algebra techniques.

EXAMPLE 6-8   A ball is thrown into the air with an initial velocity v_0 of 128 feet per second. The height h of the ball after t seconds is given by the formula

$$h = v_0 t - 16t^2$$

(a) How long does it take the ball to come back to the ground?
(b) How high does the ball go?

   Hint: The upward flight takes the same amount of time as the downward flight.

Solution

(a) *Step 1:* Let t be the whole amount of time it takes for the ball to reach the ground.
   Step 2: The initial velocity is given, $v_0 = 128$. And since the ground level is height zero, you can set $h = 0$.
   Step 3: Write the equation by substituting the known values into the formula:

$$128t - 16t^2 = 0$$

$$16t(8 - t) = 0$$

   Step 4: Solve for t:

$$16t = 0 \quad \text{or} \quad -t + 8 = 0$$

$$t = 0 \quad \text{or} \quad t = 8$$

Step 5: The solution values are $t = 0$, $t = 8$. The value $t = 0$ is the value of time t at the instant the ball is thrown. This value is not useful, so you disregard it. The value $t = 8$ is the amount of time it takes for the height to become zero again, i.e., the time the ball takes to reach the ground again.

Check: $128(8) - 16(8^2) \overset{?}{=} 0$

$1024 - 1024 \overset{\checkmark}{=} 0$

(b) *Step 1:* Let h be the highest point the ball reaches.

Step 2: Since part **(a)** showed the flight to take 8 seconds, the time for the upward flight is 4 seconds and the time for the downward flight is 4 seconds. Hence at $t = 4$ the ball is at the top.

Step 3: Find the maximum height by using the value $t = 4$ in the given height equation:

$$h = 128(4) - 16(4^2)$$

Step 4: $\qquad\qquad\qquad h = 512 - 16 \cdot 16 = 512 - 256$

Step 5: $\qquad\qquad\qquad h = 256$ feet at the highest point

Check: $256 \overset{?}{=} 128(4) - 16(4^2)$

$\overset{?}{=} 512 - 256$

$256 \overset{\checkmark}{=} 256$

6-5. Word Problems for Quadratic Inequalities

EXAMPLE 6-9   The roots of a quadratic equation $Ax^2 + Bx + C = 0$ are real and distinct if the discriminant $B^2 - 4AC$ is greater than zero. Find the values of the real number k for which the roots of $4x^2 - 8x = k^2 - 13$ are real and different.

Solution   The equation is $4x^2 - 8x + (13 - k^2) = 0$. The discriminant of this quadratic is

$$B^2 - 4AC = (-8)^2 - 4(4)(13 - k^2)$$

$$= 64 - 16(13 - k^2)$$

$$= 64 - (16)(13) + 16k^2$$

For this to be greater than zero, you write

$$64 - (16)(13) + 16k^2 > 0$$

$$16k^2 > (16)(13) - 64$$

$$k^2 > \frac{(16)(13)}{16} - \frac{64}{16}$$

$$k^2 > 13 - 4$$

$$k^2 > 9$$

The solution of this quadratic inequality gives your answer:

$$k^2 - 9 > 0$$

$$(k - 3)(k + 3) > 0$$

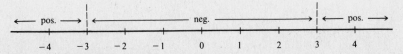

Hence we must have $k > 3$ or $k < -3$.

EXAMPLE 6-10 If a cannon ball is shot upward with initial velocity $v_0 = 64$ feet/second from the top of a building $h_0 = 20$ feet high, the equation giving the height of the ball above the ground at any time $t > 0$ is

$$h = h_0 + v_0 t - 16t^2$$

During what positive time interval is the ball more than 68 feet above the ground?

Solution   For the initial conditions of this problem the height equation is

$$h = 20 + 64t - 16t^2$$

For this to be more than 68 feet, the inequality is

$$20 + 64t - 16t^2 > 68$$

$$20 + 64t - 16t^2 - 68 > 0$$

$$-16t^2 + 64t - 48 > 0$$

$$16t^2 - 64t + 48 < 0$$

$$t^2 - 4t + 3 < 0$$

$$(t - 1)(t - 3) < 0$$

The time interval is $1 < t < 3$.

SOLVED PROBLEMS

Word Problems for First-Order Equations

PROBLEM 6-1 A basketball player, "Sorcerer" Jones, is preparing to play some exhibition ball in Europe. To complete his visa application, he needs his height in centimeters. If his height is 6 feet 7 inches, what number should he record on the form?

Solution

Step 0: Before you can begin, you need a *conversion formula*, which you can get from a math book: 1 in. = 2.54 cm.

Step 1: Let H denote height.

Step 2: Now his height is 6 ft 7 in., but in total inches that is $(6 \cdot 12 + 7)$ in. Thus

Step 3:
$$H = [(6 \cdot 12 + 7)\ \text{in.}\left(\frac{2.54\ \text{cm}}{1\ \text{in.}}\right)$$

Step 4:
$$H = (6 \cdot 12 + 7)(2.54\ \text{cm})$$

$$= (72 + 7)(2.54\ \text{cm})$$

$$= (79)(2.54\ \text{cm})$$

$$= 200.66\ \text{cm}$$

Step 5: Sorcerer should record 201 cm (rounded to three significant figures).

PROBLEM 6-2   An automatic coin counter at a bank showed that $33 in quarters and dimes had been processed in one transaction. If there were 213 coins in all, how many were quarters and how many were dimes?

Solution

Step 1: Let x be the number of quarters.

Step 2: Since there were 213 coins in all, the rest of them, $213 - x$, must be dimes. The value of the quarters is x times 25 (in cents), which is $25x$; and the value of the dimes is $(213 - x)$ times 10 (in cents), which is $10(213 - x)$. The total value is $33 or 3300 cents. Thus

Step 3:
$$25x + 10(213 - x) = 3300$$

Step 4:
$$25x + 2130 - 10x = 3300$$
$$25x - 10x = 3300 - 2130$$
$$15x = 1170$$
$$x = 78$$

Step 5: Thus, there were 78 quarters. The rest of the 213 coins were dimes, so there were $213 - x = 213 - 78 = 135$ dimes.

Check: $78 \cdot 25 + 135 \cdot 10 \overset{?}{=} 1950 + 1350 \overset{?}{=} 3300$

PROBLEM 6-3   A large rectangular crate (see Figure 6-1) has a base whose length is $3\frac{1}{2}$ times the height of the crate and whose width is $1\frac{1}{2}$ times the height. The perimeter of the smaller vertical face is 30 feet. What are the dimensions of the box?

Solution

Step 1: Let x be the height.

Step 2: Then the length is $3\frac{1}{2}$ times x or $\frac{7}{2}x$, and the width is $1\frac{1}{2}$ times x or $\frac{3}{2}x$. The small end has height $= x$ and width $= \frac{3}{2}x$, so its perimeter is $x + \frac{3}{2}x + x + \frac{3}{2}x$.

Step 3: The perimeter of the small end is equal to 30 feet. Thus

$$2(x + \tfrac{3}{2}x) = 30$$

Step 4:
$$5x = 30$$
$$x = 6$$

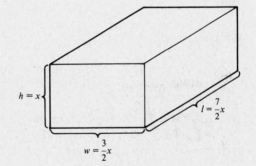

Figure 6-1

Step 5: The crate is $x = 6$ feet high. Then its length is $\frac{7}{2}x = \frac{7}{2}(6) = 21$ feet and its width is $\frac{3}{2}x = \frac{3}{2}(6) = 9$ feet. So the crate is $21' \times 9' \times 6'$.

PROBLEM 6-4   A painting is twice as tall as it is wide. If you need 15 feet of framing material to build a frame for it, what are the dimensions of the painting?

Solution

Step 1: Let w be the width.

Step 2: Then $2w$ is the height and the perimeter is the length of the framing material. Thus

Step 3:
$$w + 2w + w + 2w = 15 \text{ ft}$$

Step 4:
$$6w = 15$$
$$w = 15/6 = 2\tfrac{1}{2} \text{ ft}$$

Step 5: So the width of the painting is $w = 2\frac{1}{2}$ ft and the height is $2w = 5$ ft.

PROBLEM 6-5 The sum of three consecutive integers if 69. What are the numbers?

Solution

Step 1: Let x be the first integer.
Step 2: Then the next larger integer is $x + 1$ and the largest is $x + 2$. Thus

Step 3:
$$x + (x + 1) + (x + 2) = 69$$

Step 4:
$$3x + 3 = 69$$
$$3x = 66$$
$$x = 22$$

Step 5: Thus the consecutive numbers are $x = 22$, $x + 1 = 23$, and $x + 2 = 24$, whose sum is 69.

PROBLEM 6-6 The formula for temperature conversion from Fahrenheit (T_F) to Celsius (T_C) is

$$T_C = \tfrac{5}{9}(T_F - 32)$$

At what value do the two scales have the same reading?

Solution

Step 1: Let S be the temperature for which the Celsius reading is the same as the Fahrenheit reading.
Step 2: Then $T_C = T_F = S$. Thus

Step 3:
$$S = \tfrac{5}{9}(S - 32)$$

Step 4:
$$9S = 5(S - 32) = 5S - 160$$
$$4S = -160$$
$$S = -40$$

Step 5: Thus $S = T_F = T_C = -40°F = -40°C$—and this is the *only* value for which this condition is true.

PROBLEM 6-7 On a celery farm, the packers are paid 25 dollars per day, while their supervisors earn 40 dollars per day. The farm's daily payroll for 29 employees is 815 dollars. How many packers are employed? how many supervisors?

Solution

Step 1: Let x be the number of packers.
Step 2: Then the number of supervisors is $29 - x$.
Step 3: The payroll equation then is

$$(25)(x) + 40(29 - x) = 815$$

Step 4:
$$25x + 1160 - 40x = 815$$
$$-15x = 815 - 1160 = -345$$
$$15x = 345$$
$$x = \tfrac{345}{15} = 23$$

Step 5: So there are $x = 23$ packers and $29 - x = 6$ supervisors.

PROBLEM 6-8 A truck leaves a turnpike rest stop at 12:15 going east at 50 mph. A car leaves the same stop at 12:30 and goes east at 65 mph. How long does it take the car to overtake the truck? [*Hint:* Use the familiar relationship (distance) = (rate) × (time).]

Solution

Step 1: Both the truck and car will travel the same distance from the rest stop to the overtake point: Call that distance D. Then let T be the travel time (in minutes) of the truck.

Step 2: Since the car (going 65 mph) left 15 minutes later than the truck (going 50 mph), you can use the distance formula to write

$$\text{Truck:} \quad D = (50 \text{ mph})(T) = 50T$$

$$\text{Car:} \quad D = (65 \text{ mph})(T - 15) = 65(T - 15)$$

Step 3: Setting the expressions on the right—BOTH EQUAL TO D—equal to each other, you can denote the overtake point as

$$50T = 65(T - 15)$$

Step 4:
$$50T = 65T - 65(15)$$

$$50T - 65T = -975$$

$$-15T = -975$$

$$T = -975/-15 = 65$$

Step 5: Since the truck travels for $T = 65$ minutes, the car travels for $65 - 15 = 50$ minutes before it overtakes the truck.

PROBLEM 6-9 Eli needed some cotton thread. He bought $\frac{1}{3}$ of what he needed and then later bought 10 feet more. Now he has 5 feet less than $\frac{3}{4}$ of what he needs. How much thread did he actually need?

Solution

Step 1: Let x be the number of feet of thread he needs.
Step 2: The purchases were $\frac{1}{3}x + 10$. This amount is $\frac{3}{4}x - 5$. Thus

Step 3:
$$\frac{1}{3}x + 10 = \frac{3}{4}x - 5$$

Step 4:
$$\frac{1}{3}x - \frac{3}{4}x = -5 - 10$$

$$12(\frac{1}{3}x - \frac{3}{4}x) = 12(-15)$$

$$4x - 9x = -180$$

$$-5x = -180$$

$$x = -180/-5 = 36$$

Step 5: So Eli needed $x = 36$ feet of thread.

PROBLEM 6-10 I spent $\frac{1}{3}$ of my monthly salary on a new stereo. After I had spent 70 dollars more on clothes, I found that half of my salary had been spent. What is my monthly salary?

Solution

Step 1: Let s be my monthly salary.
Step 2: I have spent $\frac{1}{3}s + 70$ dollars and this is equal to $\frac{1}{2}s$. Thus

Step 3:
$$\frac{1}{3}s + 70 = \frac{1}{2}s$$

Step 4:
$$\frac{3}{6}s - \frac{2}{6}s = \frac{1}{6}s = 70$$

$$6(\frac{1}{6}s) = 6(70)$$

$$s = 420$$

Step 5: So my monthly salary is $s = \$420$.

PROBLEM 6-11 Mary can rake a large yard in 3 hours and Jack can do the same yard in 5 hours. How much of the raking job can the two of them together finish in an hour?

Solution

Step 1: Let n be the part of the job that both working together can finish in 1 hour.
Step 2: In 1 hour Mary can do $\frac{1}{3}$ of the job and Jack can do $\frac{1}{5}$ of the job. Thus

Step 3:
$$\tfrac{1}{3} + \tfrac{1}{5} = n$$

Step 4:
$$\tfrac{1}{3}(15) + \tfrac{1}{5}(15) = (15)n$$
$$5 + 3 = 15n$$
$$15n = 8$$
$$n = \tfrac{8}{15}$$

Step 5: So Mary and Jack can finish 8/15 of the job in 1 hour.

PROBLEM 6-12 Ryan and Maria can run around a quarter-mile track in 50 seconds and 60 seconds, respectively. If they start at the same time and continue at the same pace, in how many minutes will Ryan pass Maria?

Solution

Step 1: Let t = number of minutes until the pass takes place.
Step 2: The distance around the track is $\tfrac{1}{4}$ mile, so their respective speeds (in miles per minute) are

$$\text{Ryan:}\quad \text{speed} = \frac{\text{distance}}{\text{time}} = \frac{\tfrac{1}{4}\,\text{mi}}{\tfrac{5}{6}\,\text{min}} = \frac{6}{20}\,\text{mi/min}$$

$$\text{Maria:}\quad \text{speed} = \frac{\text{distance}}{\text{time}} = \frac{\tfrac{1}{4}\,\text{mi}}{1\,\text{min}} = \frac{1}{4}\,\text{mi/min}$$

They will pass when Ryan has run one track-length ($\tfrac{1}{4}$ mi) more than Maria runs, so Ryan's distance must be $\tfrac{1}{4}$ mi more than Maria's. Hence

$$(\text{Ryan's speed}) \cdot (\text{time } t) = (\text{Maria's speed}) \cdot (\text{time } t) + \tfrac{1}{4}.$$

Step 3:
$$\text{Thus}\quad (\tfrac{6}{20})(t) = (\tfrac{1}{4})(t) + \tfrac{1}{4}$$

Step 4:
$$\tfrac{6}{20}t - \tfrac{5}{20}t = \tfrac{1}{4}$$
$$\tfrac{1}{20}t = \tfrac{1}{4}$$
$$t = \tfrac{20}{4} = 5$$

Step 5: Ryan will pass Maria in $t = 5$ minutes.

PROBLEM 6-13 Jose is 4 years older than Kim. In 6 years the sum of their ages will be 46. What are their ages now?

Solution

Step 1: Let x be Kim's age now.
Step 2: In 6 years Kim will be $x + 6$. Since Jose is 4 years older than Kim, Jose is now $x + 4$ and in 6 years Jose will be $x + 10$. The sum of these ages, 6 years from now, is 46. Thus

Step 3:
$$(x + 6) + (x + 10) = 46$$

Step 4:
$$2x + 16 = 46$$
$$2x = 30$$
$$x = 15$$

Step 5: Kim is $x = 15$ years old now and Jose is $15 + 4 = 19$ years old.

PROBLEM 6-14 A car radiator holds 16 quarts of liquid. The present mixture has 20% antifreeze and 80% water. How much of the present mixture should be removed and replaced by pure antifreeze to raise the antifreeze percentage to 35%?

Solution

Step 1: Let V be the volume (in quarts) of the 20% mixture that must be removed.

Step 2: The volume of antifreeze in the final mixture must be 35% of 16 quarts, or $(.35)16 = 5.6$ quarts. After the mixture has had V quarts removed, its new volume is $16 - V$ quarts, which is 20% antifreeze—so that amount of antifreeze in the remaining mixture is 20% of $(16 - V)$, or $(.20)(16 - V)$. And, finally, when V quarts of antifreeze is added, there must be a total of 5.6 quarts of antifreeze in the 16-quart radiator. Thus

Step 3:
$$(.20)(16 - V) + V = 5.6$$

Step 4:
$$3.2 - .20V + V = 5.6$$
$$3.2 + .80V = 5.6$$
$$.80V = 2.4$$
$$V = 2.4/.80 = 3$$

Step 5: So $V = 3$ quarts of the original 20% mixture must be removed.

PROBLEM 6-15 The We-Makum-Good Corporation purchased a special Widgematic machine to manufacture widgets at an investment cost of 4000 dollars. For every 1000 widgets made on the Widgematic, the labor and materials plus overhead cost is 8 dollars.

(a) Find a linear polynomial whose value at any particular value of x gives the total cost of producing x thousand widgets.
(b) What is the cost of producing 20 000 widgets? 100 000?
(c) How many widgets can be produced for 8000 dollars?

Solution

(a) Total cost = cost of Widgematic + manufacturing costs

$$= (4000 \text{ dollars}) + (x)(8 \text{ dollars})$$

So the polynomial C is

$$C = 4000 + 8x$$

(b) For 20 000 widgets, the value of $x = 20$ (i.e., 20 thousand), so

$$C = 4000 + 8(20) = 4000 + 160 = 4160 \text{ dollars}$$

For 100 000, the value of $x = 100$, so

$$C = 4000 + 8(100) = 4800 \text{ dollars}$$

(c) For 8000, the x value is unknown, so

$$8000 = 4000 + 8x$$
$$4000 = 8x$$
$$x = \frac{4000}{8} = 500$$

Thus the corporation gets 500 000 widgets for 8000 dollars.

Word Problems for First-Order Inequalities

PROBLEM 6-16 If a rock is thrown down from the top of a tall tower with initial velocity v_0 of 8 feet/second (ft/s), the velocity it has at time t is given by $v = v_0 + 32t$. After how much time is the rock traveling more than 60 ft/s?

Solution

Step 1: The velocity equation here is given by the formula:

$$v = v_0 + 32t$$
$$v = 8 + 32t$$

Step 2: In order to have velocity more than 60 ft/s, you have

$$8 + 32t > 60$$

Step 3:
$$32t > 52$$

$$t > \tfrac{52}{32} = 1\tfrac{5}{8}$$

Step 5: After $t > 1\tfrac{5}{8}$ s, i.e., time greater than $1\tfrac{5}{8}$ s, the rock is traveling more than 60 ft/s.

PROBLEM 6-17 Company X will rent you a car for $25 per day. Company Y charges $16 per day plus a flat fee of $45. When is it cheaper to rent from Company Y?

Solution

Step 1: Let x be the number of days you rent.
Step 2: The cost to rent from X is $25x$. The cost to rent from Y is $16x + 45$.
Step 3: It is cheaper to rent from Y if

$$16x + 45 < 25x$$

Step 4:
$$16x + 45 - 16x < 25x - 16x$$

$$45 < 9x$$

$$5 < x$$

Step 5: At exactly 5 days the costs will be

$$\$25 \times 5 = \$125 \text{ from X}$$

and
$$(\$16 \times 5) + \$45 = \$80 + \$45 = \$125 \text{ from Y}$$

Any number of days over 5, the additional cost for Y is $16 per day versus the $25 per day from X. This is a savings of $9 per day.

Word Problems for Quadratic Equations

PROBLEM 6-18 One positive number is three less than twice another. The sum of their squares is 41. Find the two numbers.

Solution

Step 1: Let the second number be x.
Step 2: Then the first number, which is three less than twice that number, is represented by $2x - 3$.
Step 3: The sum of their squares is

$$(x)^2 + (2x - 3)^2 = 41$$

Step 4:
$$x^2 + 4x^2 - 12x + 9 = 41$$

$$5x^2 - 12x - 32 = 0$$

$$(5x + 8)(x - 4) = 0$$

Thus
$$x = 4 \quad \text{or} \quad x = -8/5$$

Step 5: Checking these using $x = 4$, you find that the first number is $2(4) - 3 = 5$ and the second number is $x = 4$. The sum of the squares is

$$25 + 16 = 41 \qquad \text{This one works.}$$

The second, $x = -8/5$, is not positive—as the conditions require—so it is rejected.

PROBLEM 6-19 The square of a number is 6 less than five times the number. What is the number?

Solution

Step 1: Let the number be x.

Steps 2 and 3:
$$x^2 = 5x - 6$$

Step 4:
$$x^2 - 5x + 6 = 0$$
$$(x - 3)(x - 2) = 0$$
$$x - 3 = 0 \quad \text{or} \quad x - 2 = 0$$
$$x = 3 \qquad\qquad x = 2$$

Step 5: The number is $x = 3$ or $x = 2$.

PROBLEM 6-20 You are required to construct an open box whose base is a square and whose height is 4 inches shorter than twice its length.

(a) Find the polynomial that expresses the total surface area of the box if the square base has sides of z inches.

(b) What is the total surface area of the box if the base is $10'' \times 10''$?

(c) What would the dimensions be if the total surface area of the box is 1540 in²?

Solution

(a) Since the base has dimension z, the height (in inches) is $2z - 4$. Each of the four sides of the box has area $z(2z - 4) = 2z^2 - 4z$ and the base has area $z(z) = z^2$.

$$\text{Total area} = A_T = 4(2z^2 - 4z) + z^2$$
$$= 9z^2 - 16z$$

(b) If the base is $10'' \times 10''$, then $z = 10$ in. So

$$A_T = 9(10^2) - 16(10) = 900 - 160 = 740 \text{ in}^2$$

(c) Given the total surface area A_T as 1540, you have $1540 = 9z^2 - 16z$, or

$$9z^2 - 16z - 1540 = 0$$

so that
$$z = \frac{16 \pm \sqrt{(-16)^2 - 4(9)(-1540)}}{2 \cdot 9} = \frac{16 \pm \sqrt{256 + 55\,440}}{18}$$

$$= \frac{16 \pm \sqrt{55\,696}}{18} = \frac{16 \pm 236}{18}$$

Thus
$$z = \frac{16 - 236}{18} \quad \text{or} \quad z = \frac{16 + 236}{18}$$

The first value is discarded, as the length of the side cannot be negative. So

$$z = \frac{16 + 236}{18} = \frac{252}{18} = 14$$

Hence the box is $14''$ long by $z = 14''$ wide by $2(z) - 4 = 28 - 4 = 24''$ high.

PROBLEM 6-21 An object is dropped from a hot-air balloon that is 1000 feet above ground level. The height h of the object t seconds into its fall is given by the polynomial

$$h(t) = -16t^2 + 1000$$

note: It should be clear that when time $t = 0$, then height $h = 1000$. At any OTHER time value, i.e., $t > 0$, the height must be calculated from the given polynomial.

(a) How long does it take for the object to hit the ground?

(b) At what time value (in seconds) is the object 500 feet above the ground?

Solution

(a) At the ground, $h = 0$, so

$$-16t^2 + 1000 = 0$$

$$16t^2 = 1000$$

$$t^2 = \frac{1000}{16}$$

$$t = \pm\sqrt{\frac{1000}{16}} = \pm\frac{\sqrt{100}\sqrt{10}}{\sqrt{16}} = \pm\frac{10\sqrt{10}}{4}$$

The negative value of time is discarded, so the solution is the final time (t_f):

$$t_f = \frac{10\sqrt{10}}{4} = \frac{5\sqrt{10}}{2} \simeq 7.9 \text{ seconds}$$

(b) When the height is 500 feet,

$$500 = -16t^2 + 1000$$

$$16t^2 = 1000 - 500 = 500$$

$$t^2 = \frac{500}{16}$$

$$t = \pm\sqrt{\frac{500}{16}} = \pm\frac{10\sqrt{5}}{4} = \pm\frac{5\sqrt{5}}{2}$$

So the solution is the positive time value

$$t = \frac{5\sqrt{5}}{2} \simeq 5.59 \text{ seconds}$$

Notice that the first half of the fall (from height 1000 to height 500) takes a much longer period of time than the second half (from 500 to 0). This is because the object is speeding up as it falls.

PROBLEM 6-22   Mr. Hsu had a piece of property bounded on one side by a river. When he enclosed a rectangular subportion of the property adjacent to the river, he used 500 feet of fencing for three sides of the rectangle and let the river be the fourth side.

(a) Find a polynomial to express the area of the rectangular field enclosed by the fence if the width of the enclosure is w feet.
(b) What should the dimensions of the enclosure be if the area is 30 000 ft^2?

Solution

(a) If the length of the fencing on three sides of the rectangle is 500 ft, the width of the enclosure is w ft, and the length of the enclosure is l ft, then

$$w + w + l = 500$$

$$l = 500 - 2w$$

Then the area A is

$$A = l \cdot w$$

$$= (500 - 2w)w$$

$$= 500w - 2w^2$$

(b) To find the dimensions l and w, use their relationship $2w + l = 500$ and the derived formula

$A = 500w - 2w^2$. Set $A = 30\,000$ ft^2, so that

$$30\,000 = 500w - 2w^2$$

$$2w^2 - 500w + 30\,000 = 0$$

$$w^2 - 250w + 15\,000 = 0$$

Now use the quadratic formula:

$$w = \frac{250 \pm \sqrt{(-250)^2 - 4(1)(15\,000)}}{2}$$

$$= \frac{250 \pm \sqrt{62\,500 - 60\,000}}{2}$$

$$= \frac{250 \pm \sqrt{2500}}{2}$$

$$= \frac{250 \pm 50}{2}$$

Thus

$$w = \frac{300}{2} \quad \text{or} \quad w = \frac{200}{2}$$

$$= 150 \qquad\qquad = 100$$

Then the corresponding length l for each of these is

$$l = 500 - 2w \qquad \text{or} \qquad l = 500 - 2(100)$$

$$l = 500 - 2(150) \qquad\qquad l = 500 - 200$$

$$l = 200 \qquad\qquad\qquad l = 300$$

So we find that the field can be 150 ft wide and 200 ft long, or 100 ft wide and 300 ft long.

Word Problems for Quadratic Inequalities

PROBLEM 6-23   A manufacturer of video cassettes finds that when x units are made and sold per day, the profit P (in dollars) is given by

$$P = x^2 - 130x - 3000$$

(a) At least how many cassettes must be made and sold daily to make a profit?
(b) For what value of x is the company losing money? At what value does it break even?

Solution

(a) To make a profit, P must be positive, so the company must have x as a solution to

$$x^2 - 130x - 3000 > 0$$

$$(x - 150)(x + 20) > 0$$

The inequality is satisfied when $x > 150$ or $x < -20$. The negative value is discarded as impossible, so the company makes a profit when it sells more than 150 units a day.
(b) The company loses money if production is between 0 and 150 units a day. At exactly 150 units the profit is zero.

PROBLEM 6-24 The sum of the squares of two consecutive positive integers is greater than or equal to 41. Find the number pairs for which this condition is true.

Solution

Step 1: Let one of the consecutive positive integers be x.

Step 2: Let the other integer be $x + 1$.

Step 3: The sum of their squares is

$$x^2 + (x + 1)^2 \geq 41$$

Step 4:

$$x^2 + (x^2 + 2x + 1) \geq 41$$

$$2x^2 + 2x + 1 \geq 41$$

$$2x^2 + 2x \geq 40$$

$$2x^2 + 2x - 40 \geq 0$$

$$x^2 + x - 20 \geq 0$$

$$(x + 5)(x - 4) \geq 0$$

Step 5: The solution set to the inequality is $x \leq -5$ or $x \geq 4$. But the numbers must be positive, so you discard $x \leq -5$ and retain the set $x \geq 4$. Hence the number pairs satisfying the condition are

$$(x, x + 1) = (4, 5), (5, 6), (6, 7), (7, 8), \ldots, (n, n + 1), \ldots \qquad \text{(where } n \text{ is a positive integer)}$$

PROBLEM 6-25 For what values of x will $\sqrt{35 - 17x + 2x^2}$ be an imaginary number?

Solution The expression is non-real when the quadratic quantity is less than zero:

$$35 - 17x + 2x^2 < 0$$

$$(5 - x)(7 - 2x) < 0$$

The values in the interval $\frac{7}{2} < x < 5$ will make $\sqrt{35 - 17x + 2x^2}$ an imaginary (complex) number.

Supplementary Exercises

PROBLEM 6-26 The sum of two consecutive odd integers is 32. What are the numbers?

PROBLEM 6-27 An Olympic rowing team can row at 12 mph in still water. If they row upstream in 30 minutes and return rowing downstream in 10 minutes, **(a)** what is the rate of the current in the stream? **(b)** How much distance did the team cover? [*Hint:* (distance) = (rate)(time).]

PROBLEM 6-28 If a car salesperson makes 150 dollars per week in salary and 40 dollars in commission on each car sold, how many cars must she sell in a week to have a gross income of more than 400 dollars a week?

PROBLEM 6-29 One positive integer is 4 more than twice another positive integer, and the product of the integers is 70. Find these integers.

PROBLEM 6-30   The sum of an integer and twice its reciprocal is $\frac{9}{2}$. Find the number.

PROBLEM 6-31   In a right triangle, the longest side is twice the shortest side. The third side is 8 inches. Find the lengths of the unknown sides. [*Hint:* Use the Pythagorean theorem.]

PROBLEM 6-32   An object is thrown straight up from the ground with an initial speed of 64 feet/second. The formula for the height above the ground at any time t is $h(t) = 64t - 16t^2$. **(a)** When is the object 48 feet high? **(b)** When does the object return to the ground?

PROBLEM 6-33   A boy can ride his bicycle 4 times as fast as he can walk. On a 2-hour trip to grandmother's, he rode for 12 miles and walked 3 miles. What is his bicycle-riding speed?

PROBLEM 6-34   When 5 is subtracted from 4 times a number, the result is the same as the sum of the number and 4. Find the number.

PROBLEM 6-35   Nancy is four years older than Deborah. In 10 years the sum of their ages will be 64. How old are they now?

PROBLEM 6-36   The sum of two consecutive integers is 13 less than 5 times the larger. What are the integers?

PROBLEM 6-37   The price of a rug is directly proportional to its area. If a circular rug 5 feet in diameter sells for 60 dollars, what would the price be for a rug 9 feet in diameter? [*Hint: $A = \pi r^2$.*]

PROBLEM 6-38   The sum of two numbers is 3 and the difference of their squares is 4. What are the numbers?

PROBLEM 6-39   The monthly profit from a microwave-oven sales outlet is given by $P = 12S - .04S^2 - 50$ if the sales S are under 250 units.

(a) How much profit is there on sales of 160 units in a month?
(b) How many units must be sold to avoid a loss?
(c) How many units must be sold to earn a profit of 750 dollars?

PROBLEM 6-40   If your rectangular garden is to have an area of 600 ft^2 and you have 120 ft of fence to enclose it, what should its dimensions be?

PROBLEM 6-41   The sum of three numbers is 27. The sum of the first two is 14 while the sum of the last two is 21. What are they?

PROBLEM 6-42   A farmer pays the same total amount for 6 cows as he does for 14 goats. If a cow costs 120 dollars more than a goat, what is the cost of **(a)** a cow? **(b)** a goat?

PROBLEM 6-43   The product of two integers is 496. One integer is one less than twice the other. What are they?

PROBLEM 6-44   Find two consecutive integers whose product is 272.

PROBLEM 6-45   In manufacturing a certain item, company X figures that its profit P (in hundreds of dollars) is given by $P = (8 - n)^2 - 64$, where n is the number of items produced. For what values of n will the company make a profit?

PROBLEM 6-46   A retired math professor has \$16000 to invest. She wants to put part of it in a one-year certificate of deposit (C.D.) at 14% and the remainder in a money market (M.M.) fund at 12%. If her expected profit for the first year is \$2000, how much should she put into each fund?

PROBLEM 6-47   If a pool can be filled by the intake pipe in 6 hours and can be emptied by the drain pipe in 8 hours, how long would it take to fill the pool if both pipes are open?

PROBLEM 6-48   Nolan can throw a baseball straight up with an initial velocity of 128 feet per second. The height that the ball reaches is given by the formula $h = 128t - 16t^2$. **(a)** During what time interval will the ball be more than 50 feet above the ground? **(b)** How high will the ball go?

PROBLEM 6-49   Walt and Sharon rent a car for a given day at an agency that charges $24 plus 11 cents per mile driven. They must keep the total cost for the rental under $45. What is the maximum number of miles they may drive?

PROBLEM 6-50   The denominator of a fraction is one larger than three times the numerator. If 10 is added to the numerator and 32 is added to the denominator, the value of the fraction is unchanged. What is this fraction?

PROBLEM 6-51   A certain number is 5 less than twice another. If their sum is 19, what are the two numbers?

PROBLEM 6-52   A publishing company spends $11 000 on editorial expenses, and $9 per book on printing, binding, and other expenses in the manufacture of a biology text. The book will wholesale to bookstores for $17.50. How many copies must be sold to show a profit?

PROBLEM 6-53   Find three consecutive even integers whose sum is 54.

PROBLEM 6-54   A new rectangular swimming pool has dimensions $16' \times 40' \times 5'$ and is being filled with water at the rate of 10 ft/min.

(a) What is the formula that will find the amount of water in the pool t minutes after the valve was opened?
(b) Find the volume of water in the pool after 1 hour.
(c) What is the depth of the water after 4 hours?
(d) How many minutes will be required to fill the pool?

Answers to Supplementary Exercises

6-26   15 and 17

6-27   **(a)** 6 mph
     (b) 6 mi

6-28 $x > 6$

6-29   5, 14

6-30 $x = 4$

6-31 $\dfrac{8}{\sqrt{3}}$ and $\dfrac{16}{\sqrt{3}}$

6-32   **(a)** $t = 1$ and $t = 3$
     (b) $t = 4$

6-33   12 mph

6-34 $n = 3$

6-35   24 and 20

6-36   3 and 4

6-37 $194.40

6-38 $\frac{13}{6}$ and $\frac{5}{6}$

6-39   **(a)** $846
     (b) 5
     (c) 100 or 200

6-40 $30 - 10\sqrt{3}$ by $30 + 10\sqrt{3}$

6-41   6, 8, and 13

6-42   **(a)** 210
     (b) 90

6-43   16 and 31

6-44   16 and 17

6-45 $n > 16$

6-46   C.D. = $4000
     M.M. = $12 000

6-47   24 hours

6-48   **(a)** .412 to 7.588
     (b) 256 feet

6-49   190

6-50 $\frac{5}{16}$

6-51   8 and 11

6-52   1295

6-53   16, 18, and 20

6-54   **(a)** $v = 10t$ **(c)** 3.75 ft
     (b) 600 ft^3 **(d)** 320

EXAM 2 (Chapters 4–6)

1. Solve and check
   (a) $2(x - 3) = 5(x + 4)$
   (b) $(2x - 1)^2 = 4(x - 3)^2$
   (c) $|2x - 5| = 7$
   (d) $4x^2 = -\frac{25}{4} - 10x$

2. Solve and graph
   (a) $|3 - 4x| \geq 9$
   (b) $|4x + 1| < 7$
   (c) $4(x + 3) \geq 2(5 - x) + 14$
   (d) $|\frac{1}{3} - x| < \frac{4}{3}$

3. Solve by factoring
   (a) $x^2 + 2x - 24 = 0$
   (b) $4x^3 - 6x^2 = 0$
   (c) $5(y - 1) + 8 = 8y^2$
   (d) $4t \cdot (2 + t) = -4t - 8$

4. Solve by completing the square: (a) $x^2 + 6x - 5 = 0$ and (b) $x^2 - 12x + 1 = 0$.

5. Solve by using the quadratic formula: (a) $4x^2 + 4x + 1 = 0$ and (b) $5x - 2 + 3x^2 = 0$.

6. Solve the following radical equations:
   (a) $v + 2 = 2\sqrt{4v - 7}$
   (b) $\sqrt{x + 6} = 7 - \sqrt{x - 1}$
   (c) $x - 2\sqrt{x} - 3 = 0$
   (d) $\sqrt[3]{5x + 8} = 2$

7. Solve and graph
   (a) $x^2 - 6x \geq 3x - 20$
   (b) $y^3 - 8y^2 + 16y \geq 0$
   (c) $(x + 4)(x - 3)(2x + 5) > 0$
   (d) $t^2 + 25 \leq 10t$

8. Debbie is 3 years older than Michelle. Thirty years from now, the sum of their ages will be 105. How old are they now?

9. Separate 32 into two parts so that 5 times the smaller minus three times the larger is equal to 24.

10. A piggy bank contains $4.10 in nickels and dimes. If there are 10 more nickels than dimes, how many of each kind does it contain?

11. At 3 PM a plane leaves San Diego for Chicago flying at an average speed of 600 mph. Two hours later, a plane leaves Chicago for San Diego and travels at an average speed of 400 mph. If the cities are 1700 miles apart, at what time do the planes pass each other?

12. The perimeter of a rectangular patio is 94 feet and the difference between the length and width is 17 feet. What are the dimensions of the patio?

13. If it takes Mark 3 hours to shovel the snow from the driveway and Dan $2\frac{1}{2}$ hours, how long would it take for both to do it together, one shovel each, no timeouts for snowball fights?

14. Suppose Vicki joins Mark and Dan in the snow job. Alone, Vicki can clear the driveway in 5 hours. With all three shoveling, how long will it take?

15. Little Ms. Hood, strolling at 1.5 miles per hour, sets out for Grandma's house in the forest. Forty minutes later, Mother Hood discovers that Ms. Hood forgot her basket of goodies and hikes after her at 3.5 miles per hour. How long will it take for Mother Hood to catch up?

Answers to Exam 2

1. (a) $x = -\frac{26}{3}$ (c) $x = -1, x = 6$

   (b) $x = \frac{7}{4}$ (d) $x = -\frac{5}{4}$

2. (a) $x \leq -\frac{3}{2}, x \geq 3$

   (b) $-2 < x < \frac{3}{2}$

   (c) $x \geq 2$

   (d) $-1 < x < \frac{5}{3}$

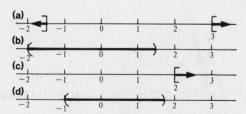

Figure E2-1

3. (a) $x = -6, x = 4$ (c) $y = -\frac{3}{8}, y = 1$

   (b) $x = 0, x = \frac{3}{2}$ (d) $t = -1, t = -2$

4. (a) $x = -3 \pm \sqrt{14}$ (b) $x = 6 \pm \sqrt{35}$

5. (a) $x = -\frac{1}{2}$ (b) $x = -2, x = \frac{1}{3}$

6. (a) $v = 4, v = 8$ (c) $x = 9$

   (b) $x = 10$ (d) $x = 0$

7. (a) $x \leq 4, x \geq 5$ (c) $-4 < x < -\frac{5}{2}, x > 3$

   (b) $y \geq 0$ (d) $t = 5$

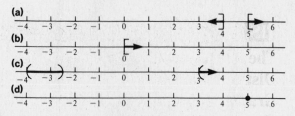

Figure E2-2

8. 21 and 24

9. 15 and 17

10. 24 dimes, 34 nickels

11. 5:30 PM

12. 15 ft by 32 ft

13. $\frac{15}{11}$ hours

14. $\frac{15}{14}$ hours

15. $\frac{1}{2}$ hour

7 GRAPHING: Lines, Equations, and Inequalities

THIS CHAPTER IS ABOUT

☑ **The Cartesian Coordinate System**
☑ **Distance Formulas**
☑ **Graphing Equations in Two Variables**
☑ **Graphing First-Degree Equations in Two Variables**
☑ **Graphing First-Degree Inequalities in Two Variables**

7-1. The Cartesian Coordinate System

A. Plotting points

A set of real numbers is graphed by locating points on the **real number line**.

EXAMPLE 7-1   Graph the set $A = \{-1, 1, 2, 3\}$ on the real number line.

Solution   Locate the points $-1, 1, 2,$ and 3:

But there can be more than one number line. If we draw one number line horizontally and one number line vertically, we obtain the **rectangular coordinate plane**. The point at which the vertical and horizontal lines cross is called the **origin** (O or 0) of the coordinate plane, and the lines are called **axes**. On the horizontal axis, called the **x-axis**, the positive numbers are to the right of the origin. On the vertical axis, called the **y-axis**, the positive numbers are above the origin. (See Figure 7-1.)

The points that make up the coordinate plane are labeled by a pair of real numbers (x, y)—such as $(3, 2)$ or $(1, 4)$—called an **ordered pair**. For every point in the plane, there is a corresponding ordered pair of real numbers, and for every ordered pair of real numbers, there is a corresponding point in the plane. The first member of the pair indicates position with respect to the horizontal axis and is called the **x-coordinate** or **abscissa** of the point. The second member of the pair indicates position with respect to the vertical axis and is called the **y-coordinate** or **ordinate** of the point.

- The association of points in a plane with ordered pairs of real numbers is called the **Cartesian coordinate system** and the rectangular coordinate plane is called the **Cartesian plane**.

To *graph* or *plot* a point in the Cartesian plane, we draw the axes, label them x and y, and assign positive and negative numbers to points on the axes (see Figure 7-1). Next, we locate the point from its x- and y-coordinates. We begin at the origin and move x units to the right or left and y units upward or downward. Then we draw a dot where we end up and label the point with its ordered pair (x, y).

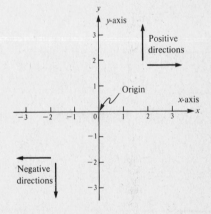

Figure 7-1. The rectangular coordinate plane.

note: The unit distances on the two axes do not have to be the same. Although we usually draw the axes on the same scale (i.e., make the distance between 0 and 1, 1 and 2, etc., the same on both axes), we can adjust the scales to fit our needs. [Imagine trying to plot the point $P(1, 100)$ on the graph in Figure 7-1!]

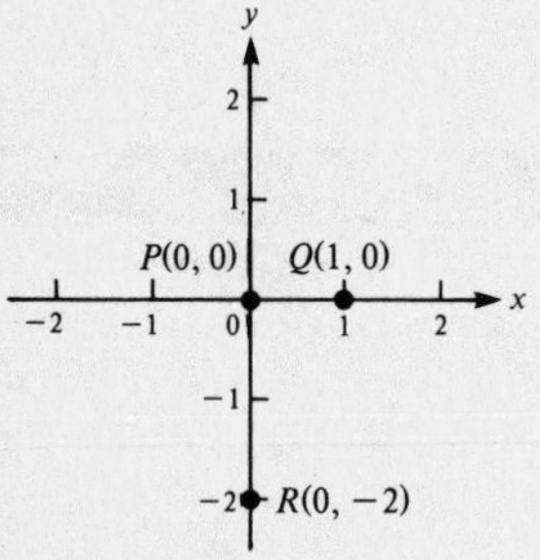

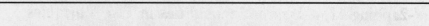

EXAMPLE 7-2   Plot the points $P(0, 0)$, $Q(1, 0)$, and $R(0, -2)$.

Solution   See Figure 7-2.

$P(0, 0)$ is the origin (we move 0 units—in any direction—from the origin).
$Q(1, 0)$ is located on the x-axis (we move 1 unit to the right from the origin but do not move upward or downward at all).
$R(0, -2)$ is located on the y-axis (we move 2 units downward from the origin but do not move to the left or right at all).

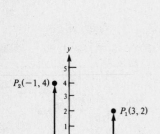

Figure 7-2

We can now make three generalizations about points in the Cartesian plane:

(1) The origin is denoted by the ordered pair $(0, 0)$.
(2) Any point ON the x-axis has a y-coordinate of zero and is of the form $(x, 0)$.
(3) Any point ON the y-axis has an x-coordinate of zero and is of the form $(0, y)$.

EXAMPLE 7-3   Plot the points $P_1(3, 2)$ and $P_2(-1, 4)$.

Solution   For $P_1(3, 2)$ we begin at the origin and move 3 units to the right and 2 units upward. For $P_2(-1, 4)$ we begin at the origin and move 1 unit to the left and 4 units upward (see Figure 7-3).

Figure 7-3

EXAMPLE 7-4   Give the ordered pairs (x, y) for the points labeled A, B, C, and D in Figure 7-4.

Solution   The pairs are $A(2, 1)$, $B(-2, -3)$, $C(-\frac{5}{2}, 2)$, and $D(4, -2)$.

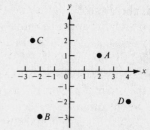

Figure 7-4

B. Quadrants

The coordinate axes divide the plane into four parts, called **quadrants**, which are numbered by roman numerals I, II, III, and IV, beginning at the upper right and numbering counterclockwise. The axes themselves are not included in any quadrant; they just form the boundaries.

note: The quadrants can also be characterized by the signs of the coordinates:

Quadrant I	$x > 0, y > 0$
Quadrant II	$x < 0, y > 0$
Quadrant III	$x < 0, y < 0$
Quadrant IV	$x > 0, y < 0$

EXAMPLE 7-5 Determine in which quadrant each of the following points lies:
$A(4, -2)$ $B(-3, 1)$ $C(-1, -1)$ $D(3, \frac{1}{3})$ $E(-2.6, 3.7)$ $F(5, 0)$ $G(0, -2)$

Solution See Figure 7-5.

$A(4, -2)$ lies in Quadrant IV, since its x-coordinate is positive and its y-coordinate is negative.

$B(-3, 1)$ lies in Quadrant II, since its x-coordinate is negative and its y-coordinate is positive.

$C(-1, -1)$ lies in Quadrant III, since both its x- and y-coordinates are negative.

$D(3, \frac{1}{3})$ lies in Quadrant I, since both its x- and y-coordinates are positive.

$E(-2.6, 3.7)$ lies in Quadrant II, since its x-coordinate is negative and its y-coordinate is positive.

$F(5, 0)$ does not lie in any quadrant, since its position is on the x-axis.

$G(0, -2)$ does not lie in any quadrant, since its position is on the y-axis.

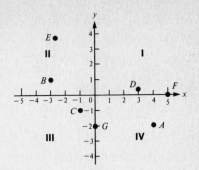

Figure 7-5

7-2. Distance Formulas

A. Finding the distance between two points

Now we can derive a formula for calculating the distance d between two points in the Cartesian coordinate plane. We already know that we can find the distance between two points on the number line by taking the absolute value of their difference (see Chapter 4). We can extend this formula to points in the plane.

First let's look at the distance between points that fall on the same horizontal or vertical line in the plane (see Figure 7-6).

- The distance between two points that fall on the *same horizontal line* is the absolute value of the difference of their x-coordinates: $d = |x_2 - x_1|$.
- The distance between two points that fall on the *same vertical line* is the absolute value of the difference of their y-coordinates: $d = |y_2 - y_1|$.

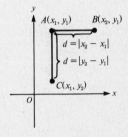

Figure 7-6. Distance between two points on the same horizontal or vertical line.

EXAMPLE 7-6 Find the distance between **(a)** $A(6, 1)$ and $B(2, 1)$ and **(b)** $C(3, -1)$ and $D(3, -3)$.

Solution

(a) Since points A and B have the same y-coordinates, they must lie on the same horizontal line (see Figure 7-7). So the distance between them is

$$d = |x_2 - x_1| = |2 - 6| = |-4| = 4$$

(b) We can see from Figure 7-7 that points C and D lie on the same vertical line. So the distance between them is

$$d = |y_2 - y_1| = |-3 - (-1)| = |-2| = 2$$

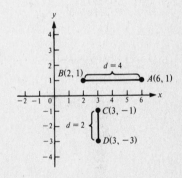

Figure 7-7

To compute the distance between two points $A(x_1, y_1)$ and $B(x_2, y_2)$ in the plane that do *not* lie on the same vertical or horizontal line, we have to use a little geometry. We begin by plotting a point C that completes the right triangle ABC, as in Figure 7-8. The coordinates of point C are (x_2, y_1). Now we can see that the horizontal distance between A and C is $|x_2 - x_1|$ and the vertical distance between B and C is $|y_2 - y_1|$, and that these two distances are the lengths of two sides of a right triangle. So we can use the Pythagorean theorem to find the length of the third—*hypotenuse*—side:

$$(AB)^2 = (AC)^2 + (BC)^2 \quad \text{(Pythagorean theorem)}$$

$$d^2 = (x_2 - x_1)^2 + (y_2 - y_1)^2$$

DISTANCE BETWEEN TWO POINTS $d = \sqrt{(x_2 - x_1)^2 + (y_2 - y_1)^2}$ **(7.1)**

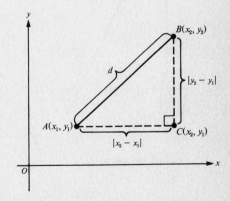

Figure 7-8. Distance between two points:

$$d = \sqrt{(x_2 - x_1)^2 + (y_2 - y_1)^2}$$

Be sure you understand the following facts:

$$(|x_2 - x_1|)^2 = (x_2 - x_1)^2$$

$$(|y_2 - y_1|)^2 = (y_2 - y_1)^2$$

$$(x_2 - x_1)^2 = (x_1 - x_2)^2$$

$$(y_2 - y_1)^2 = (y_1 - y_2)^2$$

Try some numbers if you need verification.

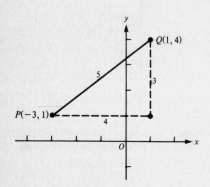

Figure 7-9

note: The formula for the distance between two points A and B is true, no matter where A and B lie in the plane. It doesn't matter if A lies in Quadrant II while B lies in Quadrant IV, or if B lies in I while A lies in III. Nor does it matter which point you choose to label A and which B, i.e., which point has coordinates (x_1, y_1) and which has (x_2, y_2).

EXAMPLE 7-7   Find the distance between $P(-3, 1)$ and $Q(1, 4)$.

Solution   Let $(x_1, y_1) = (-3, 1)$ and $(x_2, y_2) = (1, 4)$. Then

$$d = \sqrt{(x_2 - x_1)^2 + (y_2 - y_1)^2}$$

$$= \sqrt{[1 - (-3)]^2 + (4 - 1)^2}$$

$$= \sqrt{(4)^2 + (3)^2} = \sqrt{16 + 9} = \sqrt{25}$$

$$= 5$$

note: If we let $(x_1, y_1) = (1, 4)$ and $(x_2, y_2) = (-3, 1)$, we obtain the same result, since

$$d = \sqrt{(-3 - 1)^2 + (1 - 4)^2}$$

$$= \sqrt{(-4)^2 + (-3)^2} = \sqrt{16 + 9} = \sqrt{25}$$

$$= 5$$

(See Figure 7-9.)

B. Finding the midpoint of a line segment

Let $P(x_1, y_1)$ and $Q(x_2, y_2)$ be distinct points in the plane and join them to form line segment PQ. The coordinates of the midpoint $M(x, y)$ of that segment are determined by the coordinates of the endpoints:

MIDPOINT OF A LINE SEGMENT $M(x, y) = \left(\dfrac{x_1 + x_2}{2}, \dfrac{y_1 + y_2}{2} \right)$ **(7.2)**

Note that the coordinates of the midpoint are *averages* of the endpoint coordinates.

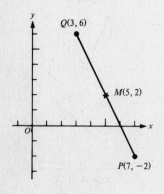

Figure 7-10

EXAMPLE 7-8   Find the midpoint of the segment joining $P(7, -2)$ to $Q(3, 6)$.

Solution   Let $(x_1, y_1) = (7, -2)$ and $(x_2, y_2) = (3, 6)$. Then

$$M(x, y) = \left(\frac{x_1 + x_2}{2}, \frac{y_1 + y_2}{2} \right)$$

$$= \left(\frac{7 + 3}{2}, \frac{-2 + 6}{2} \right)$$

$$= (5, 2)$$

(See Figure 7-10.)

EXAMPLE 7-9   If the points $P(3, 7)$ and $Q(-5, 1)$ are at opposite ends of a diameter of a circle, where is the center of the circle?

Solution   Since the center of a circle is the midpoint of any diameter, finding the midpoint of the segment joining P and Q will give us the center of the circle.

$$M(x, y) = \left(\frac{3 + (-5)}{2}, \frac{7 + 1}{2} \right)$$

$$= \left(\frac{-2}{2}, \frac{8}{2} \right)$$

$$= (-1, 4)$$

(See Figure 7-11.)

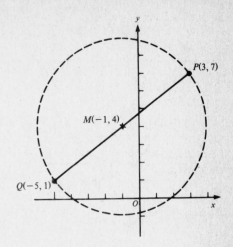

Figure 7-11

7-3. Graphing Equations in Two Variables

The Cartesian coordinate system provides a means of drawing the graph of an equation.

- The graph of an equation in two variables x and y is the set of all points whose coordinates satisfy the equation.

To find the set of points (ordered pairs) whose coordinates satisfy an equation in two variables, we can (1) *choose* values for one of the variables and then (2) *calculate* the corresponding value of the other variable by solving the equation in terms of our chosen value. Then we plot these points on the Cartesian plane, smoothly connecting them in a curve whose every point satisfies the equation.

EXAMPLE 7-10 Graph the equation $y = -2x + 1$.

Solution In order to graph this equation, we need to find ordered pairs (x, y) whose coordinates satisfy $y = -2x + 1$. We can find the pairs that satisfy the equation by picking arbitrary values of x and calculating the corresponding value of y. If, for example, we choose 2 as the x-coordinate, we can calculate the corresponding y-coordinate:

$$y = -2x + 1 \qquad (x = 2)$$

$$= -2(2) + 1 = -4 + 1$$

$$= -3$$

Then the pair $(2, -3)$ is a solution of the equation. If we do this for several values, we can construct a table of solution pairs:

$y = -2x + 1$		Solution Pairs
x	y	(x, y)
-3	$(-2)(-3) + 1 = \quad 7 \longrightarrow$	$(-3, 7)$
-2	$(-2)(-2) + 1 = \quad 5 \longrightarrow$	$(-2, 5)$
-1	$(-2)(-1) + 1 = \quad 3 \longrightarrow$	$(-1, 3)$
0	$(-2)(0) + 1 = \quad 1 \longrightarrow$	$(0, 1)$
1	$(-2)(1) + 1 = -1 \longrightarrow$	$(1, -1)$
2	$(-2)(2) + 1 = -3 \longrightarrow$	$(2, -3)$
3	$(-2)(3) + 1 = -5 \longrightarrow$	$(3, -5)$

After we have formed a collection of solution pairs, we plot the points in the Cartesian plane. We smoothly connect the points to form the graph of the equation. In this case the curve is a straight line, as shown in Figure 7-12.

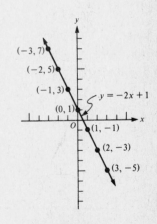

Figure 7-12

The shape of a graph is determined by the type of its equation. First-order equations in two variables have straight lines as their graphs, but other types of equations have other types of curves as their graphs. Plotting several points and looking for patterns can help you determine the shape of a particular graph.

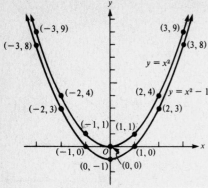

Figure 7-13

EXAMPLE 7-11 Graph the equations (**a**) $y = x^2$ and (**b**) $y = x^2 - 1$ on the same set of coordinate axes.

Solution For each equation, make a table of solution pairs, plot the points, and join them to form a smooth curve (see Figure 7-13).

(a)

$$y = x^2$$

x	y	(x, y)
-3	$(-3)^2 = 9 \longrightarrow$	$(-3, 9)$
-2	$(-2)^2 = 4 \longrightarrow$	$(-2, 4)$
-1	$(-1)^2 = 1 \longrightarrow$	$(-1, 1)$
0	$(0)^2 = 0 \longrightarrow$	$(0, 0)$
1	$(1)^2 = 1 \longrightarrow$	$(1, 1)$
2	$(2)^2 = 4 \longrightarrow$	$(2, 4)$
3	$(3)^2 = 9 \longrightarrow$	$(3, 9)$

(b)

$$y = x^2 - 1$$

x	y	(x, y)
-3	$(-3)^2 - 1 = \mathbf{8} \longrightarrow$	$(-3, 8)$
-2	$(-2)^2 - 1 = \mathbf{3} \longrightarrow$	$(-2, 3)$
-1	$(-1)^2 - 1 = \mathbf{0} \longrightarrow$	$(-1, 0)$
0	$(0)^2 - 1 = \mathbf{-1} \longrightarrow$	$(0, -1)$
1	$(1)^2 - 1 = \mathbf{0} \longrightarrow$	$(1, 0)$
2	$(2)^2 - 1 = \mathbf{3} \longrightarrow$	$(2, 3)$
3	$(3)^2 - 1 = \mathbf{8} \longrightarrow$	$(3, 8)$

EXAMPLE 7-12 Graph the equation $x = y^2 - 2$.

Solution For this equation, assign values to y and compute the corresponding x values:

$$x = y^2 - 2$$

x	y	(x, y)
$(-3)^2 - 2 = 9 - 2 = \mathbf{7}$	$-3 \longrightarrow$	$(7, -3)$
$(-2)^2 - 2 = 4 - 2 = \mathbf{2}$	$-2 \longrightarrow$	$(2, -2)$
$(-1)^2 - 2 = 1 - 2 = \mathbf{-1}$	$-1 \longrightarrow$	$(-1, -1)$
$(0)^2 - 2 = 0 - 2 = \mathbf{-2}$	$0 \longrightarrow$	$(-2, 0)$
$(1)^2 - 2 = 1 - 2 = \mathbf{-1}$	$1 \longrightarrow$	$(-1, 1)$
$(2)^2 - 2 = 4 - 2 = \mathbf{2}$	$2 \longrightarrow$	$(2, 2)$
$(3)^2 - 2 = 9 - 2 = \mathbf{7}$	$3 \longrightarrow$	$(7, 3)$

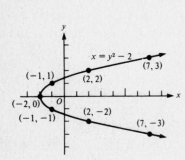

Figure 7-14

Plot the solution pairs from the table and join them to form a smooth curve, as in Figure 7-14.

EXAMPLE 7-13 Graph $2y = x^3$.

Solution First divide each side of the equation by 2 to isolate the variable y. Then assign values to x, compute y, and plot the resulting pairs.

$$y = \tfrac{1}{2}x^3$$

x	y	(x, y)
-3	$\frac{1}{2}(-3)^3 = \frac{1}{2}(-27) = \mathbf{-13\frac{1}{2}} \longrightarrow$	$(-3, -13\frac{1}{2})$
-2	$\frac{1}{2}(-2)^3 = \frac{1}{2}(-8) = \mathbf{-4} \longrightarrow$	$(-2, -4)$
-1	$\frac{1}{2}(-1)^3 = \frac{1}{2}(-1) = \mathbf{-\frac{1}{2}} \longrightarrow$	$(-1, -\frac{1}{2})$
0	$\frac{1}{2}(0)^3 = \frac{1}{2}(0) = \mathbf{0} \longrightarrow$	$(0, 0)$
1	$\frac{1}{2}(1)^3 = \frac{1}{2}(1) = \mathbf{\frac{1}{2}} \longrightarrow$	$(1, \frac{1}{2})$
2	$\frac{1}{2}(2)^3 = \frac{1}{2}(8) = \mathbf{4} \longrightarrow$	$(2, 4)$
3	$\frac{1}{2}(3)^3 = \frac{1}{2}(27) = \mathbf{13\frac{1}{2}} \longrightarrow$	$(3, 13\frac{1}{2})$

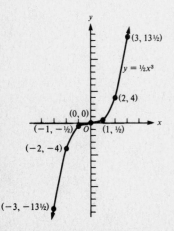

Figure 7-15

Figure 7-15 shows what the graph looks like when we connect the points. Notice how we have adjusted the scale on the y-axis to accommodate the larger y values found in this equation.

7-4. Graphing First-Degree Equations in Two Variables

A. Standard form of the equation of a line

- **A first-degree equation in two variables** x and y is an equation that can be written in the form

$$ax + by = c$$

  where a, b, and c are constants and a and b are not both zero.
- The graph of a first-degree equation in two variables is always a straight line, so $ax + by = c$ is also called a **linear equation** in two variables.
- The form $ax + by = c$ is called the **standard form** of the equation of a line.

note: For the sake of simplicity, we'll often use the expression "the line $ax + by = c$" rather than the more accurate expressions "the line described by the equation $ax + by = c$" or "the line whose equation is $ax + by = c$." Notice that we've treated points in a similar way: We say "the point (x, y)" when, strictly speaking, we mean "the point corresponding to the ordered pair (x, y)."

B. Properties of straight lines

Now that we know that a first-degree—or linear—equation in two variables describes a straight line, we don't need an extensive table of solution pairs to graph it. In fact, we need only *two points*, since *any two points determine a straight line*.

EXAMPLE 7-14 Graph the linear equations $y = x$ and $y = 3x$ on the same set of coordinate axes.

Solution We know that the equations $y = x$ and $y = 3x$ are linear equations since they can be written in standard form as $x - y = 0$ and $3x - y = 0$, respectively. But for purposes of graphing it is easier to leave them in their original form. So, choosing x values of 1 and 2 for $y = x$, we draw a line through the two points $(1, 1)$ and $(2, 2)$. For $y = 3x$ we can set $x = 1$ and $x = 2$ and draw a line through the points $(1, 3)$ and $(2, 6)$. The lines are shown in Figure 7-16. As a check, we can find and plot a third point for each line. The ordered pair $(0, 0)$ is a solution of both equations, so both lines should pass through the origin—which they do. *Voilà!*

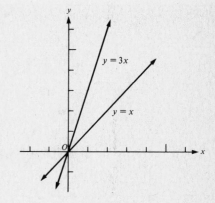

Figure 7-16

EXAMPLE 7-15 Graph **(a)** $y = 3$ and **(b)** $x = -2$ on the same set of coordinate axes.

Solution See Figure 7-17.

(a) The equation $y = 3$ is simply a short form of $0x + y = 3$. So no matter what value we assign to x, the value of y will be 3. Any ordered pair with a y-coordinate of 3 will satisfy the equation—and the corresponding point will therefore be on the line, which is *horizontal*.
(b) We can think of $x = -2$ as a short form of $x + 0y = -2$. In this case the value of x will be -2 no matter what value we assign to y, so all points having an x-coordinate of -2 will be on the line, which is *vertical*.

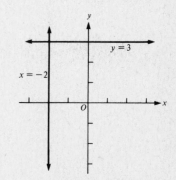

Figure 7-17

The solutions to Example 7-15 lead us to the following conclusions:

- The graph of the equation $x = c$, where c is a constant, is a vertical line drawn through the point $(c, 0)$.
- The graph of the equation $y = c$, where c is a constant, is a horizontal line drawn through the point $(0, c)$.

1. Intercepts

Any line (except vertical and horizontal lines) must cross both the *x*-axis and the *y*-axis.

caution: A *line segment* has two endpoints (so it might not cross either axis), but a *line* has no endpoints—it goes on forever in both directions.

- The point at which a line intersects the *x*-axis is called the **x-intercept**. The point at which a line intersects the *y*-axis is called the **y-intercept**.

You can find the *x*-intercept by substituting 0 for *y* in the standard equation (since the *y*-coordinate of a point on the *x*-axis is 0). Similarly, you can find the *y*-intercept by substituting 0 for *x* in the standard equation. You can use the *x*- and *y*-intercepts to graph linear equations.

EXAMPLE 7-16 Use the *x*- and *y*-intercepts to graph (**a**) $-5x + 2y = 10$ and (**b**) $2x + 2y = 3$.

Solution

(**a**) To find the *x*-intercept, let $y = 0$. Then

$$-5x + 2y = 10$$
$$-5x + 2(0) = 10$$
$$-5x = 10$$
$$x = -2$$

To find the *y*-intercept, let $x = 0$. Then

$$-5(0) + 2y = 10$$
$$2y = 10$$
$$y = 5$$

Now use the *x*-intercept $(-2, 0)$ and the *y*-intercept $(0, 5)$ to graph the line (see Figure 7-18).

(**b**) Setting $y = 0$ in the equation $2x + 2y = 3$ gives us $x = \frac{3}{2}$. Likewise, setting $x = 0$ gives us $y = \frac{3}{2}$. Graph the line through the *x*-intercept $(\frac{3}{2}, 0)$ and the *y*-intercept $(0, \frac{3}{2})$, as shown in Figure 7-18.

note: You can write the intercepts either as ordered pairs or as single numbers. For example, the *y*-intercept of a line that crosses the *y*-axis at the point $(0, 2)$ can be written as $(0, 2)$ or simply 2.

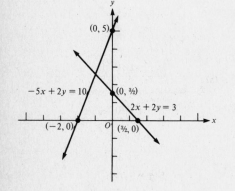

Figure 7-18

2. Slope

Every nonvertical line rises, falls, or remains horizontal as we move from left to right across the plane. The **slope** of a line—denoted by *m*—is a measure of the line's steepness or inclination. We can compute the slope by comparing the amount of vertical change to the amount of horizontal change.

Consider the line drawn in Figure 7-19, which passes through the origin and the points $P(1, 2)$ and $Q(3, 6)$. Notice how the line goes up 2 units when there is 1 unit of horizontal change and goes up 4 units for 2 units of horizontal change. The ratio of vertical to horizontal change is the same in each case: $\frac{2}{1} = \frac{4}{2}$. In fact, we would come up with the same ratio if we were to measure the vertical and horizontal change between *any* two

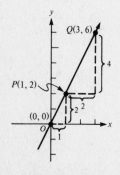

Figure 7-19

points on the line. This ratio is the slope of the line, which we can define formally as

SLOPE OF A LINE $$m = \frac{\text{vertical change}}{\text{horizontal change}} = \frac{\text{change in } y}{\text{change in } x} = \frac{y_2 - y_1}{x_2 - x_1} \qquad (7.3)$$

(See Figure 7-20.)

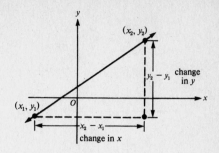

Figure 7-20. Slope: $m = \dfrac{y_2 - y_1}{x_2 - x_1}$

EXAMPLE 7-17 Compute the slopes m of the lines passing through the origin and the following points: (a) $P(4, 2)$, (b) $Q(3, 3)$, and (c) $R(1, 4)$. Draw the lines.

Solution See Figure 7-21.

(a) Let the origin $(0, 0) = (x_1, y_1)$ and $P(4, 2) = (x_2, y_2)$. Then

$$m = \frac{y_2 - y_1}{x_2 - x_1} = \frac{2 - 0}{4 - 0} = \frac{2}{4} = \frac{1}{2}$$

(b) Let the origin $(0, 0) = (x_1, y_1)$ and $Q(3, 3) = (x_2, y_2)$. Then

$$m = \frac{y_2 - y_1}{x_2 - x_1} = \frac{3 - 0}{3 - 0} = \frac{3}{3} = 1$$

(c) Let the origin $(0, 0) = (x_1, y_1)$ and $R(1, 4) = (x_2, y_2)$. Then

$$m = \frac{y_2 - y_1}{x_2 - x_1} = \frac{4 - 0}{1 - 0} = \frac{4}{1} = 4$$

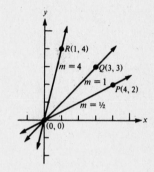

Figure 7-21

Notice the relationship between the lines in Figure 7-21 and their slopes. You can see that the greater the value of m, the steeper the line.

EXAMPLE 7-18 For each pair of points, draw the line they determine and find its slope:

(a) $P_1(0, -1)$ and $P_2(-4, 0)$

(b) $P_1(0, 0)$ and $P_2(-2, 2)$

(c) $P_1(2, -4)$ and $P_2(-1, 5)$

Solution See Figure 7-22 for the graphs of the lines. The slopes are

(a) $m = \dfrac{0 - (-1)}{-4 - 0} = -\dfrac{1}{4}$

(b) $m = \dfrac{2 - 0}{-2 - 0} = -1$

(c) $m = \dfrac{5 - (-4)}{-1 - 2} = -3$

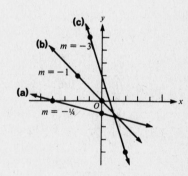

Figure 7-22

Compare the slopes of the lines in Figure 7-22 and Figure 7-21: The slopes of the latter are positive, while those of the former are negative. Then, looking at the lines in these figures, we can conclude the following:

- A line that RISES as we move from left to right has POSITIVE SLOPE.
- A line that FALLS as we move from left to right has NEGATIVE SLOPE.

The slope of a falling line also tells us something about the steepness of the line. You can see from Figure 7-22 that the line with slope -3 is steeper than the one with slope -1. Now we can say that for any line—rising or falling—the greater the absolute value of the slope ($|m|$), the steeper the line.

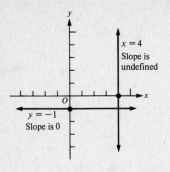

Figure 7-23

EXAMPLE 7-19 For each equation, draw the line it describes and find its slope: (a) $y = -1$ (b) $x = 4$

Solution See Figure 7-23 for the graphs of the lines.

(a) The graph of $y = -1$ is a horizontal line passing through the point $(0, -1)$. To find the slope, we choose a second point $(2, -1)$. Then

$$m = \frac{-1 - (-1)}{2 - 0} = \frac{0}{2} = 0$$

In fact, since every point on a horizontal line has the same y-coordinate, the numerator will always be $-1 - (-1) = 0$. A slope of zero means that a line neither rises nor falls—and a horizontal line fits that description.

(b) The graph of $x = 4$ is a vertical line passing through the point $(4, 0)$. To find the slope, we choose a second point $(4, 1)$. Then

$$m = \frac{1 - 0}{4 - 4} = \frac{1}{0} \to \text{Undefined}$$

In this case, since the x-coordinate of every point on a vertical line is the same, the denominator will always be zero—and division by zero is undefined.

- A horizontal line has a **slope of zero**.
- A vertical line has **no slope** (it is undefined).

note: A slope of *zero* is NOT the same as *no* slope.

C. Other forms of the equation of a line

The standard form—$ax + by = c$—is just one way of writing the equation of a line. Now that we know something about the properties of straight lines, we can use these properties to find other forms of the equation of a line that can be written in terms of x- and y-coordinates. If we know *any two* of the following three things about a line—any point, the slope, an intercept—we can write its equation and plot its graph.

note: All of the forms of the equation for a given line are equivalent, and we can derive one form from any other by algebraic manipulation. For example, the standard form $2x + 4y = 8$ can be rewritten as $4y = -2x + 8$ or as $y = -\frac{1}{2}x + 2$ or as $\frac{x}{4} + \frac{y}{2} = 1$, and so on.

1. The point-slope form

Suppose we know the slope of a line and a point on the line. Let (x, y) represent the coordinates of any point on the line and let (x_1, y_1) be the coordinates of the known point. Then we can derive an equation of the line that uses the definition of slope: If

$$m = \frac{y - y_1}{x - x_1}$$

then $m(x - x_1) = y - y_1$

POINT-SLOPE FORM $y - y_1 = m(x - x_1)$ **(7.4)**

The equation $y - y_1 = m(x - x_1)$ is called the **point-slope form** of the equation of a line and can be used to find the equation of a line with slope m passing through point (x_1, y_1).

EXAMPLE 7-20 Find the equation of the line that passes through the point $(1, -4)$ and has slope -3.

Solution   Let $m = -3$ and $(x_1, y_1) = (1, -4)$. Then use the point-slope formula:

$$y - y_1 = m(x - x_1)$$
$$y - (-4) = -3(x - 1)$$
$$y + 4 = -3x + 3$$
$$y = -3x - 1$$

Now, you can rewrite the equation in standard form:

$$3x + y = -1$$

EXAMPLE 7-21   Find the equation of the line that passes through the point $(5, 2)$ and has slope 0.

Solution   We know that any line that has a slope of 0 is horizontal. We also know that the y-coordinate of every point on a horizontal line is the same. So if $(5, 2)$ is on the line, the equation of the line must be $y = 2$. We can also derive this equation algebraically as follows:

$$y - y_1 = m(x - x_1)$$
$$y - 2 = 0(x - 5)$$
$$y - 2 = 0$$
$$y = 2$$

We can use the point-slope form of the equation of a line in another way—to help us in sketching lines in the plane. We have already graphed lines using two points (and, in particular, the x- and y-intercepts). Now let's suppose that we know the slope m of a line and one point (x_1, y_1) on the line. Since the point (x_1, y_1) is on the line, its coordinates satisfy the equation

$$y - y_1 = m(x - x_1)$$

or $$y = y_1 + m(x - x_1)$$

Then the y-coordinate of a point on the line whose x-coordinate is $x_1 + 1$ is

$$y = y_1 + m((x_1 + 1) - x_1)$$

or $$y = y_1 + m$$

Now we know another point $(x_1 + 1, y_1 + m)$ on the line. Geometrically, this means that if we are on the line at a point (x_1, y_1) and we move 1 unit to the right in the plane, we must move m units up or down. In other words, if x changes from x_1 to $x_1 + 1$, then y changes from y_1 to $y_1 + m$ (see Figure 7-24).

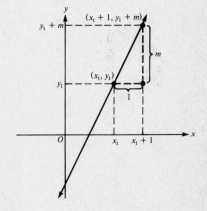

Figure 7-24

EXAMPLE 7-22   Graph the line that contains the point $(0, 2)$ and has slope 2.

Solution   Plot the point $(0, 2)$. Since m is 2, a movement of 1 unit to the right requires a movement of 2 units up (we go up since m is positive). We have arrived at a second point $(1, 4)$ on the line, so we plot it and draw the line through the two points (see Figure 7-25).

EXAMPLE 7-23   Sketch the line that contains the point $(-2, 4)$ and has slope -2. Find three additional points on the line.

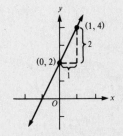

Figure 7-25

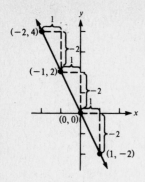

Figure 7-26

Solution Begin at the point $(-2, 4)$. Move 1 unit to the right and 2 units down (since m is -2). Now we can draw the line through this point $(-1, 2)$ and our original point $(-2, 4)$. We find two more points $(0, 0)$ and $(1, -2)$ in the same way (see Figure 7-26).

2. The two-point form

Now suppose we know the coordinates of two distinct points in the plane and want to find the equation of the line passing through them. First we use the coordinates of the points to determine the slope of the line. Then we can use the point-slope formula since we know the slope and a point on the line. (In fact, we know two points on the line, so we can take our pick!) If the points are (x_1, y_1) and (x_2, y_2), the slope is

$$m = \frac{y_2 - y_1}{x_2 - x_1}$$

Using this value for m in the point-slope formula, we get

TWO-POINT FORM $y - y_1 = \left(\dfrac{y_2 - y_1}{x_2 - x_1}\right)(x - x_1)$ (7.5)

where (x_1, y_1) and (x_2, y_2) are distinct points and $x_1 \neq x_2$.

note: Since both (x_1, y_1) and (x_2, y_2) are on the line, we would come up with an equivalent equation if we used (x_2, y_2) in the point-slope formula:

$$y - y_2 = \left(\frac{y_2 - y_1}{x_2 - x_1}\right)(x - x_2)$$

EXAMPLE 7-24   Find the equation of the line passing through the points $(-2, 5)$ and $(6, 1)$.

Solution   Let $(-2, 5) = (x_1, y_1)$ and $(6, 1) = (x_2, y_2)$. Then use the two-point formula:

$$y - y_1 = \left(\frac{y_2 - y_1}{x_2 - x_1}\right)(x - x_1)$$

$$y - 5 = \left(\frac{1 - 5}{6 - (-2)}\right)(x - (-2))$$

$$y - 5 = -\tfrac{1}{2}(x + 2) = -\tfrac{1}{2}x - 1$$

$$y = -\tfrac{1}{2}x + 4$$

Now, you can rewrite the equation in standard form:

$$\tfrac{1}{2}x + y = 4$$

If we use the other point, we come up with the same equation:

$$y - y_2 = \left(\frac{y_2 - y_1}{x_2 - x_1}\right)(x - x_2)$$

$$y - 1 = \left(\frac{1 - 5}{6 - (-2)}\right)(x - 6)$$

$$y - 1 = -\tfrac{1}{2}(x - 6)$$

$$y - 1 = -\tfrac{1}{2}x + 3$$

$$y = -\tfrac{1}{2}x + 4$$

$$\tfrac{1}{2}x + y = 4$$

3. The slope-intercept form

We saw earlier that every nonvertical line must cross the y-axis at some point and that this point is called the y-intercept. By convention, we give this y-intercept the coordinates $(x_1, y_1) = (0, b)$. Now let's suppose we have a line with slope m and y-intercept $(0, b)$. We can use the point-slope formula to find an equation of the line:

$$y - y_1 = m(x - x_1)$$
$$y - b = m(x - 0)$$
$$y - b = mx$$

SLOPE-INTERCEPT FORM $y = mx + b$ (7.6)

slope y-intercept

The equation $y = mx + b$ is called the **slope-intercept form** of the equation of a line, where m is the coefficient of x and b is the constant term. When the equation of a line is in this form (where the coefficient of y is 1), the slope is always the coefficient of x and the y-intercept is always the constant term. This makes the slope-intercept form a very useful one. Just from looking at the equation, we know how steep a line is, whether it rises or falls (or is horizontal), and where it crosses the y-axis. We can also use the slope and y-intercept to graph the line.

note: Don't confuse the constant term b in the slope-intercept form $y = mx + b$ with the coefficient of y (also b) in the standard form $ax + by = c$. It just happens that the same letter of the alphabet is used to represent two different things.

EXAMPLE 7-25   Rewrite each of the following linear equations in slope-intercept form. Give the slope and y-intercept for each line.

(a) $2x + 4y = 3$ **(b)** $x + 3y = 7$ **(c)** $y - 7 = 6x + 2$ **(d)** $y - 4 = 0$

Solution   Remember that all equations of a particular line are equivalent. You just have to do some algebraic manipulation to get the form you want.

(a) $2x + 4y = 3$

$$4y = -2x + 3$$
$$y = -\tfrac{2}{4}x + \tfrac{3}{4}$$
$$y = -\tfrac{1}{2}x + \tfrac{3}{4}$$

The slope is $-\tfrac{1}{2}$.
The y-intercept is $\tfrac{3}{4}$.

(b) $x + 3y = 7$

$$3y = -x + 7$$
$$y = -\tfrac{1}{3}x + \tfrac{7}{3}$$

The slope is $-\tfrac{1}{3}$.
The y-intercept is $\tfrac{7}{3}$.

(c) $y - 7 = 6x + 2$

$$y = 6x + 9$$

The slope is 6.
The y-intercept is 9.

(d) $y - 4 = 0$

$$y = 4$$
$$y = 0x + 4$$

The slope is 0.
The y-intercept is 4.

EXAMPLE 7-26   Find the equation of the line with slope -1 and y-intercept 4 and graph the line.

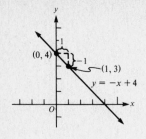

Figure 7-27

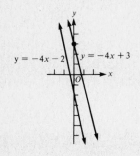

Figure 7-28

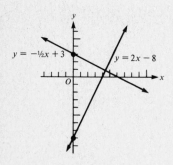

Figure 7-29

Solution   Using $y = mx + b$, the equation of the line is $y = (-1)x + 4$ or $y = -x + 4$. To graph the line, we begin at the point $(0, 4)$. Moving 1 unit to the right and 1 unit down, we arrive at the point $(1, 3)$ and sketch the line through the two points (see Figure 7-27).

D. Parallel and perpendicular lines

We can use slope to tell us whether two lines are parallel, perpendicular, or neither.

- Two nonvertical lines are *parallel* if and only if their *slopes are equal.*
- Two nonvertical lines are *perpendicular* if and only if the *product of their slopes is -1.*

EXAMPLE 7-27   Show that the lines described by the equations (**a**) $4x + y = 3$ and (**b**) $-32x - 8y = 16$ are parallel.

Solution   Rewrite each equation in slope-intercept form:

(**a**) $4x + y = 3$ $\qquad\qquad$ (**b**) $-32x - 8y = 16$

$\qquad\quad y = -4x + 3$ $\qquad\qquad\qquad\quad -8y = 32x + 16$

$\qquad\qquad\qquad\qquad\qquad\qquad\qquad\qquad\quad y = -4x - 2$

The value of m in each equation is -4, so the slope of each line is -4. The lines are parallel since their slopes are equal (see Figure 7-28).

EXAMPLE 7-28   Show that the lines described by the equations $y = 2x - 8$ and $y = -\frac{1}{2}x + 3$ are perpendicular.

Solution   The equations are already in slope-intercept form, so we can see that the lines have slopes of 2 and $-\frac{1}{2}$, respectively. The product of their slopes is $2(-\frac{1}{2}) = -1$, so the lines are perpendicular (see Figure 7-29).

7-5. Graphing First-Degree Inequalities in Two Variables

A. Definition of a first-degree inequality in two variables

- A **first-degree inequality in two variables** (or **linear inequality in two variables**) is an inequality that can be written in one of the following forms:

$$ax + by > c$$

$$ax + by \geq c$$

$$ax + by < c$$

$$ax + by \leq c$$

We know that the solution set of a linear *equation* in two variables is a line. Now we want to know what the solution set of a linear *inequality* in two variables describes.

A straight line (whose equation is $ax + by = c$) divides the Cartesian plane into two half-planes, one on each side of the line. One half-plane contains the solutions to the inequality $ax + by > c$, while the other contains the solutions to $ax + by < c$. If the points on the dividing line—or **boundary line**—are part of the solution set, we write the inequalities as $ax + by \geq c$ and $ax + by \leq c$, respectively.

B. Graphing linear inequalities

To graph a linear inequality, we begin by drawing its boundary line. Then we choose a **test point** from one of the half-planes. If the coordinates of the test point satisfy the inequality, we shade in the half-plane containing that point. If the coordinates of the test point do not satisfy the inequality, we shade in the other half-plane. If the boundary line is part of the solution set, we draw it as a solid line; if not, we draw it as a dashed line.

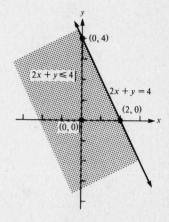

EXAMPLE 7-29   Graph the linear inequality $2x + y \leq 4$.

Solution   We can use the x- and y-intercepts to graph the equation $2x + y = 4$. When $x = 0$, $y = 4$, so the y-intercept is $(0, 4)$. When $y = 0$, $x = 2$, so the x-intercept is $(2, 0)$. We draw the boundary line through these two points, as shown in Figure 7-30. We can choose $(0, 0)$ as a test point, since it does not fall on the line. Substituting 0 for x and 0 for y in the inequality, we get

$$2(0) + 0 \overset{?}{\leq} 4$$

$$0 \overset{\checkmark}{\leq} 4$$

The test point $(0, 0)$ does satisfy the inequality, so we shade in the half-plane containing $(0, 0)$. The boundary line is solid, since it is part of the solution set.

Figure 7-30

EXAMPLE 7-30   Graph the linear inequality $x > 3$.

Solution   The solution set of $x > 3$ consists of all points in the plane whose x-coordinate is greater than 3; i.e., the half-plane to the right of—but not including—the line $x = 3$ (see Figure 7-31).

For linear inequalities other than the type found in Example 7-30 (where the boundary line is vertical), the slope-intercept form saves us a step in graphing. The following is true for linear inequalities in slope-intercept form:

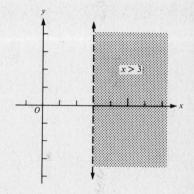

- The graph of $y > mx + b$ consists of the half-plane above the line $y = mx + b$.
- The graph of $y \geq mx + b$ consists of the half-plane above the line $y = mx + b$ and the line itself.
- The graph of $y < mx + b$ consists of the half-plane below the line $y = mx + b$.
- The graph of $y \leq mx + b$ consists of the half-plane below the line $y = mx + b$ and the line itself.

This way we don't need a test point to decide which half-plane is the solution set.

Figure 7-31

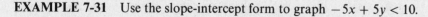

EXAMPLE 7-31   Use the slope-intercept form to graph $-5x + 5y < 10$.

Solution   Rewrite the inequality in slope-intercept form:

$$-5x + 5y < 10$$

$$5y < 5x + 10$$

$$y < x + 2$$

Sketch the line $y = x + 2$, using the y-intercept and the slope to find another point on the line. Since the inequality is of the form $y < mx + b$, we shade in the half-plane below the boundary line (see Figure 7-32).

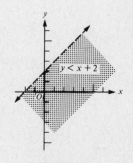

Figure 7-32

SUMMARY

1. The Cartesian coordinate system identifies points in the two-dimensional plane with ordered pairs of real numbers (x, y). The first number is called the x-coordinate, or abscissa, and the second number is called the y-coordinate, or ordinate.
2. The formula for finding the distance between two points (x_1, y_1) and (x_2, y_2) is

$$d = \sqrt{(x_2 - x_1)^2 + (y_2 - y_1)^2}$$

3. The formula for finding the coordinates of the midpoint M of a line segment joining (x_1, y_1) and (x_2, y_2) is

$$M(x, y) = \left(\frac{x_1 + x_2}{2}, \frac{y_1 + y_2}{2} \right)$$

4. A first-degree (or linear) equation in two variables is written in standard form as

$$ax + by = c$$

and is graphed as a straight line.
5. The point at which a line intersects the x-axis is called the x-intercept; the point at which a line intersects the y-axis is called the y-intercept.
6. The slope of a line is a measure of its steepness and is defined as

$$m = \frac{y_2 - y_1}{x_2 - x_1}$$

where (x_1, y_1) and (x_2, y_2) are any two distinct points on the line.
7. The point-slope form of the equation of a line is

$$y - y_1 = m(x - x_1)$$

where m is the slope and (x_1, y_1) is a known point on the line.
8. The two-point form of the equation of a line is

$$y - y_1 = \left(\frac{y_2 - y_1}{x_2 - x_1} \right)(x - x_1)$$

where (x_1, y_1) and (x_2, y_2) are two known distinct points on the line.
9. The slope-intercept form of the equation of a line is

$$y = mx + b$$

where m is the slope and b is the y-intercept.
10. Two nonvertical lines are parallel if and only if their slopes are equal. Two nonvertical lines are perpendicular if and only if the product of their slopes is -1.
11. The graph of a first-degree—or linear—inequality is a half-plane and the line that forms the edge of the half-plane is called the boundary line. In particular,

the graph of $y > mx + b$ consists of the half-plane above the line $y = mx + b$;
the graph of $y \geq mx + b$ consists of the half-plane above the line $y = mx + b$ and the line itself;
the graph of $y < mx + b$ consists of the half-plane below the line $y = mx + b$;
the graph of $y \leq mx + b$ consists of the half-plane below the line $y = mx + b$ and the line itself.

RAISE YOUR GRADES

Can you...?

☑ explain the construction of the Cartesian coordinate plane
☑ plot a point anywhere in the plane, given its x-coordinate and y-coordinate
☑ find the distance between two points in the plane
☑ find the coordinates of the midpoint of a line segment, given its endpoints
☑ use the table construction method to graph an equation in two variables

☑ find the *x*-intercept and *y*-intercept of a line from its equation
☑ find the slope of a line, given two points on the line
☑ find the equation of a line, given the slope and a point on the line
☑ graph a line using its slope and a point on the line
☑ find the equation of a line, given two points on the line
☑ write the equation of a line in slope-intercept form
☑ describe the tests for determining whether two lines are parallel, perpendicular, or neither
☑ graph a linear inequality in two variables

SOLVED PROBLEMS

The Cartesian Coordinate System

PROBLEM 7-1   Plot the points $A(4, -2)$, $B(-3, 3)$, $C(2, 5)$, $D(-1, 0)$, $E(0, -3)$, and $F(-2.3, 4.6)$.

Solution   Begin at the origin in each case. Then

$A(4, -2)$ is 4 units to the right and 2 units down.
$B(-3, 3)$ is 3 units to the left and 3 units up.
$C(2, 5)$ is 2 units to the right and 5 units up.
$D(-1, 0)$ is 1 unit to the left and 0 units up or down.
$E(0, -3)$ is 0 units to the right or left and 3 units down.
$F(-2.3, 4.6)$ is 2.3 units to the left and 4.6 units up.

See Figure 7-33.

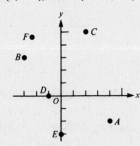

Figure 7-33

PROBLEM 7-2   Find the quadrant in which each point lies for $M(-3, 2)$, $N(-2, -5)$, $P(4, -3)$, and $Q(6, 0)$.

Solution   See Figure 7-34.

$M(-3, 2)$ lies in Quadrant II since its *x*-coordinate is negative and its *y*-coordinate is positive.
$N(-2, -5)$ lies in Quadrant III since both its *x*-coordinate and *y*-coordinate are negative.
$P(4, -3)$ lies in Quadrant IV since its *x*-coordinate is positive and its *y*-coordinate is negative.
$Q(6, 0)$ does not lie in any quadrant since it is on the *x*-axis.

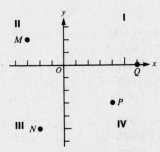

Figure 7-34

Distance Formulas

PROBLEM 7-3   Find the distance between

(a) $P_1(1, 3)$ and $P_2(5, 1)$ (b) $P_3(4, 0)$ and $P_4(0, 3)$ (c) $P_5(0, 0)$ and $P_6(5, 12)$

Solution

(a) Let $(x_1, y_1) = (1, 3)$ and $(x_2, y_2) = (5, 1)$. Then

$$d = \sqrt{(x_2 - x_1)^2 + (y_2 - y_1)^2}$$
$$= \sqrt{(5 - 1)^2 + (1 - 3)^2} = \sqrt{(4)^2 + (-2)^2} = \sqrt{16 + 4} = \sqrt{20} = \sqrt{4} \cdot \sqrt{5} = 2\sqrt{5}$$

(b) Let $(x_1, y_1) = (0, 3)$ and $(x_2, y_2) = (4, 0)$. (Remember that it doesn't matter which point you call (x_1, y_1) and which (x_2, y_2).) Then

$$d = \sqrt{(4 - 0)^2 + (0 - 3)^2} = \sqrt{(4)^2 + (-3)^2} = \sqrt{16 + 9} = \sqrt{25} = 5$$

(c) Let $(x_1, y_1) = (0, 0)$ and $(x_2, y_2) = (5, 12)$. Then

$$d = \sqrt{(5 - 0)^2 + (12 - 0)^2} = \sqrt{(5)^2 + (12)^2} = \sqrt{25 + 144} = \sqrt{169} = 13$$

PROBLEM 7-4   Suppose you connect the points $A(0, 5)$, $B(3, 2)$, $C(2, -2)$, and $D(-1, 1)$ in order so that they form a quadrilateral. Is this quadrilateral a square, a rectangle, a parallelogram, or what?

Solution   First use the distance formula to find the lengths of the sides. (*Note:* d_{AB} means "the distance between points A and B.")

$$d_{AB} = \sqrt{(3 - 0)^2 + (2 - 5)^2} = \sqrt{(3)^2 + (-3)^2} = \sqrt{9 + 9} = \sqrt{18} = \sqrt{9} \cdot \sqrt{2} = 3\sqrt{2}$$
$$d_{BC} = \sqrt{(2 - 3)^2 + (-2 - 2)^2} = \sqrt{(-1)^2 + (-4)^2} = \sqrt{1 + 16} = \sqrt{17}$$

You know it's not a square, since two of the sides aren't equal. Now find the lengths of the other two sides:

$$d_{CD} = \sqrt{(-1 - 2)^2 + (1 - (-2))^2}$$
$$= \sqrt{(-3)^2 + (3)^2} = \sqrt{9 + 9} = \sqrt{18}$$
$$= 3\sqrt{2}$$
$$d_{DA} = \sqrt{(-1 - 0)^2 + (1 - 5)^2}$$
$$= \sqrt{(-1)^2 + (-4)^2} = \sqrt{1 + 16} = \sqrt{17}$$

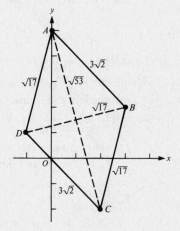

Figure 7-35

The quadrilateral has opposite sides of equal length, so it's either a rectangle or a parallelogram. If it's a rectangle, the diagonals must be equal. Find the lengths of the diagonals:

$$d_{AC} = \sqrt{(2 - 0)^2 + (-2 - 5)^2}$$
$$= \sqrt{(2)^2 + (-7)^2} = \sqrt{4 + 49} = \sqrt{53}$$
$$d_{BD} = \sqrt{(-1 - 3)^2 + (1 - 2)^2}$$
$$= \sqrt{(-4)^2 + (-1)^2} = \sqrt{16 + 1} = \sqrt{17}$$

Since the diagonals aren't equal, it's not a rectangle but a parallelogram. (See Figure 7-35.)

PROBLEM 7-5   Use your hand-held calculator to find the distance between $(-23.1, 12.62)$ and $(5.47, 19.18)$.

Solution   Let $x_1 = -23.1$, $y_1 = 12.62$, $x_2 = 5.47$, and $y_2 = 19.18$. Then

$$d = \sqrt{(5.47 - (-23.1))^2 + (19.18 - 12.62)^2}$$

Perform the following operations:

Calculate $5.47 - (-23.1) = 5.47 + 23.1 = 28.57$.

Square this value and store it: $(28.57)^2 = 816.2449$.

Calculate $19.18 - 12.62 = 6.56$.

Square this value: $(6.56)^2 = 43.0336$.

Add this to your stored value: $43.0336 + 816.2449 = 859.2785$.

Find the square root of the value: $\sqrt{859.2785} = 29.313453$.

The distance d is 29.313453.

PROBLEM 7-6 Find the coordinates of the midpoint of the segment joining

(a) $(4, -2)$ and $(-6, 8)$ **(b)** $(-3, -1)$ and $(5, 9)$

(c) $(4.2, -1.8)$ and $(2.6, 7.3)$ **(d)** $(8, 0)$ and $(-3, 0)$

Solution Use the midpoint formula in each case:

$$M(x, y) = \left(\frac{x_1 + x_2}{2}, \frac{y_1 + y_2}{2} \right)$$

(a) $M(x, y) = \left(\dfrac{4 + (-6)}{2}, \dfrac{-2 + 8}{2} \right) = \left(\dfrac{-2}{2}, \dfrac{6}{2} \right) = (-1, 3)$

(b) $M(x, y) = \left(\dfrac{-3 + 5}{2}, \dfrac{-1 + 9}{2} \right) = \left(\dfrac{2}{2}, \dfrac{8}{2} \right) = (1, 4)$

(c) $M(x, y) = \left(\dfrac{4.2 + 2.6}{2}, \dfrac{-1.8 + 7.3}{2} \right) = \left(\dfrac{6.8}{2}, \dfrac{5.5}{2} \right) = (3.4, 2.75)$

(d) $M(x, y) = \left(\dfrac{8 + (-3)}{2}, \dfrac{0 + 0}{2} \right) = \left(\dfrac{5}{2}, \dfrac{0}{2} \right) = \left(\dfrac{5}{2}, 0 \right)$

PROBLEM 7-7 The infield of a baseball field is a perfect square with a distance of 90 ft between consecutive bases. How far is it from home plate to second base?

Solution If you place home plate at the origin on a graph, as in Figure 7-36, then second base is at the point (90 ft, 90 ft). Then the distance between $(0, 0)$ and $(90, 90)$ is

$$d = \sqrt{(90 - 0)^2 + (90 - 0)^2}$$

$$= \sqrt{8100 + 8100}$$

$$= \sqrt{16\,200}$$

$$= 127.28$$

A catcher must throw 127.28 ft to catch a player trying to steal second!

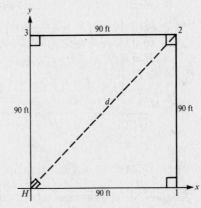

Figure 7-36

PROBLEM 7-8 Three points in the plane are collinear if they can be arranged in an order PQR such that $d_{PQ} + d_{QR} = d_{PR}$. Determine if $(-3, 1)$, $(0, -1)$, and $(6, -5)$ are collinear.

Solution

$$d_{PQ} = \sqrt{(0 - (-3))^2 + (-1 - 1)^2} = \sqrt{9 + 4} = \sqrt{13}$$

$$d_{QR} = \sqrt{(6 - 0)^2 + (-5 - (-1))^2} = \sqrt{36 + 16} = 2\sqrt{13}$$

$$d_{PR} = \sqrt{(6 - (-3))^2 + (-5 - 1)^2} = \sqrt{81 + 36} = \sqrt{117} = \sqrt{9} \cdot \sqrt{13} = 3\sqrt{13}$$

The points are collinear since

$$d_{PQ} + d_{QR} = \sqrt{13} + 2\sqrt{13} = 3\sqrt{13} = d_{PR}$$

PROBLEM 7-9 Find the coordinates of the points that lie a distance of 3 units from the origin.

Solution There is no unique solution to this problem since there are at least four obvious points that satisfy the condition: the intercepts $(3, 0)$, $(0, 3)$, $(-3, 0)$, and $(0, -3)$. In general, ANY point (x, y)

lying 3 units from the origin $(0, 0)$ must satisfy the distance formula:

$$3 = \sqrt{(x - 0)^2 + (y - 0)^2}$$

$$3 = \sqrt{x^2 + y^2} \quad \text{[Square both sides.]}$$

$$9 = x^2 + y^2$$

$$x^2 + y^2 = 9$$

Congratulations! You have just found the equation of a circle with its center at the origin and radius 3! (Remember that a circle consists of all the points that are the same distance from the center of the circle—and that distance is called the radius. See Figure 7-37.)

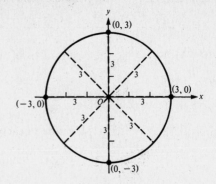

Figure 7-37

Graphing Equations in Two Variables

PROBLEM 7-10   Graph the equation $2y = x^2$.

Solution   First divide each side by 2 to isolate y and then assign values to x. Complete the table by finding the corresponding y values, plot the points, and join them to form the curve.

$y = \frac{1}{2}x^2$

x	y	(x, y)
0	$\frac{1}{2}(0)^2 = 0 \longrightarrow$	$(0, 0)$
1	$\frac{1}{2}(1)^2 = \frac{1}{2} \longrightarrow$	$(1, \frac{1}{2})$
-1	$\frac{1}{2}(-1)^2 = \frac{1}{2} \longrightarrow$	$(-1, \frac{1}{2})$
2	$\frac{1}{2}(2)^2 = 2 \longrightarrow$	$(2, 2)$
-2	$\frac{1}{2}(-2)^2 = 2 \longrightarrow$	$(-2, 2)$
3	$\frac{1}{2}(3)^2 = \frac{9}{2} \longrightarrow$	$(3, \frac{9}{2})$
-3	$\frac{1}{2}(-3)^2 = \frac{9}{2} \longrightarrow$	$(-3, \frac{9}{2})$

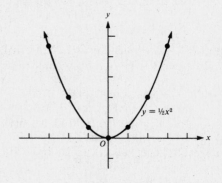

Figure 7-38

See Figure 7-38.

PROBLEM 7-11   Sketch the graph of $x = y^3$.

Solution   In this case it is better to assign values to y and compute the corresponding x values from the equation.

$x = y^3$

x	y	(x, y)
$0^3 = 0$	$0 \longrightarrow$	$(0, 0)$
$1^3 = 1$	$1 \longrightarrow$	$(1, 1)$
$2^3 = 8$	$2 \longrightarrow$	$(8, 2)$
$(-1)^3 = -1$	$-1 \longrightarrow$	$(-1, -1)$
$(-2)^3 = -8$	$-2 \longrightarrow$	$(-8, -2)$

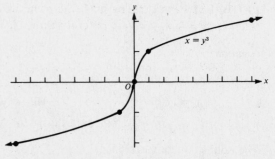

Figure 7-39

Connect the points as shown in Figure 7-39.

PROBLEM 7-12   Graph the equation $x^2 + y^2 = 4$.

Solution   Assign values to x first and compute the corresponding y values; then assign values to y and compute the x values. The pattern that emerges should help you draw the graph.

$$x^2 + y^2 = 4$$

x	y	(x, y)
0	$y^2 = 4 \Rightarrow y = \pm 2$	$\longrightarrow (0, 2)$ and $(0, -2)$
1	$y^2 = 3 \Rightarrow y = \pm\sqrt{3}$	$\longrightarrow (1, \sqrt{3})$ and $(1, -\sqrt{3})$
2	$y^2 = 0 \Rightarrow y = 0$	$\longrightarrow (2, 0)$
-1	$y^2 = 3 \Rightarrow y = \pm\sqrt{3}$	$\longrightarrow (-1, \sqrt{3})$ and $(-1, -\sqrt{3})$
-2	$y^2 = 0 \Rightarrow y = 0$	$\longrightarrow (-2, 0)$
$x^2 = 4 \Rightarrow x = \pm 2$	0	$\longrightarrow (2, 0)$ and $(-2, 0)$
$x^2 = 3 \Rightarrow x = \pm\sqrt{3}$	± 1	$\longrightarrow (\sqrt{3}, 1), (-\sqrt{3}, 1), (\sqrt{3}, -1), (-\sqrt{3}, -1)$

(See Figure 7-40.)

note: See Problem 7-9. The equation $x^2 + y^2 = 4$ is a circle whose center is at $(0, 0)$ and radius is 2.

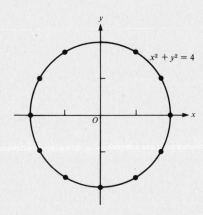

Figure 7-40

Graphing First-Degree Equations in Two Variables

PROBLEM 7-13 Graph the equations (a) $y = 3$, (b) $x = -\frac{3}{2}$, (c) $y = -4$, and (d) $x = \frac{7}{2}$.

Solution

(a) The graph of $y = 3$ is a horizontal line drawn through the point $(0, 3)$.

(b) The graph of $x = -\frac{3}{2}$ is a vertical line drawn through the point $(-\frac{3}{2}, 0)$.

(c) The graph of $y = -4$ is a horizontal line drawn through the point $(0, -4)$.

(d) The graph of $x = \frac{7}{2}$ is a vertical line drawn through the point $(\frac{7}{2}, 0)$.

(See Figure 7-41.)

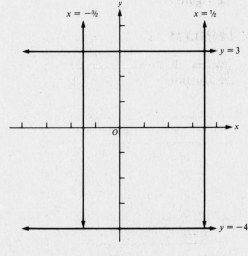

Figure 7-41

PROBLEM 7-14 Find the x- and y-intercepts for the following lines:

(a) $y = -3x + 9$

(b) $6x - 2y + 5 = 0$

(c) $3(x + 1) - 8(y - 2) = 20$

(d) $x - 3 = 5$

Solution Remember that you can find the x-intercept by setting $y = 0$ and the y-intercept by setting $x = 0$.

(a) $y = -3x + 9$

Let $y = 0$. Then

$0 = -3x + 9$

$3x = 9$

$x = 3$

Let $x = 0$. Then

$y = -3(0) + 9$

$y = 9$

The x-intercept is $(3, 0)$; the y-intercept is $(0, 9)$.

(b) $6x - 2y + 5 = 0$

Let $y = 0$. Then

$6x - 2(0) + 5 = 0$

$6x = -5$

$x = -\frac{5}{6}$

Let $x = 0$. Then

$6(0) - 2y + 5 = 0$

$-2y = -5$

$y = \frac{5}{2}$

The x-intercept is $(-\frac{5}{6}, 0)$; the y-intercept is $(0, \frac{5}{2})$.

(c) $3(x + 1) - 8(y - 2) = 20$

Let $y = 0$. Then

$3(x + 1) - 8(0 - 2) = 20$

$3x + 3 + 16 = 20$

$3x = 1$

$x = \frac{1}{3}$

Let $x = 0$. Then

$3(0 + 1) - 8(y - 2) = 20$

$3 - 8y + 16 = 20$

$-8y = 1$

$y = -\frac{1}{8}$

The x-intercept is $(\frac{1}{3}, 0)$; the y-intercept is $(0, -\frac{1}{8})$.

(d) $x - 3 = 5$

Let $y = 0$. Then

$x - 3 = 5$

$x = 8$

The graph of $x - 3 = 5$ (or $x = 8$) is a vertical line, so it never intersects the y-axis.

The x-intercept is $(8, 0)$; there is no y-intercept.

PROBLEM 7-15 Find the slopes of

(a) the line containing $(3, 5)$ and $(-2, -7)$.
(b) the line that crosses the y-axis at -2 and the x-axis at -5.
(c) the line passing through $(-1, 1)$ and $(1, -1)$.

Solution Use the formula $m = \dfrac{y_2 - y_1}{x_2 - x_1}$ to find the slope of each line.

(a) Let $(3, 5) = (x_1, y_1)$ and $(-2, -7) = (x_2, y_2)$. Then

$$m = \frac{-7 - 5}{-2 - 3} = \frac{-12}{-5} = \frac{12}{5}$$

(b) The coordinates of the y-intercept are $(0, -2)$ and those of the x-intercept are $(-5, 0)$. Let $(0, -2) = (x_1, y_1)$ and $(-5, 0) = (x_2, y_2)$. Then

$$m = \frac{0 - (-2)}{-5 - 0} = \frac{2}{-5} = -\frac{2}{5}$$

(c) Let $(-1, 1) = (x_1, y_1)$ and $(1, -1) = (x_2, y_2)$. Then

$$m = \frac{-1 - 1}{1 - (-1)} = \frac{-2}{2} = -1$$

PROBLEM 7-16 Use the point-slope form to find the equations of

(a) the line with slope $m = 2$ passing through point $(-1, -2)$.
(b) the line with slope $m = -1$ passing through point $(2, 3)$.
(c) the line with slope $m = 0$ passing through point $(-3, 1)$.
(d) the line with slope $m = \frac{1}{4}$ passing through point $(-4, -2)$.

Solution In each case, use the point-slope form of the equation of a line:

$$y - y_1 = m(x - x_1)$$

(a) $y - (-2) = 2(x - (-1))$

$\qquad y + 2 = 2(x + 1)$

$\qquad y + 2 = 2x + 2$

$\qquad\qquad y = 2x$

(b) $y - 3 = -1(x - 2)$

$\qquad y - 3 = -x + 2$

$\qquad\qquad y = -x + 5$

(c) $y - 1 = 0(x - (-3))$

$\qquad y - 1 = 0$

$\qquad\qquad y = 1$

(d) $y - (-2) = \frac{1}{4}(x - (-4))$

$\qquad y + 2 = \frac{1}{4}(x + 4)$

$\qquad y + 2 = \frac{1}{4}x + 1$

$\qquad\qquad y = \frac{1}{4}x - 1$

PROBLEM 7-17 Find an equation of the line that contains the following points:

(a) $(1, -4)$ and $(-3, 2)$

(b) $(0, 8)$ and $(4, 0)$

(c) $(-5, -\frac{1}{2})$ and $(\frac{3}{2}, 4)$

(d) $(3, 2)$ and $(9, 5)$

Solution In each case, use the two-point formula:

$$y - y_1 = \left(\frac{y_2 - y_1}{x_2 - x_1}\right)(x - x_1)$$

(a) $y - (-4) = \left(\dfrac{2 - (-4)}{-3 - 1}\right)(x - 1)$

$\qquad y - (-4) = -\frac{3}{2}(x - 1)$

$\qquad\qquad y + 4 = -\frac{3}{2}x + \frac{3}{2}$

$\qquad\qquad\qquad y = -\frac{3}{2}x - \frac{5}{2}$

(b) $y - 8 = \left(\dfrac{0 - 8}{4 - 0}\right)(x - 0)$

$\qquad y - 8 = -2(x - 0)$

$\qquad y - 8 = -2x$

$\qquad\qquad y = -2x + 8$

(c) $y - (-\frac{1}{2}) = \left(\dfrac{4 - (-\frac{1}{2})}{\frac{3}{2} - (-5)}\right)(x - (-5))$

$\qquad y - (-\frac{1}{2}) = \frac{9}{13}(x - (-5))$

$\qquad\qquad y + \frac{1}{2} = \frac{9}{13}x + \frac{45}{13}$

$\qquad\qquad\qquad y = \frac{9}{13}x + \frac{77}{26}$

(d) $y - 2 = \left(\dfrac{5 - 2}{9 - 3}\right)(x - 3)$

$\qquad y - 2 = \frac{1}{2}(x - 3)$

$\qquad y - 2 = \frac{1}{2}x - \frac{3}{2}$

$\qquad\qquad y = \frac{1}{2}x + \frac{1}{2}$

PROBLEM 7-18 If line ℓ_1 contains points $(-5, 1)$ and $(2, 7)$ and line ℓ_2 contains points $(-4, -3)$ and $(2, 3)$, which line is steeper?

Solution Find the slope of each line and compare them to see which line is steeper. Remember that the greater the absolute value of m, the steeper the line.

The slope of ℓ_1 is $\quad m = \dfrac{7-1}{2-(-5)} = \dfrac{6}{7}$

The slope of ℓ_2 is $\quad m = \dfrac{3-(-3)}{2-(-4)} = \dfrac{6}{6} = 1$

Line ℓ_2 is slightly steeper than line ℓ_1.

PROBLEM 7-19 Find the slope and the y-intercept for the line corresponding to each equation:

(a) $-5x + 15y = 20$

(b) $3(x + 3) - 5(y - 1) = 6$

(c) $8x = 2(y - x + 1)$

(d) $3(y + x) = 3(x - 4)$

Solution Rewrite each equation in slope-intercept form:

(a) $-5x + 15y = 20$

$15y = 5x + 20$

$y = \frac{5}{15}x + \frac{20}{15}$

$= \frac{1}{3}x + \frac{4}{3}$

The slope is $\frac{1}{3}$. The y-intercept is $\frac{4}{3}$.

(b) $3(x + 3) - 5(y - 1) = 6$

$3x + 9 - 5y + 5 = 6$

$3x - 5y + 14 = 6$

$3x - 5y = -8$

$-5y = -3x - 8$

$y = \frac{3}{5}x + \frac{8}{5}$

The slope is $\frac{3}{5}$. The y-intercept is $\frac{8}{5}$.

(c) $8x = 2(y - x + 1)$

$8x = 2y - 2x + 2$

$10x = 2y + 2$

$5x = y + 1$

$y = 5x - 1$

The slope is 5. The y-intercept is -1.

(d) $3(y + x) = 3(x - 4)$

$3y + 3x = 3x - 12$

$3y = -12$

$y = -4$

or $\quad y = 0x - 4$

The slope is 0. The y-intercept is -4.

PROBLEM 7-20 Graph the lines of Problem 7-19.

Solution For each line, plot the y-intercept first and use the slope to find another point on the line. Then draw the line through the two points.

(a) Plot the y-intercept $(0, \frac{4}{3})$. The slope is $\frac{1}{3}$, so an increase in x of 1 unit means an increase in y of $\frac{1}{3}$ unit; thus the line also passes through $(0 + 1, \frac{4}{3} + \frac{1}{3}) = (1, \frac{5}{3})$.

(b) Plot the y-intercept $(0, \frac{8}{5})$. The slope is $\frac{3}{5}$, so y increases by $\frac{3}{5}$ when x increases by 1; thus the line also passes through $(0 + 1, \frac{8}{5} + \frac{3}{5}) = (1, \frac{11}{5})$.

> *note:* You can also fall back on the original definition of slope to find the second point. You know, for example, that $m = \dfrac{\text{vertical change}}{\text{horizontal change}} = \dfrac{3}{5}$ in part (b), so in this case a horizontal increase of 5 and a vertical increase of 3 will also give you a second point on the line, $(0 + 5, \frac{8}{5} + 3) = (5, \frac{23}{5})$.

(c) Plot the y-intercept $(0, -1)$. Since the slope is 5, a movement of 1 unit to the right and 5 units upward gives you another point, $(1, 4)$.

(d) Plot the y-intercept $(0, -4)$. The slope is 0, so the line is horizontal and passes through every point with y-coordinate -4.

See Figure 7-42 (facing page).

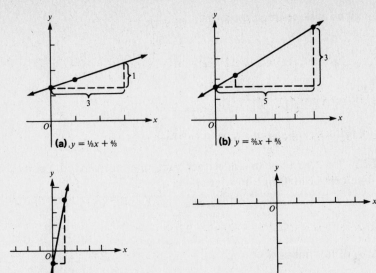

Figure 7-42

PROBLEM 7-21 In each case, determine from the equations whether the two lines are parallel, perpendicular, or neither:

(a) $y = x$; $y = -x$

(b) $y = 3(x + 1)$; $y - 3x = 4$

(c) $x = 0$; $y = 0$

(d) $2x + 3y = 7$; $6x + 9y = 2$

(e) $x - 5y = 1$; $2x + y = 4$

Solution Determine the slopes of the lines in each pair and use the following test:

If $m_1 = m_2$, the lines are parallel.

If $m_1 \cdot m_2 = -1$, the lines are perpendicular.

(a) The equations $y = x$ and $y = -x$ are already in slope-intercept form, so you know that the slopes of the lines are 1 and -1, respectively. Then

$$m_1 \cdot m_2 = 1 \cdot (-1) = -1$$

The lines are perpendicular.

(b) Put the equations in slope-intercept form:

$y = 3(x + 1)$ $\qquad$ $y - 3x = 4$

$y = 3x + 3$ $\qquad\qquad$ $y = 3x + 4$

$$m_1 = 3 \text{ and } m_2 = 3, \text{ so } m_1 = m_2$$

The lines are parallel.

(c) The equation $x = 0$ describes a vertical line—the y-axis. The equation $y = 0$ describes a horizontal line—the x-axis. You can't use the test for these lines, since m is not defined for $x = 0$, but you know that by definition the x-axis and y-axis are perpendicular.

(d) Put the equations in slope-intercept form:

$2x + 3y = 7$ $\qquad\qquad$ $6x + 9y = 2$

$3y = -2x + 7$ $\qquad\qquad$ $9y = -6x + 2$

$y = -\frac{2}{3}x + \frac{7}{3}$ $\qquad\qquad$ $y = -\frac{2}{3}x + \frac{2}{9}$

$$m_1 = -\frac{2}{3} \text{ and } m_2 = -\frac{2}{3}, \text{ so } m_1 = m_2$$

The lines are parallel.

(e) Put the equations in slope-intercept form:

$$x - 5y = 1 \qquad\qquad 2x + y = 4$$
$$-5y = -x + 1 \qquad\qquad y = -2x + 4$$
$$y = \tfrac{1}{5}x - \tfrac{1}{5}$$

$$m_1 = \tfrac{1}{5} \text{ and } m_2 = -2, \text{ so } m_1 \neq m_2 \text{ and } m_1 \cdot m_2 \neq -1$$

The lines are neither parallel nor perpendicular.

PROBLEM 7-22 Let ℓ be a line that is neither horizontal nor vertical. Let its x-intercept be at $(a, 0)$ and its y-intercept be at $(0, b)$. Show for any (x, y) on ℓ that

$$\frac{x}{a} + \frac{y}{b} = 1$$

and that the slope of this line is $m = -\dfrac{b}{a}$.

Solution Since you know two points on the line, you can find the slope from

$$m = \frac{y_2 - y_1}{x_2 - x_1} = \frac{b - 0}{0 - a} = -\frac{b}{a}$$

Then with the point-slope form using $(a, 0)$ as the point and $m = -b/a$, you get

$$y - 0 = \left(-\frac{b}{a}\right)(x - a)$$

$$y = -\frac{b}{a}x - \left(\frac{b}{a}\right)(-a)$$

$$y = -\frac{b}{a}x + b$$

$$\frac{y}{b} = -\frac{b}{a}\cdot\frac{x}{b} + \frac{b}{b}$$

$$\frac{y}{b} = -\frac{x}{a} + 1$$

$$\frac{x}{a} + \frac{y}{b} = 1$$

PROBLEM 7-23 Find the equation of the line that contains the point $(1, 4)$ and is parallel to the line $x - 3y = 5$.

Solution If the line is parallel to $x - 3y = 5$, it must have the same slope. Put $x - 3y = 5$ in slope-intercept form to find the slope:

$$x - 3y = 5$$
$$-3y = -x + 5$$
$$y = \tfrac{1}{3}x - \tfrac{5}{3}$$

So the slope is $m = \tfrac{1}{3}$. Use the point-slope form to find the equation of the line:

$$y - 4 = \tfrac{1}{3}(x - 1)$$
$$y - 4 = \tfrac{1}{3}x - \tfrac{1}{3}$$
$$y = \tfrac{1}{3}x + \tfrac{11}{3}$$

PROBLEM 7-24 Find the equation of the line that passes through the point $(-2, 4)$ and is perpendicular to the line $y = \tfrac{1}{3}x - 1$.

Solution If the line is perpendicular to $y = \frac{1}{3}x - 1$, then the product of the two slopes must be -1. The slope of $y = \frac{1}{3}x - 1$ is $\frac{1}{3}$, so the slope of the line we're looking for must be -3 (since $-3 \cdot \frac{1}{3} = -1$). Now use the given point $(-2, 4)$ and the computed slope -3 in point-slope form to find the equation of the line:

$$y - 4 = -3(x - (-2))$$
$$y - 4 = -3x - 6$$
$$y = -3x - 2$$

Graphing First-Degree Inequalities in Two Variables

PROBLEM 7-25 Sketch the graph of the inequality $3x + 6y \geq 12$.

Solution First graph the boundary line $3x + 6y = 12$. Then choose a test point from one side of the line and see if it satisfies the inequality. If you choose $(0, 0)$ as your test point, you get

$$3(0) + 6(0) \overset{?}{\geq} 12$$
$$0 \not\geq 12$$

Zero is not greater than or equal to twelve, so you can shade in the other half-plane as the solution set. The boundary line is part of the solution set (since the inequality is $\geq$), so it is a solid line.

You get the same result if you use the slope-intercept form of the inequality to graph the solution set. Rewrite the inequality in slope-intercept form:

$$3x + 6y \geq 12$$
$$6y \geq -3x + 12$$
$$y \geq -\tfrac{1}{2}x + 2$$

Then the boundary line is a line with slope of $-\frac{1}{2}$ and y-intercept $(0, 2)$ and the graph of the inequality is the half-plane above—and including—that line. Figure 7-43 shows the solution set.

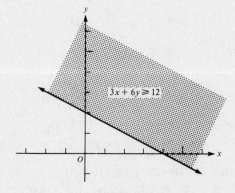

$3x + 6y \geq 12$

Figure 7-43

PROBLEM 7-26 Graph $y < 3x - 5$.

Solution The inequality is already in slope-intercept form, so graph the boundary line $y = 3x - 5$ and shade in the half-plane below the line. The line is dashed to show that it is not part of the solution set (see Figure 7-44).

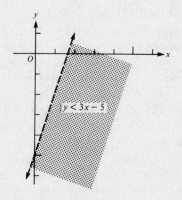

$y < 3x - 5$

Figure 7-44

Supplementary Exercises

PROBLEM 7-27 Plot the points $A(-2, 7)$, $B(\frac{5}{2}, -\frac{1}{2})$, $C(-4, -4)$, $D(8, 1)$, and $E(0, 4)$. Indicate which quadrant each is in.

PROBLEM 7-28 Find the distance **(a)** between the origin and $(5, 8)$ and **(b)** between the origin and $(\sqrt{10}, -10)$.

PROBLEM 7-29   Given $P(-1, 2)$, find a point that has a first coordinate of 2 and is 5 units from P.

PROBLEM 7-30   Without plotting the following points, name the quadrant in which each lies:

(a) $(-5, -1)$ (b) $(3, 7)$ (c) $(\pi, \sqrt{8})$

(d) $(1.5, -20)$ (e) $(\frac{13}{4}, -\frac{1}{8})$ (f) $(-0.3, 0.6)$

(g) $(20, 20)$

PROBLEM 7-31   The perimeter of a triangle is the sum of the lengths of its three sides. Find the perimeter of the triangle formed by $(1, 1)$, $(7, 5)$, and $(3, -4)$.

PROBLEM 7-32   Find the perimeter of the triangle whose vertices are at $(-1, 2)$, $(3, -1)$, and $(3, 2)$.

PROBLEM 7-33   A rectangle with a height of 7 units has the points $(-2, -3)$ and $(5, -3)$ as its two lowest corners. (a) What is the length of the diagonal of the rectangle? (b) Where is its geometric center?

PROBLEM 7-34   A triangle has vertices at $(0, 0)$, $(6, 0)$, and $(3, 8)$. Find its perimeter and its area.

PROBLEM 7-35   Find x so that the distance between $(5, 0)$ and $(x, 8)$ is 10.

PROBLEM 7-36   A rhombus is a polygon whose four sides are of equal length. If a polygon has the points $(0, 0)$, $(0, 5)$, $(4, 3)$, and $(4, 8)$ as vertices, is that polygon a rhombus?

PROBLEM 7-37   Find the points on the x-axis that are 4 units from the point $(3, 1)$.

PROBLEM 7-38   Show that the diagonals of a square intersect at their midpoints. [*Hint:* Select four convenient points as vertices such as $(0, 0)$, $(0, c)$, $(c, 0)$, and (c, c).]

PROBLEM 7-39   The infield of a softball field is a square whose sides are each 60 feet. (a) How far is it from home plate to second base? (b) If the pitching distance is 46 feet, on which side of the imaginary line connecting first base to third base does the pitcher stand?

PROBLEM 7-40   Herb, who lives in city H, needs to ship a computer to his friend Gordon, who lives in city G. The cities are situated as shown on Figure 7-45, with one road joining H to G. But there is an alternate route—through city P—which forms a right triangle with H and G. Ace Trucking uses only the direct route and charges $6.50 per mile; Bruce Trucking uses only the H-to-P-to-G route and charges $4.00 per mile. Both firms guarantee same-day delivery. Which company should Herb use if he is watching his budget?

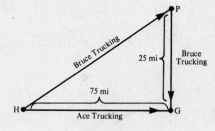

Figure 7-45

PROBLEM 7-41   The points $(5, 2)$ and $(-3, 2)$ are two vertices of a square. Find three additional pairs of points that could serve as the other two vertices.

PROBLEM 7-42   A *normal* line to a given line is a line that is perpendicular to the given line. If the given line has slope m, then the normal line has slope $-1/m$. (If $m = 0$, then the slope of the normal is undefined.) For each of the following, find the slope of the normal line. Then, at the indicated point, find the equation of the normal line and write it in slope-intercept form:

(a) $x + 2y = 2$ Point $(0, 1)$

(b) $4x - y = 2$ Point $(1, 2)$

(c) $2(x - 1) + 3y = -7$ Point $(2, -3)$

PROBLEM 7-43 Use the information given in (**a**)–(**e**) to find the equation of each line and write the equation in standard form:

(**a**) $m = 2$, contains point $(1, -4)$
(**c**) intercepts at $(0, 5)$ and $(2, 0)$
(**e**) $m = 4$, x-intercept $(-2, 0)$

(**b**) contains $(7, 1)$ and $(-1, 5)$
(**d**) $m = -\frac{3}{2}$, y-intercept $(0, -1)$

PROBLEM 7-44 If the point $(1, -3)$ lies on a line whose slope is 2, which of the following points also lie on that line? $(1, 1)$, $(2, -1)$, $(11, -7)$, $(3, 7)$, $(3, 1)$, $(4, 3)$

PROBLEM 7-45 If the population of Gotham City grows linearly and if in 1970 the population was 21 000 and in 1980 it was 28 500, what will the population be in 1990? in 2000?

PROBLEM 7-46 Find the equation of the line parallel to $-4x + y = 7$ and passing through the point $(-2, 1)$.

PROBLEM 7-47 Fahrenheit temperature and Celsius temperature are related linearly. Using the facts that $32°F = 0°C$ and $212°F = 100°C$, find the equation that expresses C in terms of F; then find the equation that expresses F in terms of C.

PROBLEM 7-48 Using the model developed in Problem 7-47, find C if $F = 68°$ and find F if $C = 24°$.

PROBLEM 7-49 A line has slope $-\frac{1}{2}$ and y-intercept 2. Does the point $(8, -2)$ lie on this line?

PROBLEM 7-50 In an electric circuit the current and the voltage are related linearly. If the current i is 6 amperes, then the voltage V is 3 volts. Also, when $i = 20$, then $V = 10$. Find the equation relating i to V. Use this to find the voltage when the current is 12 amperes.

PROBLEM 7-51 Assume a \$16 000 automobile depreciates 10% of its original cost each year. Find an equation to represent its current value V after t years. When does it become worthless?

PROBLEM 7-52 Find the point of intersection of the lines $y = -3x$ and $x - y = 4$.

PROBLEM 7-53 The following lines form a rectangle:

$$y = \tfrac{1}{2}x + 4, \quad y = \tfrac{1}{2}x - 1, \quad y = -2x + 3, \quad y = -2x + 6$$

What is the area of the rectangle? What are the vertices of the rectangle?

PROBLEM 7-54 Use the given information to write the equation of the line that

(**a**) passes through $(4, 1)$ and $(-2, 7)$
(**b**) passes through $(-1, 4)$ and is horizontal
(**c**) has slope 3 and x-intercept 5
(**d**) passes through $(0, 4)$ and $(-4, 0)$
(**e**) passes through $(7, 4)$ and is perpendicular to $y = 2x - 3$
(**f**) passes through $(-4, 2)$ and is parallel to the y-axis

PROBLEM 7-55 Graph the solution set of $y \le 4$.

PROBLEM 7-56 Graph the inequality $-x + 3y > -2$.

Answers to Supplementary Exercises

7-27 See Figure 7-46.

A is in Quadrant II.
B is in Quadrant IV.
C is in Quadrant III.

D is in Quadrant I.
E is not in any quadrant.

7-28 (**a**) $\sqrt{89}$ (**b**) $\sqrt{110}$

7-29 $(2, 6)$ or $(2, -2)$

7-30 (a) III (b) I (c) I
(d) IV (e) IV (f) II
(g) I

7-31 $\sqrt{29} + \sqrt{97} + \sqrt{52}$

7-32 12

7-33 (a) $7\sqrt{2}$ (b) $(\frac{3}{2}, \frac{1}{2})$

7-34 Perimeter $= 6 + 2\sqrt{73}$
Area $= 24$

7-35 $x = 11$ or $x = -1$

7-36 Yes—The length of each side is 5.

7-37 $(3 + \sqrt{15}, 0)$ and $(3 - \sqrt{15}, 0)$

7-38 The diagonals intersect at $\left(\frac{c}{2}, \frac{c}{2}\right)$.

7-39 (a) $d = 60\sqrt{2}$ ft $\simeq 84.84$ ft
(b) The pitcher stands *behind* the imaginary line connecting first base to third base.

7-40 Herb should use Bruce Trucking: Ace charges $487.50; Bruce charges $416.23.

7-41 $(-3, -6)$ and $(5, -6)$
$(-3, 10)$ and $(5, 10)$
$(1, 6)$ and $(1, -2)$

7-42 (a) $m = 2;\ y = 2x + 1$
(b) $m = -\frac{1}{4};\ y = -\frac{1}{4}x + \frac{9}{4}$
(c) $m = \frac{3}{2};\ y = \frac{3}{2}x - 6$

7-43 (a) $2x - y = 6$
(b) $x + 2y = 9$
(c) $\frac{x}{2} + \frac{y}{5} = 1$ or $5x + 2y = 10$
(d) $3x + 2y = -2$
(e) $4x - y = -8$

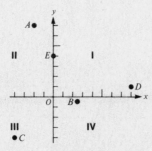

Figure 7-46

7-44 $(2, -1), (3, 1),$ and $(4, 3)$

7-45 36 000 in 1990; 43 500 in 2000

7-46 $4x - y = -9$

7-47 $C = \frac{5}{9}(F - 32); F = \frac{9}{5}C + 32$

7-48 $C = 20; F = 75.2$

7-49 Yes

7-50 $i - 2V = 0; V = 6$

7-51 $V = 16\,000 - 1600t$; 10 years

7-52 $(1, -3)$

7-53 Area $= 6$
Vertices: $\left(-\frac{2}{5}, \frac{19}{5}\right), \left(\frac{4}{5}, \frac{22}{5}\right), \left(\frac{8}{5}, -\frac{1}{5}\right), \left(\frac{14}{5}, \frac{2}{5}\right)$

7-54 (a) $x + y = 5$ (b) $y = 4$
(c) $y = 3x - 15$ (d) $-x + y = 4$
(e) $x + 2y = 15$ (f) $x = -4$

7-55 See Figure 7-47.

7-56 See Figure 7-48.

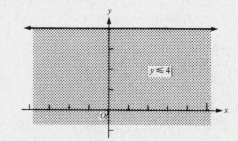

Figure 7-47

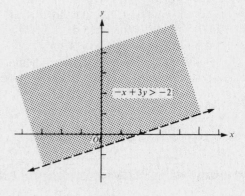

Figure 7-48

8 RELATIONS, FUNCTIONS, GRAPHS

THIS CHAPTER IS ABOUT

☑ **Relations**

☑ **Functions**

☑ **Special Functions**

☑ **Graphing Functions**

☑ **Composite and Inverse Functions**

☑ **Variation**

8-1. Relations

A. Definition of a relation

- A **relation** is any set of ordered pairs.

Suppose we have sets $A = \{x_1, x_2, x_3, x_4\}$ and $B = \{y_1, y_2, y_3, y_4\}$. We can form a relation by *pairing* the elements of set A with those of set B.

EXAMPLE 8-1 Form two relations using sets $A = \{x_1, x_2, x_3, x_4\}$ and $B = \{y_1, y_2, y_3, y_4\}$.

Solution One possible pairing is

$$x_1 \longrightarrow y_1$$
$$x_2 \longrightarrow y_2$$
$$x_3 \longrightarrow y_3$$
$$x_4 \longrightarrow y_4$$

And we can represent this relation as the set of ordered pairs $\{(x_1, y_1), (x_2, y_2), (x_3, y_3), (x_4, y_4)\}$.

Another possible pairing is

$$x_1 \longrightarrow y_3$$
$$x_2 \longrightarrow y_4$$
$$x_3 \longrightarrow y_1$$
$$x_4 \longrightarrow y_2$$

So, using ordered-pair notation, we have the relation $\{(x_1, y_3), (x_2, y_4), (x_3, y_1), (x_4, y_2)\}$.

Relations may be *finite* or *infinite*. The two relations formed in Example 8-1 are finite relations—they contain a finite number of ordered pairs. The set of ordered pairs satisfying the equation $3x + 5y = 2$ is an example of an infinite relation—there are infinitely many points on the line described by $3x + 5y = 2$.

B. Domain and range

Given that a relation is a set of ordered pairs (x, y):

- The set of first components (or x-coordinates) is called the **domain** of the relation.
- The set of second components (or y-coordinates) is called the **range** of the relation.

EXAMPLE 8-2   Find the domain and range of the relation

$$\{(1, -2), (2, -4), (2, 5), (3, 0), (3, 7)\}$$

Solution   The domain is the set of first components in the relation:

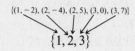

$$\{1, 2, 3\}$$

note: Elements that occur in more than one pair (such as 2 and 3 here) are listed only once.

The range is the set of second components in the relation:

$$\{(1, -2), (2, -4), (2, 5), (3, 0), (3, 7)\}$$

$$\{-4, -2, 0, 5, 7\}$$

EXAMPLE 8-3   Find the domain and range of the relation

$$\{(3, 1), (-3, 2), (-3, 1), (-1, 3), (-1, 2), (2, 2)\}$$

Sketch the graph of the relation.

Solution   The domain is the set $\{-3, -1, 2, 3\}$. The range is the set $\{1, 2, 3\}$. To graph the relation, we simply plot the set of ordered pairs, as shown in Figure 8-1.

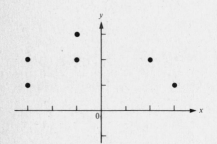

Figure 8-1

We can also find the domain and range of a relation from its graph. To find the domain, we project all of the points up or down to the x-axis. The collection of values on the x-axis is the domain. To find the range, we project all of the points onto the y-axis. The collection of values on the y-axis is the range.

EXAMPLE 8-4   Find the domain and range of the relation in Example 8-3 from its graph.

Solution   As Figure 8-2 shows, the values projected onto the x-axis are $-3, -1$, 2, and 3, and the values projected onto the y-axis are 1, 2, and 3. Thus the domain is the set $\{-3, -1, 2, 3\}$ and the range is the set $\{1, 2, 3\}$—which is exactly what we found in Example 8-3.

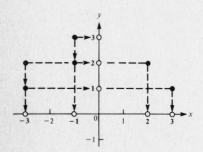

Figure 8-2

8-2. Functions

A. Definition of a function

- A **function** is a relation with the property that, for each value of the first component in any ordered pair, there is *only one* value of the second component.

In other words, in order for a relation to be a function, no two different ordered pairs can have the same x-coordinate.

EXAMPLE 8-5 **(a)** Is the relation $\{(3, 5), (4, 6), (5, 7)\}$ a function? **(b)** Is the set $\{(3, 5), (3, 6), (5, 7)\}$ a function?

Solution

(a) Yes. The relation $\{(3, 5), (4, 6), (5, 7)\}$ is a function, since no two of its ordered pairs have the same x-coordinate.
(b) No. The set $\{(3, 5), (3, 6), (5, 7)\}$ is a relation but not a function, since two different ordered pairs—$(3, 5)$ and $(3, 6)$—have the same x-coordinate. Or, to put it another way, the set cannot be a function since, for the value 3 in the first component, there are *two* values—5 and 6—in the second component.

> *note:* For a set of ordered pairs to be a function, different values of y MAY NOT have the same value of x, but different values of x MAY have the same value of y.

EXAMPLE 8-6 Which of the following relations is a function?

(a) $\{(-2, 1), (2, -4), (1, 3), (3, 3)\}$ **(b)** $\{(-1, 4), (2, 2), (2, 6), (3, 5)\}$

Solution In each case, check the ordered pairs in the set.

(a)

x	y

$-2 \longrightarrow 1$
$2 \longrightarrow -4$ This relation IS a function, since different values of x may
$\left.\begin{array}{c}1 \\ 3\end{array}\right\rangle 3$ have the same value of y.

(b)

x	y

$-1 \longrightarrow 4$
$2 \left\langle\begin{array}{c}2 \\ 6\end{array}\right.$ This relation is NOT a function, since different values of y
$3 \longrightarrow 5$ may not have the same value of x.

We have described a function as a set of ordered pairs, but we can also say that a function consists of a *domain* and a *rule*:

- The *domain* is a collection of real numbers and the *rule* assigns to each number in that domain a unique number.
- The total collection of numbers that the function assigns to members of the domain is called the *range* of the function.

EXAMPLE 8-7 Consider the function defined by the following rule:

$$\{(x, y): \ y = 3x - 1, \text{ where } x \text{ is any real number}\}$$

Find some members of the function and graph the function.

Solution We can make a table of ordered pairs for the function just as we do for equations (see Chapter 7). We find members of the function by assigning any real number to x and calculating the corresponding y by the "rule."

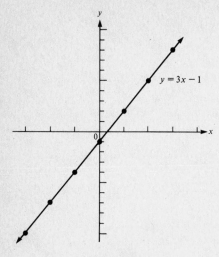

Figure 8-3

$y = 3x - 1$		Function pairs
x	y	(x, y)
-3	-10 ⟶	$(-3, -10)$
-2	-7 ⟶	$(-2, -7)$
-1	-4 ⟶	$(-1, -4)$
0	-1 ⟶	$(0, -1)$
1	2 ⟶	$(1, 2)$
2	5 ⟶	$(2, 5)$
3	8 ⟶	$(3, 8)$

The graph is shown in Figure 8-3.

note: Since we are free to choose any x from the domain, we call x the **independent variable**. But, since the y value that is matched with any x "depends on" the value of that x, we call y the **dependent variable**. We're not restricted to x and y, though—for the function with the rule $s = \frac{1}{2}t^2 + 3t + 5$, the independent variable is t and the dependent variable is s.

B. Vertical line test

When a relation is given as a set of ordered pairs, we can determine if it is a function by comparing the x-coordinates of each pair—if no two pairs have the same x-coordinate, the relation is a function. If the relation is given in terms of a domain and a rule, we can use the graph of the relation to determine if it is a function.

Since no two distinct points on the graph of a function can have the same x-coordinate and since a vertical line has the property that all of its points have the same x-coordinate, it follows that no vertical line intersects the graph of a function more than once. Thus

- A given relation is a function if its graph passes the **vertical line test**, i.e., if no vertical line intersects the graph more than once.

EXAMPLE 8-8   Which of the graphs in Figure 8-4 are graphs of functions?

(a) $y = x^2 - 1$

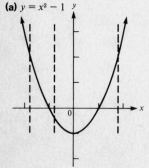

(b) $x = y^2$

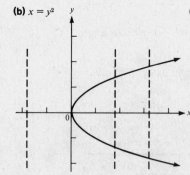

(c) $x^2 + \frac{1}{4}y^2 = 1$

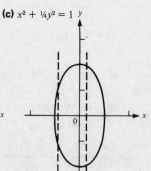

(d) $y = \frac{1}{2}x + 1$

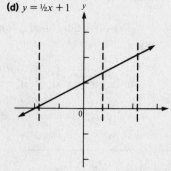

Figure 8-4

Solution

(a) The graph of $y = x^2 - 1$ is a graph of a function, since all vertical lines intersect the graph exactly once.

(b) The graph of $x = y^2$ is not a graph of a function, since many vertical lines intersect the graph more than once.

(c) The graph of $x^2 + \frac{1}{4}y^2 = 1$ is not a graph of a function, since many vertical lines intersect the graph more than once.

(d) The graph of $y = \frac{1}{2}x + 1$ is a graph of a function, since all vertical lines intersect the graph exactly once.

C. Domain and range

The definition of the domain and range of a relation holds for functions: The domain of a function is the set of all x-coordinates and the range is the set of all y-coordinates.

EXAMPLE 8-9 Find the domain and range of each function:

(a) $\{(-1, 2), (1, -3), (2, 5), (4, 1)\}$ (b) $\{(0, 5), (1, 5), (2, 5), (3, 5)\}$

Solution

(a) The domain is the set $\{-1, 1, 2, 4\}$. The range is the set $\{-3, 1, 2, 5\}$.

(b) The domain is the set $\{0, 1, 2, 3\}$. The range is the set $\{5\}$.

Notice that the set in (b) describes a function, since it is the y-coordinate—not the x-coordinate—that is repeated.

When a function (or relation) is given in terms of a domain and a rule, we can use the graph of the function (or relation) to find the domain and range. Obviously, we can't project *all* of the points onto the x- and y-axes (as we can for finite sets of ordered pairs), but the curve itself usually provides enough clues.

EXAMPLE 8-10 Find the domain and range of each function or relation in Example 8-8.

Solution See Figure 8-4.

(a) The domain is the set of all real numbers. The range is the set $\{y \geq -1\}$.

(b) The domain is the set $\{x \geq 0\}$. The range is the set of all real numbers.

(c) The domain is the set $\{-1 \leq x \leq 1\}$. The range is the set $\{-2 \leq y \leq 2\}$.

(d) The domain is the set of all real numbers and the range is also the set of all real numbers.

---

note: If it is not specified, the domain of a function or relation defined by a rule (equation) is assumed to be the set of all real numbers—provided that substituting a member of the domain into the equation does not result in an undefined term. Consider, for example, the equation $x = y^2$ or $y = \pm\sqrt{x}$. In this case, the domain must be restricted to $x \geq 0$, since the square root of a negative number is not defined. Similarly, the domain of $\dfrac{1}{x - 3}$ includes all real numbers except $x = 3$, since division by zero is undefined.

D. Function notation and terminology

Functions are normally denoted by the letters f, g, h, F, G, or H, and the value in the range assigned by the function f to a value x in the domain is denoted by $f(x)$, read "f of x." If, for example, the function f assigns to domain value 5 the range value 3, we write $f(5) = 3$ and read "f of $5 = 3$."

- If x is in the domain of a function f, we say that f is *defined* at x. If x is not in the domain of f, we say that f is *undefined* at x.
- When a function is defined by an equation or rule such as $y = 2x + 1$, we use $f(x)$ in place of the letter y and write $f(x) = 2x + 1$.
- To evaluate a function f at a particular value of x, we find the value $f(x)$ for that x.

EXAMPLE 8-11   If $f(x) = x^2 + x - 2$, find $f(0)$, $f(2)$, and $f(a)$.

Solution   In each case, substitute the value in the parentheses for x in the equation $f(x) = x^2 + x - 2$:

$$f(0) = (0)^2 + 0 - 2 = 0 + 0 - 2 = -2$$

$$f(2) = (2)^2 + 2 - 2 = 4 + 2 - 2 = 4$$

$$f(a) = (a)^2 + a - 2 = a^2 + a - 2$$

EXAMPLE 8-12   If $F(x) = x^3 + 5x - 3$, find $F(-1)$, $F(3)$, $F(t)$, $F(r^2)$, and $F(a + h)$.

Solution $F(-1) = (-1)^3 + 5(-1) - 3 = -1 - 5 - 3 = -9$

$$F(3) = (3)^3 + 5(3) - 3 = 27 + 15 - 3 = 39$$

$$F(t) = (t)^3 + 5(t) - 3 = t^3 + 5t - 3$$

$$F(r^2) = (r^2)^3 + 5(r^2) - 3 = r^6 + 5r^2 - 3$$

$$F(a + h) = (a + h)^3 + 5(a + h) - 3$$

Beware of the following traps when dealing with function notation:

(1) $f(x)$ is NOT $f \times x$: $f(x)$ is a notation for a value in the range of a function.
(2) $f(x^2)$ is NOT $[f(x)]^2$: $f(x^2)$ is the notation for the value in the range corresponding to the domain value x^2.
(3) $f(4x)$ is NOT $4 \times f(x)$: $f(4x)$ is the notation for the value in the range corresponding to the domain value $4x$.
(4) $f(x + 2)$ is NOT $f(x) + f(2)$; NOR is it $f(x) + 2$: $f(x + 2)$ is the notation for the value in the range corresponding to the domain value $(x + 2)$.

EXAMPLE 8-13   Given the function $g(x) = x^2 + 5$, find $g(2)$, $g(x^2)$, $g(4x)$, $g(x + 2)$, and $g(x + a) - g(a)$.

Solution

$$g(2) = (2)^2 + 5 = 4 + 5 = 9$$

$$g(x^2) = (x^2)^2 + 5 = x^4 + 5$$

$$g(4x) = (4x)^2 + 5 = 16x^2 + 5$$

$$g(x + 2) = (x + 2)^2 + 5 = x^2 + 4x + 4 + 5$$

$$= x^2 + 4x + 9$$

$$g(x + a) - g(a) = [(x + a)^2 + 5] - [(a)^2 + 5]$$

$$= x^2 + 2ax + a^2 + 5 - a^2 - 5$$

$$= x^2 + 2ax$$

8-3. Special Functions

There are some special functions whose properties and shapes it's useful to know.

- A function is called a **one-to-one function** if each member of the range is paired with exactly one member of the domain.

We can tell whether a function is one-to-one by looking at its graph—if no horizontal line crosses the graph more than once, the function is one-to-one. This test is called the **horizontal line test**.

(a) $f(x) = 2x + 4$

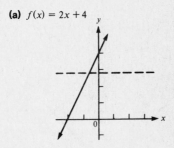

EXAMPLE 8-14 Which of the following is a one-to-one function?

(a) $f(x) = 2x + 4$ **(b)** $g(x) = x^2$

Graph each function.

Solution

(a) $f(x) = 2x + 4$ is a one-to-one function, since each member of the range is paired with exactly one member of the domain; i.e., different values of x will always produce different values of y (see Figure 8-5a).

(b) $g(x) = x^2$ is not a one-to-one function, because each member of the range—excluding 0—is paired with two members of the domain; e.g., if $g(x) = 4$, then $x = 2$ and $x = -2$ (see Figure 8-5b).

Notice that the graph of $f(x) = 2x + 4$ passes the horizontal line test, but the graph of $g(x) = x^2$ fails the test—any horizontal line drawn above the x-axis crosses the graph in two points.

(b) $g(x) = x^2$

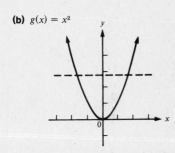

Figure 8-5

- The function $f(x) = x$ is called the **identity function**.
- If a function has the property that only one range value c is paired with every value in the domain, that function is called a **constant function** and is denoted by $f(x) = c$. The graph of a constant function is a horizontal line crossing the y-axis at $y = c$.

If a function is not constant, it is either *increasing* or *decreasing*. We usually say that a function is increasing or decreasing over an interval or subset of the domain.

- A function is **increasing** over an interval I if $f(x_1) < f(x_2)$ whenever $x_1 < x_2$ for any x_1 and x_2 in I.
- A function is **decreasing** over an interval I if $f(x_1) > f(x_2)$ whenever $x_1 < x_2$ for any x_1 and x_2 in I.

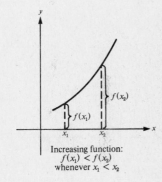

Increasing function:
$f(x_1) < f(x_2)$
whenever $x_1 < x_2$

EXAMPLE 8-15 Tell where the function is increasing or decreasing for **(a)** $f(x) = 2x + 4$ and **(b)** $f(x) = x^2$.

Solution   See the graphs in Figure 8-5.

(a) The function f defined by $f(x) = 2x + 4$ is increasing over the entire domain, since for all x in the domain $f(x_1) < f(x_2)$ whenever $x_1 < x_2$. Notice that the graph of $f(x) = 2x + 4$ is a rising line.

note: In general, a rising line indicates an increasing function and a falling line indicates a decreasing function.

(b) The function f defined by $f(x) = x^2$ is decreasing over the interval $(-\infty, 0)$ and increasing over the interval $(0, \infty)$. As Figure 8-5b shows, the left side of the graph (where $x < 0$) is "falling" and the right side (where $x > 0$) is "rising."

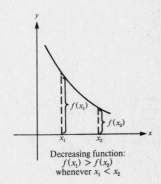

Decreasing function:
$f(x_1) > f(x_2)$
whenever $x_1 < x_2$

- A function that can be written in the form $f(x) = mx + b$, where m and b are real numbers ($m \neq 0$), is called a **linear function**. The graph of a linear function is a straight line.
- A function that can be written in the form $f(x) = ax^2 + bx + c$, where $a, b,$ and c are real numbers ($a \neq 0$), is called a **quadratic function**. The graph of a quadratic function is a *parabola*. (We discuss parabolas in detail in Chapter 9, but they're not new to us—the curve in Figure 8-5b is a parabola.)
- A function of the form $f(x) = |x|$, with domain all real numbers x, is called the **absolute value function**.

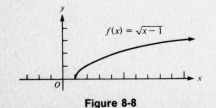

$f(x) = |x|$

Figure 8-6

EXAMPLE 8-16 Given the absolute value function $f(x) = |x|$, (a) graph the function, (b) find the domain and range of the function, and (c) tell whether the function is one-to-one.

Solution

(a) If $f(x) = |x|$, then $f(0) = 0$, $f(1) = 1$, $f(-1) = 1$; $f(2) = 2$, $f(-2) = 2$; $f(3) = 3, f(-3) = 3$; and so on. The graph is drawn in Figure 8-6.
(b) The domain is the set of all real numbers. The range is the set of all nonnegative real numbers, since the absolute value of any number is always positive—or 0.
(c) The graph fails the horizontal line test, so it is not a one-to-one function. This is what we would expect, since both $|x| = x$ and $|-x| = x$.

- The function defined by $f(x) = \sqrt{x}$, where x is a nonnegative real number, is called the **square root function**.
- A function of the form $f(x) = b^x$, where b is a positive real number and $b \neq 1$, is called an **exponential function**.

(a)

$f(x) = \sqrt{x}$

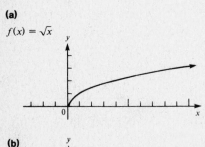

(b)

$f(x) = 2^x$

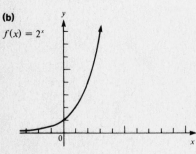

Figure 8-7

EXAMPLE 8-17 (a) Graph $f(x) = \sqrt{x}$ and find its domain and range.
  (b) Graph $f(x) = 2^x$ and find its domain and range.

Solution

(a) If $f(x) = \sqrt{x}$, then $f(0) = 0$, $f(1) = 1$, $f(4) = 2$, $f(9) = 3$, and so on. The graph is drawn in Figure 8-7a. The domain is the set of nonnegative real numbers, and so is the range.
(b) If $f(x) = 2^x$, then $f(0) = 1$, $f(1) = 2$, $f(2) = 4$, $f(3) = 8$, $f(-1) = \frac{1}{2}$, $f(-2) = \frac{1}{4}, f(-3) = \frac{1}{8}$, and so on. The curve is shown in Figure 8-7b. The domain is the set of all real numbers, but the range includes only positive real numbers (since 2 raised to any power can't be zero or negative). Notice also that the curve approaches the x-axis but never crosses it—there is no x such that $2^x = 0$.

8-4. Graphing Functions

We can graph a function by constructing a table of solution pairs—just as we can for equations (Chapter 7). But there are more efficient, and easier, ways. We can use our knowledge of functions and their properties to help us in graphing. The special functions discussed in the last section are important because we'll see their properties and shapes again and again—with various modifications. Thus, for example, graphing $f(x) = \sqrt{x - 1}$ (Figure 8-8) becomes easier when we remember the shape of the basic curve described by $f(x) = \sqrt{x}$ (Figure 8-7a).

A. Symmetry

A graph of a function is *symmetric with respect to the y-axis* if that portion of the graph to the left of the vertical axis is the exact mirror image of that

$f(x) = \sqrt{x - 1}$

Figure 8-8

portion of the graph to the right of the vertical axis. The graph of $f(x) = x^2$ (Figure 8-5b) displays this type of symmetry. Notice that if the point (x, y) is on the graph, then the point $(-x, y)$ is also on the graph. In function notation, this means that $f(-x) = f(x)$.

- Any function with the property that $f(-x) = f(x)$ for all x in the domain of f is an **even function** and its graph is symmetric with respect to the y-axis.

Other examples of even functions are $f(x) = x^2 - 2$, $f(x) = x^4$, $f(x) = 1/x^2$, $f(x) = -x^6$, and $f(x) = |x|$.

The graph of $f(x) = x^3$ (Figure 8-9) is *symmetric with respect to the origin*. Notice that if the point (x, y) is on the graph, then the point $(-x, -y)$ is also on the graph. In function notation, this means that $f(-x) = -f(x)$.

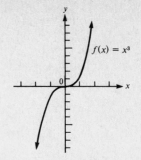

Figure 8-9

- Any function with the property that $f(-x) = -f(x)$ for all x in the domain of f, is an **odd function** and its graph is symmetric with respect to the origin.

Other odd functions include $f(x) = x$, $f(x) = x^5$, $f(x) = x^{-3}$, and $f(x) = x^3 - x$.

note: Polynomial functions that have only even exponents of x are even functions and those with only odd exponents of x are odd functions. (The absolute value function is an exception.)

B. Shifting

1. Vertical shifts

- If we have a function that is related to another by the formula $g(x) = f(x) + c$, where c is a constant, then we say that the graph of the function g is a **vertical shift** of the graph of the original function f.

This is extremely helpful in graphing when we are familiar with the graph of f. To graph g, we simply take the graph of f and move it up c units if $c > 0$ and down $|c|$ units if $c < 0$.

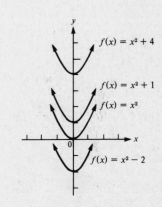

Figure 8-10

EXAMPLE 8-18 Graph $f(x) = x^2$. Then on the same set of axes graph $f(x) = x^2 - 2$, $f(x) = x^2 + 1$, and $f(x) = x^2 + 4$.

Solution The graph of $f(x) = x^2$ is the same as the simple upward-opening parabola shown in Figure 8-5b.

The graph of $f(x) = x^2 - 2$ is the graph of $f(x) = x^2$ shifted down $|-2| = 2$ units, since $c < 0$.
The graph of $f(x) = x^2 + 1$ is the graph of $f(x) = x^2$ shifted up 1 unit, since $c > 0$.
The graph of $f(x) = x^2 + 4$ is the graph of $f(x) = x^2$ shifted up 4 units, since $c > 0$.

The four curves are sketched in Figure 8-10. Notice that if the point (x, y) is on the graph of $f(x) = x^2$, then the points $(x, y - 2)$, $(x, y + 1)$, and $(x, y + 4)$ are on the graphs of $f(x) = x^2 - 2$, $f(x) = x^2 + 1$, and $f(x) = x^2 + 4$, respectively.

EXAMPLE 8-19 Graph $f(x) = |x|$, $f(x) = |x| + 2$, and $f(x) = |x| - 3$ on the same set of axes.

Solution See Figure 8-11.

The absolute value function $f(x) = |x|$ was graphed earlier in Figure 8-6.
To graph $f(x) = |x| + 2$, we move the graph of $f(x) = |x|$ up 2 units.
To graph $f(x) = |x| - 3$, we move the graph of $f(x) = |x|$ down 3 units.

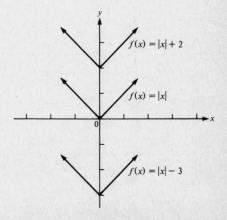

Figure 8-11

2. Horizontal shifts

- If we have a function that is related to another by the formula $g(x) = f(x - c)$, where c is a constant, then we say that the graph of g is a **horizontal shift** of the graph of f.

  To graph g, we simply take the graph of f and move it c units to the right if $c > 0$ and $|c|$ units to the left if $c < 0$.

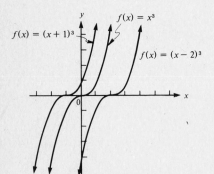

Figure 8-12

EXAMPLE 8-20   Graph $f(x) = x^3$. Then graph $f(x) = (x - 2)^3$ and $f(x) = (x + 1)^3$ on the same set of axes.

Solution   See Figure 8-12.

The graph of the odd function $f(x) = x^3$ was sketched in Figure 8-9.

The graph of $f(x) = (x - 2)^3$ is the graph of $f(x) = x^3$ shifted 2 units to the right, since $c > 0$.

The graph of $f(x) = (x + 1)^3$ is the graph of $f(x) = x^3$ shifted 1 unit to the left, since $c < 0$. [Note that $x + 1 = x - (-1)$.]

C. Piecewise functions

In the definition of some functions, explicit restrictions are placed on the domain, so that different rules are used over distinct portions of the domain. These functions are often referred to as **piecewise functions**.

EXAMPLE 8-21   Sketch the graph of the function defined by

$$f(x) = \begin{cases} 1 - 2x & \text{if } x < 2 \\ -1 + x & \text{if } x \geq 2 \end{cases}$$

Solution   We need to compute $f(x)$ in two separate batches—one for x when $x < 2$ and one for x when $x \geq 2$:

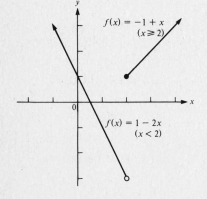

Figure 8-13

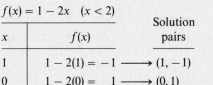

$f(x) = 1 - 2x$	$(x < 2)$	Solution pairs		$f(x) = -1 + x$	$(x \geq 2)$	Solution pairs
x	$f(x)$			x	$f(x)$	
1	$1 - 2(1) = -1 \longrightarrow$	$(1, -1)$		2	$-1 + 2 = 1 \longrightarrow$	$(2, 1)$
0	$1 - 2(0) = \ \ \ 1 \longrightarrow$	$(0, 1)$		4	$-1 + 4 = 3 \longrightarrow$	$(4, 3)$

Notice that since both parts of the function are *linear* functions, we need to find only two points on the graph of each. Since the domain for $f(x) = 1 - 2x$ is $x < 2$, we draw a line through $(1, -1)$ and $(0, 1)$ but "stop" the line at $x = 2$, as shown in Figure 8-13. (We draw an open circle at $x = 2$ since 2 is not included in the domain.) Next we sketch a line through $(2, 1)$ and $(4, 3)$ with an endpoint at $(2, 1)$, since the domain of $f(x) = -1 + x$ includes only those values of x greater than or equal to 2. (We draw a closed circle at $x = 2$ to indicate that the point is included in the domain.)

EXAMPLE 8-22   Sketch the graph of the function defined by

$$f(x) = \begin{cases} 1 & \text{if } x < 0 \\ x - 1 & \text{if } 0 < x < 2 \\ 2 & \text{if } x > 2 \end{cases}$$

Solution   See Figure 8-14.

Over the domain $x < 0$, f is a constant function—its graph is the horizontal line $y = 1$, restricted to the second quadrant.

Figure 8-14

Over the domain $0 < x < 2$, f is a linear function—its graph is a rising line segment between the points $(0, -1)$ and $(2, 1)$.

Over the domain $x > 2$, f is a constant function—its graph is the horizontal line $y = 2$, restricted to the first quadrant with $x > 2$.

The function is undefined at $x = 0$ and $x = 2$.

D. Restricted domains

When certain functions are defined by a rule, the domain of the function must be restricted to a specific subset of the real number line. Recognizing this restriction can help us in graphing those functions.

EXAMPLE 8-23   Given $f(x) = \sqrt{x - 3}$, find the domain of the function and graph the function.

Solution   Since the square root of a negative number is undefined, the domain of f must be restricted so that the quantity under the radical sign is not negative. This means that $x - 3$ must not be negative, so $x - 3 \geq 0$ or $x \geq 3$. The domain is the set $\{x \geq 3\}$. We can now construct a table of solution pairs, choosing values of x that make our computations simpler:

$$f(x) = \sqrt{x - 3}$$

x	$f(x)$	Solution pairs
3	$\sqrt{3 - 3} = \sqrt{0} = 0 \longrightarrow$	$(3, 0)$
4	$\sqrt{4 - 3} = \sqrt{1} = 1 \longrightarrow$	$(4, 1)$
7	$\sqrt{7 - 3} = \sqrt{4} = 2 \longrightarrow$	$(7, 2)$
12	$\sqrt{12 - 3} = \sqrt{9} = 3 \longrightarrow$	$(12, 3)$

The points are connected to form a smooth curve, as in Figure 8-15.

> *note:* (1) The rule of the function also tells us the range of the function. Since the square root of a number is always nonnegative, the range cannot include negative real numbers. Thus we know that the graph of f is restricted to the first quadrant with $x \geq 3$.
>
> (2) We can use an alternative method of graphing if we recognize $f(x) = \sqrt{x - 3}$ as a variation of the square root function. We could simply draw the square root function shifted 3 units to the right.

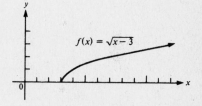

Figure 8-15

8-5. Composite and Inverse Functions

A. Composite functions

• If f and g are two functions, then the **composite function** $g \circ f$ is defined by the rule

$$(g \circ f)(x) = g(f(x))$$

note: The composition of two functions may be denoted by either $(g \circ f)(x)$—using the special operational symbol "$\circ$"—or $g(f(x))$. Both are read as "g of f of x."

EXAMPLE 8-24   Let functions f and g be defined by $f(x) = x^2$ and $g(x) = 2x$, respectively. Find the rules for the composite functions $(g \circ f)(x)$ and $(f \circ g)(x)$. Evaluate each function at $x = 3$.

Solution First find $(g \circ f)(x)$. Since $(g \circ f)(x) = g(f(x))$, we simply substitute the entire function $f(x)$ for x in $g(x)$. Then, for $f(x) = x^2$ and $g(x) = 2x$,

$$g(f(x)) = g(x^2) = 2(x^2) = 2x^2$$

$$\text{If } x = 3, \text{ then } (g \circ f)(x) = 2x^2 = 2(3)^2 = 18.$$

Now find $(f \circ g)(x)$.

Since $(f \circ g)(x) = f(g(x))$, we simply substitute the function $g(x)$ for x in $f(x)$. Then, for $g(x) = 2x$ and $f(x) = x^2$,

$$f(g(x)) = f(2x) = (2x)^2 = 4x^2$$

If $x = 3$, then $(f \circ g)(x) = 4x^2 = 4(3)^2 = 36$.

note: We arrive at the same value of $(f \circ g)(x)$ if we compute it the "long" way: If $x = 3$, then $g(x) = 6$. Then $f(g(x)) = f(6) = 36$.

- The domain of the composite function $g \circ f$ consists of all numbers x in the domain of f such that $f(x)$ is in the domain of g.

We need to be careful when specifying the domain of a composite function, since the domain may not be apparent from the rule of the composite function alone.

EXAMPLE 8-25 Let $f(x) = \sqrt{2x}$ and $g(x) = x^2 + 1$. Find the domain and rule for each of the composite functions $(g \circ f)(x)$ and $(f \circ g)(x)$.

Solution First find $(g \circ f)(x)$.

For $f(x) = \sqrt{2x}$ and $g(x) = x^2 + 1$,

$$(g \circ f)(x) = g(f(x)) = g(\sqrt{2x}) = (\sqrt{2x})^2 + 1 = 2x + 1$$

Notice that there are no *apparent* restrictions on the domain of $g \circ f$; i.e., the rule $2x + 1$ does not place any restrictions on the domain. We know from the definition above that the domain of $g \circ f$ consists of all numbers x in the domain of f such that $f(x)$ is in the domain of g. The domain of g includes all real numbers, so it follows that $f(x)$ can be any real number. Thus the only condition on the domain of $g \circ f$ is that x be in the domain of f. But, since $f(x) = \sqrt{2x}$, the domain of f must exclude negative numbers, and so the domain of $g \circ f$ is the set of all nonnegative real numbers.

Now find $(f \circ g)(x)$.

For $f(x) = \sqrt{2x}$ and $g(x) = x^2 + 1$,

$$(f \circ g)(x) = f(g(x)) = f(x^2 + 1) = \sqrt{2(x^2 + 1)} = \sqrt{2x^2 + 2}$$

The domain of $f \circ g$ consists of all numbers x in the domain of g such that $g(x)$ is in the domain of f. The domain of f is the set of nonnegative real numbers and $g(x)$ is in that domain since $g(x)$ will always be positive (for any x, the quantity $x^2 + 1$ is greater than or equal to 1). So the domain of $f \circ g$ is the same as the domain of g, which is the set of all real numbers.

B. Inverse functions

1. Definition of an inverse function

Consider a function f defined by $y = f(x)$, which takes a value x and produces a value y to form the solution pair (x, y). It is often useful to reverse this process, i.e., start with a value y and produce the corresponding value of x.

EXAMPLE 8-26   The function that converts degrees Fahrenheit to degrees Celsius is defined by the rule $C = \frac{5}{9}(F - 32)$. Find the function for the reverse process, i.e., for converting degrees Celsius to degrees Fahrenheit.

Solution   In the given equation, C is the dependent variable and F is the independent variable. In order to find the equation in which F is the dependent variable and C is the independent variable, we solve the given equation for F:

$$C = \tfrac{5}{9}(F - 32)$$

$$9C = 5(F - 32) = 5F - 160$$

$$5F = 9C + 160$$

$$F = \tfrac{9}{5}C + 32$$

Let's explore the results of Example 8-26 further. We now have one function that takes an F value and converts it to C, creating the pair (F, C), and a second function that takes a C value and converts it to F, creating the pair (C, F). We'll define these functions by $f(x) = \frac{5}{9}(x - 32)$ and $g(x) = \frac{9}{5}x + 32$, respectively.

If $y = f(x)$, then $y = \frac{5}{9}(x - 32)$. Then

$$g\big(f(x)\big) = g(y) = g\big(\tfrac{5}{9}(x - 32)\big) = \tfrac{9}{5}\big(\tfrac{5}{9}(x - 32)\big) + 32$$

$$= (\tfrac{9}{5} \cdot \tfrac{5}{9})(x - 32) + 32$$

$$= x - 32 + 32$$

$$= x$$

Thus, if $y = f(x)$, then $x = g(y)$.

The converse is also true: If $x = g(y)$, then $y = f(x)$. We can express this relationship between the two functions f and g by saying that the function g is the inverse of the function f (and that the function f is the inverse of the function g). In general,

- A function f has an **inverse** if there is a function g such that

  (1) the domain of g is the range of f
  (2) $f(x) = y$ if and only if $g(y) = x$ for all x in the domain of f and all y in the range of f.
- The function g is called the inverse of f and is usually denoted as f^{-1}.

Caution: $f^{-1}(x)$ *does not* mean $\dfrac{1}{f(x)}$; it is the notation for the inverse of a function f. If we wish to denote the reciprocal of $f(x)$, we write $[f(x)]^{-1}$.

To find the inverse of a function given in set notation, we simply interchange x and y in each pair of the set.

EXAMPLE 8-27   Find the inverse of the function $f = \{(1, 0), (3, 1), (4, 2)\}$.

Solution   Interchange x and y in each pair:

$$f = \{(1, 0), (3, 1), (4, 2)\} \longrightarrow f^{-1} = \{(0, 1), (1, 3), (2, 4)\}$$

To find the inverse of a function defined by a rule $y = f(x)$, we solve the equation for x and then interchange x and y in the equation.

EXAMPLE 8-28 Given the function f defined by $y = f(x) = 2x + 3$, find its inverse function f^{-1}.

Solution

$$y = 2x + 3 \qquad \text{[Solve the given equation for } x.\text{]}$$
$$2x = y - 3$$
$$x = \tfrac{1}{2}y - \tfrac{3}{2}$$
$$y = \tfrac{1}{2}x - \tfrac{3}{2} \qquad \text{[Interchange the variables.]}$$

Thus if $f(x) = 2x + 3$, then $f^{-1}(x) = \tfrac{1}{2}x - \tfrac{3}{2}$.

> *note:* If we substitute f^{-1} for g in our definition of an inverse function, we see that $f(x) = y$ if and only if $f^{-1}(y) = x$. Then, since $f(x) = y$, we have $f^{-1}(f(x)) = x$; and since $f^{-1}(y) = x$, we have $f(f^{-1}(y)) = y$. We can use these properties to check the formula for the inverse of a function.

EXAMPLE 8-29 Check to see that the functions of Example 8-28 satisfy the formulas $f^{-1}(f(x)) = x$ and $f(f^{-1}(y)) = y$.

Solution From Example 8-28, we have $f(x) = 2x + 3$ and $f^{-1}(x) = \tfrac{1}{2}x - \tfrac{3}{2}$. Then

$$f^{-1}(f(x)) = f^{-1}(2x + 3) = \tfrac{1}{2}(2x + 3) - \tfrac{3}{2} = x + \tfrac{3}{2} - \tfrac{3}{2} = x \quad \checkmark$$
$$f(f^{-1}(y)) = f(\tfrac{1}{2}y - \tfrac{3}{2}) = 2(\tfrac{1}{2}y - \tfrac{3}{2}) + 3 = y - 3 + 3 = y \quad \checkmark$$

Thus we can be sure that $f^{-1}(x) = \tfrac{1}{2}x - \tfrac{3}{2}$ defines the inverse function of $f(x) = 2x + 3$.

EXAMPLE 8-30 If $f(x) = 3 - x^3$, find $f^{-1}(x)$ and verify the formula.

Solution Let $f(x) = y = 3 - x^3$. Then

$$y = 3 - x^3 \qquad \text{[Solve the equation for } x.\text{]}$$
$$x^3 = 3 - y$$
$$x = \sqrt[3]{3 - y}$$
$$y = \sqrt[3]{3 - x} \qquad \text{[Interchange the variables.]}$$

Using $f(x) = 3 - x^3$ and $f^{-1}(x) = \sqrt[3]{3 - x}$, we have

$$f^{-1}(f(x)) = f^{-1}(3 - x^3) = \sqrt[3]{3 - (3 - x^3)} = \sqrt[3]{x^3} = x \quad \checkmark$$
$$f(f^{-1}(y)) = f(\sqrt[3]{3 - y}) = 3 - (\sqrt[3]{3 - y})^3 = 3 - (3 - y) = y \quad \checkmark$$

2. Graphs of inverses

- If the point (a, b) is on the graph of f, then (b, a) is on the graph of f^{-1}, and vice versa.
- The graphs of a function and its inverse are symmetric with respect to the line $y = x$.

EXAMPLE 8-31 Given $f(x) = -3x + 1$, find $f^{-1}(x)$ and graph both functions on the same set of axes.

Solution Solve $y = -3x + 1$ for x and then interchange x and y:

$$y = -3x + 1$$
$$3x = -y + 1$$
$$x = -\tfrac{1}{3}y + \tfrac{1}{3}$$
$$y = -\tfrac{1}{3}x + \tfrac{1}{3}$$
$$f^{-1}(x) = -\tfrac{1}{3}x + \tfrac{1}{3}$$

The graphs of both functions are straight lines (see Figure 8-16). Notice that they are symmetric with respect to the line $y = x$; i.e., if we folded the entire plane along the line $y = x$, the graphs of f and f^{-1} would coincide.

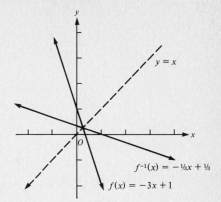

Figure 8-16

- Not every function has an inverse; only functions that are *one-to-one* have inverses.

note: This restriction is not an arbitrary one. Remember that the range of a function is the domain of its inverse. Then, since a function that is not one-to-one has more than one range value for each domain value, its inverse would have more than one domain value for each range value—which means that it is not a function!

EXAMPLE 8-32 Which of the following functions have inverses?

(a) $f(x) = x^2$ **(b)** $f(x) = x - 1$ **(c)** $f(x) = |x|$ **(d)** $f(x) = \sqrt{x}$

Solution The graphs of these functions are well-known to us. We can use the horizontal line test to determine if the function is one-to-one.

(a) $f(x) = x^2$ does not have an inverse, since it is not one-to-one. (See Fig. 8-5b.)
(b) $f(x) = x - 1$ has an inverse, since it is one-to-one.
(c) $f(x) = |x|$ does not have an inverse, since it is not one-to-one. (See Fig. 8-6.)
(d) $f(x) = \sqrt{x}$ has an inverse, since it is one-to-one. (See Fig. 8-7a.)

EXAMPLE 8-33 Given $f(x) = \sqrt{x}$, find $f^{-1}(x)$ and graph both functions on the same set of axes.

Solution Solve $y = \sqrt{x}$ for x and interchange the variables:

$$y = \sqrt{x}$$
$$\sqrt{x} = y$$
$$x = y^2 \qquad \text{[Square both sides.]}$$
$$y = x^2 \qquad \text{[Interchange the variables.]}$$

Then $f^{-1}(x) = x^2$, where $x \geq 0$. The domain of f^{-1} must be restricted to $x \geq 0$, since the range of f includes only nonnegative numbers. Also, without this restriction, the graphs would not be symmetric to the line $y = x$ (see Figure 8-17).

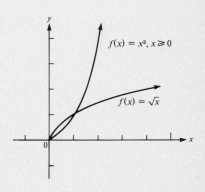

Figure 8-17

If we restrict the domain of a function that is *not* one-to-one such that the function *is* one-to-one over that domain, we can find the inverse of that function.

EXAMPLE 8-34 Given $f(x) = (x - 2)^2$, $x \geq 2$, find $f^{-1}(x)$ and graph both functions on the same set of axes.

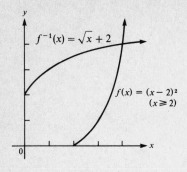

Figure 8-18

Solution   Solve $y = (x - 2)^2$ for x and interchange the variables:

$$y = (x - 2)^2$$

$$(x - 2)^2 = y$$

$$x - 2 = \sqrt{y} \qquad \text{[Take the square root of both sides.]}$$

$$x = \sqrt{y} + 2$$

$$y = \sqrt{x} + 2 \qquad \text{[Interchange the variables.]}$$

Thus $f^{-1}(x) = \sqrt{x} + 2$. The graphs of $f(x) = (x - 2)^2$, $x \geq 2$, and $f^{-1}(x) = \sqrt{x} + 2$ are shown in Figure 8-18.

8-6. Variation

Another way of describing the relationship between a given dependent variable and one or more independent variables is with the language of *variation and proportion*.

A. Direct variation

- If k is a nonzero constant and if x and y are related by the formula $y = kx$, then we say that *y* **varies directly as** *x* or *y* **is directly proportional to** *x*.

In this case, k is called the **constant of variation** or the **constant of proportionality**.

EXAMPLE 8-35   Find the formulas for the variation of x and y if

(a) x is length in feet and y is length in yards;
(b) x is liquid volume in ounces and y is liquid volume in quarts.

Solution

(a) We begin with $y = kx$, since the relationship between length in feet and yards is a direct proportionality. But we know that when $x = 3$ (feet), $y = 1$ (yard), so we substitute these values into the equation:

$$y = kx$$

$$1 = k(3)$$

$$k = \tfrac{1}{3}$$

Thus the correct equation is $y = \tfrac{1}{3}x$.

(b) Since there are 32 ounces in a quart, $x = 32$ when $y = 1$. This gives us the formula $1 = k(32)$ or $k = \tfrac{1}{32}$. The equation relating quarts and ounces is $y = \tfrac{1}{32}x$.

EXAMPLE 8-36   Find the formulas for the following relationships:

(a) y varies directly as x and $y = 12$ when $x = 2$;
(b) y varies directly as the square of x and $y = 18$ when $x = 3$;
(c) The pressure P in a liquid varies directly as the distance d below the surface and the pressure is 45 lb/in² at a depth of 3 ft.

Solution

(a) If y varies directly as x, then $y = kx$. If $y = 12$ when $x = 2$, then $12 = k(2)$ or $k = 6$. The formula is $y = 6x$.
(b) If y varies directly as the square of x, then $y = kx^2$. If $y = 18$ when $x = 3$, then $18 = k(3)^2$ or $k = 2$. The formula is $y = 2x^2$.
(c) If P varies directly as d, then $P = kd$. If $P = 45$ when $d = 3$, then $45 = k(3)$ or $k = 15$. The formula is $P = 15d$.

B. Inverse variation

- If k is a nonzero constant and if x and y are related by the formula $y = k/x$, then we say that y **varies inversely as** x or y **is inversely proportional to** x. (Again, k is called the constant of variation.)

EXAMPLE 8-37   Find the formulas for the following relationships:

(a) y varies inversely as x and $y = 4$ when $x = 8$;
(b) F varies inversely as the square of d;
(c) s varies inversely as the square root of t and $s = 32$ when $t = 4$.

Solution

(a) Since $y = k/x$ and $y = 4$ when $x = 8$, we have $4 = k/8$ or $k = 32$. The formula is $y = 32/x$.
(b) Since F varies inversely as the square of d, the formula is $F = k/d^2$.
(c) Since $s = k/\sqrt{t}$ and $s = 32$ when $t = 4$, we have $32 = k/\sqrt{4}$ or $k = 64$. The formula is $s = 64/\sqrt{t}$.

C. Joint variation

Frequently some physical quantity depends on more than one other quantity.

- If k is a nonzero constant and x, y, and z are related by the formula $z = kxy$, then we say that z **varies jointly as** x **and** y or z **is jointly proportional to** x **and** y.
- If k is a nonzero constant and x, y, and z are related by the formula $z = kx/y$, then we say that z **varies directly as** x **and inversely as** y.

EXAMPLE 8-38   The volume V of a cylinder is jointly proportional to the square of the radius of the base and the height. Given that $V = 20\pi$ in^3 when the radius $r = 2$ in. and the height $h = 5$ in., find the formula for the relationship.

Solution   Since this is a joint proportionality, we have $V = kr^2h$. Then $20\pi = k(2)^2(5)$ or $k = \pi$. Thus $V = \pi r^2 h$ is the formula for the volume of any right circular cylinder.

EXAMPLE 8-39   Find the formulas for the following relationships:

(a) z is jointly proportional to x and the square root of y and $z = 12$ when $x = 3$ and $y = 16$;
(b) t is jointly proportional to q and r and inversely proportional to the square of s and $t = 6$ when $q = -6$, $r = 4$, and $s = -2$;
(c) j is directly proportional to the square of h and inversely proportional to f and g and $j = 15$ when $h = 10$, $f = 25$, and $g = 8$.

Solution

(a) The relationship is $z = kx\sqrt{y}$. Substituting the values $z = 12$, $x = 3$, and $y = 16$ gives us $12 = k(3)\sqrt{16}$ or $k = 1$. Then the formula is $z = x\sqrt{y}$.
(b) The relationship is $t = kqr/s^2$. Solving for k, we have $6 = k(-6)(4)/(-2)^2$ or $k = -1$. The formula is $t = -qr/s^2$.
(c) The relationship is $j = kh^2/fg$. Solving for k, we have $15 = k(10)^2/(25)(8)$ or $k = 30$. The formula is $j = 30h^2/fg$.

SUMMARY

1. A relation is a set of ordered pairs.
2. A function is a relation with the property that no two pairs have the same first component (or x-coordinate).
3. The domain of a relation or function is the set of all x-coordinates and the range is the set of all y-coordinates.
4. The notation $f(x)$ is read "f of x" and is defined to be the value of the function f at x.
5. Functions can be classified by type. Some of the most common classifications are one-to-one, identity, constant, increasing, decreasing, linear, quadratic, absolute value, square root, and exponential.
6. If $f(-x) = f(x)$, then f is an even function and its graph is symmetric with respect to the y-axis. If $f(-x) = -f(x)$, then f is an odd function and its graph is symmetric with respect to the origin.
7. If f and g are two functions, then the composite function $f \circ g$ is defined by $(f \circ g)(x) = f(g(x))$.
8. If f is a function, then f has an inverse f^{-1} if

   • f is one-to-one
   • the domain of f^{-1} is the range of f
   • $f^{-1}(f(x)) = x$ for all x in the domain of f^{-1}
   • $f(f^{-1}(y)) = y$ for all y in the range of f

   The graphs of a function and its inverse are symmetric with respect to the line $y = x$.
9. A variation is a special type of function that relates a dependent variable to one or more independent variables. Variations can be direct, inverse, or joint—or a combination of the three types.

RAISE YOUR GRADES

Can you...?

☑ define a relation and explain how to find its domain and range
☑ define a function and explain how it differs from a relation
☑ explain the difference between a dependent variable and an independent variable
☑ explain the vertical line test for determining whether a relation is a function
☑ define a constant function, an increasing function, and a decreasing function
☑ define the absolute value function and sketch its graph
☑ explain the difference between direct variation, inverse variation, and joint variation
☑ explain how symmetry and shifting help in graphing functions
☑ define an even function, an odd function, and a piecewise function and describe their graphs
☑ define the composition of two functions
☑ explain how to find the inverse of a function and how its graph relates to the graph of the original function

SOLVED PROBLEMS

Relations

PROBLEM 8-1 Given set $P = \{1, 4, 6, 8\}$ and set $Q = \{2, 3, 6, 9, 10, 12\}$, form two relations using elements of P and Q.

Solution Even though Q has more elements than P, you need only pair the elements to form a relation. Thus, $R_1 = \{(1, 2), (4, 3), (6, 9), (8, 10)\}$ is one relation, $R_2 = \{(1, 10), (4, 6), (6, 2), (8, 12), (4, 2)\}$ is another, and many others are possible.

PROBLEM 8-2 Find the domain and range of the relations formed in Problem 8-1.

Solution For relation R_1 the domain is the set of first components, or $\{1, 4, 6, 8\}$, and the range is the set of second components, or $\{2, 3, 9, 10\}$. For relation R_2, the domain is $\{1, 4, 6, 8\}$ and the range is $\{2, 6, 10, 12\}$.

PROBLEM 8-3 Sketch the graphs of the relations $R_1 = \{(1, 2), (4, 3), (6, 9), (8, 10)\}$ and $R_2 = \{(1, 10), (4, 6), (6, 2), (8, 12), (4, 2)\}$. Check the domain and range of the relations.

Solution See Figure 8-19. The domain and range are shown on the x- and y-axes, respectively.

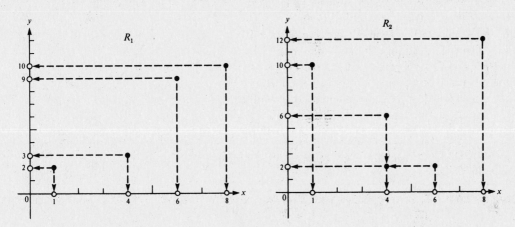

Figure 8-19

Functions

PROBLEM 8-4 Are the two relations R_1 and R_2 given in Problem 8-1 functions?

Solution The relation R_1 is a function since no two pairs have the same first element. The relation R_2 is not a function since there are two pairs with the same first element $(4, 6)$ and $(4, 2)$. The graphs in Problem 8-3 verify this: In relation R_1, no vertical line contains two points of the relation, and in relation R_2, a vertical line through the point $(4, 0)$ contains two points of the graph of the relation.

PROBLEM 8-5 Let $R = \{(2, -1), (3, 2), (-4, 1), (2, 8), (-1, 5)\}$. Is R a function?

Solution Inspecting the pairings:

$$
\begin{array}{cc}
x & y \\
\hline
2 & \begin{cases} -1 \\ 8 \end{cases} \\
3 \longrightarrow & 2 \\
-4 \longrightarrow & 1 \\
-1 \longrightarrow & 5
\end{array}
$$

One element of the domain is paired with two different elements of the range, *so* the given R is not a function.

PROBLEM 8-6 Sketch the graph of the function that is defined by the rule $f(x) = x^2 - 5x + 6$ and has a domain of all real numbers. Determine the range of the function.

Solution $f(x) = y = x^2 - 5x + 6$:

$y = x^2 - 5x + 6$		Function Pairs
x	y	(x, y)
-1	$(-1)^2 - 5(-1) + 6 = 12 \longrightarrow$	$(-1, 12)$
0	$0^2 - \quad 0 + 6 = 6 \longrightarrow$	$(0, 6)$
1	$1^2 - \quad 5(1) + 6 = 2 \longrightarrow$	$(1, 2)$
2	$2^2 - \quad 5(2) + 6 = 0 \longrightarrow$	$(2, 0)$
3	$3^2 - \quad 5(3) + 6 = 0 \longrightarrow$	$(3, 0)$
4	$4^2 - \quad 5(4) + 6 = 2 \longrightarrow$	$(4, 2)$
5	$5^2 - \quad 5(5) + 6 = 6 \longrightarrow$	$(5, 6)$
$\frac{5}{2}$	$(\frac{5}{2})^2 - \quad 5(\frac{5}{2}) + 6 = -\frac{1}{4} \longrightarrow$	$(\frac{5}{2}, -\frac{1}{4})$

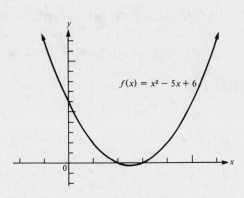

The range consists of all the y values that pair with any x values in the domain. You can see from Figure 8-20 that the graph is an upward-opening parabola, a symmetric curve, whose y values are always equal to or greater than $-\frac{1}{4}$. (That's why it's important to add the last point of the table—it determines the minimum point on the graph, which is assured by symmetry.) Thus the y values that are members of some pair in the function are $\{y : y \geq -\frac{1}{4}\}$, which is the range.

Figure 8-20

PROBLEM 8-7   Which of the graphs in Figure 8-21 are graphs of functions?

(a)
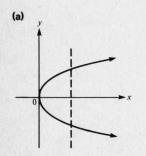

(b)
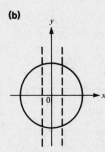

(c)
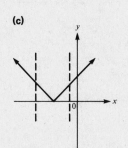

(d)
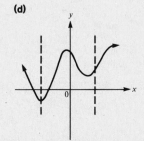

(e)
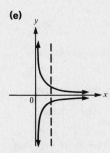

Figure 8-21

Solution

(a) A vertical line drawn at any positive x value will cut the graph twice, so this is not the graph of a function.
(b) The vertical line $x = 0$ (as well as several others) cuts the graph twice, so this is not the graph of a function.
(c) Every vertical line cuts this graph exactly once, so this is the graph of a function.
(d) Every vertical line cuts the graph exactly once, so this is the graph of a function.
(e) A vertical line at any positive value of x cuts the graph twice, so this is not the graph of a function. [*Note:* Despite the fact that there are two parts to the curve in Figure 8-21e, it is considered a *single* curve].

PROBLEM 8-8   Identify the dependent variable and the independent variable in (a) $B = \frac{1}{3}t^3$ and (b) $d = 15z + 9$.

Solution

(a) In $B = \frac{1}{3}t^3$, the value of B "depends on" the value of t, so B is the dependent variable and t is the independent variable.
(b) In $d = 15z + 9$, the dependent variable is d and independent variable is z.

PROBLEM 8-9 Let f be a function such that $f(x) = 2x^2 - 5$. Find $f(2)$, $f(-3)$, $f(t)$, $f(b^3)$, and $f(x + h)$.

Solution Since $f(x) = 2x^2 - 5$:

$$f(2) = 2(2)^2 - 5 = 8 - 5 = 3$$

$$f(-3) = 2(-3)^2 - 5 = 18 - 5 = 13$$

$$f(t) = 2(t)^2 - 5 = 2t^2 - 5$$

$$f(b^3) = 2(b^3)^2 - 5 = 2b^6 - 5$$

$$f(x + h) = 2(x + h)^2 - 5 = 2(x^2 + 2xh + h^2) - 5 = 2x^2 + 4xh + 2h^2 - 5$$

PROBLEM 8-10 Let $F(t) = \sqrt{t - 1}$. Find $F(1)$, $F(5)$, $F(x + h)$, and $F(t + 5)$.

Solution Since $F(t) = \sqrt{t - 1}$:

$$F(1) = \sqrt{1 - 1} = \sqrt{0} = 0$$

$$F(5) = \sqrt{5 - 1} = \sqrt{4} = 2$$

$$F(x + h) = \sqrt{x + h - 1}$$

$$F(t + 5) = \sqrt{(t + 5) - 1} = \sqrt{t + 4}$$

PROBLEM 8-11 Let $G(x) = \dfrac{x + 2}{x + 5}$. Find $G(4)$, $G(-7)$, $G(x + h)$, and $G(x + h) - G(x)$.

Solution Since $G(x) = \dfrac{x + 2}{x + 5}$:

$$G(4) = \frac{4 + 2}{4 + 5} = \frac{6}{9} = \frac{2}{3} \qquad G(-7) = \frac{-7 + 2}{-7 + 5} = \frac{-5}{-2} = \frac{5}{2} \qquad G(x + h) = \frac{x + h + 2}{x + h + 5}$$

$$\begin{aligned}
G(x + h) - G(x) &= \frac{x + h + 2}{x + h + 5} - \frac{x + 2}{x + 5} \\[2mm]
&= \frac{x + h + 2}{x + h + 5} \cdot \frac{x + 5}{x + 5} - \frac{x + 2}{x + 5} \cdot \frac{x + h + 5}{x + h + 5} \\[2mm]
&= \frac{(x + h + 2)(x + 5) - (x + 2)(x + h + 5)}{(x + h + 5)(x + 5)} \\[2mm]
&= \frac{(x^2 + hx + 7x + 5h + 10) - (x^2 + hx + 7x + 2h + 10)}{(x + h + 5)(x + 5)} \\[2mm]
&= \frac{3h}{(x + h + 5)(x + 5)}
\end{aligned}$$

PROBLEM 8-12 Let $P(y) = \dfrac{(y - 2)^2}{5}$. Find $P(-3)$, $P(\frac{1}{2})$, and $P(z^2 + 2)$.

Solution Since $P(y) = \dfrac{(y - 2)^2}{5}$:

$$P(-3) = \frac{(-3 - 2)^2}{5} = \frac{(-5)^2}{5} = \frac{25}{5} = 5$$

$$P(\tfrac{1}{2}) = \frac{(\frac{1}{2} - 2)^2}{5} = \frac{(-\frac{3}{2})^2}{5} = \frac{\frac{9}{4}}{5} = \frac{9}{20}$$

$$P(z^2 + 2) = \frac{(z^2 + 2 - 2)^2}{5} = \frac{(z^2)^2}{5} = \frac{z^4}{5}$$

PROBLEM 8-13   Find the value of each function at the given values in the domain:

$$\text{(a) } f(x) = |x + 4| \quad \text{at } x = 3, x = 0, \text{ and } x = -7$$

$$\text{(b) } f(x) = 5 \quad \text{at } x = 0 \text{ and } x = 2$$

$$\text{(c) } f(x) = \sqrt{x - 2} \quad \text{at } x = 0, x = 3, \text{ and } x = 10$$

Solution   For each function, substitute the value in the domain for x in the equation.

(a) Since $f(x) = |x + 4|$, $f(3) = |3 + 4| = |7| = 7$, $f(0) = |4| = 4$, and $f(-7) = |-3| = 3$.

(b) For any x in the domain of f, $f(x) = 5$, so $f(0) = 5$ and $f(2) = 5$.

(c) If $f(x) = \sqrt{x - 2}$, then $f(0) = \sqrt{-2}$. But the square root of a negative number is undefined, so the function is undefined at $x = 0$. At $x = 3$, $f(x) = \sqrt{3 - 2} = \sqrt{1} = 1$ and at $x = 10$, $f(x) = \sqrt{10 - 2} = \sqrt{8} = 2\sqrt{2}$.

Special Functions

PROBLEM 8-14   Determine if the functions (a) $f(x) = 2x - 1$ and (b) $f(x) = x^2 + 2$ are one-to-one.

Solution   In a one-to-one function, each value in the domain corresponds to exactly one value in the range.

(a) The graph of $f(x) = 2x - 1$ is a straight line with slope 2 and y-intercept -1. Each value x in the domain corresponds to exactly one value y in the range, so this is a one-to-one function. You can also tell that the function is one-to-one from the graph in Figure 8-22a, since no horizontal line crosses the graph in more than one point.

(b) Since $f(x) = x^2 + 2$, there are clearly two values in the domain corresponding to each value in the range—a number and its opposite. Therefore the function is not one-to-one. As Figure 8-22b shows, the graph of the function fails the horizontal line test, since any horizontal line drawn above $y = 2$ crosses the graph in two points.

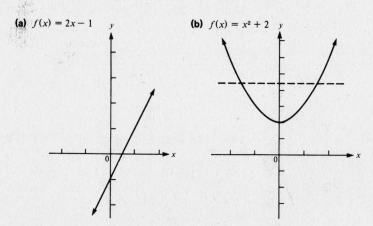

Figure 8-22

PROBLEM 8-15   Determine the values of x where the following functions are increasing, decreasing, or constant:

$$\text{(a) } f(x) = x^2 - 2x \quad \text{(b) } f(x) = \begin{cases} x & \text{if } x \le 3 \\ 3 & \text{if } x > 3 \end{cases} \quad \text{(c) } f(x) = (x + 1)^3$$

Solution   First graph each function, as shown in Figure 8-23. Then use the graph to find the areas over which the function is increasing, decreasing, or constant.

(a) The function $f(x) = x^2 - 2x$ is decreasing over the domain $x < 1$ and increasing over the domain $x > 1$.

(a) $f(x) = x^2 - 2x$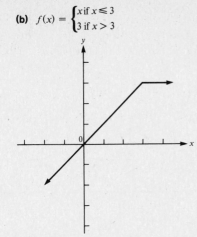

(b) $f(x) = \begin{cases} x & \text{if } x \le 3 \\ 3 & \text{if } x > 3 \end{cases}$

(c) $f(x) = (x+1)^3$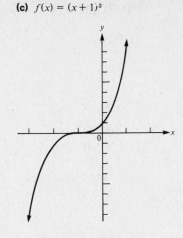

Figure 8-23

(b) The function $f(x) = \begin{cases} x & \text{if } x \le 3 \\ 3 & \text{if } x > 3 \end{cases}$ is increasing over the domain $x < 3$ and constant over the domain $x \ge 3$.

(c) The function $f(x) = (x+1)^3$ is increasing over the entire domain of the function.

Graphing Functions

PROBLEM 8-16   Without graphing the function, determine whether each function is even or odd or neither, and comment on its symmetry:

   (a) $f(x) = x^3 + x$ **(b)** $f(x) = -x^2 + 16$ **(c)** $f(x) = -3x - 7$ **(d)** $f(x) = -\frac{1}{2}x$

Solution   If $f(-x) = f(x)$, the function is even and symmetric with respect to the y-axis. If $f(-x) = -f(x)$, the function is odd and symmetric with respect to the origin.

(a) Since $f(x) = x^3 + x$,

$$f(-x) = (-x)^3 + (-x) = -x^3 - x = -(x^3 + x) = -f(x)$$

The function is odd, so the graph of the function is symmetric with respect to the origin.

(b) Since $f(x) = -x^2 + 16$,

$$f(-x) = -(-x)^2 + 16 = -x^2 + 16 = f(x)$$

This function is even and the graph of the function is symmetric with respect to the y-axis.

(c) Since $f(x) = -3x - 7$,

$$f(-x) = -3(-x) - 7 = 3x - 7$$

The result $3x - 7$ is neither $f(x)$ nor $-f(x)$, so this function is neither even nor odd and the graph displays no symmetry.

(d) Since $f(x) = -\frac{1}{2}x$,

$$f(-x) = -\frac{1}{2}(-x) = \frac{1}{2}x = -f(x)$$

The function is odd and its graph is symmetric with respect to the origin.

PROBLEM 8-17   Sketch the graphs of **(a)** $f(x) = |x| - 2$ and **(b)** $f(x) = |x - 2|$.

Solution   See Figure 8-24.

(a) The graph of $f(x) = |x| - 2$ is the graph of $f(x) = |x|$ shifted downward 2 units.

(b) The graph of $f(x) = |x - 2|$ is the graph of $f(x) = |x|$ shifted 2 units to the right.

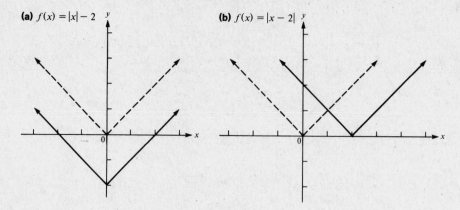

Figure 8-24

PROBLEM 8-18   Sketch the graphs of the following functions on the same set of axes:

$$\text{(a)}\ f(x) = x + 3 \qquad \text{(b)}\ f(x) = x + 1 \qquad \text{(c)}\ f(x) = x - 2$$

Solution   You can find all of the graphs easily by shifting the graph of the identity function $f(x) = x$. (See Figure 8-25.)

(a) The graph of $f(x) = x + 3$ is the graph of $f(x) = x$ shifted upward 3 units.

(b) The graph of $f(x) = x + 1$ is the graph of $f(x) = x$ shifted upward 1 unit.

(c) The graph of $f(x) = x - 2$ is the graph of $f(x) = x$ shifted downward 2 units.

PROBLEM 8-19   Sketch the graph of the piecewise function $f(x) = \begin{cases} x & \text{if } x \le -1 \\ 2 & \text{if } -1 < x < 2. \\ -x + 4 & \text{if } x \ge 2 \end{cases}$

Solution   The pieces consist of a linear function, a constant function, and a linear function. Graph each function independently—on a single set of axes—as in Figure 8-26.

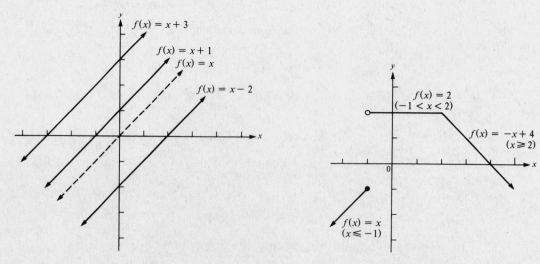

Figure 8-25   **Figure 8-26**

PROBLEM 8-20   Find the domain for (a) $f(x) = \dfrac{\sqrt{x}}{x - 1}$ and (b) $f(x) = \dfrac{2x}{4 - \sqrt{x - 1}}$.

Solution   Remember that, unless specified in the rule, the domain of a function is the set of all real numbers x for which $f(x)$ is defined.

(a) In order for the numerator of $\sqrt{x}/(x-1)$ to be well defined, x must be greater than or equal to 0. The domain must be further restricted, since if x is equal to 1, the denominator is 0 and division by 0 is undefined. Thus the domain of $f(x) = \sqrt{x}/(x-1)$ is the set $\{x: x \neq 1 \text{ and } x \geq 0\}$.

(b) For $f(x) = 2x/(4 - \sqrt{x-1})$ to be well defined, two restrictions are needed. First, the quantity under the radical must not be negative; i.e., $x - 1 \geq 0$ or $x \geq 1$. Also the denominator $4 - \sqrt{x-1}$ must not be equal to 0:

$$4 - \sqrt{x-1} \neq 0$$

$$\sqrt{x-1} \neq 4 \qquad \text{[Square both sides.]}$$

$$x - 1 \neq 16$$

$$x \neq 17$$

Thus the domain of $f(x) = 2x/(4 - \sqrt{x-1})$ is the set $\{x: x \neq 17 \text{ and } x \geq 1\}$.

Composite and Inverse Functions

PROBLEM 8-21   Let $f(x) = x^2 + 4$ and $g(x) = x + 6$. Find the formula and domain for **(a)** $(f \circ g)(x)$ and **(b)** $(g \circ f)(x)$.

Solution

(a) $(f \circ g)(x) = f[g(x)]$

$$= f[x+6] = (x+6)^2 + 4$$

$$= x^2 + 12x + 36 + 4$$

$$= x^2 + 12x + 40$$

The domain is the set of all real numbers.

(b) $(g \circ f)(x) = g[f(x)] = g[x^2 + 4] = (x^2 + 4) + 6 = x^2 + 10$

The domain is the set of all real numbers.

PROBLEM 8-22   Suppose $f(x) = 5x + 1$ and $g(x) = x^2 + x - 2$. Find **(a)** $(f \circ g)(2)$ and **(b)** $(g \circ f)(-3)$.

Solution

(a) First find $(f \circ g)(x) = f[g(x)]$

$$= f[x^2 + x - 2]$$

$$= 5(x^2 + x - 2) + 1$$

$$= 5x^2 + 5x - 9$$

Then $(f \circ g)(2) = 5(2)^2 + 5(2) - 9$

$$= 20 + 10 - 9 = 21$$

(b) First find $(g \circ f)(x) = g[f(x)]$

$$= g[5x + 1]$$

$$= (5x + 1)^2 + (5x + 1) - 2$$

$$= 25x^2 + 10x + 1 + 5x + 1 - 2$$

$$= 25x^2 + 15x$$

$$\text{Then } (g \circ f)(-3) = 25(-3)^2 + 15(-3)$$
$$= 225 - 45$$
$$= 180$$

PROBLEM 8-23 Determine the inverse of the function given by $f = \{(1, 5), (2, 2), (3, -4)\}$.

Solution The inverse is the function given by $f^{-1} = \{(5, 1), (2, 2), (-4, 3)\}$. Note that the domain of f has become the range of f^{-1} and the range of f has become the domain of f^{-1}.

PROBLEM 8-24 Find the inverse function for $f(x) = \frac{1}{3}x + \frac{5}{3}$. Sketch f and f^{-1} on the same set of axes.

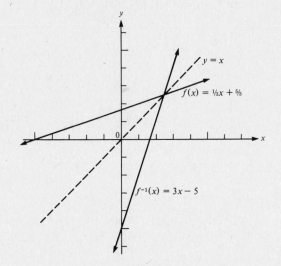

Solution Let $y = f(x) = \frac{1}{3}x + \frac{5}{3}$. Then

$$y = \tfrac{1}{3}x + \tfrac{5}{3} \qquad \text{[Solve the equation for } x.\text{]}$$
$$\tfrac{1}{3}x = y - \tfrac{5}{3}$$
$$x = 3y - 5$$
$$y = 3x - 5 \qquad \text{[Interchange the variables.]}$$

Thus

$$f(x) = \tfrac{1}{3}x + \tfrac{5}{3} \quad \text{and} \quad f^{-1}(x) = 3x - 5$$

You can see from Figure 8-27 that the graphs of the two functions are symmetric with respect to the line $y = x$.

Figure 8-27

PROBLEM 8-25 Find the expression for $f^{-1}(x)$ if **(a)** $f(x) = \dfrac{1}{x + 2}$ and **(b)** $f(x) = \sqrt{x - 4}$.

Solution

(a) Let $y = f(x) = \dfrac{1}{x + 2}$. Then

$$y = \frac{1}{x + 2} \qquad \text{[Solve the equation for } x.\text{]}$$
$$x + 2 = \frac{1}{y}$$
$$x = \frac{1}{y} - 2$$
$$y = \frac{1}{x} - 2 \qquad \text{[Interchange the variables.]}$$

Thus $\qquad\qquad f(x) = \dfrac{1}{x + 2} \quad$ and $\quad f^{-1}(x) = \dfrac{1}{x} - 2$

(b) Let $y = f(x) = \sqrt{x - 4}$. Then

$$y = \sqrt{x - 4} \qquad \text{[Solve the equation for } x.\text{]}$$
$$x - 4 = y^2$$
$$x = y^2 + 4$$
$$y = x^2 + 4 \qquad \text{[Interchange the variables.]}$$

Thus

$$f(x) = \sqrt{x - 4} \quad \text{and} \quad f^{-1}(x) = x^2 + 4$$

But wait! The range of a function f is the domain of its inverse (and vice versa). The range of $f(x) = \sqrt{x-4}$ is $f(x) \geq 0$, so the domain of $f^{-1}(x) = x^2 + 4$ must also be $x \geq 0$. This must be specified with the rule; otherwise, we would assume that the domain of $x^2 + 4$ is any real number x. So $f^{-1}(x) = x^2 + 4$, $x \geq 0$.

PROBLEM 8-26 Check to see that the formulas $f^{-1}(f(x)) = x$ and $f(f^{-1}(y)) = y$ hold for the two function pairs in Problem 8-25.

Solution

(a) If $f(x) = \dfrac{1}{x+2}$ and $f^{-1}(x) = \dfrac{1}{x} - 2$, then

$$f^{-1}(f(x)) = f^{-1}\left(\frac{1}{x+2}\right) = \frac{1}{\left(\dfrac{1}{x+2}\right)} - 2 = (x+2) - 2 = x \quad \checkmark$$

$$f(f^{-1}(y)) = f\left(\frac{1}{y} - 2\right) = \frac{1}{\left(\dfrac{1}{y} - 2\right) + 2} = \frac{1}{1/y} = y \quad \checkmark$$

(b) If $f(x) = \sqrt{x-4}$ and $f^{-1}(x) = x^2 + 4$, $x \geq 0$, then

$$f^{-1}(f(x)) = f^{-1}(\sqrt{x-4}) = (\sqrt{x-4})^2 + 4 = x - 4 + 4 = x \quad \checkmark$$

$$f(f^{-1}(y)) = f(y^2 + 4) = \sqrt{(y^2 + 4) - 4} = \sqrt{y^2} = y \quad \checkmark$$

PROBLEM 8-27 Let $f(x) = \dfrac{x+1}{x-1}$. Find $f^{-1}(x)$.

Solution Begin with $y = \dfrac{x+1}{x-1}$ and solve for x:

$$y = \frac{x+1}{x-1}$$

$$(x-1)y = x + 1$$

$$xy - y = x + 1$$

$$xy - x = y + 1$$

$$x(y-1) = y + 1$$

$$x = \frac{y+1}{y-1}$$

$$y = \frac{x+1}{x-1} \qquad \text{[Interchange x and y.]}$$

Thus $f(x) = \dfrac{x+1}{x-1}$ and $f^{-1}(x) = \dfrac{x+1}{x-1}$. Amazing! This is a function that is its *own inverse*. As a check, make sure that the formula $f^{-1}(f(x)) = x$ holds for these "two" functions:

$$f^{-1}(f(x)) = f^{-1}\left(\frac{x+1}{x-1}\right) = \frac{\dfrac{x+1}{x-1} + 1}{\dfrac{x+1}{x-1} - 1} = \frac{\dfrac{x+1}{x-1} + \dfrac{x-1}{x-1}}{\dfrac{x+1}{x-1} - \dfrac{x-1}{x-1}} = \frac{\dfrac{(x+1) + (x-1)}{x-1}}{\dfrac{(x+1) - (x-1)}{x-1}} = \frac{\dfrac{2x}{x-1}}{\dfrac{2}{x-1}} = \frac{2x}{2} = x$$

The graph of $f(x) = (x+1)/(x-1)$ is drawn in Figure 8-28 (on p. 202). Notice that the graph is symmetric with respect to the line $y = x$, satisfying the requirement of an inverse function. [*note:* Chapter 9 discusses the techniques of graphing a function like $f(x) = (x+1)/(x-1)$.]

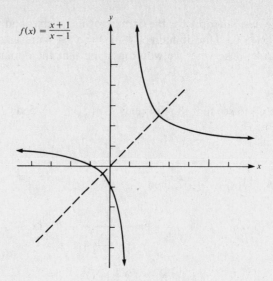

$$f(x) = \frac{x+1}{x-1}$$

Figure 8-28

Variation

PROBLEM 8-28   Find the equations for the following relationships:

(a) y varies directly as the square of x and $y = 8$ when $x = 4$;
(b) V varies inversely as the cube of w and $V = \frac{1}{4}$ when $w = 2$;
(c) z is directly proportional to s and inversely proportional to the square of t and $z = 6$ when $s = 1$ and $t = 3$.

Solution

(a) Since y varies directly as the square of x, then $y = kx^2$. But $8 = k(4)^2$, so $8 = 16k$ or $k = \frac{1}{2}$. The equation for this relationship is $y = \frac{1}{2}x^2$.
(b) Since V varies inversely as the cube of w, then $V = k/w^3$. But $1/4 = k/(2)^3$, so $1/4 = k/8$ or $k = 2$. The equation is $V = 2/w^3$.
(c) Since z is directly proportional to s and inversely proportional to the square of t, then $z = ks/t^2$. But $6 = k(1)/(3)^2$, so $6 = k/9$ or $k = 54$. The equation is $z = 54s/t^2$.

PROBLEM 8-29   Suppose r varies jointly as y and the square root of z and $r = 16$ when $y = 2$ and $z = 4$. Find the formula for the dependent variable r.

Solution   Since r varies jointly as y and the square root of z, then $r = ky\sqrt{z}$. But $16 = k(2)(\sqrt{4})$, so $16 = 4k$ or $k = 4$. The formula is $r = 4y\sqrt{z}$.

PROBLEM 8-30   Suppose that the amount of diesel fuel that your truck uses varies jointly as the distance traveled and the cube root of the average speed. If you used 40 gallons on a 500-mile trip going an average speed of 64 mph, how many gallons would you use on an 800-mile trip averaging 54 mph?

Solution   The formula for the variation is $F = kd\sqrt[3]{s}$.

Since the fuel $F = 40$ when the distance $d = 500$ and the speed $s = 64$, you have

$$40 = k(500)(\sqrt[3]{64}) \quad \text{or} \quad 40 = 2000k \quad \text{or} \quad k = \tfrac{1}{50}$$

The formula then becomes $F = \frac{1}{50}d\sqrt[3]{s}$ and you can use it with the new conditions, $d = 800$ and $s = 54$:

$$F = \tfrac{1}{50}(800)(\sqrt[3]{54}) = 16(\sqrt[3]{54}) = 16(\sqrt[3]{27})(\sqrt[3]{2}) = 48\sqrt[3]{2}$$

You would use $48\sqrt[3]{2}$ or approximately 60 gallons of fuel on an 800-mile trip averaging 54 mph.

Supplementary Exercises

PROBLEM 8-31 List the domain and range of the relation $R_1 = \{(-3, 1), (-2, 4), (0, 1), (4, 2), (-2, 1), (5, 4)\}$. Sketch the graph of R_1.

PROBLEM 8-32 List the domain and range of the relation $R_2 = \{(4, 1), (-2, 3), (1, 5), (-3, 4), (2, -2)\}$. Is R_2 a function?

PROBLEM 8-33 Is R_2 in Problem 8-32 a one-to-one function?

PROBLEM 8-34 Determine the domain of each of the following functions and identify the independent variable and the dependent variable:

$$\textbf{(a)}\; y = 2 - x - x^2 \qquad \textbf{(b)}\; s = \frac{3}{t + 2} \qquad \textbf{(c)}\; F = \frac{4.721}{m^2} \qquad \textbf{(d)}\; z = \frac{1}{x^2 - 4}$$

PROBLEM 8-35 Use the vertical line test to determine which of the following are graphs of functions and which are graphs of relations.

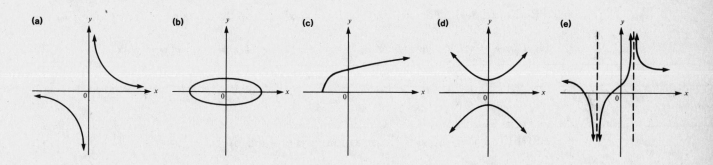

PROBLEM 8-36 Given $f(x) = x^2 + 5x$, find **(a)** $f(1)$, **(b)** $f(-5)$, **(c)** $f(a)$, **(d)** $f(a + h)$, and **(e)** $\dfrac{f(a + h) - f(a)}{h}$.

PROBLEM 8-37 Let $f(x) = 5 - 2x$. Find **(a)** $f(4)$, **(b)** $f(-4)$, **(c)** $f(t)$, **(d)** $f(-t)$, **(e)** $f(x + h)$, and **(f)** $\dfrac{f(x + h) - f(x)}{h}$.

PROBLEM 8-38 Let $g(x) = \dfrac{3x}{x^2 + 1}$. Find **(a)** $g(2)$, **(b)** $g(0)$, **(c)** $g(t^2)$, and **(d)** $[g(t)]^2$.

PROBLEM 8-39 **(a)** If $y = 5x + 2$, what number is paired with $y = 7$? **(b)** If $y = \sqrt{x - 5}$, what number is paired with $y = 8$?

PROBLEM 8-40 Are the following functions one-to-one?

$$\textbf{(a)}\; f(x) = -9x + \tfrac{1}{2} \qquad \textbf{(b)}\; f(x) = x^3 + 4 \qquad \textbf{(c)}\; f(x) = |x|$$

PROBLEM 8-41 If the temperature of a gas in an enclosed container remains constant, the pressure of the gas is inversely proportional to its volume. If a certain gas has a pressure of 12 lb/in^2 in a cubical box of edge 6 inches, what would the pressure be if the box is expanded **(a)** to a cube of edge 8 inches? **(b)** to a cube of edge 12 inches?

PROBLEM 8-42 Write the equation to describe each of the following variational relationships:

(a) P is inversely proportional to the sum of x and y and $P = \tfrac{1}{4}$, when $x = 5$ and $y = 3$.
(b) Z varies jointly as the cube of x and the square of y and $Z = 10$ when $x = 1$ and $y = 2$.
(c) V varies directly as s and inversely as the square root of t; when $s = 4$ and $t = 8$, $V = 1$.

PROBLEM 8-43 If an object is dropped from the top of a tall building, the distance it falls in t seconds is directly proportional to the square of t. If height $s = 100$ feet when $t = 2\frac{1}{2}$ seconds, what is the formula for s as a function of t? If a ball is dropped from 800 feet, how long does it take to reach the ground?

PROBLEM 8-44 Graph the functions

$$\text{(a) } f(x) = |x| + 4 \quad \text{(b) } f(x) = |x + 4| \quad \text{(c) } f(x) = \sqrt{x + 4}$$

PROBLEM 8-45 Determine if the given function is odd, even, or neither:

$$\text{(a) } f(x) = x^3 + 1 \quad \text{(b) } g(x) = \frac{1}{x^2} \quad \text{(c) } h(x) = 2x^4 - x^2 + 3$$

PROBLEM 8-46 Sketch the graphs of

(a) $f(x) = \begin{cases} -x \text{ if } x < -2 \\ 2 \text{ if } -2 \le x \le 3 \\ x - 1 \text{ if } x > 3 \end{cases}$
(b) $f(x) = \begin{cases} |x| \text{ if } x \le 4 \\ 8 - x \text{ if } x > 4 \end{cases}$

PROBLEM 8-47 For each pair of functions given, find the domain and the rule for $g \circ f$:

(a) $f(x) = -x + 4, \quad g(x) = 2x^2 + 5$
(b) $f(x) = \frac{1}{3x}, \quad g(x) = \sqrt{3x}$

(c) $f(x) = \frac{x^2 - 1}{x^2 + 1}, \quad g(x) = \frac{1}{x}$

PROBLEM 8-48 Find the inverse of (a) $f(x) = \frac{1}{x}$ and (b) $f(x) = \frac{x}{x - 1}$.

PROBLEM 8-49 Given $f(x) = 2x^2 + x - 1$ and $g(x) = x^2 + 9$, find (a) $(f \circ g)(2)$ and (b) $(g \circ f)(4)$.

PROBLEM 8-50 Given $f(x) = 3x^3$, find $f^{-1}(x)$.

PROBLEM 8-51 Find the formula for the inverse of each function:

$$\text{(a) } f(x) = -\frac{1}{3}(x + 2) \quad \text{(b) } g(t) = t^5 \quad \text{(c) } h(x) = \frac{3 - x}{2x - 5}$$

PROBLEM 8-52 On the same coordinate plane, sketch the graphs of the given function and its inverse:

$$\text{(a) } f(x) = 4x \quad \text{(b) } g(x) = \sqrt{x + 3} \quad \text{(c) } h(x) = 10 - \frac{1}{2}x$$

PROBLEM 8-53 If an equilateral triangle has side length x, then its area is given by $A = \frac{\sqrt{3}}{4}x^2$. (a) Derive a formula for the side length in terms of the area. (b) If the area is 24 in^2, what is the side length?

PROBLEM 8-54 Given $f(a) = a^3 + 3a$, compute $\dfrac{f(a + h) - f(a)}{h}$.

Answers to Supplementary Exercises

8-31 domain $= \{-3, -2, 0, 4, 5\}$
range $= \{1, 2, 4\}$
See Figure 8-29.

8-32 domain $= \{-3, -2, 1, 2, 4\}$
range $= \{-2, 1, 3, 4, 5\}$
Yes, R_2 is a function.

8-33 Yes, R_2 is a one-to-one function.

8-34 (a) domain = all real numbers
x is independent, y is dependent

(b) domain = $\{t:\ t \ne -2\}$
t is independent, s is dependent

(c) domain = $\{m:\ m \ne 0\}$;
m is independent and F is dependent

(d) domain = $\{x:\ x \ne 2, x \ne -2\}$,
x is independent, z is dependent

8-35 (a) function (d) relation

(b) relation (e) function

(c) function

8-36 (a) 6 (b) 0 (c) $a^2 + 5a$

(d) $a^2 + 2ah + h^2 + 5a + 5h$

(e) $2a + 5 + h$

8-37 (a) -3 (b) 13 (c) $5 - 2t$

(d) $5 + 2t$ (e) $5 - 2x - 2h$ (f) -2

8-38 (a) $\dfrac{6}{5}$ (b) 0 (c) $\dfrac{3t^2}{t^4 + 1}$

(d) $\dfrac{9t^2}{(t^2 + 1)^2}$

8-39 (a) $x = 1$ (b) $x = 69$

8-40 (a) yes (b) yes (c) no

8-41 (a) $\dfrac{81}{16}$ lb/in^2 (b) $\dfrac{3}{2}$ lb/in^2

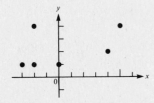

Figure 8-29

8-42 (a) $P = \dfrac{2}{x + y}$

(b) $Z = \frac{5}{2}x^3 y^2$

(c) $V = \left(\dfrac{\sqrt{2}}{2}\right)\left(\dfrac{s}{\sqrt{t}}\right)$

8-43 $s = 16t^2$; $t = 5\sqrt{2}$ seconds

8-44 See Figure 8-30.

8-45 (a) neither (b) even (c) even

8-46 See Figure 8-31.

8-47 (a) $(g \circ f)(x) = 2x^2 - 16x + 37$
domain = all real numbers

(b) $(g \circ f)(x) = \dfrac{1}{\sqrt{x}}$
domain = $\{x:\ x > 0\}$

(c) $(g \circ f)(x) = \dfrac{x^2 + 1}{x^2 - 1}$
domain = $\{x:\ x \ne 1, x \ne -1\}$

8-48 (a) $f^{-1}(x) = \dfrac{1}{x}$ (b) $f^{-1}(x) = \dfrac{x}{x - 1}$

8-49 (a) 350 (b) 1234

8-50 $f^{-1}(x) = \sqrt[3]{\dfrac{x}{3}}$

8-51 (a) $f^{-1}(x) = -3x - 2$

(b) $g^{-1}(t) = t^{1/5}$

(c) $h^{-1}(x) = \dfrac{5x + 3}{2x + 1}$

8-52 See Figure 8-32.

8-53 (a) $x = \sqrt{\dfrac{4A}{\sqrt{3}}}$ (b) $4\sqrt{2\sqrt{3}}$

8-54 $3a^2 + 3ah + h^2 + 3$

(a) (b) (c)

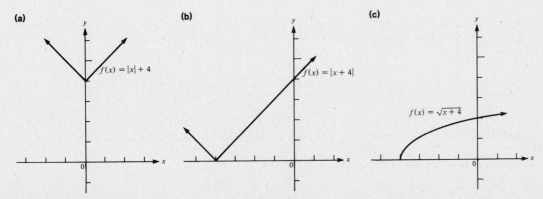

Figure 8-30

(a)

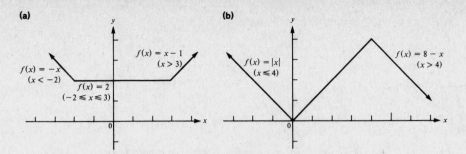

(b)

Figure 8-31

(a)

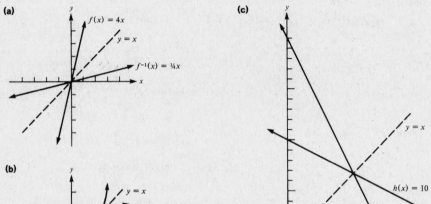

(b)

(c)

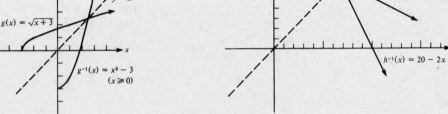

Figure 8-32

9 POLYNOMIAL FUNCTIONS, RATIONAL FUNCTIONS, AND CONIC SECTIONS

THIS CHAPTER IS ABOUT

☑ **Quadratic Functions**
☑ **Higher-Degree Polynomial Functions**
☑ **Rational Functions**
☑ **Conic Sections**

9-1. Quadratic Functions

A. Definition of a quadratic function

A **polynomial function of degree n** is a function given by the expression

$$f(x) = a_n x^n + a_{n-1} x^{n-1} + a_{n-2} x^{n-2} + \cdots + a_2 x^2 + a_1 x + a_0$$

where the coefficients are real numbers, the exponents are nonnegative integers, and $a_n \neq 0$. For example, a constant function $f(x) = a_0$ is a polynomial function of degree 0, and a linear function $f(x) = a_1 x + a_0$ is a polynomial function of degree 1.

- Polynomial functions of degree 2, where $f(x) = a_2 x^2 + a_1 x + a_0$, are called **quadratic functions** and are of the general form

QUADRATIC FUNCTION (general form)
$$f(x) = ax^2 + bx + c$$

where a, b, and c are constants and $a \neq 0$.

B. Properties of a quadratic function

- The graph of every quadratic function is a parabola whose axis of symmetry is on or parallel to the y-axis.

We can begin to describe the properties of a quadratic function in terms of the graph of its simplest case. If $b = c = 0$ and $a \neq 0$, the function $f(x) = ax^2 + bx + c$ becomes

QUADRATIC FUNCTION (parabola, $V = (0, 0)$)
$$f(x) = ax^2$$

which is a parabola whose axis of symmetry is the y-axis ($x = 0$) and whose vertex is the origin, $V = (0, 0)$.

(1) If $a > 0$, the vertex $(0, 0)$ is the *lowest* point of the graph and the parabola opens *upward*.
(2) If $a < 0$, the vertex $(0, 0)$ is the *highest* point of the graph and the parabola opens *downward*.

EXAMPLE 9-1 Graph the parabolas described by

      (a) $f(x) = x^2$ **(b)** $f(x) = -x^2$

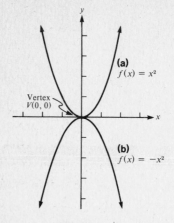

Figure 9-1

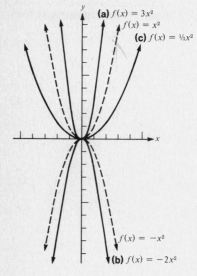

Figure 9-2

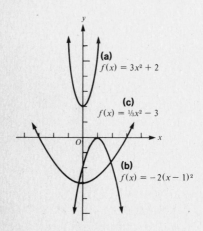

Figure 9-3

Solution See Figure 9-1.

(a) In $f(x) = x^2$, the constant $a = 1$, so the graph is an upward-opening parabola with vertex at the origin.

(b) In $f(x) = -x^2$, the constant $a = -1$, so the graph is a downward-opening parabola with vertex at the origin.

Notice that the graph of $f(x) = -x^2$ is simply a reflection of the graph of $f(x) = x^2$ over the x-axis through the origin.

- If $f(x) = ax^2$ and $a \neq 1$ and $a \neq -1$, the graph has the same general shape as $f(x) = x^2$ or $f(x) = -x^2$, with the following differences:

  (1) If $|a| < 1$, the graph of $f(x) = ax^2$ is "fatter" than the graph of $f(x) = x^2$.

  (2) If $|a| > 1$, the graph of $f(x) = ax^2$ is "thinner" than the graph of $f(x) = x^2$.

EXAMPLE 9-2 Graph the parabolas described by **(a)** $f(x) = 3x^2$, **(b)** $f(x) = -2x^2$, and **(c)** $f(x) = \frac{1}{3}x^2$. Compare each with the graph of $f(x) = \pm x^2$.

Solution See Figure 9-2.

(a) Since $a > 0$, the graph of $f(x) = 3x^2$ is an upward-opening parabola with vertex at the origin. Since $|a| > 1$, it is thinner than the graph of $f(x) = x^2$.

(b) Since $a < 0$, the graph of $f(x) = -2x^2$ is a downward-opening parabola with vertex at the origin. Since $|a| > 1$, it is thinner than the graph of $f(x) = -x^2$.

(c) Since $a > 0$, the parabola opens upward and has vertex at the origin. Since $|a| < 1$, it is fatter than the graph of $f(x) = x^2$.

- If $f(x) = ax^2 + k$, where k is a nonzero constant, the graph is a *vertical shift* of the graph of $f(x) = ax^2$. The parabola is the same shape as $f(x) = ax^2$, but the vertex has been shifted from $(0,0)$ to $(0,k)$. The axis of symmetry does not change.

- If $f(x) = a(x - h)^2$, where h is a nonzero constant, the graph is a *horizontal shift* of the graph of $f(x) = ax^2$. The parabola is the same shape as $f(x) = ax^2$, but the vertex has been shifted from $(0,0)$ to $(h,0)$ so that the axis of symmetry shifts from $x = 0$ (the y-axis) to $x = h$ (a vertical line).

  note: If h is positive, the graph is shifted to the *right*. If h is negative, the graph is shifted to the *left*.

  recall: If two functions are related such that $g(x) = f(x) + k$, where k is a constant, then their graphs are vertical shifts of each other.
  If two functions are related such that $f(x) = g(x - h)$, where h is a nonzero real number, then their graphs are horizontal shifts of each other. (See Chapter 8.)

EXAMPLE 9-3 Graph the parabolas described by

(a) $f(x) = 3x^2 + 2$ **(b)** $f(x) = -2(x - 1)^2$ **(c)** $f(x) = \frac{1}{3}x^2 - 3$

Solution See Figure 9-3.

(a) The graph of $f(x) = 3x^2 + 2$ is the same as that of the parabola $f(x) = 3x^2$ (Figure 9-2a), but shifted upward 2 units. The vertex is at $(0, 2)$.

(b) The graph of $f(x) = -2(x - 1)^2$ is the same as that of the parabola $f(x) = -2x^2$ (Figure 9-2b), but shifted 1 unit to the right. The vertex is at $(1, 0)$.

(c) The graph of $f(x) = \frac{1}{3}x^2 - 3$ is the same as that of the parabola $f(x) = \frac{1}{3}x^2$ (Figure 9-2c), but shifted downward 3 units. The vertex is at $(0, -3)$.

The vertex of a parabola doesn't always lie on the *x*- or *y*-axis; it can be anywhere in the plane.

- In general, the graph of

QUADRATIC FUNCTION
(parabola, V = (h, k)) $f(x) = a(x - h)^2 + k$

is the graph of $f(x) = ax^2$ shifted vertically and horizontally away from the origin: The vertex is at the point (h, k) and the axis of symmetry is $x = h$.

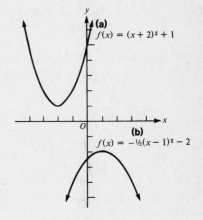

EXAMPLE 9-4   Graph (a) $f(x) = (x + 2)^2 + 1$ and (b) $f(x) = -\frac{1}{2}(x - 1)^2 - 2$.

Solution   See Figure 9-4.

(a) The graph of $f(x) = (x + 2)^2 + 1$ is simply the basic parabola $f(x) = x^2$ shifted 2 units to the left and 1 unit upward. The vertex is at $(-2, 1)$.

(b) The graph of $f(x) = -\frac{1}{2}(x - 1)^2 - 2$ is a downward-opening parabola with vertex at $(1, -2)$. It is fatter than $f(x) = -x^2$.

Figure 9-4

C. Techniques for graphing quadratic functions

1. Completing the square

Given a quadratic function in the form $f(x) = ax^2 + bx + c$, we can use algebra to rewrite it in the form $f(x) = a(x - h)^2 + k$. This latter expression makes graphing a parabola easier, since we know the vertex to be the point (h, k).

The process we use to go from $f(x) = y = ax^2 + bx + c$ to $f(x) = y = a(x - h)^2 + k$ is called **completing the square** (see also Chapter 5):

Step 1: Rewrite the equation as $y = (ax^2 + bx + \square) + c$, leaving room inside the parentheses for the number that completes the square.

Step 2: Factor the *a* out of $ax^2 + bx$ so that the coefficient of x^2 is 1.

Step 3: Compute the number equal to one-half the coefficient of *x*; then square it.

Step 4: Add the value computed in Step 3 to the quantity inside the parentheses (to complete the square) and subtract *a* times that value from the quantity outside the parentheses.

Step 5: Rewrite the value in the parentheses as a squared binomial.

Step 6: Simplify the quantity outside the parentheses.

note: From now on, we'll use the symbols $f(x)$ and *y* interchangeably when dealing with functions. The equations $y = ax^2 + bx + c$ and $f(x) = ax^2 + bx + c$ are equivalent: $f(x)$ is really just the function notation for *y*.

EXAMPLE 9-5   Express the quadratic equations (a) $y = x^2 + 4x + 5$ and (b) $y = x^2 - 2x - 4$ in the form $y = a(x - h)^2 + k$. Then sketch each parabola.

Solution   Complete the square on each equation.

(a) $y = x^2 + 4x + 5$

$\qquad = (x^2 + 4x + \square) + 5$ *Step 1:* Rewrite the equation.

$\qquad = (x^2 + 4x + \square) + 5$ *Step 2:* The coefficient of *x* is already 1, so we can omit this step.

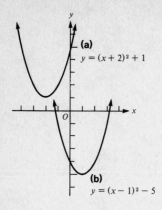

(a)
$y = (x + 2)^2 + 1$

(b)
$y = (x - 1)^2 - 5$

Figure 9-5

Find $\frac{1}{2}(4) = 2$; then $(2)^2 = 4$ *Step 3:* Find one-half the coefficient of x and square it.

$y = (x^2 + 4x + 4) + 5 - 4$ *Step 4:* Add 4 and subtract 4.

$= (x + 2)^2 + 5 - 4$ *Step 5:* Rewrite $x^2 + 4x + 4$ as $(x + 2)^2$.

$= (x + 2)^2 + 1$ *Step 6:* Simplify.

The graph of $y = (x + 2)^2 + 1$ is the "basic" parabola ($y = x^2$) shifted 2 units to the left and 1 unit upward. Its vertex, then, is at $(-2, 1)$. (See Figure 9-5a).

(b)
$$y = x^2 - 2x - 4$$

$$= (x^2 - 2x + \quad) - 4 \qquad \textit{Step 1}$$

Find $\frac{1}{2}(-2) = -1$; then $(-1)^2 = 1$ *Step 3*

$$y = (x^2 - 2x + 1) - 4 - 1 \qquad \textit{Step 4}$$

$$= (x - 1)^2 - 4 - 1 \qquad \textit{Step 5}$$

$$= (x - 1)^2 - 5 \qquad \textit{Step 6}$$

The graph of $y = (x - 1)^2 - 5$ is the basic parabola shifted 1 unit to the right and 5 units downward. Its vertex is at $(1, -5)$. (See Figure 9-5b.)

EXAMPLE 9-6 Complete the square on the general quadratic equation $y = ax^2 + bx + c$.

Solution

$y = ax^2 + bx + c$

$$= (ax^2 + bx + \quad) + c \qquad \textit{Step 1}$$

$$= a\left(x^2 + \frac{b}{a}x + \quad\right) + c \qquad \textit{Step 2}$$

Find $\dfrac{1}{2}\left(\dfrac{b}{a}\right) = \dfrac{b}{2a}$; then $\left(\dfrac{b}{2a}\right)^2 = \dfrac{b^2}{4a^2}$ *Step 3*

$$y = a\left(x^2 + \frac{b}{a}x + \frac{b^2}{4a^2}\right) + c - \frac{b^2}{4a} \qquad \textit{Step 4}\ \left[\text{We subtract } \frac{b^2}{4a}, \text{ since } \frac{a}{1}\left(\frac{b^2}{4a^2}\right) = \frac{b^2}{4a}.\right]$$

$$= a\left(x + \frac{b}{2a}\right)^2 + c - \frac{b^2}{4a} \qquad \textit{Step 5}$$

$$= a\left(x + \frac{b}{2a}\right)^2 + \left(c - \frac{b^2}{4a}\right) \qquad \textit{Step 6}$$

We know from completing the square in Example 9-6 that the equations $y = ax^2 + bx + c$ and $y = a(x + b/2a)^2 + (c - b^2/4a)$ are equivalent. But the latter equation is of the form $y = a(x - h)^2 + k$, where $h = -b/2a$ and $k = c - b^2/4a$. It follows, then, that

- The graph of an equation of the form $y = ax^2 + bx + c$ is a parabola whose vertex is $(h, k) = \left(-\dfrac{b}{2a}, c - \dfrac{b^2}{4a}\right)$ and whose axis of symmetry is the vertical line $x = h = -\dfrac{b}{2a}$.

EXAMPLE 9-7 Complete the square on **(a)** $y = 2x^2 + 8x + 1$ and **(b)** $y = -x^2 + 3x + 4$ to find the vertex and axis of symmetry for each parabola. Use the original values of a, b, and c to check your answer.

Solution   See Figure 9-6.

(a)
$$y = 2x^2 + 8x + 1 = (2x^2 + 8x +) + 1$$
$$= 2(x^2 + 4x +) + 1 = 2(x^2 + 4x + 4) + 1 - 8$$
$$= 2(x + 2)^2 - 7$$

The vertex is $(h, k) = (-2, -7)$ and the axis of symmetry is the line $x = -2$.

check: In the original equation $y = 2x^2 + 8x + 1$, $a = 2$, $b = 8$, and $c = 1$.

Then the vertex is $\left(-\dfrac{b}{2a}, c - \dfrac{b^2}{4a}\right)$ or $\left(\dfrac{-8}{2(2)}, 1 - \dfrac{(8)^2}{4(2)}\right) = (-2, -7)$

and the axis of symmetry is the line $x = -\dfrac{b}{2a} = -2$.

(b)
$$y = -x^2 + 3x + 4 = (-x^2 + 3x +) + 4$$
$$= (-1)(x^2 - 3x +) + 4 = -1(x^2 - 3x + \tfrac{9}{4}) + 4 - (-\tfrac{9}{4})$$
$$= -1(x - \tfrac{3}{2})^2 + \tfrac{25}{4}$$

The vertex is $(\tfrac{3}{2}, \tfrac{25}{4})$ and the axis of symmetry is the line $x = \tfrac{3}{2}$. The parabola is the basic parabola opening downward.

check: In the original equation $y = -x^2 + 3x + 4$, $a = -1$, $b = 3$, and $c = 4$. Then the vertex is $\left(\dfrac{-3}{2(-1)}, 4 - \dfrac{(3)^2}{4(-1)}\right)$ or $\left(\dfrac{3}{2}, \dfrac{25}{4}\right)$ and the axis of symmetry is the line $x = -\dfrac{3}{2(-1)} = \dfrac{3}{2}$.

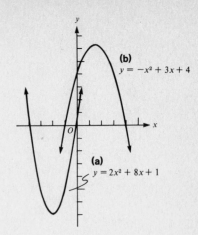

(b) $y = -x^2 + 3x + 4$

(a) $y = 2x^2 + 8x + 1$

Figure 9-6

2. Intercepts

When graphing any function $y = f(x)$, it's useful to locate its intercepts. The value of x at which the graph crosses the x-axis is an x-intercept and the value of y at which the graph crosses the y-axis is a y-intercept. We find the x-intercept(s) of a quadratic function by setting $y = 0$ in the expression $y = f(x)$ and solving the equation for x. We find the y-intercept by setting $x = 0$ and solving the equation for y.

note: The graph of a quadratic function has one and only one y-intercept. (If there were more than one, there would be more than one point on the vertical line $x = 0$—and that means the graph is not the graph of a function!) The graph of a quadratic function may have one, two, or no x-intercepts.

EXAMPLE 9-8   Find the x-intercept(s) and the y-intercept for **(a)** $y = x^2 - 4$, **(b)** $y = -4x^2$, and **(c)** $y = 2x^2 + 1$. Sketch each parabola.

Solution   See Figure 9-7.

(a) To find the x-intercepts, set $y = 0$ and solve for x:

$$y = x^2 - 4$$
$$0 = x^2 - 4$$
$$x^2 = 4$$
$$x = \pm 2$$

There are two x-intercepts: $(2, 0)$ and $(-2, 0)$.

To find the y-intercept, set $x = 0$ and solve for y:

$$y = x^2 - 4$$
$$y = -4$$

The y-intercept is $(0, -4)$, and this point is also the vertex.

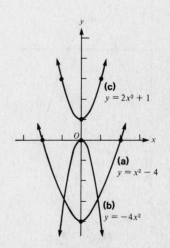

(c) $y = 2x^2 + 1$

(a) $y = x^2 - 4$

(b) $y = -4x^2$

Figure 9-7

(b) Set $y = 0$ in $y = -4x^2$ and solve for x:

$$y = -4x^2$$

$$0 = -4x^2$$

$$x = 0$$

Thus the only x-intercept is the point $(0, 0)$. We don't need to look any further for a y-intercept, since the origin $(0, 0)$ is also on the y-axis—and there can be only one y-intercept. As a check, set $x = 0$ and solve for y in $y = -4x^2$.

$$y = -4x^2 = -4(0)^2 = 0$$

The only intercept of either axis is the point $(0, 0)$, which is also the vertex.

(c) Set $y = 0$ in $y = 2x^2 + 1$ and solve for x:

$$y = 2x^2 + 1$$

$$0 = 2x^2 + 1$$

$$2x^2 = -1$$

$$x^2 = -\tfrac{1}{2}$$

But there is no real value of x such that x^2 is negative, so there are no x-intercepts.

Set $x = 0$ and solve for y:

$$y = 2(0)^2 + 1 = 1$$

The y-intercept is $(0, 1)$, which is also the vertex.

EXAMPLE 9-9   Find the intercepts and the vertex for $y = -2x^2 + 12x - 15$. Graph the parabola.

Solution   If $x = 0$, $y = -15$, so the y-intercept is $(0, -15)$. To find the x-intercept(s), set $y = 0$ and solve the equation for x:

$$y = -2x^2 + 12x - 15$$

$$0 = -2x^2 + 12x - 15$$

But $-2x^2 + 12x - 15$ doesn't appear to factor over the set of integers, so we'll use the quadratic formula to solve for x, using $a = -2$, $b = 12$, and $c = -15$:

$$x = \frac{-b \pm \sqrt{b^2 - 4ac}}{2a} = \frac{-12 \pm \sqrt{(12)^2 - 4(-2)(-15)}}{2(-2)}$$

$$= \frac{-12 \pm \sqrt{24}}{-4} = 3 \pm \frac{\sqrt{24}}{4} = 3 \pm \sqrt{\frac{3}{2}}$$

Using the formula for the vertex, we have

$$\left(-\frac{b}{2a}, c - \frac{b^2}{4a}\right) = \left(\frac{-12}{2(-2)}, -15 - \frac{(12)^2}{4(-2)}\right) = (3, 3)$$

The graph is shown in Figure 9-8.

note: You can also find the vertex by completing the square:

$$y = -2x^2 + 12x - 15$$

$$= -2(x^2 - 6x + \quad) - 15$$

$$= -2(x^2 - 6x + 9) - 15 + 18$$

$$= -2(x - 3)^2 + 3$$

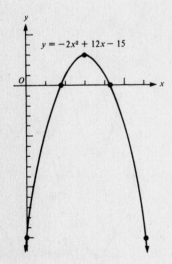

$y = -2x^2 + 12x - 15$

Figure 9-8

D. Maximum and minimum values of a quadratic function

Given the quadratic function $f(x) = ax^2 + bx + c$:

- If $a > 0$, then the vertex is the lowest point on the graph of $f(x) = ax^2 + bx + c$ and the y-coordinate of the vertex is called the **minimum value** of the function.
- If $a < 0$, then the vertex is the highest point on the graph of $f(x) = ax^2 + bx + c$ and the y-coordinate of the vertex is called the **maximum value** of the function.

EXAMPLE 9-10   Given $f(x) = 3x^2 + 6x - 5$, find the minimum value of $f(x)$.

Solution   We know that it is a minimum value we're looking for, since $a > 0$. Then the y-coordinate of the vertex is the minimum value of the function. Use $a = 3$, $b = 6$, and $c = -5$ in the formula for the vertex:

$$\left(\frac{-b}{2a}, c - \frac{b^2}{4a} \right) = \left(\frac{-6}{2(3)}, -5 - \frac{(6)^2}{4(3)} \right) = (-1, -8)$$

The y-coordinate is -8, so the minimum value of $f(x) = 3x^2 + 6x - 5$ is -8.

Maximum and minimum values have applications in the real world. Frequently, we wish to maximize or minimize a variable quantity. If we can express that quantity as a quadratic function of some other variable quantity, we can use the techniques of this section.

EXAMPLE 9-11   Rancher Myk wishes to fence in a rectangular pasture for her sheep but needs to fence only three sides, since the remaining side will be along a straight river. She has 2 miles of fencing. What should the dimensions of the pasture be in order to get the largest area?

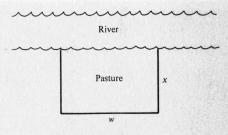

Figure 9-9

Solution   Let the sides of the pasture be x and w, as shown in Figure 9-9. Then the total of the lengths of the three sides must be equal to 2 miles, since Rancher Myk has only 2 miles of fencing. We can represent this as $2x + w = 2$. Now the area of the pasture is $A = xw$. But if we solve for w in the first equation, we can substitute that value for w in the second equation:

$$2x + w = 2$$

$$w = -2x + 2 \qquad \text{Solve for } w.$$

$$A = xw = x(-2x + 2) \qquad \text{Replace } w \text{ with } -2x + 2.$$

$$= -2x^2 + 2x$$

This makes A a function of x only—and the function is a quadratic, so its graph is a parabola. Moreover, the value of a is -2, so the parabola opens downward and the y-coordinate of the vertex locates the maximum value of the function. The value of x at the vertex will be the dimension for x to be used to maximize the area. Use $a = -2$, $b = 2$, and $c = 0$ in the formula $x = -b/2a$ to get the x-coordinate of the vertex:

$$x = -\frac{b}{2a} = -\frac{2}{2(-2)} = \frac{1}{2}$$

The vertex is at $x = \frac{1}{2}$. The corresponding value for w is

$$w = -2x + 2 = -2(\tfrac{1}{2}) + 2 = -1 + 2 = 1$$

The dimensions of the pasture should be $x = \frac{1}{2}$ mile and $w = 1$ mile for Rancher Myk to have the largest pasture possible.

9-2. Higher-Degree Polynomial Functions

Although it's fairly easy to use the properties of quadratic functions to determine the vertex, axis, intercepts, and general shape of a given parabola, it's not so easy to do this for polynomials of higher degree. In fact, the higher the degree of a polynomial, the less accuracy we have in graphing it. But there are some techniques that allow us to sketch the general shape of many higher-degree polynomial functions.

A. Graphing monomial functions of degree greater than 2

Let's begin with the monomial functions of $f(x) = ax^n$ for $n \geq 3$.

EXAMPLE 9-12   Graph (a) $f(x) = x^3$, (b) $f(x) = x^4$, and (c) $f(x) = x^5$.

Solution

(a) Since $f(-x) = (-x)^3 = -x^3 = -f(x)$, the function $f(x) = x^3$ is an odd function. Its graph is symmetric with respect to the origin. (See Figure 9-10a.)

(b) Since $f(-x) = (-x)^4 = x^4 = f(x)$, the function $f(x) = x^4$ is an even function. Its graph is symmetric with respect to the y-axis. (See Figure 9-10b.)

(c) Since $f(-x) = (-x)^5 = -x^5 = -f(x)$, the function $f(x) = x^5$ is an odd function. Its graph is symmetric with respect to the origin. (See Figure 9-10c.)

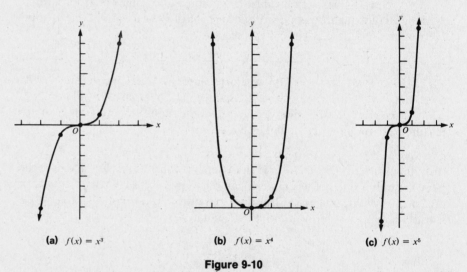

(a) $f(x) = x^3$ (b) $f(x) = x^4$ (c) $f(x) = x^5$

Figure 9-10

Now we can make the following generalizations:

(1) The graphs of $x^5, x^7, x^9, \ldots$, where the exponent of x is odd, are similar in shape to the graph of $f(x) = x^3$. They're all odd functions and symmetric with respect to the origin.

(2) The graphs of $x^6, x^8, x^{10}, \ldots$, where the exponent of x is even, are similar in shape to the graph of $f(x) = x^4$. They're all even functions and symmetric with respect to the y-axis.

> *note:* Although $f(x) = x^4$ and other monomials of even degree *look* a little like $f(x) = x^2$ (Fig. 9-1) they are NOT parabolas. Only graphs of QUADRATIC functions are parabolas.

(3) As the exponent of x—whether even or odd—becomes larger, the graph of $y = x^n$ becomes increasingly flat around the origin and increasingly steep for values of x greater than 1.

We can see the difference between the curves $f(x) = y = x^n$ for $n = 2, 3, 4, \ldots$, if we graph the portion of each curve between 0 and 1 on the x-axis.

EXAMPLE 9-13 On the same set of coordinate axes, graph $y = x^2$, $y = x^3$, $y = x^4$, and $y = x^5$ over the domain $0 \le x \le 1$.

Solution Since the computations are more difficult for values of x less than 1, we'll make a table of x and y values:

x	0	.2	.4	.6	.8	1.0
$y = x^2$	0	.04	.16	.36	.64	1.0
$y = x^3$	0	.008	.064	.216	.512	1.0
$y = x^4$	0	.0016	.0256	.1296	.4096	1.0
$y = x^5$	0	.00032	.01024	.07776	.32768	1.0

The curves are drawn in Figure 9-11. Notice that $y = x^5$ is flatter than $y = x^4$, $y = x^4$ is flatter than $y = x^3$, and so on.

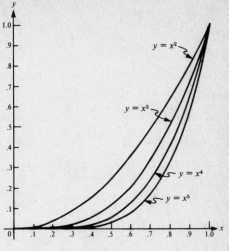

Figure 9-11

Some of the properties of quadratic functions can be applied to higher-degree polynomial functions:

(1) If $a < 0$, the graph of $y = ax^n$ is a reflection through the x-axis of the graph of $y = |a|x^n$.

(2) If $|a| > 1$, the graph of $y = ax^n$ is similar to the graph of $y = x^n$ but "pulled away" from the x-axis.

(3) If $|a| < 1$, the graph of $y = ax^n$ is similar to the graph of $y = x^n$ but "pulled toward" the x-axis.

EXAMPLE 9-14 Draw the graphs of **(a)** $y = -\frac{1}{4}x^3$ and **(b)** $y = 2x^4$.

Solution See Figure 9-12.

(a) The graph of $y = -\frac{1}{4}x^3$ is similar to the graph of $y = x^3$ (Figure 9-10a), but it is pulled toward the x-axis and reflected through the x-axis.

(b) The graph of $y = 2x^4$ is similar to the graph of $y = x^4$ (Figure 9-10b), but pulled away from the x-axis.

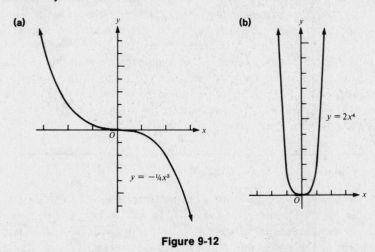

Figure 9-12

Sometimes the graph of a higher-degree polynomial function can be found by shifting the graph of ax^n vertically and/or horizontally:

• If $y = a(x - h)^n + k$ and n is even, the graph is the same shape as $y = ax^n$, but the vertex is shifted from $(0, 0)$ to (h, k) and the axis of symmetry to the vertical line $x = h$.

(a)

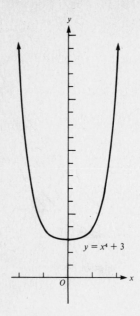

$y = x^4 + 3$

(b)

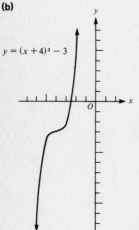

$y = (x + 4)^3 - 3$

(c)

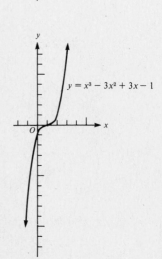

$y = x^3 - 3x^2 + 3x - 1$

Figure 9-13

• If $y = a(x - h)^n + k$ and n is odd, the graph is the same shape as $y = ax^n$, but the point of symmetry is shifted from the origin $(0, 0)$ to the point (h, k).

EXAMPLE 9-15   Draw the graphs of **(a)** $y = x^4 + 3$, **(b)** $y = (x + 4)^3 - 3$, and **(c)** $y = x^3 - 3x^2 + 3x - 1$.

Solution   See Figure 9-13.

(a) The graph of $y = x^4 + 3$ is simply the graph of $y = x^4$ (Figure 9-10b) shifted upward 3 units, with no change in the axis of symmetry.

(b) The graph of $y = (x + 4)^3 - 3$ is the graph of $y = x^3$ (Figure 9-10a) shifted 4 units to the left and 3 units downward, so that the point of symmetry is $(-4, -3)$.

(c) This is not as bad as it looks. The polynomial $x^3 - 3x^2 + 3x - 1$ is the expanded version of $(x - 1)^3$. Thus the graph of $y = x^3 - 3x^2 + 3x - 1 = (x - 1)^3$ is the graph of $y = x^3$ (Figure 9-10a) shifted 1 unit to the right, so that the point of symmetry is $(1, 0)$.

B. Zeros of higher-degree polynomials

The graphs of all polynomial functions of degree greater than 1 are smooth curves with varying numbers of "hills" and "valleys." A close analysis of higher-degree polynomials and their graphs requires methods of calculus— but if we can factor the polynomial, we can usually determine the general shape of the graph.

• If a polynomial function

$$f(x) = a_n x^n + a_{n-1} x^{n-1} + a_{n-2} x^{n-2} + \cdots + a_1 x + a_0$$

can be factored as

$$f(x) = a_n (x - c_1)(x - c_2) \times \cdots \times (x - c_n)$$

then $c_1, c_2, \ldots, c_n$ are called the **zeros** of the polynomial, since $f(x) = 0$ if and only if $x = c_1$ or $x = c_2$ or $\ldots x = c_n$.

The zeros of a polynomial function divide the number line into intervals in which the function does not change its sign. Thus the zeros are the points at which the graph crosses or touches the x-axis (the x-intercepts), since at those points $y = f(x) = 0$. When the graph is NOT crossing or touching the x-axis, it must be either above or below it. To determine the shape of the graph, then, we must find out where the graph is above the x-axis (where $f(x) > 0$) and where it is below the x-axis (where $f(x) < 0$). To do this, we choose a test value x from each interval determined by the x-intercepts. If $f(x) > 0$, the graph is above the x-axis over all of the interval; if $f(x) < 0$, the graph is below the x-axis over all of the interval.

EXAMPLE 9-16   Graph the function $y = f(x) = (x - 1)(x + 1)(x - 3)$.

Solution   Since this function is given in factored form, it is easy to see the zeros— those values of x for which $y = f(x) = 0$. The zeros are $x = 1$, $x = -1$, and $x = 3$, so the x-intercepts are at $x = 1$, $x = -1$, and $x = 3$. We draw a number line (representing the x-axis) and divide it into four intervals, I_1, I_2, I_3, and I_4:

We choose a test value x from each interval and compute $f(x)$:

Interval	Test value	$f(x) = (x - 1)(x + 1)(x - 3)$
I_1	$x = -2$	$f(x) = (-2 - 1)(-2 + 1)(-2 - 3) = (-3)(-1)(-5) = -15$
I_2	$x = 0$	$f(x) = (0 - 1)(0 + 1)(0 - 3) = (-1)(1)(-3) = 3$
I_3	$x = 2$	$f(x) = (2 - 1)(2 + 1)(2 - 3) = (1)(3)(-1) = -3$
I_4	$x = 4$	$f(x) = (4 - 1)(4 + 1)(4 - 3) = (3)(5)(1) = 15$

The table shows that $f(x)$ is negative on the interval I_1, positive on I_2, negative on I_3, and positive on I_4. This means that the graph is below the x-axis when $x < -1$, above the x-axis when $-1 < x < 1$, below the x-axis when $1 < x < 3$, and above the x-axis when $x > 3$.

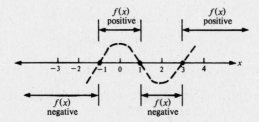

We can see that there must be at least one "hill" on the interval $-1 < x < 1$ when $f(x)$ changes from increasing to decreasing and at least one "valley" on the interval $1 < x < 3$ when $f(x)$ changes from decreasing to increasing. Although we need the calculus to find the exact points where these occur, for now, we calculate some additional solution pairs and approximate the graph (Figure 9-14) through all the known points.

$$y = (x - 1)(x + 1)(x - 3)$$

x	y
-3	$(-4)(-2)(-6) = -48$
$-\frac{1}{2}$	$(-\frac{3}{2})(\frac{1}{2})(-\frac{7}{2}) = \frac{21}{8}$
$\frac{1}{2}$	$(-\frac{1}{2})(\frac{3}{2})(-\frac{5}{2}) = \frac{15}{8}$
$\frac{3}{2}$	$(\frac{1}{2})(\frac{5}{2})(-\frac{3}{2}) = -\frac{15}{8}$
$\frac{5}{2}$	$(\frac{3}{2})(\frac{7}{2})(-\frac{1}{2}) = -\frac{21}{8}$
$\frac{7}{2}$	$(\frac{5}{2})(\frac{9}{2})(\frac{1}{2}) = \frac{45}{8}$

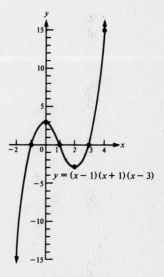

Figure 9-14

note: Every polynomial of degree 3 has an S-shaped graph similar to the one drawn in Figure 9-14. The graph may be squeezed or elongated and moved up or down or to the right or left, but it is always a basic S shape.

EXAMPLE 9-17 Sketch the graph of $y = f(x) = x^3 + 4x^2 + 4x$.

Solution First factor the polynomial:

$$f(x) = x^3 + 4x^2 + 4x$$
$$= x(x^2 + 4x + 4)$$
$$= x(x + 2)(x + 2)$$
$$= x(x + 2)^2$$

The value of $f(x)$ is equal to 0 only if $x = 0$ or $x = -2$, so the only x-intercepts are at $x = 0$ and $x = -2$. As before, we draw a number line and divide it into intervals:

We choose a test value from each interval and compute $f(x)$:

Interval	Test value	$f(x) = x(x + 2)^2$
I_1	$x = -3$	$f(x) = (-3)(-3 + 2)^2 = (-3)(-1)^2 = -3$
I_2	$x = -1$	$f(x) = (-1)(-1 + 2)^2 = (-1)(1)^2 = -1$
I_3	$x = 1$	$f(x) = (1)(1 + 2)^2 = (1)(3)^2 = 9$

As the table shows, $f(x)$ is negative on the intervals I_1 and I_2 and positive on I_3. This means that the graph is below the x-axis when $x < 0$ and above the x-axis when $x > 0$:

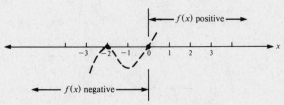

Since the graph remains below the x-axis on both I_1 and I_2, it must not CROSS the axis at the intercept $x = -2$ but only TOUCH it.

note: In general, if a factor of a polynomial $(x - c_n)$ has an exponent of 2 (or any other even integer)—i.e., if a zero is of even degree—then the graph of the polynomial touches but does not cross the x-axis at $x = c_n$.

We now calculate some additional points that can be plotted on the graph, as in Figure 9-15.

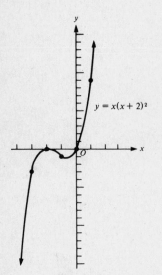

Figure 9-15

$$y = x(x + 2)^2$$

x	y
-4	$-4(-4 + 2)^2 = -16$
$-\frac{5}{2}$	$-\frac{5}{2}(-\frac{5}{2} + 2)^2 = -\frac{5}{8}$
$-\frac{3}{2}$	$-\frac{3}{2}(-\frac{3}{2} + 2)^2 = -\frac{3}{8}$
$-\frac{1}{2}$	$-\frac{1}{2}(-\frac{1}{2} + 2)^2 = -\frac{9}{8}$
2	$2(2 + 2)^2 = 32$

9-3. Rational Functions

A. Definition

- A **rational function** is a quotient of two polynomial functions. In other words, if $P(x)$ and $Q(x)$ are polynomial functions, we can write a rational function as

RATIONAL FUNCTION $\qquad f(x) = \dfrac{P(x)}{Q(x)}$

The domain of $f(x)$ consists of all real numbers x such that $Q(x) \neq 0$ (since division by zero is undefined).

B. Asymptotes and graphing

The zeros of the numerator and denominator play an important part in graphing a rational function. The zeros of the numerator $P(x)$ are the same as the zeros of $f(x)$, since $f(x) = 0$ if and only if $P(x) = 0$. The zeros of the numerator, then, will give us the x-intercepts (if any) of the graph of $f(x)$. The zeros of the denominator help us in a different way. If b is a zero of the denominator $Q(x)$, then $f(b)$ is undefined (since division by zero is undefined). We are interested in what happens to the graph of $f(x)$ when x is NEAR b.

EXAMPLE 9-18 Sketch the graph of $f(x) = 1/x$.

Solution The function $f(x) = 1/x$ is a rational function with $P(x) = 1$, a constant polynomial function, and $Q(x) = x$, a first-degree polynomial function. The domain of $f(x)$ is the set of all real numbers except 0. (If $x = 0$, the denominator is zero, and the function $f(x)$ is undefined.) The graph of $f(x)$ has no x-intercepts, since the numerator $P(x)$ can never be equal to zero—and thus $f(x)$ can never be equal to 0. We also know that the function is odd—and that its graph is symmetric with respect to the origin—since $f(-x) = 1/(-x) = -(1/x) = -f(x)$.

We wish to see what happens to the graph of $f(x)$ when x is near 0 (since the zero of the denominator is 0). If we construct a table of values in which x gets closer to 0, we see that $f(x)$ grows larger and larger. If, however, we use larger values of x, then $f(x)$ grows smaller and smaller.

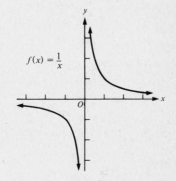

$f(x) = \frac{1}{x}$

x	$f(x)$	x	$f(x)$
1	1	1	1
$\frac{1}{10}$	10	10	$\frac{1}{10}$
$\frac{1}{100}$	100	100	$\frac{1}{100}$
$\frac{1}{1000}$	1000	1000	$\frac{1}{1000}$

The graph is shown in Figure 9-16. Notice that no matter how small $f(x)$ becomes, it is never zero, since the graph never crosses the x-axis. And no matter how small x becomes, it is never zero, since zero is excluded from the domain. Symmetry assures us of the shape of the graph in the third quadrant.

Figure 9-16

When a curve constantly approaches but never touches a line, that line is called an *asymptote*:

- A function $f(x)$ has the line $x = b$ as a **vertical asymptote** if, as x approaches b from the left or right or both, $f(x)$ grows without bound; i.e., as $x \to b$, $f(x) \to \infty$ or $f(x) \to -\infty$.
  A function $f(x) = P(x)/Q(x)$ has a vertical asymptote at $x = b$ if $Q(b) = 0$ (i.e., b is a zero of the denominator) but $P(b) \neq 0$.
- A function $f(x)$ has the line $y = c$ as a **horizontal asymptote** if, as x grows without bound, $f(x)$ approaches c; i.e., as $x \to \infty$ or as $x \to -\infty$, $f(x) \to c$.

EXAMPLE 9-19 Find the asymptotes for the function $f(x) = 1/x$ in Example 9-18.

Solution The function has a vertical asymptote at $x = 0$ (the y-axis), since as $x \to 0$, $f(x) \to \infty$ and $f(x) \to -\infty$. Notice that $Q(0) = 0$, so 0 is a zero of the denominator. The function has a horizontal asymptote at $y = 0$, since as $x \to \infty$ and $x \to -\infty$, $f(x) \to 0$. (See Figure 9-16.)

EXAMPLE 9-20 Determine the asymptotes for $f(x) = 1/x^2$. Graph the function.

Solution When $x = 0$, the denominator is zero and the numerator is not, so the line $x = 0$ is a vertical asymptote. Also, as x approaches 0, $f(x)$ grows without bound.

As $|x|$ grows without bound, $f(x)$ approaches zero, so the line $y = 0$ is a horizontal asymptote. There are no x-intercepts, since $f(x)$ is never zero. Finally, since $f(-x) = 1/(-x)^2 = 1/x^2 = f(x)$, the graph is symmetric with respect to the y-axis. (See Figure 9-17.)

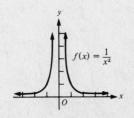

$f(x) = \frac{1}{x^2}$

Figure 9-17

EXAMPLE 9-21   Find the asymptotes for $f(x) = \dfrac{1}{(x-1)(x+3)}$ and sketch the graph of $f(x)$.

Solution   The zeros of the denominator $(x-1)(x+3)$ are $x = 1$ and $x = -3$, so there are vertical asymptotes at $x = 1$ and $x = -3$. These asymptotes determine three regions or intervals: I_1, where $x < -3$; I_2, where $-3 < x < 1$; and I_3, where $x > 1$.

note: Since there are no x-intercepts ($f(x)$ is never zero), the graph never crosses the x-axis. Therefore, if the graph is above the x-axis, for example, in a particular region, it stays above the axis throughout the entire region.

In order to find out where the graph of $f(x)$ is above or below the x-axis, we make a table of test values:

Interval	Test value	$f(x) = \dfrac{1}{(x-1)(x+3)}$
I_1	$x = -4$	$f(x) = \dfrac{1}{(-4-1)(-4+3)} = \dfrac{1}{(-5)(-1)} = \dfrac{1}{5}$
I_2	$x = 0$	$f(x) = \dfrac{1}{(0-1)(0+3)} = \dfrac{1}{(-1)(3)} = -\dfrac{1}{3}$
I_3	$x = 2$	$f(x) = \dfrac{1}{(2-1)(2+3)} = \dfrac{1}{(1)(5)} = \dfrac{1}{5}$

The value of the function is positive on I_1, negative on I_2, and positive on I_3. This means that the graph is above the x-axis when $x < -3$, below the x-axis when $-3 < x < 1$, and above the axis when $x > 1$.

The graph has a horizontal asymptote of $y = 0$, since as $x \to \infty$ and as $x \to -\infty$, $f(x) \to 0$.

We can sketch the graph in intervals I_1 and I_3, knowing the asymptotes and plotting a few points. To sketch the curve in the middle interval, we set $x = 0$ and solve for y to find the y-intercept.

$$y = \frac{1}{(x-1)(x+3)} = \frac{1}{(0-1)(0+3)} = -\frac{1}{3}$$

We have a curve that crosses the y-axis at $y = -\frac{1}{3}$ and approaches the lines $x = -3$ and $x = 1$ (the asymptotes) as $f(x) \to -\infty$. The complete rough graph of $f(x) = 1/(x-1)(x+3)$ is shown in Figure 9-18.

note: The y-intercept $-\frac{1}{3}$ is not the maximum value of the curve in I_2, since, for example, $f(-1) = -\frac{1}{4}$. We can only *approximate* the curve here—we need calculus to determine the exact spot at which the curve is highest in an interval such as I_2.

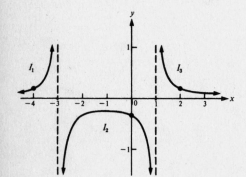

Figure 9-18: $f(x) = \dfrac{1}{(x-1)(x+3)}$

The graph of a rational function doesn't always have the x-axis as a horizontal asymptote. We can use the following **degree-comparison test** to tell when a function has a horizontal asymptote:

Given $f(x) = \dfrac{P(x)}{Q(x)} = \dfrac{a_m x^m + a_{m-1} x^{m-1} + \cdots + a_2 x^2 + a_1 x + a_0}{b_n x^n + b_{n-1} x^{n-1} + \cdots + b_2 x^2 + b_1 x + b_0}$,

(1) If $m < n$, the line $y = 0$ (the x-axis) is a horizontal asymptote.
(2) If $m = n$, the line $y = a_m/b_n$ is a horizontal asymptote.
(3) If $m > n$, there is no horizontal asymptote.

EXAMPLE 9-22   Find the asymptotes for $f(x) = \dfrac{2x^2 + 1}{(x - 3)(x + 2)}$ and sketch the graph of the function.

Solution   The zeros of the denominator are $x = 3$ and $x = -2$, so there are vertical asymptotes at $x = 3$ and $x = -2$. The degree of the numerator polynomial is 2 and the degree of the denominator polynomial is 2. Thus we have case (**2**) of the degree-comparison test given above, since $m = n = 2$. The value of a_m is 2 and the value of b_n is 1, so the line $y = a_m/b_n = 2/1 = 2$ is the horizontal asymptote.

note: We can also use our original definition of a horizontal asymptote here. Given $f(x) = \dfrac{2x^2 + 1}{(x - 3)(x + 2)}$, we divide both the numerator and the denominator by x^2. Then

$$\frac{2x^2 + 1}{(x - 3)(x + 2)} = \frac{2x^2 + 1}{x^2 - x - 6} = \frac{2 + 1/x^2}{1 - 1/x - 6/x^2}$$

As $x \to \infty$ and $x \to -\infty$, the terms $1/x^2$, $1/x$, and $6/x^2$ all approach 0, so $f(x) \to 2/1$ or 2. ✓

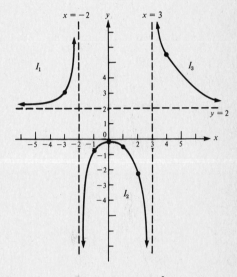

Figure 9-19: $f(x) = \dfrac{2x^2 + 1}{(x - 3)(x + 2)}$

Our next step in graphing the function is to find where the graph is above the x-axis and where it is below it. Since the function has no zeros on the horizontal axis ($2x^2 + 1 \neq 0$), the vertical asymptotes give us these intervals: I_1, where $x < -2$; I_2, where $-2 < x < 3$; and I_3, where $x > 3$. For I_1, we choose the test value $x = -3$. Since $f(-3) = \frac{19}{6}$, $f(x)$ is positive on all of I_1. For I_2, we choose $x = 0$. Since $f(0) = -\frac{1}{6}$, $f(x)$ is negative on all of I_2. For I_3, we choose $x = 4$. The value of $f(4)$ is $\frac{33}{6}$, so $f(x)$ is positive on all of I_3. Finally, since $f(0) = -\frac{1}{6}$, the y-intercept is at $(0, -\frac{1}{6})$. The rough graph is shown in Figure 9-19. [Extra points $f(-1) = -\frac{3}{4}$ and $f(2) = -\frac{9}{4}$.]

Not all asymptotes are vertical or horizontal.

- If the degree of the numerator of a rational function is exactly one more than the degree of the denominator, the graph of the function has a **slant** or **oblique asymptote**.

This asymptote is a slanted straight line toward which the function gets very close when the x values are very large positive or very large negative. To find the equation of the line that is the oblique asymptote, use long division to simplify the rational expression (see Chapter 2). Then see what happens as $|x| \to \infty$.

EXAMPLE 9-23   Graph $f(x) = \dfrac{x^2 - 3x + 3}{x - 1}$.

Solution   The zero of the denominator is 1, so there is a vertical asymptote of $x = 1$. Since the degree of the numerator is exactly one more than the degree of the denominator, there is an oblique asymptote. We simplify the expression, using long division:

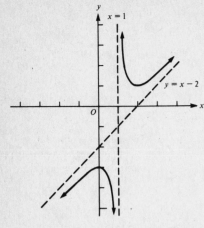

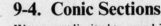

$$f(x) = x - 1 \overline{) x^2 - 3x + 3}$$
$$\underline{-(x^2 - \;x)}$$
$$-2x + 3$$
$$\underline{-(-2x + 2)}$$
$$1$$

As $|x| \to \infty$, the remainder $1/(x - 1)$ approaches 0, so $f(x) \to x - 2$; i.e., the graph of the function approaches the line $y = x - 2$, which is an oblique asymptote with slope 1 and y-intercept -2.

Finally, let's check the intercepts. There is an x-intercept when the numerator $x^2 - 3x + 3$ is equal to zero (since $f(x) = 0$ if and only if the numerator is 0). We use the quadratic formula to solve for x:

$$x = \frac{-b \pm \sqrt{b^2 - 4ac}}{2a} = \frac{-(-3) \pm \sqrt{(-3)^2 - 4(1)(3)}}{2(1)} = \frac{3 \pm \sqrt{-3}}{2}$$

There is no real number that is the square root of -3, so there are no x-intercepts. To find the y-intercept, set $x = 0$ and solve for y:

$$y = \frac{x^2 - 3x + 3}{x - 1} = \frac{(0)^2 - 3(0) + 3}{0 - 1} = \frac{3}{-1} = -3$$

We plot the y-intercept at $(0, -3)$ and sketch an approximation of the graph, as in Figure 9-20.

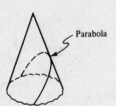

Figure 9-20: $y = \dfrac{x^2 - 3x + 3}{x - 1}$

9-4. Conic Sections

We are not limited to graphing functions. We can also graph conic sections. **Conic sections** are curves that can be obtained by cutting a right circular cone with a plane. The four major conic sections are the *parabola*, the *circle*, the *ellipse*, and the *hyperbola*, as shown in Figure 9-21.

note: If a plane slices a cone such that the intersection is a single point, a line, or two intersecting lines, that intersection is called a **degenerate conic section**.

A. Parabolas

Geometrically, we define a **parabola** as the set of all points (x, y) that are equidistant from a fixed point F and a fixed line L. The line is called the *directrix* and the point F is called the *focus*. Algebraically, however, we can think of parabolas as graphs of their equations.

The graph of an equation of the form $y = ax^2 + bx + c$, $a \neq 0$, is an "up-and-down" parabola that opens upward if $a > 0$ and downward if $a < 0$ (see Section 9-1). By the process of completing the square, we can rewrite this equation as

- $y = a(x - h)^2 + k$, where (h, k) is the vertex of the parabola and $x = h$ is the axis of symmetry.

The graph of an equation of the form $x = ay^2 + by + c$, $a \neq 0$, is a "sideways" parabola that opens to the right if $a > 0$ and to the left if $a < 0$. By completing the square, this equation can be rewritten as

- $x = a(y - k)^2 + h$, where (h, k) is the vertex of the parabola and $y = k$ is the axis of symmetry.

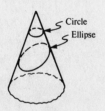

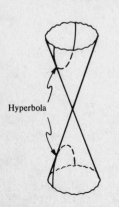

Figure 9-21: Conic sections

EXAMPLE 9-24 Find the vertex and axis of symmetry for the graphs of (**a**) $y = 2x^2 + 2x - 12$ and (**b**) $x = 2y^2 + 2y - 12$. Sketch each parabola.

Solution To find the vertex and axis of symmetry, rewrite each equation in the form $y = a(x - h)^2 + k$ or $x = a(y - k)^2 + h$.

(a) Complete the square on the right side of the equation.

$$y = 2x^2 + 2x - 12 = 2(x^2 + x + \quad) - 12$$

$$= 2(x^2 + x + \tfrac{1}{4}) - 12 - \tfrac{1}{2} = 2(x + \tfrac{1}{2})^2 - 12 - \tfrac{1}{2}$$

$$= 2(x + \tfrac{1}{2})^2 - \tfrac{25}{2}$$

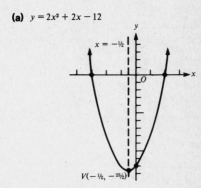

(a) $y = 2x^2 + 2x - 12$

Now we can see that $h = -\tfrac{1}{2}$ and $k = -\tfrac{25}{2}$, so the vertex of this upward-opening $(a > 0)$ parabola is at $(h, k) = (-\tfrac{1}{2}, -\tfrac{25}{2})$ and the axis of symmetry is $x = h = -\tfrac{1}{2}$.

To help shape the graph, we find the intercepts. Setting $x = 0$ in $y = 2x^2 + 2x - 12$ gives us $y = -12$, so the y-intercept is at $(0, -12)$. Setting $y = 0$ gives us

$$0 = 2x^2 + 2x - 12$$

$$= 2(x^2 + x - 6) = 2(x - 2)(x + 3)$$

The x-intercepts are at $(2, 0)$ and $(-3, 0)$. The graph is shown in Figure 9-22a.

(b) Since the equation $x = 2y^2 + 2y - 12$ is the same equation as $y = 2x^2 + 2x - 12$—with the variables switched—we don't need to repeat the computations. We simply switch x and y in $y = 2(x + \tfrac{1}{2})^2 - \tfrac{25}{2}$ to obtain $x = 2(y + \tfrac{1}{2})^2 - \tfrac{25}{2}$. Then, since in this case $h = -\tfrac{25}{2}$ and $k = -\tfrac{1}{2}$, the vertex of the parabola is at $(-\tfrac{25}{2}, -\tfrac{1}{2})$ and the axis of symmetry is $y = -\tfrac{1}{2}$.

The intercepts are $(-12, 0)$, $(0, -3)$, and $(0, 2)$. The parabola, shown in Figure 9-22b, has the same general shape as that in Figure 9-22a, but it lies on its side and opens to the right.

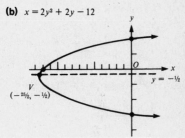

(b) $x = 2y^2 + 2y - 12$

Figure 9-22

EXAMPLE 9-25 Graph $x = -y^2 - 3y - 2$.

Solution Complete the square on the right side of the equation:

$$x = -y^2 - 3y - 2 = -1(y^2 + 3y + \quad) - 2$$

$$= -1(y^2 + 3y + \tfrac{9}{4}) - 2 + \tfrac{9}{4}$$

$$= -1(y + \tfrac{3}{2})^2 + \tfrac{1}{4}$$

Since $h = \tfrac{1}{4}$ and $k = -\tfrac{3}{2}$, the vertex of the parabola is at $(\tfrac{1}{4}, -\tfrac{3}{2})$ and the axis of symmetry is $y = -\tfrac{3}{2}$. Since $a = -1$, the parabola opens to the left.

Setting $y = 0$ in $x = -y^2 - 3y - 2$ gives us $x = -2$ so we have an x-intercept of $(-2, 0)$. Setting $x = 0$ gives us

$$0 = -y^2 - 3y - 2$$

$$= -1(y^2 + 3y + 2) = -1(y + 2)(y + 1)$$

so the y-intercepts are at $(0, -2)$ and $(0, -1)$. See Figure 9-23.

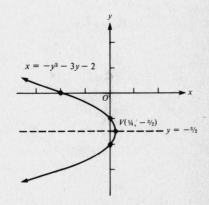

$x = -y^2 - 3y - 2$

Figure 9-23

B. Circles

A **circle** is the set of all points (x, y) that are a fixed distance r from a fixed point (h, k). The value r is called the *radius* and the point (h, k) is called the *center*. We can use this definition to write an equation:

- The graph of an equation of the form

CIRCLE
($C = (h, k)$)
$$(x - h)^2 + (y - k)^2 = r^2$$

is a circle with center $C = (h, k)$ and radius r.

note: This equation follows directly from the formula for the distance between the center (h, k) and a point r units away from it on the circle.

EXAMPLE 9-26 Find the equations of the circles with (**a**) center at $(2, 1)$ and radius 4 and (**b**) center at $(-3, 2)$ and radius 3.

Solution

(**a**) Since the center of the circle is $(2, 1)$ and the radius is 4, we let $h = 2, k = 1,$ and $r = 4$. Then the equation of the circle is

$$(x - h)^2 + (y - k)^2 = r^2$$
$$(x - 2)^2 + (y - 1)^2 = (4)^2$$
$$(x - 2)^2 + (y - 1)^2 = 16$$

(**b**) Since $h = -3, k = 2,$ and $r = 3$, the equation of the circle is

$$\left(x - (-3)\right)^2 + (y - 2)^2 = (3)^2$$
$$(x + 3)^2 + (y - 2)^2 = 9$$

EXAMPLE 9-27 Show that the graph of $x^2 + y^2 - 2x + 6y - 6 = 0$ is a circle. Graph the circle.

Solution Group the variables and complete the square on each group:

$$x^2 + y^2 - 2x + 6y - 6 = 0$$
$$(x^2 - 2x + \quad) + (y^2 + 6y + \quad) = 6$$
$$(x^2 - 2x + 1) + (y^2 + 6y + 9) = 6 + 1 + 9$$
$$(x - 1)^2 + (y + 3)^2 = 16$$

This is the equation of a circle with center $(h, k) = (1, -3)$ and radius $r = 4$. (See Figure 9-24.)

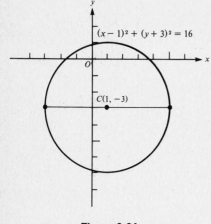

Figure 9-24

C. Ellipses

1. Geometry of ellipses

- An **ellipse** is the set of all points (x, y) the SUM of whose distances from two fixed points F_1 and F_2 is a constant.

The two fixed points are called the *foci* of the ellipse, and the midpoint of the line segment joining these foci is the *center* of the ellipse. The distance from one focus to the center of the ellipse is c units.

Ellipses have two axes of symmetry:

(**1**) a **major axis**, along which the foci lie;
(**2**) a **minor axis**, perpendicular to the major axis.

The endpoints of the major axis are called *vertices* V_1 and V_2. The distance between the center of the ellipse and one vertex is a units. The endpoints of the minor axis are sometimes called covertices B_1 and B_2. The distance between one covertex and the center of the ellipse is b units.

The distances between the center and a focus (c), the center and a vertex (a), and the center and a covertex (b) are related: $a^2 = b^2 + c^2$.

2. Equations of ellipses

The geometry of an ellipse allows us to write equations for fat ellipses, thin ellipses, and round ellipses. We'll take these one by one, beginning with the simplest cases, where the center is at the origin.

- The graph of an equation of the form

X-ELLIPSE $\dfrac{x^2}{a^2} + \dfrac{y^2}{b^2} = 1$ $(a > b > 0)$

is an ellipse whose center is at $(0, 0)$ and whose foci $F_1(-c, 0)$ and $F_2(c, 0)$ are on the x-axis. Thus

(1) the major axis is on the x-axis and the vertices V_1 and V_2 are the x-intercepts $(\pm a, 0)$;

(2) the minor axis is on the y-axis and the covertices B_1 and B_2 are the y-intercepts $(0, \pm b)$;

(3) the ellipse is wider than it is tall.

See Figure 9-25a.

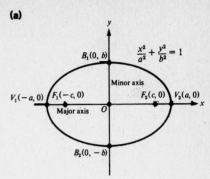

- The graph of an equation of the form

   Y-ELLIPSE $\qquad \dfrac{x^2}{b^2} + \dfrac{y^2}{a^2} = 1 \qquad (a > b > 0)$

is an ellipse whose center is at $(0, 0)$ and whose foci $F(0, \pm c)$ are on the y-axis. Thus

(1) the major axis is on the y-axis and the vertices V_1 and V_2 are the y-intercepts $(0, \pm a)$;

(2) the minor axis is on the x-axis and the covertices B_1 and B_2 are the x-intercepts $(\pm b, 0)$;

(3) the ellipse is taller than it is wide.

See Figure 9-25b.

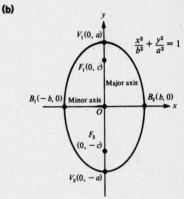

- The graph of $x^2/a^2 + y^2/b^2 = 1$, where $a = b$, becomes

$$\frac{x^2}{a^2} + \frac{y^2}{a^2} = 1 \quad \text{or} \quad x^2 + y^2 = a^2$$

which is the graph of a circle (a "round ellipse") whose center is $(0, 0)$ and whose radius $r = a$.

Figure 9-25

EXAMPLE 9-28 Show that the ellipse defined by $\dfrac{x^2}{a^2} + \dfrac{y^2}{b^2} = 1$ has x-intercepts $(a, 0)$ and $(-a, 0)$ and y-intercepts $(0, b)$ and $(0, -b)$.

Solution

If $y = 0$:

$$\frac{x^2}{a^2} + \frac{0}{b^2} = 1$$

$$\frac{x^2}{a^2} = 1$$

$$x^2 = a^2$$

$$x = \pm a$$

Since $x = \pm a$ when $y = 0$, the x-intercepts are $(a, 0)$ and $(-a, 0)$.

If $x = 0$:

$$\frac{0}{a^2} + \frac{y^2}{b^2} = 1$$

$$\frac{y^2}{b^2} = 1$$

$$y^2 = b^2$$

$$y = \pm b$$

Since $y = \pm b$ when $x = 0$, the y-intercepts are $(0, b)$ and $(0, -b)$.

EXAMPLE 9-29 Find the vertices and covertices for the ellipse defined by $\dfrac{x^2}{25} + \dfrac{y^2}{9} = 1$. Sketch the ellipse.

Solution In this equation, $25 > 9$, so $a^2 = 25$; thus the vertices are the x-intercepts $(\pm a, 0) = (\pm 5, 0)$. Then, since $b^2 = 9$, the covertices are the y-intercepts $(0, \pm b) = (0, \pm 3)$. We plot these four points and use symmetry to sketch the ellipse, as in Figure 9-26.

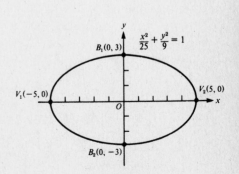

Figure 9-26

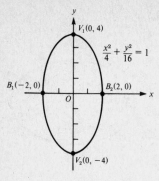

Figure 9-27

EXAMPLE 9-30   Sketch the ellipse defined by $4x^2 + y^2 = 16$.

Solution   First we divide both sides of the equation by 16 to put the equation in standard form:

$$4x^2 + y^2 = 16$$

$$\frac{4x^2}{16} + \frac{y^2}{16} = \frac{16}{16}$$

$$\frac{x^2}{4} + \frac{y^2}{16} = 1$$

Since $16 > 4, b^2 = 4$; so the covertices are the x-intercepts $(2, 0)$ and $(-2, 0)$. Then, since $a^2 = 16$, the vertices are the y-intercepts $(0, 4)$ and $(0, -4)$. The ellipse is shown in Figure 9-27. Notice that the ellipse is taller than it is wide.

An ellipse is said to be in *standard position* when its axes of symmetry are on the x- and y-axes and its center is at the origin. But we can also write equations for ellipses that are not in standard position.

- If the ellipse does not have its center at the origin, but at some point in the plane whose coordinates are (h, k), then the equation of the ellipse is

ELLIPSES   or

$$\frac{(x - h)^2}{a^2} + \frac{(y - k)^2}{b^2} = 1$$

$$\frac{(x - h)^2}{b^2} + \frac{(y - k)^2}{a^2} = 1 \qquad \text{where } a > b > 0$$

If the major axis is parallel to the x-axis, then

(1) The vertices are $(h - a, k)$ and $(h + a, k)$.
(2) The endpoints of the minor axis—the covertices—are $(h, k - b)$ and $(h, k + b)$.

If the major axis is parallel to the y-axis, then

(1) The vertices are $(h, k - a)$ and $(h, k + a)$.
(2) The endpoints of the minor axis—the covertices—are $(h - b, k)$ and $(h + b, k)$.

EXAMPLE 9-31   Graph the equation $4x^2 + y^2 + 24x - 10y + 45 = 0$.

Solution   Complete the square on each variable and rearrange the terms so that the right side of the equation is 1:

$$4x^2 + y^2 + 24x - 10y + 45 = 0$$

$$4(x^2 + 6x + \quad) + (y^2 - 10y + \quad) = -45$$

$$4(x^2 + 6x + 9) + (y^2 - 10y + 25) = -45 + 36 + 25$$

$$4(x + 3)^2 + (y - 5)^2 = 16$$

$$\frac{(x + 3)^2}{4} + \frac{(y - 5)^2}{16} = 1$$

The center of the ellipse is (h, k) or $(-3, 5)$.
Since $16 > 4, a^2 = 16$ and $b^2 = 4$; so $a = \sqrt{16} = 4$ and $b = \sqrt{4} = 2$. Thus

the vertices are $(h, k - a)$ and $(h, k + a)$ or $(-3, 1)$ and $(-3, 9)$;
the covertices are $(h - b, k)$ and $(h + b, k)$ or $(-5, 5)$ and $(-1, 5)$.

See Figure 9-28.

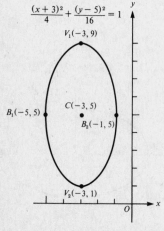

Figure 9-28

D. Hyperbolas

- A **hyperbola** is the set of all points (x, y) the DIFFERENCE of whose distances from two fixed points F_1 and F_2 is a constant.

The two fixed points are called the *foci* of the hyperbola, and the midpoint of the line segment joining these foci is the *center* of the hyperbola. Then it starts to get complicated, so we'll describe a hyperbola in standard position in order to see its general characteristics.

A hyperbola is said to be in standard position if the foci are either on the x-axis or on the y-axis and are equidistant ($\pm c$ units) from the origin. The origin is then the center of the hyperbola.

- When the foci are on the x-axis, the equation of the hyperbola is

X-HYPERBOLA (standard form)
$$\frac{x^2}{a^2} - \frac{y^2}{b^2} = 1 \qquad (C = (0,0))$$

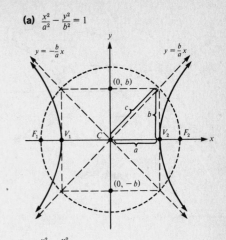

The standard x-hyperbola (shown in Figure 9-29a) has the following properties:

(1) The hyperbola consists of two nonintersecting parts or *branches*, each of which looks a little like a parabola opening away from the origin.

(2) The points at which the branches cross the x-axis on either side of the origin are $(a, 0)$ and $(-a, 0)$. These points are called the *vertices* V_1 and V_2 of the hyperbola.

(3) The line segment joining the two vertices is called the **transverse axis** of the hyperbola and the midpoint of this axis—the origin $(0,0)$—is the *center* C of the hyperbola.

(4) The hyperbola is shaped by the lines $y = \frac{b}{a}x$ and $y = -\frac{b}{a}x$. These are the **asymptotes** of the hyperbola. (See Section 9-3B.)

(5) The hyperbola is symmetric with respect to both axes and to the origin.

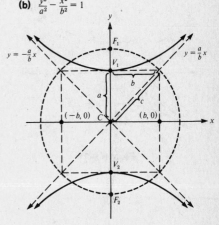

Figure 9-29

- When the foci are on the y-axis, the equation of the hyperbola is

Y-HYPERBOLA (standard form)
$$\frac{y^2}{a^2} - \frac{x^2}{b^2} = 1 \qquad (C = (0,0))$$

The standard y-hyperbola (shown in Figure 9-29b) is the same as the x-hyperbola, except for the following differences:

(1) The vertices of the hyperbola—$(0, a)$ and $(0, -a)$—are the points at which the branches cross the y-axis on either side of the origin.

(2) The asymptotes of the y-hyperbola are the lines $y = \frac{a}{b}x$ and $y = -\frac{a}{b}x$.

note: The foci do *not* lie on the curves forming the hyperbola. They lie on the extended transverse axis a distance of $\pm c$ units from the center. The vertices *do* lie on the curves, a distance of $\pm a$ units from the center. The factor b in the equations of the hyperbola is related to the distances a and c by the Pythagorean relation $a^2 + b^2 = c^2$. Thus $b^2 = c^2 - a^2$, or $b = \pm\sqrt{c^2 - a^2}$.

EXAMPLE 9-32 Find the vertices and asymptotes for the hyperbola defined by $\dfrac{x^2}{4} - \dfrac{y^2}{1} = 1$. Graph the hyperbola.

Solution The equation $x^2/4 - y^2/1 = 1$ is of the form $x^2/a^2 - y^2/b^2 = 1$, so we have a standard x-hyperbola with $a = \sqrt{4} = 2$ and $b = \sqrt{1} = 1$. The vertices are $(a, 0)$ and $(-a, 0)$ or $(2, 0)$ and $(-2, 0)$. The asymptotes are $y = \frac{b}{a}x$ and $y = -\frac{b}{a}x$ or $y = \frac{1}{2}x$ and $y = -\frac{1}{2}x$. We could plot these asymptotes by setting up a table of values—but there's an easier way: Construct a rectangle whose dimensions are $2a$ by $2b$. Mark off $a = 2$ units on both sides of the origin on

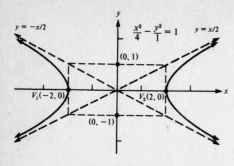

$$y = -x/2 \qquad \frac{x^2}{4} - \frac{y^2}{1} = 1 \qquad y = x/2$$

Figure 9-30

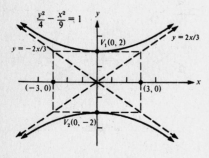

$$\frac{y^2}{4} - \frac{x^2}{9} = 1$$

$$y = -2x/3 \qquad y = 2x/3$$

Figure 9-31

the x-axis, which contains the vertices. Mark off $b = 1$ unit on both sides of the opposite axis. Construct the box of dimensions $2a = 4$ by $2b = 2$ centered at the origin. The diagonals extended are the asymptotes.

note: If we draw a circle whose center is the center of the hyperbola and whose radius is $\sqrt{a^2 + b^2} = \sqrt{5}$, we can locate the foci $(\sqrt{5}, 0)$ and $(-\sqrt{5}, 0)$ on the x-axis.

Now draw the two branches of the hyperbola—with vertices at $(\pm 2, 0)$—within the bounds of the asymptotes. See Figure 9-30.

EXAMPLE 9-33 Find the vertices and asymptotes for $9y^2 - 4x^2 = 36$. Sketch the curve.

Solution First we put the equation into standard form:

$$9y^2 - 4x^2 = 36$$

$$\frac{9y^2}{36} - \frac{4x^2}{36} = \frac{36}{36}$$

$$\frac{y^2}{4} - \frac{x^2}{9} = 1$$

The equation is of the form $y^2/a^2 - x^2/b^2 = 1$, so we have a standard y-hyperbola with $a = 2$ and $b = 3$. The vertices are $(0, a)$ and $(0, -a)$ or $(0, 2)$ and $(0, -2)$. The asymptotes are $y = \frac{a}{b}x = \frac{2}{3}x$ and $y = -\frac{a}{b}x = -\frac{2}{3}x$. This hyperbola is shown in Figure 9-31.

If a hyperbola does not have its center at the origin but at some point in the plane whose coordinates are (h, k), the equation of the hyperbola is of one of the following forms:

x-Hyperbola	y-Hyperbola
$\dfrac{(x - h)^2}{a^2} - \dfrac{(y - k)^2}{b^2} = 1$	$\dfrac{(y - k)^2}{a^2} - \dfrac{(x - h)^2}{b^2} = 1$
Vertices: $(h - a, k)$ and $(h + a, k)$	Vertices: $(h, k - a)$ and $(h, k + a)$
Asymptotes: $y - k = \dfrac{b}{a}(x - h)$	Asymptotes: $y - k = \dfrac{a}{b}(x - h)$
$y - k = -\dfrac{b}{a}(x - h)$	$y - k = -\dfrac{a}{b}(x - h)$
Center: (h, k)	Center: (h, k)
Foci: $(h - c, k)$ and $(h + c, k)$	Foci: $(h, k - c)$ and $(h, k + c)$

EXAMPLE 9-34 Sketch the hyperbola $x^2 - 4y^2 - 2x - 8y - 7 = 0$.

Solution First, put the equation into standard form. Complete the square on both the x and y variables:

$$x^2 - 4y^2 - 2x - 8y - 7 = 0$$

$$(x^2 - 2x + \quad) - 4(y^2 + 2y + \quad) = 7$$

$$(x^2 - 2x + 1) - 4(y^2 + 2y + 1) = 7 + 1 - 4$$

$$(x - 1)^2 - 4(y + 1)^2 = 4$$

$$\frac{(x - 1)^2}{4} - \frac{(y + 1)^2}{1} = 1$$

This equation is of the form $\dfrac{(x-h)^2}{a^2} - \dfrac{(y-k)^2}{b^2} = 1$, so we have an x-hyperbola with center $(h, k) = (1, -1)$, whose branches open to the left and right. The value of a is 2 and the value of b is 1. Thus

The vertices are $(h - a, k) = (1 - 2, -1) = (-1, -1)$

$$\text{and } (h + a, k) = (1 + 2, -1) = (3, -1).$$

The asymptotes are

$$y - k = \frac{b}{a}(x - h) \quad \text{and} \quad y - k = -\frac{b}{a}(x - h)$$

$$y - (-1) = \frac{1}{2}(x - 1) \qquad\qquad y + 1 = -\frac{1}{2}(x - 1)$$

$$y + 1 = \frac{1}{2}x - \frac{1}{2} \qquad\qquad y + 1 = -\frac{1}{2}x + \frac{1}{2}$$

$$y = \frac{1}{2}x - \frac{3}{2} \qquad\qquad y = -\frac{1}{2}x - \frac{1}{2}$$

We sketch the hyperbola using the vertices and the asymptotes, as shown in Figure 9-32.

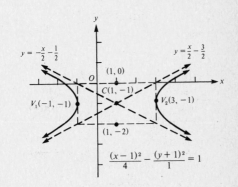

Figure 9-32

E. General equation of a conic section

The equations of the four conic sections may all be obtained from one general second-degree equation in two variables. In fact, the graph of any equation of the following form is a conic section.

CONIC SECTION (general form) $\qquad Ax^2 + By^2 + Cx + Dy + E = 0$

And we can easily recognize the particular conic represented by an equation in general form by looking at A and B:

(1) If $A \neq 0$ and $B = 0$ or if $A = 0$ and $B \neq 0$, the graph is a parabola.
(2) If A and B have the same sign, the graph is an ellipse. (If A and B have the same sign and $A = B$, the graph is a special case of the ellipse—it is a circle.)
(3) If A and B have different signs, the graph is a hyperbola.

EXAMPLE 9-35 Determine the type of conic section represented by the equations

(a) $9y^2 - x^2 - 36y + 12x - 9 = 0$ $\qquad$ (b) $4x^2 + 16x + y^2 - 4y = -31$

(c) $x^2 - y + 6x + 16 = 0$ $\qquad$ (d) $4x^2 + 16x + 4y^2 - 4y = -31$

Solution

(a) In $9y^2 - x^2 - 36y + 12x - 9 = 0$, the value of A is -1 and the value of B is 9. Since A and B have different signs, the graph of this equation is a hyperbola.
(b) In $4x^2 + 16x + y^2 - 4y = -31$, $A = 4$ and $B = 1$. Since A and B have the same sign, the graph is an ellipse. It is not a circle, since $A \neq B$.
(c) In $x^2 - y + 6x + 16 = 0$, $A = 1$ and $B = 0$. Since $A \neq 0$ and $B = 0$, the graph is a parabola.
(d) In $4x^2 + 16x + 4y^2 - 4y = -31$, $A = B = 4$. Since A and B have the same sign and are equal, the graph is a circle.

SUMMARY

1. The graph of a second-degree polynomial function $f(x) = ax^2 + bx + c$ (where $a \neq 0$), called a quadratic function, is a parabola whose axis of symmetry is on or parallel to the y-axis. If $a > 0$, the parabola opens upward and if $a < 0$, the parabola opens downward; if $|a| > 1$, the parabola is thin and if $|a| < 1$, the parabola is fat.

2. By completing the square, the function $f(x) = ax^2 + bx + c$ may be written as $f(x) = a(x - h)^2 + k$, where (h, k) is the vertex of the parabola. The vertex (h, k) is the minimum point if $a > 0$ and the maximum point if $a < 0$.

3. Quadratic functions may be graphed by finding their x- and y-intercepts and/or by writing their equations in the form $f(x) = a(x - h)^2 + k$ to find the vertex (h, k). The location of the vertex shows the amount of horizontal (h) and vertical (k) shift from a vertex at the origin.

4. The zeros of a polynomial function divide the number line into intervals in which the function does not change its sign. In general, a zero of even degree produces a point at which the graph touches the axis but does not cross.

5. A rational function is the quotient of two polynomial functions.

6. A function $f(x)$ has the line $x = b$ as a vertical asymptote if, as $x \to b$, $f(x) \to \pm \infty$. The vertical asymptotes of a rational function occur at the zeros of the denominator polynomial. A function $f(x)$ has the line $y = c$ as a horizontal asymptote if, as $x \to \pm \infty$, $f(x) \to c$. The horizontal asymptote of a rational function may be found by the degree-comparison test.

7. The conic sections are the four curves—parabola, circle, ellipse, hyperbola—that are formed when a right circular cone is cut by a plane. In standard form, the equations of these curves may be written as follows:

Parabola: $y = a(x - h)^2 + k$ ($a \neq 0$; axis of symmetry $x = h$)
((h, k) = vertex) $x = a(y - k)^2 + h$ ($a \neq 0$; axis of symmetry $y = k$)

Circle: $(x - h)^2 + (y - k)^2 = r^2$ (r = radius)
((h, k) = center)

Ellipse: $\dfrac{(x - h)^2}{a^2} + \dfrac{(y - k)^2}{b^2} = 1$ ($a > b > 0$; major axis $y = k$)
((h, k) = center)

$\dfrac{(x - h)^2}{b^2} + \dfrac{(y - k)^2}{a^2} = 1$ ($a > b > 0$; major axis $x = h$)

Hyperbola: $\dfrac{(x - h)^2}{a^2} - \dfrac{(y - k)^2}{b^2} = 1$ (transverse axis $y = k$)
((h, k) = center)

$\dfrac{(y - k)^2}{a^2} - \dfrac{(x - h)^2}{b^2} = 1$ (transverse axis $x = h$)

8. The conics are all defined by the general equation

$$Ax^2 + By^2 + Cx + Dy + E = 0$$

such that

- if $A \neq 0$ and $B = 0$, or if $A = 0$ and $B \neq 0$, the graph is a parabola
- if A and B have the same sign, the graph is an ellipse
- if $A = B$, the graph is a circle (a special ellipse)
- if A and B have different signs, the graph is a hyperbola

RAISE YOUR GRADES

Can you . . . ?

☑ define a polynomial function, a quadratic function, and a rational function
☑ explain the location of the maximum or minimum point of a quadratic function
☑ explain the relation between two functions if they are vertical or horizontal shifts of one another

☑ explain the importance of the sign of the constant a in the equation of a parabola
☑ show how to find the vertex of a parabola given its equation in any form
☑ sketch the functions $y = ax^n$ for $n \geq 3$.
☑ explain how the zeros of a polynomial function are important to its graph
☑ explain the processes of finding vertical and horizontal asymptotes for a rational function
☑ write the equations in standard form of the four conic sections
☑ explain the relationship between the vertices and the foci for ellipses and hyperbolas
☑ give the procedure for finding the asymptotes for a hyperbola in standard position
☑ describe which conic section is represented when values of A, B, C, D, and E are given in $Ax^2 + By^2 + Cx + Dy + E = 0$

SOLVED PROBLEMS

Quadratic Functions

PROBLEM 9-1   Determine the degree of the following functions. Are any of them quadratic?

$$\text{(a) } f(x) = 5x - 14 \qquad \text{(b) } f(x) = \frac{6x^2 - 5x + 1}{2x^3 - 8x + 3} \qquad \text{(c) } f(x) = -2x^2 + 5$$

Solution

(a) $f(x) = 5x - 14$ is a first-degree polynomial function, so it is a linear function.
(b) $f(x) = \dfrac{6x^2 - 5x + 1}{2x^3 - 8x + 3}$ is a rational function whose numerator is of second degree and denominator is of third degree.
(c) $f(x) = -2x^2 + 5$ is a second-degree polynomial function, so it is a quadratic function.

PROBLEM 9-2   Describe the shape of the graph for the following functions:

$$\text{(a) } y_1 = -\tfrac{1}{2}x + 3 \qquad \text{(b) } y_2 = -3x^2 \qquad \text{(c) } y_3 = x^2 + 6x + 1$$

Solution

(a) $y_1 = -\tfrac{1}{2}x + 3$ is a linear polynomial function, so its graph is a straight line.
(b) $y_2 = -3x^2$ is a quadratic polynomial function, so its graph is a parabola. Since $bx = c = 0$, the axis of symmetry is on the y-axis and the vertex is at the origin. Since $-3 < 0$, the parabola opens downward; and since $|-3| > 1$, the parabola is thin.
(c) $y_3 = x^2 + 6x + 1$ is a quadratic polynomial function, so its graph is a parabola—opening upward, since the coefficient of x^2 is positive. The characteristics of the parabola can be determined algebraically by completing the square:

$$y_3 = x^2 + 6x + 1 = (x^2 + 6x + \square) + 1 = (x^2 + 6x + 9) + 1 - 9 = (x + 3)^2 - 8$$

So the parabola is the same shape as the basic parabola $y = x^2$, but its vertex is at $(-3, -8)$ and its axis of symmetry is $x = -3$.

PROBLEM 9-3   Sketch the graphs of (a) $y = -2x^2 + 2$ and (b) $y = \tfrac{1}{2}x^2 - 2$.

Solution   Both of these functions are parabolas of the form $y = ax^2 + k$, so their axes of symmetry are on the y-axis. Construct these graphs in steps.

(a) **(1)** Graph the parabola $y = -x^2$, which opens downward.
(2) Graph $y = -2x^2$, which is thinner than $y = -x^2$.
(3) Shift two units upward to get $y = -2x^2 + 2$, whose vertex is $(0, 2)$.

See Figure 9-33.

(b) **(1)** Graph the parabola $y = x^2$, which opens upward.
(2) Graph $y = \frac{1}{2}x^2$, which is fatter than $y = x^2$.
(3) Shift two units downward to get $y = \frac{1}{2}x^2 - 2$, whose vertex is $(0, -2)$.

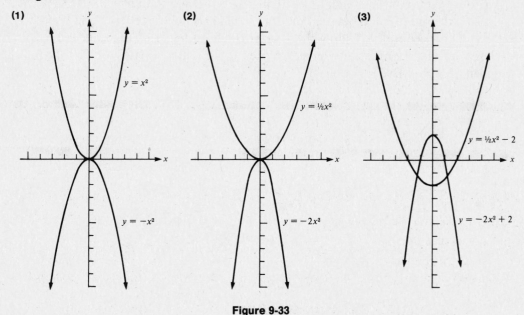

Figure 9-33

PROBLEM 9-4 Find the x- and y-intercepts for the quadratic functions **(a)** $y = 2x^2 - 6$, **(b)** $y = -x^2 + 5$, and **(c)** $y = -2x^2 - 4$. Sketch each graph, using your intercept information.

Solution To find x-intercepts, set $y = 0$ and solve the resulting equation for x. To find the y-intercepts, set $x = 0$ and solve the resulting equation for y.

(a) If $y = 0$: $\quad 0 = 2x^2 - 6$

$$2x^2 = 6$$
$$x^2 = 3$$
$$x = \pm\sqrt{3}$$

If $x = 0$: $\quad y = 2(0)^2 - 6$
$$y = -6$$

The graph contains the intercepts $(\sqrt{3}, 0)$ $(-\sqrt{3}, 0)$, $(0, -6)$. See Figure 9-34a.

(b) If $y = 0$: $\quad 0 = -x^2 + 5$

$$x^2 = 5$$
$$x = \pm\sqrt{5}$$

If $x = 0$: $\quad y = -0^2 + 5$
$$y = 5$$

The intercepts are at $(\sqrt{5}, 0)$, $(-\sqrt{5}, 0)$, $(0, 5)$. See Figure 9-34b.

(c) If $y = 0$: $\quad 0 = -2x^2 - 4$

$$2x^2 = -4$$
$$x^2 = -2$$

No x-intercepts.

If $x = 0$: $\quad y = -2(0)^2 - 4$
$$y = -4$$

The y-intercept is $(0, -4)$. See Figure 9-34c.

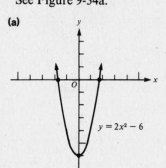

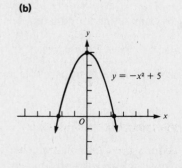

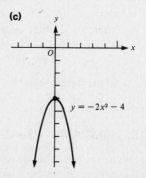

Figure 9-34

PROBLEM 9-5 Sketch the graphs of (a) $y = (x + 3)^2$ and (b) $y = \frac{1}{2}(x - 4)^2$.

Solution

(a) The graph of $y = (x + 3)^2$ has the same shape as $y = x^2$ but is shifted three units to the *left*. See Figure 9-35a.

(b) The graph of $y = \frac{1}{2}(x - 4)^2$ has the same shape as $y = \frac{1}{2}x^2$ but is shifted four units to the *right*. See Figure 9-35b.

(a)

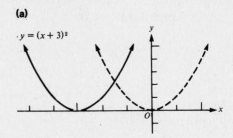

(b)

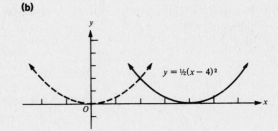

Figure 9-35

PROBLEM 9-6 Sketch the graphs of (a) $y = -(x + 1)^2 + 2$ and (b) $y = 3(x - 1)^2 + 1$.

Solution We form the sketches in steps.

(a) (1) Graph the parabola $y = -x^2$, which opens downward.
    (2) By shifting one unit to the left, we get $y = -(x + 1)^2$.
    (3) By shifting two units up, we get $y = -(x + 1)^2 + 2$.
See Figure 9-36a.

(b) (1) Graph the parabola $y = 3x^2$, which opens upward.
    (2) Shift one unit to the right to get $y = 3(x - 1)^2$.
    (3) Shift one unit up to get $y = 3(x - 1)^2 + 1$.
See Figure 9-36b.

(a) **(1)** **(2)** **(3)**

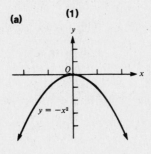
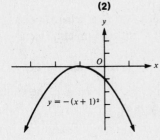
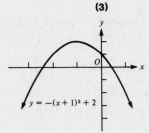

(b)

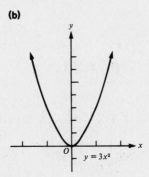
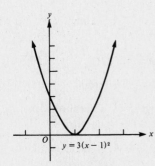
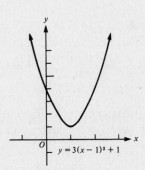

Figure 9-36

PROBLEM 9-7 Complete the square on the functions

    (a) $y = x^2 + 6x - 2$ (b) $y = -x^2 + x + \frac{5}{4}$

Describe the graph of each function.

Solution

(a)
$$y = x^2 + 6x - 2$$
$$= (x^2 + 6x + \quad) - 2 \qquad \text{Rewrite the equation.}$$
$$y = (x^2 + 6x + 9) - 2 - 9 \qquad \text{Square half of the } x \text{ coefficient;}$$
$$= (x + 3)^2 - 11 \qquad \begin{array}{l}\text{add it inside and subtract it outside}\\ \text{the parentheses.}\end{array}$$

In this form, the equation obviously represents the basic upward-opening parabola $y = x^2$ shifted 3 units to the left and 11 units down.

(b)
$$y = -x^2 + x + \tfrac{5}{4}$$
$$= -1(x^2 - x + \quad) + \tfrac{5}{4}$$
$$= -1(x^2 - x + \tfrac{1}{4}) + \tfrac{5}{4} + \tfrac{1}{4}$$
$$= -1(x - \tfrac{1}{2})^2 + \tfrac{3}{2}$$

The graph of this function has the same shape as the basic downward-opening parabola $y = -x^2$ shifted $\tfrac{1}{2}$ unit to the right and $\tfrac{3}{2}$ units upward.

PROBLEM 9-8 Locate the vertex and the symmetry line for the quadratic polynomial function $y = 3x^2 + 12x + 2$.

Solution Complete the square:
$$y = 3x^2 + 12x + 2$$
$$= (3x^2 + 12x + \quad) + 2$$
$$= 3(x^2 + 4x + \quad) + 2$$
$$= 3(x^2 + 4x + 4) + 2 - 12$$
$$= 3(x + 2)^2 - 10$$

This function is in the form $y = a(x - h)^2 + k$, where $h = -2$ and $k = -10$. This would put the vertex at $(h, k) = (-2, -10)$ and its vertical symmetry line at $x = h = -2$.

PROBLEM 9-9 Find the vertex of the parabola $y = 4x^2 - 2x + 11$.

Solution From the standard form $y = ax^2 + bx + c$, the vertex is located at $(-\tfrac{b}{2a}, c - \tfrac{b^2}{4a})$. Here $a = 4, b = -2, c = 11$, so the vertex is at
$$\left(-\frac{-2}{8}, 11 - \frac{4}{16}\right) = \left(\frac{1}{4}, \frac{43}{4}\right)$$

PROBLEM 9-10 Find two numbers whose sum is 60 and whose product is as large as possible. What is the maximum product value?

Solution Let the numbers be x and y. Then $x + y = 60$, or $y = 60 - x$. Their product is
$$P = xy = x(60 - x) = 60x - x^2$$

This is a quadratic function whose graph is a parabola opening downward. The maximum value of P occurs at the vertex. We complete the square on P:
$$P = -x^2 + 60x$$
$$= -(x^2 - 60x + \quad)$$
$$= -(x^2 - 60x + 900) + 900$$
$$= -(x - 30)^2 + 900$$

The x-coordinate of the vertex is $x = h = 30$, for which $P = k = 900$. So the numbers are $x = 30$ and $y = 60 - 30 = 30$, whose maximum product value is $k = 900$.

PROBLEM 9-11   You are producing and selling personally designed skateboards. Your total revenue per week (in dollars) is given by

$$R = -\tfrac{15}{2}x^2 + 75x$$

How many should you make to maximize the total revenue?

Solution   The revenue function is a parabola opening downward, so the maximum value of R is obtained at the vertex of the parabola. Complete the square to find it:

$$R = -\tfrac{15}{2}x^2 + 75x = -\tfrac{15}{2}(x^2 - 10x + \quad)$$

$$= -\tfrac{15}{2}(x^2 - 10x + 25) + \tfrac{15}{2}(25) = -\tfrac{15}{2}(x - 5)^2 + \tfrac{375}{2}$$

The value $x = 5$, with corresponding $R = \tfrac{375}{2}$, locates the vertex. Thus you should produce 5 boards per week to maximize your revenue of $(\tfrac{375}{2}) = \$187.50$.

PROBLEM 9-12   An isosceles triangle with base 6 and sides 5 is to have a rectangle inscribed so that one of the sides of the rectangle lies on the base of the triangle. Find the dimensions of the largest rectangle that may be constructed.

Solution   Draw the rectangle within the triangle and label the rectangle so that its area is $A = (2x)(y) = 2xy$, as in Figure 9-37a. This area is the value to be maximized. But you need a relationship between x and y, which you obtain by noting that the small right triangle to the right of the rectangle is similar to the large right triangle making up half of the original isosceles triangle (see Figure 9-37b). By labeling the sides of the small and large right triangles, you can see the similarity relationship $\tfrac{4}{3} = \tfrac{y}{3-x}$, so that

note: The height of the perpendicular bisector of the isosceles triangle is $\sqrt{5^2 - 3^2} = \sqrt{16} = 4$.

(a)

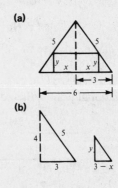

(b)

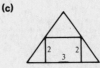

$$3y = 4(3 - x) \quad \text{or} \quad y = \tfrac{4}{3}(3 - x)$$

Now you have

$$A = 2x\big(\tfrac{4}{3}(3 - x)\big) = \tfrac{8}{3}x(3 - x)$$

$$= 8x - \tfrac{8}{3}x^2$$

This is a quadratic function whose graph is a parabola and whose vertex will determine the maximum point. So you find the vertex by completing the square:

$$A = -\tfrac{8}{3}x^2 + 8x$$

$$= -\tfrac{8}{3}(x^2 - 3x + \quad)$$

$$= -\tfrac{8}{3}(x^2 - 3x + \tfrac{9}{4}) + \tfrac{8}{3}\cdot\tfrac{9}{4}$$

$$= -\tfrac{8}{3}(x - \tfrac{3}{2})^2 + 6$$

(c)

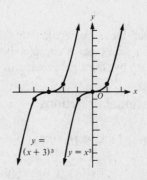

Figure 9-37

The vertex occurs at $(\tfrac{3}{2}, 6)$, and the maximum value of A is then 6. The dimensions of the rectangle are $2x$ by y, so you make it $2(\tfrac{3}{2})$ by $\tfrac{4}{3}(3 - \tfrac{3}{2})$, or 3 units wide and 2 units high, as in Figure 9-37c.

Higher-Degree Polynomial Functions

PROBLEM 9-13   Sketch the graph of $y = (x + 3)^3$.

Solution   This function is a translation of the odd function $y = x^3$. Use the previously determined graph of $y = x^3$ (Figure 9-10a) and simply shift it to the left three units, as in Figure 9-38.

Figure 9-38

PROBLEM 9-14   Sketch the graph of $y = (x - 4)(x^2 - 4)$.

Solution   First factor the function completely:

$$y = (x - 4)(x^2 - 4) = (x - 4)(x - 2)(x + 2)$$

The zeros of this function are at $x = -2$, $x = 2$, and $x = 4$. The intervals on the number line are thus defined: I_1 is $x < -2$, I_2 is $-2 < x < 2$, I_3 is $2 < x < 4$, and I_4 is $x > 4$.

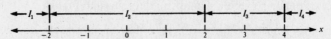

Now pick representative points in each interval to determine the sign of the function throughout the interval.

In I_1, pick $x = -3$. Here

$$y = (-3 - 4)((-3)^2 - 4)$$
$$= (-7)(5) = -35$$

So $f(x)$ is negative for all values in I_1.

In I_2, pick $x = 0$. Here

$$y = (0 - 4)(0^2 - 4)$$
$$= (-4)(-4) = 16$$

So $f(x)$ is positive for all values in I_2.

In I_3, pick $x = 3$. Here

$$y = (3 - 4)(3^2 - 4)$$
$$= (-1)(9 - 4) = -5$$

So $f(x)$ is negative for all values in I_3.

In I_4, pick $x = 5$. Here

$$y = (5 - 4)(5^2 - 4)$$
$$= 1(21) = 21$$

So $f(x)$ is positive for all values in I_4.

You now have four number pairs on the graph as well as the three x-intercepts. Calculate a few additional points, then draw a rough sketch, as in Figure 9-39.

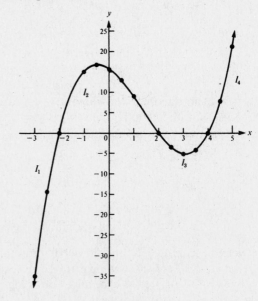

$y = (x - 4)(x^2 - 4)$	
x	y
-1	15
1	9
$\frac{1}{2}$	$\frac{105}{8} = 13.125$
$-\frac{1}{2}$	$\frac{135}{8} = 16.875$
$-\frac{5}{2}$	$-\frac{117}{8} = -14.625$
$\frac{5}{2}$	$-\frac{27}{8} = -3.375$
$\frac{7}{2}$	$-\frac{33}{8} = -4.125$
$\frac{9}{2}$	$\frac{65}{8} = 8.125$

For values of x less than -3, the values of $f(x)$ are very large and negative; for values of x larger than 5, the values of $f(x)$ are very large and positive.

Figure 9-39

note: Unless you calculate *all* the points, you can't get a completely accurate graph by this algebraic method.

Rational Functions

PROBLEM 9-15   Find the domain ($\mathscr{D}$) and range of these rational functions:

(a) $f(x) = \dfrac{x + 3}{(x - 2)(x + 1)}$ (b) $g(x) = \dfrac{x}{2x^2 + x - 6}$

Solution   The domain of a rational function consists of all real numbers x except those values that make the denominator zero.

(a) This function has denominator zeros at $x = 2$ and $x = -1$. Thus $\mathscr{D}_f = \{x : x \neq 2, x \neq -1\}$. The range consists of all y-values.

(b) The domain includes all real numbers x except those for which $2x^2 + x - 6 = 0$. This is $(2x - 3)(x + 2) = 0$. Hence $\mathscr{D}_g = \{x : x \neq \frac{3}{2}, x \neq -2\}$. The range is the set of all real values.

PROBLEM 9-16   Determine the vertical asymptotes for these rational functions:

$$\text{(a) } y = \frac{x + 6}{x^2 - 4x - 5} \qquad \text{(b) } y = \frac{x^2 + 1}{x^3 - x^2 - 12x}$$

Solution   On a graph the vertical asymptotes occur at those values of x for which the denominator polynomial is zero but the numerator polynomial is not simultaneously zero.

(a) Factor:
$$y = \frac{x + 6}{x^2 - 4x - 5} = \frac{x + 6}{(x - 5)(x + 1)}$$

The vertical asymptotes occur at $x = 5$ and $x = -1$.

(b) Factor:
$$y = \frac{x^2 + 1}{x(x^2 - x - 12)} = \frac{x^2 + 1}{x(x - 4)(x + 3)}$$

The vertical asymptotes occur at $x = 0$, $x = 4$, and $x = -3$.

PROBLEM 9-17   Find the horizontal asymptotes for these rational functions:

$$\text{(a) } f(x) = \frac{2x + 1}{x - 3} \qquad \text{(b) } f(x) = \frac{x - 4}{x^2 + 6}$$

Solution   The line $y = c$ is a horizontal asymptote for a function if, as the values of x get larger and larger, the value of the function gets arbitrarily close to c.

(a) For $f(x) = (2x + 1)/(x - 3)$ a table shows

x	10	100	1000	10 000	Very large
$f(x)$	$\dfrac{21}{7} = 3$	$\dfrac{201}{97} = 2.07$	$\dfrac{2001}{997} = 2.007$	$\dfrac{20\,001}{9997} = 2.000\,7$	$\to 2$

Thus the line $y = 2$ is a horizontal asymptote.

(b) For $f(x) = (x - 4)/(x^2 + 6)$ a table shows

x	10	100	1000	Very large
$f(x)$	$\dfrac{6}{106} = .056\,6$	$\dfrac{96}{10\,006} = .009\,59$	$\dfrac{996}{1\,000\,006} = .000\,995$	$\to 0$

Thus the line $y = 0$ is a horizontal asymptote.

PROBLEM 9-18   Use the degree-comparison test to determine the horizontal asymptotes for

$$\text{(a) } y = \frac{3x^2 + x - 2}{-x^2 + 2x - 5} \qquad \text{(b) } y = \frac{5x - 12}{13x^2 + x - 8}$$

Solution

(a) Since the degree of the numerator is the same as the degree of the denominator, the horizontal asymptote occurs at $y = c$ where c is the quotient of the leading coefficients. Here $y = c = 3/(-1) = -3$ is the horizontal asymptote.

(b) The degree of the numerator is less than the degree of the denominator. Hence $y = 0$ is the horizontal asymptote.

PROBLEM 9-19   Construct the graph of $y = \dfrac{-2x^2 + 1}{(x+1)(x-2)}$.

Solution   Begin by gathering all the information you can concerning symmetry, asymptotes, intercepts, etc. Thus, checking for symmetry in this function, you see that if x is replaced by $(-x)$, you get

$$y = \frac{-2(-x)^2 + 1}{(-x+1)(-x-2)} = \frac{-2x^2 + 1}{(x-1)(x+2)}$$

which is not the same as the original; and if y is replaced by $-y$, the sign of the function is changed. So there is no symmetry in this function. The vertical asymptotes occur at $x = -1$ and $x = 2$, and the horizontal asymptote occurs at $y = -2$. The x-intercepts are at the values where $-2x^2 + 1 = 0$ or $x^2 = 1/2$, so $x = \pm\sqrt{1/2}$. The y-intercept is at $y = (0+1)/(0+1)(0-2) = -1/2$. Finally, by determining the values of y at some strategically picked values of x, you can make a rough sketch, as in Figure 9-40.

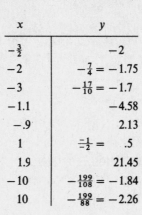

x	y
$-\frac{3}{2}$	-2
-2	$-\frac{7}{4} = -1.75$
-3	$-\frac{17}{10} = -1.7$
-1.1	-4.58
$-.9$	2.13
1	$\frac{-1}{-2} = .5$
1.9	21.45
-10	$-\frac{199}{108} = -1.84$
10	$-\frac{199}{88} = -2.26$

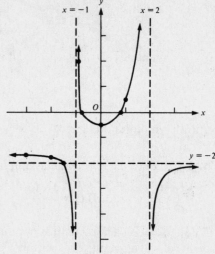

Figure 9-40

Conic Sections

PROBLEM 9-20   Find the equation of the circle whose center is $(-4, -2)$ and whose radius is 3.

Solution   Here, you're given $h = -4, k = -2, r = 3$, so the standard form of the equation of the circle is

$$(x - h)^2 + (y - k)^2 = r^2$$

giving

$$(x + 4)^2 + (y + 2)^2 = 9$$

The expanded—or general—form of this equation is

$$x^2 + 8x + 16 + y^2 + 4y + 4 = 9$$

or

$$x^2 + y^2 + 8x + 4y + 11 = 0$$

PROBLEM 9-21   Show that the graph of $x^2 + y^2 + 8x - 2y - 8 = 0$ is a circle, and find the radius and the location of the center.

Solution   Regroup the variables and complete the square on each:

$$(x^2 + 8x + \quad) + (y^2 - 2y + \quad) = 8$$

$$(x^2 + 8x + 16) + (y^2 - 2y + 1) = 8 + 16 + 1$$

$$(x + 4)^2 + (y - 1)^2 = 25$$

This is a circle of radius $r = \sqrt{25} = 5$ centered at $(h, k) = (-4, 1)$.

PROBLEM 9-22 The following equations are all parabolas. In which direction does each parabola open?

(a) $y - 3 = -x^2 + 4x$

(b) $-6x = -y^2 + 2y$

(c) $2x + y^2 = 8$

(d) $y - 4x^2 + 4x = 0$

Solution Put each in the form $y = ax^2 + bx + c$, which opens up and down, or $x = ay^2 + by + c$, which opens left and right.

(a) $y = -x^2 + 4x + 3$ $\Rightarrow$ Opens downward, since $a < 0$.

(b) $x = \frac{1}{6}y^2 - \frac{1}{3}y$ $\Rightarrow$ Opens to the right, since $a > 0$.

(c) $x = -\frac{1}{2}y^2 + 4$ $\Rightarrow$ Opens to the left, since $a < 0$.

(d) $y = 4x^2 - 4x$ $\Rightarrow$ Opens upward, since $a > 0$.

PROBLEM 9-23 Find the vertex and the axis of symmetry of each parabola (a) $x = 3y^2 - 12y + 1$ and (b) $y = -2x^2 + 6x - 2$.

Solution Complete the square on each quadratic expression to rewrite the equation in the standard form $x = a(y - k)^2 + h$ or $y = a(x - h)^2 + k$. Then identify the vertex (h, k).

(a) $x = 3y^2 - 12y + 1$

$\quad = 3(y^2 - 4y + \quad) + 1$

$\quad = 3(y^2 - 4y + 4) + 1 - 12$

$\quad = 3(y - 2)^2 - 11$

Thus the vertex is located at $(h, k) = (-11, 2)$ and the axis of symmetry is $y = 2$.

(b) $y = -2x^2 + 6x - 2$

$\quad = -2(x^2 - 3x + \quad) - 2$

$\quad = -2(x^2 - 3x + \frac{9}{4}) - 2 + 2 \cdot \frac{9}{4}$

$\quad = -2(x - \frac{3}{2})^2 + \frac{5}{2}$

Thus the vertex is at $(\frac{3}{2}, \frac{5}{2})$ and the axis of symmetry is $x = \frac{3}{2}$.

PROBLEM 9-24 Sketch the parabola $2y = 6x^2 + 15x$.

Solution First, get as much information as possible by completing the square on the right:

$$2y = 6x^2 + 15x$$

$$= 6(x^2 + \tfrac{5}{2}x + \quad)$$

$$= 6(x^2 + \tfrac{5}{2}x + \tfrac{25}{16}) - 6 \cdot \tfrac{25}{16}$$

$$= 6(x + \tfrac{5}{4})^2 - \tfrac{75}{8}$$

$$\text{or} \quad y = 3(x + \tfrac{5}{4})^2 - \tfrac{75}{16}$$

This locates the vertex at $(-\frac{5}{4}, -\frac{75}{16})$ and tells you that the parabola opens upward (because of the $+3$ in front of $(x + \frac{5}{4})^2$). You can also see that two x-intercepts should exist, so set $y = 0$ and solve for x:

$$0 = 6x^2 + 15x = 6x(x + \tfrac{5}{2})$$

Thus $x = 0$ or $x = -\frac{5}{2}$ when $y = 0$. Finally, with only a couple of carefully chosen additional points to find from the equation $2y = 6x^2 + 15x$, you get the parabola shown in Figure 9-41.

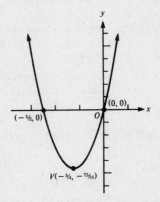

Figure 9-41

PROBLEM 9-25 Given the ellipse $\dfrac{x^2}{16} + \dfrac{y^2}{4} = 1$, find the location of the vertices, the endpoints of the minor axis, and the foci.

Solution The standard form of the equation for an ellipse is $\dfrac{x^2}{a^2} + \dfrac{y^2}{b^2} = 1$, where $a > b > 0$ and where the center is at $(0, 0)$ and the foci lie on the x-axis. The given equation is of this form, since the larger number is under the x^2 term. So the value of $a^2 = 16$ and the value of $b^2 = 4$. Hence the vertices, which lie on the x-axis at $(-a, 0)$ and $(a, 0)$, are located at $(-4, 0)$ and $(4, 0)$; and the ends of the minor axis, which lie on the y-axis at $(0, -b)$ and $(0, b)$, are located at $(0, -2)$ and $(0, 2)$. Then, since $a^2 = b^2 + c^2$ for every ellipse,

$$16 = 4 + c^2 \quad \text{or} \quad c^2 = 12 \longrightarrow c = \pm\sqrt{12} = \pm 2\sqrt{3}$$

Thus the foci, which occur at $(-c, 0)$ and $(c, 0)$, are located at $(-2\sqrt{3}, 0)$ and $(2\sqrt{3}, 0)$.

PROBLEM 9-26   Find the intercepts of the ellipse $12x^2 + 5y^2 = 60$.

Solution   You can find the intercepts directly without going to the standard form:

Let $y = 0$: $12x^2 = 60$ Let $x = 0$: $5y^2 = 60$

$\qquad\qquad\quad x^2 = 5$ $\qquad\qquad\qquad\qquad\qquad\qquad y^2 = 12$

$\qquad\qquad\quad x = \pm\sqrt{5}$ $\qquad\qquad\qquad\qquad\qquad\quad y = \pm\sqrt{12} = \pm 2\sqrt{3}$

PROBLEM 9-27   Find the center, the vertices, and the lengths of the major and minor axes of the ellipse $9x^2 + 4y^2 - 36x + 8y + 4 = 0$. Then sketch the graph.

Solution   First, complete the square on both the x and y variables:

$$9x^2 + 4y^2 - 36x + 8y + 4 = 0$$
$$(9x^2 - 36x) + (4y^2 + 8y) = -4$$
$$9(x^2 - 4x) + 4(y^2 + 2y) = -4$$
$$9(x^2 - 4x + 4) + 4(y^2 + 2y + 1) = -4 + 36 + 4$$
$$9(x - 2)^2 + 4(y + 1)^2 = 36$$
$$\tfrac{9}{36}(x - 2)^2 + \tfrac{4}{36}(y + 1)^2 = \tfrac{36}{36}$$
$$\frac{(x - 2)^2}{4} + \frac{(y + 1)^2}{9} = 1$$

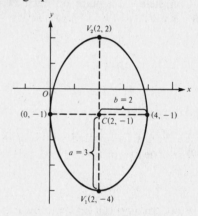

Figure 9-42

Now you can see that the given equation is of the form $\dfrac{(x - h)^2}{b^2} + \dfrac{(y - k)^2}{a^2} = 1$, so the center is at

the point $(h, k) = (2, -1)$ and the major axis is vertical. The vertices are 3 units above and below the center on the major axis of length $2\sqrt{a^2} = 2\sqrt{9} = 6$. Thus $V_1 = (h, k - a) = (2, -1 - 3) = (2, -4)$ and $V_2 = (h, k + a) = (2, -1 + 3) = (2, 2)$. The end-points of the minor axis, whose length is $2b = 4$, are 2 units to the right and left of the center. These points are $(h - b, k) = (2 - 2, -1) = (0, -1)$ and $(h + b, k) = (2 + 2, -1) = (4, -1)$. The ellipse is shown in Figure 9-42.

PROBLEM 9-28   Find the vertices and the asymptotes of the hyperbola $\dfrac{x^2}{16} - \dfrac{y^2}{9} = 1$.

Solution   This hyperbola has x-intercepts at $x = \pm 4$ (since $x^2/16 - 0 = 1$ or $x^2 = 16$) but no y-intercepts (since $-y^2/9 = 1$ or $y^2 = -9$ is not real). Its vertices are thus on the y-axis at $(-a, 0) = (-4, 0)$ and $(a, 0) = (4, 0)$. The asymptotes, then, are the lines $y = bx/a$ and $y = -bx/a$, where $a = \sqrt{16} = 4$ and $b = \sqrt{9} = 3$; so the asymptotes are $y = 3x/4$ and $y = -3x/4$.

PROBLEM 9-29   Find the vertices and the asymptotes for the hyperbola $4y^2 - 5x^2 = 20$.

Solution   First put the equation into standard form:

$$\frac{4y^2}{20} - \frac{5x^2}{20} = \frac{20}{20}$$

$$\frac{y^2}{5} - \frac{x^2}{4} = 1$$

Here the value of a is $\sqrt{5}$ and the value of b is $\sqrt{4} = 2$, and the hyperbola has y-intercepts (its vertices) at $(0, \pm a) = (0, -\sqrt{5})$ and $(0, \sqrt{5})$. The asymptotes have equations $y = ax/b = \sqrt{5}x/2$ and $y = -ax/b = -\sqrt{5}x/2$.

PROBLEM 9-30 For the hyperbola $\dfrac{x^2}{25} - \dfrac{y^2}{4} = 1$, find the values of a and b and the dimensions of the asymptote box.

Solution Since the positive term is the x^2 term, the value of $a = \sqrt{25} = 5$ and $b = \sqrt{4} = 2$. Hence, from the center point (the origin), the asymptote box extends 5 units to the right and left and 2 units up and down and has dimensions $2a \times 2b = 10 \times 4$.

PROBLEM 9-31 For the hyperbola $2x^2 - 9y^2 + 8x + 36y - 46 = 0$, find the center, the vertices, and the equations of the asymptotes. Then sketch its graph.

Solution First complete the square on both variables:

$$2x^2 - 9y^2 + 8x + 36y - 46 = 0$$
$$(2x^2 + 8x) - (9y^2 - 36y) = 46$$
$$2(x^2 + 4x) - 9(y^2 - 4y) = 46$$
$$2(x^2 + 4x + 4) - 9(y^2 - 4y + 4) = 46 + 8 - 36$$
$$2(x + 2)^2 - 9(y - 2)^2 = 18$$
$$\tfrac{2}{18}(x + 2)^2 - \tfrac{9}{18}(y - 2)^2 = \tfrac{18}{18}$$
$$\frac{(x + 2)^2}{9} - \frac{(y - 2)^2}{2} = 1$$

This hyperbola is centered at $(h, k) = (-2, 2)$ and the vertices are on the horizontal line $y = 2$. Thus the vertices are located 3 units to the right and left of the center point at $(-5, 2)$ and $(1, 2)$, since the value of a is $\sqrt{9} = 3$. Then, since the value of b is $\sqrt{2}$, the asymptote lines are

$$y - k = \frac{b}{a}(x - h) \longrightarrow y - 2 = \frac{\sqrt{2}}{3}(x + 2)$$

and

$$y - k = -\frac{b}{a}(x - h) \longrightarrow y - 2 = -\frac{\sqrt{2}}{3}(x + 2)$$

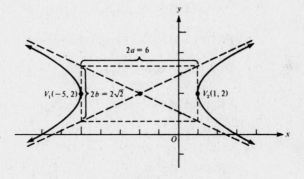

Figure 9-43

The hyperbola is sketched in Figure 9-43.

PROBLEM 9-32 Just by inspecting the following equations, decide which conic section each represents:

(a) $3x^2 + 5x - 3y + 10 = 0$ **(b)** $4x^2 + 3 - y^2 + 6y = 12$

(c) $3y^2 + 4x^2 + 5y - 2x - 7 = 0$ **(d)** $(2x - 1)^2 + 5y^2 = 4x^2 + 9$

(e) $x^2 + y^2 + 4x + 6y + 4 = 0$

Solution In each case, rearrange the terms to put the equation in the form

$$Ax^2 + By^2 + Cx + Dy + E = 0$$

Then inspect the values of A and B specifically.

(a) $3x^2 + 0y^2 + 5x - 3y + 10 = 0$ has $A = 3$, $B = 0$. Since $D \neq 0$, this is a parabola. (A nondegenerate parabola contains an x^2 term and a y term, or a y^2 term and an x term).

(b) $4x^2 - y^2 + 0x + 6y - 9 = 0$ has $A = 4, B = -1$, so this is a hyperbola because A and B have different signs.

(c) $4x^2 + 3y^2 - 2x + 5y - 7 = 0$ has $A = 4, B = 3$, so this is an ellipse because A and B have the same sign and are not equal.

(d)
$$(2x - 1)^2 + 5y^2 = 4x^2 + 9$$
$$4x^2 - 4x + 1 + 5y^2 = 4x^2 + 9$$
$$5y^2 - 4x - 8 = 0$$

$0x^2 + 5y^2 - 4x + 0y - 8 = 0$ has $A = 0, B = 5$, so this is a parabola (it has a y^2 term and an x term).

(e) $x^2 + y^2 + 4x + 6y + 4 = 0$ has $A = 1, B = 1$, so this is a circle—a special case of the ellipse—because A and B have the same sign and are equal.

Supplementary Exercises

PROBLEM 9-33 Identify the graph of the function $f(x) = 1 + 2x - 3x^2$ and find the intercepts and the vertex of its graph.

PROBLEM 9-34 Identify the graph of the function $f(x) = 2 + x - x^2$ and find the intercepts and the vertex of its graph.

PROBLEM 9-35 Find the vertex of the parabola $y = 8x^2 - 8x - 1$.

PROBLEM 9-36 Sketch the graph of $f(x) = -4x^2 + 2$.

PROBLEM 9-37 Sketch the graph of $y = 8x^2 - 8x - 1$ (see Problem 9-35).

PROBLEM 9-38 Sketch the graph of $y = (x - \frac{1}{2})^2$.

PROBLEM 9-39 Sketch the graph of $y = -(x + 5)^2 + 2$.

PROBLEM 9-40 Find the vertex of $y = -x^2 + 4x + 13$.

PROBLEM 9-41 Find the vertex of $y = -3x^2 + 18x - 27$.

PROBLEM 9-42 Find the dimensions of the rectangular field that will enclose the maximum pasture area if 150 meters of fence are available.

PROBLEM 9-43 A rectangular piece of tin 20 cm by 150 cm is to be formed into a 150-cm-long water trough by turning up equal segments on each edge. Find the height of the segment to turn up in order to maximize the carrying capacity.

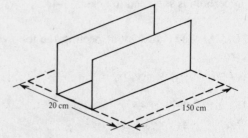

PROBLEM 9-44 If $D = \sqrt{x^2 + 1} - x$, what value of x will make D take its *minimum* value?

PROBLEM 9-45 What is the *maximum* value of the function $y = -4t^4 + 16t^2$?

PROBLEM 9-46 Among all pairs of numbers whose sum is 25, which two have the largest product?

PROBLEM 9-47 Among all rectangles having a perimeter of 20 feet, what are the dimensions of the one with the greatest area?

PROBLEM 9-48 A baseball is hit straight up, so that its height above the ground is given by $h = -16t^2 + 96t$, where t is measured in seconds and h in feet. Find the maximum height the ball reaches.

PROBLEM 9-49 What is the maximum area a right triangle can have if the sum of the lengths of the two shorter sides is 50 cm?

PROBLEM 9-50 Sketch the graph of $y = 3x^3 + 1$.

PROBLEM 9-51 Sketch the graph of $y = -\frac{1}{4}x^4$.

PROBLEM 9-52 Sketch the graph of $y = -x^3 + x$.

PROBLEM 9-53 Sketch the graph of $y = (x + 1)(x^2 - 9)$.

PROBLEM 9-54 Determine the vertical asymptotes for $y = \dfrac{x + 6}{x^2 - 9x + 20}$.

PROBLEM 9-55 Find the horizontal and vertical asymptotes for $y = \dfrac{2x^2 - 3}{x^2 - 2x + 1}$.

PROBLEM 9-56 Find the horizontal and vertical asymptotes for $y = \dfrac{-x^2 + 1}{2x^2 + 1}$.

PROBLEM 9-57 Sketch the graph of $y = \dfrac{1}{(x + 2)^3}$.

PROBLEM 9-58 Sketch the graph of $x^2 + y^2 + 2x + 4y = 11$.

PROBLEM 9-59 Find the equation of a line that is tangent to the circle $x^2 + y^2 = 25$ at $(0, 5)$. [*Hint*: If a line is tangent to a circle at a certain point, the line and the circle *share* that point.]

PROBLEM 9-60 Of all the points on the circle $x^2 + y^2 - 2x + 2y = 2$, which has the largest first coordinate and which has the largest second coordinate? How far apart are these points? [*Hint*: Sketch the circle first.]

PROBLEM 9-61 What is the equation of the circle with center at $(2, -5)$ and radius 4?

PROBLEM 9-62 Identify the conic $y = 1 - 4x - x^2$.

PROBLEM 9-63 Find the vertex of $x = y^2 - 6y + 2$.

PROBLEM 9-64 Find the intercepts and vertex of $x^2 - 2x - 4y = 7$ and sketch its graph.

PROBLEM 9-65 Find the equation of the parabola that is symmetric with respect to the x-axis, has its vertex at the origin, and passes through the point $(10, 5)$.

PROBLEM 9-66 A headlight reflector is constructed in the shape of a parabola. The light bulb must be at the focus of the parabola so that the light rays travel parallel to the axis of symmetry. For the reflector whose cross section is shown here, find the location of the bulb with respect to the vertex of the parabola.

PROBLEM 9-67 Find the ends of the axes (V_1, V_2 and B_1, B_2) and the center for the ellipse $4x^2 + 9y^2 - 16x + 36y + 16 = 0$.

PROBLEM 9-68 Find the equation of the ellipse for which

(a) center $(0, 0)$, x-intercept 5, y-intercept 2
(b) center $(2, -1)$, vertex $(2, -5)$, point $(0, 1)$ on curve
(c) center $(-2, -3)$, points $(2, -3)$ and $(-2, 2)$ on curve

PROBLEM 9-69 Find the ellipse(s) whose major axis has length 6, minor axis has length 3, and whose center is at $(2, 4)$. Sketch your result(s).

PROBLEM 9-70 Find the coordinates of the vertices and foci for the hyperbola $5y^2 - 2x^2 = 10$.

PROBLEM 9-71 What are the asymptotes for $\dfrac{x^2}{25} - \dfrac{y^2}{9} = 1$?

PROBLEM 9-72 Find the equation of the hyperbola whose center is $(0, 0)$, whose foci are on the y-axis, and whose curve contains points $(2, -5)$ and $(10, 7)$.

PROBLEM 9-73 Find the equation of the hyperbola for which

(a) center $(0, 0)$, asymptote $y = 3x$, x-intercept 4
(b) center $(1, 2)$, vertex $(1, -2)$, asymptote $y = x + 1$

PROBLEM 9-74 Find the equation of the hyperbola with vertices at $(6, 2)$ and $(-2, 2)$ and foci at $(7, 2)$ and $(-3, 2)$.

PROBLEM 9-75 Find the equation of the hyperbola with center at $(1, 2)$, vertex at $(5, 2)$, and asymptote at $y = -2x$.

PROBLEM 9-76 Determine which conic section is represented by

(a) $4x^2 = 9y^2 + 4y + 2$

(b) $x^2 + 4x + 2y^2 - 7 = 0$

(c) $5x^2 = 4y - 2x$

(d) $2(x - 3)^2 + 5y = \dfrac{4x^2 + 7y^2}{2}$

Answers to Supplementary Exercises

9-33 Parabola: $(0, 1)$, $(1, 0)$, $(-\tfrac{1}{3}, 0)$; $V(\tfrac{1}{3}, \tfrac{4}{3})$

9-42 37.5 m by 37.5 m

9-34 Parabola: $(2, 0)$, $(-1, 0)$, $(0, 2)$; $V(\tfrac{1}{2}, \tfrac{9}{4})$

9-43 5 cm

9-35 $V(\tfrac{1}{2}, -3)$

9-44 $x = \tfrac{1}{2}$

9-36 See Figure 9-44.

9-45 16

9-37 See Figure 9-45.

9-46 $x = 12.5$ and $y = 12.5$

9-38 See Figure 9-46.

9-47 5 ft by 5 ft

9-39 See Figure 9-47.

9-48 144 ft

9-40 $V(2, 17)$

9-49 312.5 cm^2

9-41 $V(3, 0)$

9-50 See Figure 9-48.

9-51 See Figure 9-49.

9-52 See Figure 9-50.

9-53 See Figure 9-51.

9-54 $x = 4, x = 5$

9-55 $y = 2, x = 1$

9-56 $y = -\frac{1}{2}$, no vertical asymptote

9-57 See Figure 9-52.

9-58 See Figure 9-53.

9-59 $y = 5$

9-60 $(3, -1), (1, 1), \sqrt{8}$

9-61 $(x - 2)^2 + (y + 5)^2 = 16$

9-62 parabola

9-63 $V(-7, 3)$

9-64 $y = -\frac{7}{4}, x = 1 \pm \sqrt{8}, V = (1, -2)$
See Figure 9-54.

9-65 $5x = 2y^2$

9-66 $\frac{25}{48}$ inch from vertex

9-67 $C(2, -2); V_1(-1, -2), V_2(5, -2); B_1(2, 0), B_2(2, -4)$

9-68 (a) $\dfrac{x^2}{25} + \dfrac{y^2}{4} = 1$

   (b) $\dfrac{3(x - 2)^2}{16} + \dfrac{(y + 1)^2}{16} = 1$

   (c) $\dfrac{(x + 2)^2}{16} + \dfrac{(y + 3)^2}{25} = 1$

9-69 $\dfrac{4(x - 2)^2}{9} + \dfrac{(y - 4)^2}{9} = 1$
or
$\dfrac{(x - 2)^2}{9} + \dfrac{4(y - 4)^2}{9} = 1$
See Figure 9-55.

9-70 $V = (0, \sqrt{2}), (0, -\sqrt{2}); F = (0, \sqrt{7}), (0, -\sqrt{7})$

9-71 $y = \pm\frac{3}{5}x$

9-72 $\dfrac{y^2}{24} - \dfrac{x^2}{96} = 1$

9-73 (a) $\dfrac{x^2}{16} - \dfrac{y^2}{144} = 1$

   (b) $\dfrac{(y - 2)^2}{16} - \dfrac{(x - 1)^2}{16} = 1$

9-74 $\dfrac{(x - 2)^2}{16} - \dfrac{(y - 2)^2}{9} = 1$

9-75 $\dfrac{(x - 1)^2}{16} - \dfrac{(y - 2)^2}{64} = 1$

9-76 (a) hyperbola

   (b) ellipse

   (c) parabola

   (d) parabola

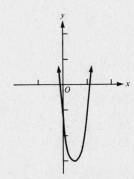

Figure 9-45

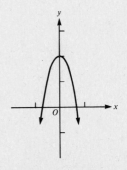

Figure 9-44

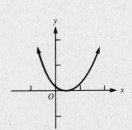

Figure 9-46

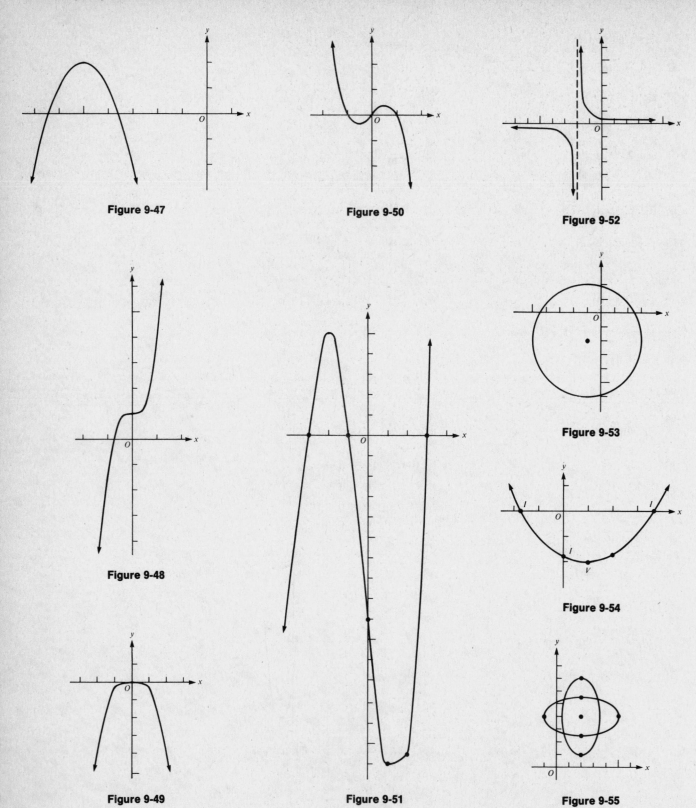

Figure 9-47

Figure 9-50

Figure 9-52

Figure 9-48

Figure 9-51

Figure 9-53

Figure 9-54

Figure 9-49

Figure 9-55

10 EXPONENTIAL AND LOGARITHMIC FUNCTIONS

THIS CHAPTER IS ABOUT

☑ **Exponential Functions**
☑ **Logarithmic Functions**
☑ **Computation with Logarithms**
☑ **Applications of Exponential and Logarithmic Functions**

10-1. Exponential Functions

A. Definitions and graphs

- An **exponential function** is a function of the form $f(x) = a^x$, where a is a positive constant $(a \neq 1)$ and the domain of the function is the set of all real numbers.
- The constant a is called the **base** of the exponential function.

EXAMPLE 10-1 Which of the following are examples of exponential functions? Find the base of each exponential function.

(a) $f(x) = 2^x$ **(b)** $f(x) = (-2)^x$ **(c)** $f(x) = x^2$

(d) $f(x) = 1^x$ **(e)** $f(x) = (\frac{1}{2})^x$

Solution

(a) $f(x) = 2^x$ is an exponential function with base 2.
(b) $f(x) = (-2)^x$ is NOT an exponential function, since the base -2 is not a positive constant.
(c) $f(x) = x^2$ is NOT an exponential function, since the base x is a variable, not a constant, and the exponent 2 is a constant, not a variable.
(d) $f(x) = 1^x$ is NOT an exponential function, since the base is $a = 1$. [*Note:* The function $f(x) = 1^x$ is a *constant function*, since $1^x = 1$ for all x.]
(e) $f(x) = (\frac{1}{2})^x$ is an exponential function with base $\frac{1}{2}$.

EXAMPLE 10-2 Sketch the graph of the exponential function $f(x) = 2^x$.

Solution If $f(x) = 2^x$, then $f(0) = 1$, $f(1) = 2$, $f(2) = 4$, $f(3) = 8$, $f(-1) = \frac{1}{2}$, $f(-2) = \frac{1}{4}$, $f(-3) = \frac{1}{9}$, and so on. The domain is the set of all real numbers, but the range is restricted to positive real numbers (since 2 raised to any power can't be negative). Also, since there is no x such that $2^x = 0$, the curve never touches the x-axis. See Figure 10-1.

note: The function $f(x) = 2^x$ has the property that if $r < s$, then $2^r < 2^s$. Thus the function $f(x) = 2^x$ is *increasing* for all x, which the graph in Figure 10-1 confirms.

In general, for any exponential function $f(x) = a^x$,

- If $a > 1$, the function is always INCREASING.
- If $0 < a < 1$, the function is always DECREASING.

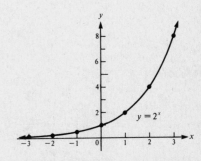

Figure 10-1: $f(x) = y = 2^x$.

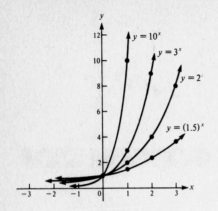

Figure 10-2

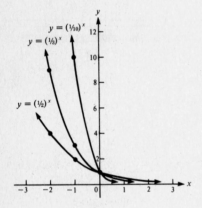

Figure 10-3

EXAMPLE 10-3   Graph $y = (1.5)^x$, $y = 2^x$, $y = 3^x$, and $y = 10^x$ on the same set of axes.

Solution   See Figure 10-2. Since $a > 1$ in each case, the graphs are all increasing.

EXAMPLE 10-4   Graph $y = (\frac{1}{2})^x$, $y = (\frac{1}{3})^x$, and $y = (\frac{1}{10})^x$ on the same set of axes.

Solution   See Figure 10-3. Since $0 < a < 1$ in each case, the graphs are all decreasing.

note: Since there is no x such that $a^x = 0$, the graph of a function of the form $y = a^x$ has no x-intercepts. (It has the x-axis as a horizontal asymptote.) However, since $a^0 = 1$ for any value of a, the graph of every function of the form $y = a^x$ has a y-intercept at $(0, 1)$.

B. Rules of exponents

The rules of exponents that apply to a^x for rational values of x (see Chapter 3) also apply for all real values of x.

If a and b are fixed positive numbers and x and y are arbitrary real numbers, then

RULES OF EXPONENTS

(1) $a^x \cdot a^y = a^{x+y}$

(2) $\dfrac{a^x}{a^y} = a^{x-y}$

(3) $(a^x)^y = a^{xy}$

(4) $(ab)^x = a^x b^x$

(5) $\left(\dfrac{a}{b}\right)^x = \dfrac{a^x}{b^x}$

(6) $a^{-x} = \dfrac{1}{a^x} = \left(\dfrac{1}{a}\right)^x$

(7) $a^0 = 1$

(8) $1^x = 1$

EXAMPLE 10-5   Use the rules of exponents to simplify

(a) $(\sqrt[3]{3})^\pi (\sqrt[3]{9})^\pi$ (b) $\dfrac{(\sqrt{2})^{\sqrt{2}}(\sqrt{3})^{\sqrt{2}}}{(\sqrt{12})^{\sqrt{2}}}$ (c) $\left(\dfrac{1}{3^{\sqrt{5}}}\right)^{\sqrt{5}}$

Solution

(a) $(\sqrt[3]{3})^\pi (\sqrt[3]{9})^\pi = (\sqrt[3]{3 \cdot 9})^\pi$ Rule 4 $[(ab)^x = a^x b^x]$

$= (\sqrt[3]{27})^\pi$

$= 3^\pi$

(b) $\dfrac{(\sqrt{2})^{\sqrt{2}}(\sqrt{3})^{\sqrt{2}}}{(\sqrt{12})^{\sqrt{2}}} = \dfrac{(\sqrt{2}\sqrt{3})^{\sqrt{2}}}{(\sqrt{12})^{\sqrt{2}}}$ Rule 4

$= \dfrac{(\sqrt{6})^{\sqrt{2}}}{(\sqrt{12})^{\sqrt{2}}} = \left(\dfrac{\sqrt{6}}{\sqrt{12}}\right)^{\sqrt{2}}$ Rule 5 $\left[\left(\dfrac{a}{b}\right)^x = \dfrac{a^x}{b^x}\right]$

$= \left(\sqrt{\dfrac{6}{12}}\right)^{\sqrt{2}} = \left(\sqrt{\dfrac{1}{2}}\right)^{\sqrt{2}}$

$= \left(\dfrac{\sqrt{1}}{\sqrt{2}}\right)^{\sqrt{2}} = \dfrac{1^{\sqrt{2}}}{(\sqrt{2})^{\sqrt{2}}}$ Rule 5

$= \dfrac{1}{(\sqrt{2})^{\sqrt{2}}}$ Rule 8 $[1^x = 1]$

$= (\sqrt{2})^{-\sqrt{2}}$ Rule 6 $\left[a^{-x} = \dfrac{1}{a^x}\right]$

(c)

$$\left(\frac{1}{3^{\sqrt{5}}}\right)^{\sqrt{5}} = \frac{1^{\sqrt{5}}}{(3^{\sqrt{5}})^{\sqrt{5}}} \qquad \text{Rule 5}$$

$$= \frac{1}{3^5} \qquad \text{Rules 3 and 8} \quad [(a^x)^y = a^{xy}; 1^x = 1]$$

$$= \frac{1}{243}$$

EXAMPLE 10-6 Sketch the graph of $y = 5^{2-x}$.

Solution First we use the rules of exponents to rewrite 5^{2-x}:

$$y = 5^{2-x} = 5^2 \cdot 5^{-x} \qquad \text{Rule 1}$$

$$= 25 \cdot 5^{-x} = 25\left(\frac{1}{5^x}\right) \qquad \text{Rule 6}$$

$$= 25\left(\frac{1}{5}\right)^x \qquad \text{Rule 5}$$

This graph has y values 25 times larger than the y values of the graph of $y = (\frac{1}{5})^x$:

$$y = 25(\tfrac{1}{5})^x$$

x	0	1	2	3	−1	−2
y	25	5	1	$\frac{1}{5}$	125	625

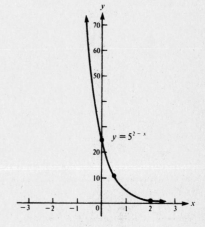

Figure 10-4

And since the base $\frac{1}{5}$ is between 0 and 1, the graph of $y = 25(\frac{1}{5})^x$, i.e., $y = 5^{2-x}$, is decreasing for all x. See Figure 10-4.

C. The exponential function $y = e^x$

Of all the possible base values that may be assigned to a in the function $y = a^x$, one value outshines all others in importance. It is an irrational number—denoted by e—whose decimal expansion begins

$$e = 2.718\,281\,828\,459\,045\,235\ldots$$

Because of its usefulness in explaining natural physical phenomena, the function $y = e^x$ is called the **natural exponential function**.

note: The number e is defined as the value that the expression $(1 + \frac{1}{n})^n$ approaches as n grows larger without bound. Notice, for example, that if $n = 10$, $(1 + \frac{1}{n})^n \approx 2.593\,742\,5$ and if $n = 10\,000$, $(1 + \frac{1}{n})^n \approx 2.718\,145\,9$.

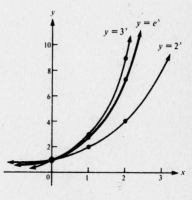

EXAMPLE 10-7 Sketch the graph of $y = e^x$.

Solution Since the base value e lies between 2 and 3, the graph of $y = e^x$ lies between the graphs of $y = 2^x$ and $y = 3^x$. We can approximate the graph by drawing a curve through the points $(0, 1)$, $(1, 2.7)$, and $(2, 7.3)$, as in Figure 10-5.

note: Many hand-held calculators have a key for e^x, which makes the computations of points on its graph much easier.

Figure 10-5

EXAMPLE 10-8 Sketch the graph of $y = e^{-(x^2)}$.

Solution We form a table of solution pairs, using a hand-held calculator to compute the powers of e:

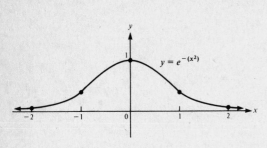

Figure 10-6

\multicolumn{2}{c}{$y = e^{-(x^2)}$}
x	y
0	$e^0 = 1$
1	$e^{-(1^2)} = \dfrac{1}{e} \approx .368$
2	$e^{-(2^2)} = \dfrac{1}{e^4} \approx .018$
-1	$e^{-(-1^2)} = \dfrac{1}{e} \approx .368$
-2	$e^{-(-2^2)} = \dfrac{1}{e^4} \approx .018$

Since $e^{-(-x^2)} = e^{-(x^2)}$, the function is even and the graph is symmetric with respect to the y-axis, as shown in Figure 10-6.

D. Rules for solving exponential equations

If $a \neq 1$, then $y = a^x$ is either increasing (if $a > 1$) or decreasing (if $0 < a < 1$) for all x in the domain. Thus the exponential function $y = a^x$ is a one-to-one function and we can state the following rules:

(1) If $a^x = a^y$, then $x = y$.
(2) If $a^x = b^x$ (where $x \neq 0$), then $a = b$.

We can use rules (1) and (2) to help in solving certain equations involving exponential functions.

EXAMPLE 10-9   Solve the equations

(a) $5^8 = 5^{2x}$ **(b)** $4^5 = (x - 1)^5$ **(c)** $2^{2x+3} = 32$ **(d)** $(\sqrt{3})^{x+1} = 3^{x-\sqrt{2}}$

Solution

(a) Since $5^8 = 5^{2x}$, then $8 = 2x$ or $x = 4$.   Rule (1)
(b) Since $4^5 = (x - 1)^5$, then $4 = x - 1$ or $x = 5$.   Rule (2)
(c) First rewrite $2^{2x+3} = 32$ as $2^{2x+3} = 2^5$. Then since
$2^{2x+3} = 2^5$, $2x + 3 = 5$ or $x = 1$.   Rule (1)
(d) Rewrite $(\sqrt{3})^{x+1} = 3^{x-\sqrt{2}}$ as $(\sqrt{3})^{x+1} = ((\sqrt{3})^2)^{x-\sqrt{2}}$
or $(\sqrt{3})^{2x-2\sqrt{2}}$. Then since $(\sqrt{3})^{x+1} = (\sqrt{3})^{2x-2\sqrt{2}}$,
$x + 1 = 2x - 2\sqrt{2}$ or $x = 1 + 2\sqrt{2}$.   Rule (1)

10-2. Logarithmic Functions

A. Definitions and graphs

Since the exponential function $y = a^x$ is a one-to-one function, we can say that for each positive real number y, there is a unique real number x such that $a^x = y$. Let $a > 0$ with $a \neq 1$. Then, for any positive number y,

• The **logarithm of y to the base a** is the unique number x such that $a^x = y$.

We denote the logarithm of y to the base a as $\log_a y$. Thus, for any x and any $y > 0$,

$$\log_a y = x \quad \text{if and only if} \quad a^x = y$$

note: In the expression $x = \log_a y$, x is called the *logarithm*, a is called the *base*, and y is called the *argument*.

EXAMPLE 10-10 Convert the following equations involving exponents into equations involving logarithms:

(a) $10^1 = 10$ (b) $10^2 = 100$ (c) $10^3 = 1000$
(d) $2^2 = 4$ (e) $27^{1/3} = 3$ (f) $25^{1/2} = 5$

Solution   Use the definition $\log_a y = x$ if and only if $a^x = y$.

(a) In $10^1 = 10$, $a = 10$, $x = 1$, and $y = 10$. Thus $\log_a y = x$ becomes $\log_{10} 10 = 1$.
(b) Since $a = 10$, $x = 2$, and $y = 100$, the equation $10^2 = 100$ is equivalent to $\log_{10} 100 = 2$.
(c) The equation $10^3 = 1000$ is equivalent to $\log_{10} 1000 = 3$.
(d) The equation $2^2 = 4$ is equivalent to $\log_2 4 = 2$.
(e) The equation $27^{1/3} = 3$ is equivalent to $\log_{27} 3 = \frac{1}{3}$.
(f) The equation $25^{1/2} = 5$ is equivalent to $\log_{25} 5 = \frac{1}{2}$.

EXAMPLE 10-11 Convert the following equations involving logarithms into equations involving exponents:

(a) $\log_3 27 = 3$ (b) $\log_4 16 = 2$ (c) $\log_{1/2} 8 = -3$

Solution

(a) In $\log_3 27 = 3$, $a = 3$, $x = 3$, and $y = 27$, so the equivalent equation $y = a^x$ is $3^3 = 27$.
(b) Since $a = 4$, $x = 2$, and $y = 16$, the equation $\log_4 16 = 2$ is equivalent to $4^2 = 16$.
(c) Since $a = \frac{1}{2}$, $x = -3$, and $y = 8$, the equation $\log_{1/2} 8 = -3$ is equivalent to $(\frac{1}{2})^{-3} = 8$.

EXAMPLE 10-12 Evaluate (a) $\log_3 81$, (b) $\log_{10} 1$, (c) $\log_{1/3} 9$.

Solution

(a) If $\log_3 81 = x$, then $3^x = 81$. We know that $3^4 = 81$, so $x = 4$.
(b) If $\log_{10} 1 = x$, then $10^x = 1$. We know that $10^0 = 1$, so $x = 0$.
(c) If $\log_{1/3} 9 = x$, then $(\frac{1}{3})^x = 9$. We can rewrite this equation as $(\frac{1}{3})^x = 3^2$ or $3^{-x} = 3^2$, so $x = -2$.

EXAMPLE 10-13 Solve the following equations for x:

(a) $\log_2 x = 3$ (b) $\log_x 16 = 2$ (c) $\log_e x = 1$

Solution

(a) Since $\log_2 x = 3$, then $2^3 = x$ or $x = 8$.
(b) Since $\log_x 16 = 2$, then $x^2 = 16$ or $x = \pm 4$. But by definition, the base of a logarithm must be positive, so the only solution is $x = 4$.
(c) Since $\log_e x = 1$, then $e^1 = x$ or $x = e$.

We can develop the following formulas from the general definition of a logarithm:

(1) $\log_a 1 = 0$ This follows from the definition, since $a^0 = 1$
(2) $\log_a a = 1$ This follows, since $a^1 = a$
(3) $a^{\log_a x} = x$ This follows, since if $a^y = x$, then $y = \log_a x$
(4) $\log_a(a^x) = x$ This follows, since if $a^x = y$, then $x = \log_a y$

EXAMPLE 10-14 Evaluate (a) $\log_3 1$, (b) $\log_5 5$, (c) $2^{\log_2 5}$, (d) $\log_2(2^3)$.

Solution   Use the formulas developed above.

(a) $\log_3 1 = 0$ Formula (1)
(b) $\log_5 5 = 1$ Formula (2)
(c) $2^{\log_2 5} = 5$ Formula (3)
(d) $\log_2(2^3) = 3$ Formula (4)

Now suppose that we have the positive numbers x and z such that $\log_a x = \log_a z$. From formula (3) above, we have $x = a^{\log_a x} = a^{\log_a z} = z$. Thus, if $\log_a x = \log_a z$, then $x = z$. The converse is also true, so we can state this result as

(5) $\log_a x = \log_a z$ if and only if $x = z$

This result is helpful in solving certain logarithmic equations.

EXAMPLE 10-15   Solve the equation $\log_2(x^2 + 3) = \log_2 4x$.

Solution   Since $\log_2(x^2 + 3) = \log_2 4x$, then $x^2 + 3 = 4x$.   Formula (5)

So
$$x^2 + 3 = 4x$$
$$x^2 - 4x + 3 = 0$$
$$(x - 1)(x - 3) = 0$$
$$x = 1 \quad \text{or} \quad x = 3$$

Check by substituting the values of x into the original equation:

$x = 1$: $\log_2(x^2 + 3) \overset{?}{=} \log_2 4x$

                $\log_2(1^2 + 3) \overset{?}{=} \log_2 4(1)$

                    $\log_2 4 = \log_2 4$

$x = 3$: $\log_2(x^2 + 3) \overset{?}{=} \log_2 4x$

                $\log_2(3^2 + 3) \overset{?}{=} \log_2 4(3)$

                   $\log_2 12 \overset{\checkmark}{=} \log_2 12$

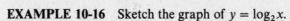

• Every logarithmic function is the *inverse* of an exponential function. We can obtain the graph of the logarithmic function by reflecting the graph of the corresponding exponential function through the line $y = x$.

EXAMPLE 10-16   Sketch the graph of $y = \log_2 x$.

Solution   Since $y = \log_2 x$, then $x = 2^y$ is the equivalent exponential equation. The inverse of $x = 2^y$ is $y = 2^x$, which we graphed earlier in Figure 10-1. We reflect the graph of $y = 2^x$ through the line $y = x$ to obtain the graph of $y = \log_2 x$, as in Figure 10-7.

EXAMPLE 10-17   Sketch the graph of $y = \log_{1/2} x$.

Solution   The exponential function corresponding to $y = \log_{1/2} x$ is $(\frac{1}{2})^y = x$; the inverse of $(\frac{1}{2})^y = x$ is $y = (\frac{1}{2})^x$. We have already graphed $y = (\frac{1}{2})^x$ (see Figure 10-3), so we simply reflect it through $y = x$ to obtain the graph of $y = \log_{1/2} x$, as in Figure 10-8.

note: Since the equation $y = \log_a x$ is equivalent to the equation $x = a^y$, we can also graph a logarithmic function by preparing a table of solution pairs. We choose arbitrary values of y, compute the corresponding values of x, and plot the points.

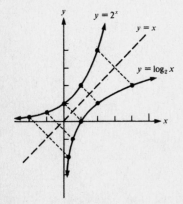

Figure 10-7

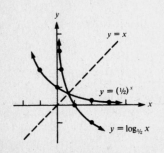

Figure 10-8

We can compare the features of the graph of the logarithmic function $y = \log_a x$ with the corresponding features of its inverse, the exponential function $y = a^x$.

$y = a^x$	$y = \log_a x$
The domain consists of all real numbers.	The domain consists of all positive real numbers.
The range consists of all positive real numbers.	The range consists of all real numbers.
The y-intercept is 1, and there is no x-intercept.	The x-intercept is 1, and there is no y-intercept.
The x-axis is a horizontal asymptote of the graph.	The y-axis is a vertical asymptote of the graph.
If $a > 1$, $y = a^x$ is increasing.	If $a > 1$, $y = \log_a x$ is increasing.
If $0 < a < 1$, $y = a^x$ is decreasing.	If $0 < a < 1$, $y = \log_a x$ is decreasing.
The value of a^x is positive for all x.	If $0 < x < 1$, the value of $\log_a x$ is negative if $a > 1$ and positive if $0 < a < 1$.
	If $x > 1$, the value of $\log_a x$ is positive if $a > 1$ and negative if $0 < a < 1$.

B. Special bases

The definition of a logarithmic function states that the base value a can be any positive number except 1. There are, however, two bases that are used more than any others.

In Section 10-1C we defined the natural exponential function by means of the equation $y = e^x$. The logarithmic function with base e is called the **natural logarithmic function** and is important in calculus and other branches of mathematics. We use **ln x** as an abbreviation for $\log_e x$ and call it the natural logarithm of x.

Computations with logarithms have traditionally made use of logarithms to the base 10. This makes sense, since our number system is base 10. The logarithmic function with base 10 is called the **common logarithmic function** and we abbreviate $\log_{10} x$ simply as **log x**.

note: When we write $\ln x$, base e is understood. Likewise, when we write $\log x$, base 10 is understood. But if the base value of a logarithm is anything other than e or 10, we must always write the base in the subscript form.

EXAMPLE 10-18   Write the following exponential equations in logarithmic form:

(a) $e^1 = e$ **(b)** $10^3 = 1000$ **(c)** $e^{1.6094} = 5$ **(d)** $10^{-4} = .0001$

Solution

(a) The equation $e^1 = e$ is equivalent to $\log_e e = 1$ or $\ln e = 1$.
(b) The equation $10^3 = 1000$ is equivalent to $\log_{10} 1000 = 3$ or $\log 1000 = 3$.
(c) The equation $e^{1.6094} = 5$ is equivalent to $\ln 5 = 1.6094$.
(d) The equation $10^{-4} = .0001$ is equivalent to $\log .0001 = -4$.

EXAMPLE 10-19   Sketch the graph of (a) $y = \ln x$ and (b) $y = \log x$.

Solution

(a) The function $y = \ln x$ is the inverse of $y = e^x$, so we can reflect the graph of $y = e^x$ (Fig. 10-5) through the line $y = x$ to obtain the graph of $y = \ln x$, as in Figure 10-9a.

(a)

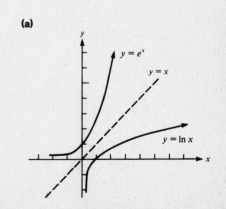

Figure 10-9a

(b)

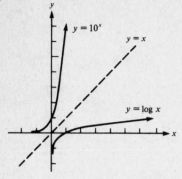

Figure 10-9b

(b) The function $y = \log x$ is the inverse of $y = 10^x$, so we can reflect the graph of $y = 10^x$ (Fig. 10-2) through the line $y = x$ to obtain the graph of $y = \log x$, as in Figure 10-9b.

C. Laws of logarithms and change of base formulas

1. Laws of logarithms

The following laws are fundamental for all work involving logarithms.

Given the positive real numbers x and y:

LAWS OF LOGARITHMS

(1) $\log_a(xy) = \log_a x + \log_a y$ — The logarithm of a product equals the sum of the logarithms of the parts

(2) $\log_a\left(\dfrac{x}{y}\right) = \log_a x - \log_a y$ — The logarithm of a quotient equals the difference of the logarithms of the parts

(3) $\log_a x^n = n \log_a x$ — The logarithm of a function to a power equals the exponent times the logarithm of the function alone

EXAMPLE 10-20 Write the following expressions in simpler logarithmic form:

(a) $\log_3(3^5)$ **(b)** $\log_{10}\left(\dfrac{421}{137}\right)$ **(c)** $\log_5\left(\dfrac{4\sqrt{10}}{\sqrt{13}}\right)$ **(d)** $\log_a\left(\dfrac{mn}{p}\right)$

Solution

(a) $\log_3(3^5) = 5\log_3 3 = 5(1) = 5$ — Law 3

(b) $\log_{10}\left(\dfrac{421}{137}\right) = \log_{10}421 - \log_{10}137$ — Law 2

(c) $\log_5\left(\dfrac{4\sqrt{10}}{\sqrt{13}}\right) = \log_5 4\sqrt{10} - \log_5\sqrt{13}$ — Law 2

$= \log_5 2^2 + \log_5(10)^{1/2} - \log_5(13)^{1/2}$ — Law 1

$= 2\log_5 2 + \tfrac{1}{2}\log_5 10 - \tfrac{1}{2}\log_5 13$ — Law 3

(d) $\log_a\left(\dfrac{mn}{p}\right) = \log_a m + \log_a n - \log_a p$ — Laws 1 and 2

2. Change of base formulas

Sometimes it is necessary to change the base of a logarithm by converting an expression from $\log_a x$ to $\log_b x$. We can do this by using two basic formulas, which can be obtained from one of the basic formulas derived from the general definition of a logarithm: $a^{\log_a x} = x$. Thus

$$a^{\log_a x} = x$$

$\log_b(a^{\log_a x}) = \log_b x$ — Take the logarithm of each side

$(\log_a x)(\log_b a) = \log_b x$ — Use law of logarithms (3)

(i) $\log_a x = \dfrac{\log_b x}{\log_b a}$ — Divide by $\log_b a$

Now let $x = b$ in formula (i). Then

$$\log_a b = \frac{\log_b b}{\log_b a}$$

(ii) $$\log_a b = \frac{1}{\log_b a}$$

EXAMPLE 10-21   Change the base of the following logarithms to base 10:
(a) $\log_2 9$ **(b)** $\log_6 10$.

Solution   Use the change of base formulas given above.

(a) $\log_2 9 = \dfrac{\log_{10} 9}{\log_{10} 2}$ Formula (i)

(b) $\log_6 10 = \dfrac{1}{\log_{10} 6}$ Formula (ii)

note: *Beware!* The expression $(\log_{10} 9)/(\log_{10} 2)$ is NOT equal to $\log_{10} 9 - \log_{10} 2$; it is already in its simplest form. The expression $\log_{10}(\frac{9}{2})$, however, IS equal to $\log_{10} 9 - \log_{10} 2$.

D. Solving logarithmic equations

We can use the definition and laws of logarithms to solve logarithmic equations. And it's often useful to convert a given logarithmic equation into its equivalent exponential form.

EXAMPLE 10-22   Solve the equation $\log(3x + 6) = 1 + \log(x - 5)$.

Solution   First rewrite the equation with the logarithmic expressions on one side:

$$\log(3x + 6) = 1 + \log(x - 5)$$

$$\log(3x + 6) - \log(x - 5) = 1$$

Now use the second law of logarithms $\log_a\left(\dfrac{x}{y}\right) = \log_a x - \log_a y$ to obtain

$$\log\left(\frac{3x + 6}{x - 5}\right) = 1$$

Convert this equation into an exponential equation and solve for x:

$$\frac{3x + 6}{x - 5} = 10^1 = 10$$

$$3x + 6 = 10(x - 5) = 10x - 50$$

$$-7x = -56$$

$$x = 8$$

Check: $\log(3 \cdot 8 + 6) \overset{?}{=} 1 + \log(8 - 5)$

$\log 30 \overset{?}{=} 1 + \log 3$

$\log(10 \cdot 3) \overset{?}{=} 1 + \log 3$

$\log 10 + \log 3 \overset{?}{=} 1 + \log 3$

$1 + \log 3 \overset{\checkmark}{=} 1 + \log 3$

EXAMPLE 10-23   Solve

(a) $2\log_3 x = \log_3 36$ **(b)** $\log_2(x - 1) + \log_2(x + 1) = 3$

Solution

(a)
$$2\log_3 x = \log_3 36$$
$$\log_3 x^2 = \log_3 36 \qquad \text{Law 3}$$

Thus
$$x^2 = 36$$
$$x = 6 \quad \text{or} \quad x \neq -6 \qquad \text{(Reject } x = -6 \text{ since } \log_3(-6) \text{ is not defined.)}$$

(b)
$$\log_2(x - 1) + \log_2(x + 1) = 3$$
$$\log_2[(x - 1)(x + 1)] = 3 \qquad \text{Law 1}$$

Converting to exponential form, we have
$$2^3 = (x - 1)(x + 1)$$
$$8 = x^2 - 1$$
$$x^2 = 9$$
$$x = 3 \quad \text{or} \quad x \neq -3 \qquad \text{(Reject } x = -3 \text{: } \log_2(-4) \text{ and } \log_2(-2) \text{ are undefined.)}$$

EXAMPLE 10-24   Solve for x in $\log(x + 15) + \log x = 2$.

Solution   Combine the terms on the left:
$$\log(x + 15) + \log x = 2$$
$$\log[(x + 15)x] = 2 \qquad \text{Law 1}$$

Convert the equation to exponential form:
$$(x + 15)x = 10^2 = 100 \qquad \text{(Base 10 is understood.)}$$
$$x^2 + 15x - 100 = 0$$
$$(x + 20)(x - 5) = 0$$
$$x \neq -20 \quad \text{or} \quad x = 5 \qquad \text{(Reject } x = -20 \text{: } \log(-5) \text{ and } \log(-20) \text{ are undefined.)}$$

Check:
$$\log(5 + 15) + \log 5 \overset{?}{=} 2$$
$$\log 20 + \log 5 \overset{?}{=} 2$$
$$\log 20 \cdot 5 \overset{?}{=} 2$$
$$\log 100 \overset{?}{=} 2$$
$$2 \overset{\checkmark}{=} 2$$

10-3. Computation with Logarithms

A. Scientific notation

When we're dealing with very large or very small numbers, it is convenient to write all numbers in a common form called *scientific notation*.

- A positive number other than zero is in **scientific notation** if it is written in the form $b \times 10^n$, where $1 \le b < 10$ and n is an integer.

EXAMPLE 10-25 Write the following numbers in scientific notation: (a) 482, (b) 23.61, (c) 105 362, (d) .000 149.

Solution In each case we rewrite the number by moving the decimal point. Multiplication by 10 moves the decimal point one place to the right, while multiplication by 10^{-1} (same as division by 10) moves the decimal point one place to the left.

(a) $482 = 4.82 \times 100 = 4.82 \times 10^2$

(b) $23.61 = 2.361 \times 10 = 2.361 \times 10^1$

(c) $105\,362 = 1.053\,62 \times 100\,000 = 1.053\,62 \times 10^5$

(d) $.000\,149 = 1.49 \times .0001 = 1.49 \times 10^{-4}$

EXAMPLE 10-26 Evaluate each of the following and write the result in scientific notation:

(a) $(2.5 \times 10^4)(6.7 \times 10^3)$ (b) $\dfrac{.002\,19}{.067}$ (c) $\dfrac{4.139 \times 10^6}{7.12 \times 10^{11}}$

Solution

(a)
$$(2.5 \times 10^4)(6.7 \times 10^3) = (2.5 \times 6.7)(10^4 \times 10^3)$$
$$= 16.75 \times 10^7$$
$$= 1.675 \times 10^8$$

(b)
$$\frac{.002\,19}{.067} = \frac{2.19 \times 10^{-3}}{6.7 \times 10^{-2}} = \frac{2.19}{6.7} \times \frac{10^{-3}}{10^{-2}}$$
$$= (.3269) \times (10^{-3+2}) = (3.269 \times 10^{-1}) \times 10^{-1}$$
$$= 3.269 \times 10^{-2}$$

(c)
$$\frac{4.139 \times 10^6}{7.12 \times 10^{11}} = \frac{4.139}{7.12} \times \frac{10^6}{10^{11}}$$
$$= (.5813) \times 10^{6-11} = (5.813 \times 10^{-1}) \times 10^{-5}$$
$$= 5.813 \times 10^{-6}$$

B. Common logarithms

When we are working with logarithms to the base 10 (common logarithms), we know that $\log 1 = 0$ and $\log 10 = 1$. If $1 < x < 10$, then we expect that $\log x$ lies between 0 and 1. Table A in the Appendix gives us four-place approximations to the common logarithms of all three-digit numbers between 1 and 10.

EXAMPLE 10-27 Use Table A to find (a) $\log 6.28$, (b) $\log 1.49$, (c) $\log 5.05$, (d) $\log 3$.

Solution

(a) In the row labeled 6.2, we move across to column 8 to find $\log 6.28 = .7980$.
(b) In the row labeled 1.4, we move across to column 9 to find $\log 1.49 = .1732$.
(c) In the row labeled 5.0, we move across to column 5 to find $\log 5.05 = .7033$.
(d) In the row labeled 3.0, we look in column 0 to find $\log 3 = .4771$.

But the table shows the common logarithms only for numbers between 1 and 10—to find the logarithms of *other* numbers, we need to use the fundamental properties of logarithms.

- If we are given any positive number x, then $x = b \times 10^n$, where $1 \leq b < 10$ and n is an integer. Then

$$\log x = \log(b \times 10^n)$$
$$= \log b + \log 10^n$$
$$= \log b + n \log 10$$

So
$$\log x = \log b + n$$

The number $\log b$ is called the **mantissa** of the number $\log x$ and the integer n is called the **characteristic** of the number $\log x$.

Since the value of b is between 1 and 10, the mantissa of $\log x$ is always between 0 and 1 and can be found in Table A.

EXAMPLE 10-28 Use Table A to find the values of (a) $\log 27.3$, (b) $\log 426$, (c) $\log .001\,51$.

Solution First write each number in scientific notation. Then use the rule $\log x = \log b + n$ and Table A.

(a) $\log 27.3 = \log(2.73 \times 10^1)$

$\qquad = \log 2.73 + \log 10^1$

$\qquad = \log 2.73 + 1(\log 10)$

$\qquad = \log 2.73 + 1 \cdot 1$

$\qquad = .4362 + 1$

$\qquad = 1.4362$

(b) $\log 426 = \log(4.26 \times 10^2)$

$\qquad = \log 4.26 + \log 10^2$

$\qquad = \log 4.26 + 2$

$\qquad = .6294 + 2$

$\qquad = 2.6294$

(c) $\log .001\,51 = \log(1.51 \times 10^{-3})$

$\qquad = \log 1.51 + \log 10^{-3} = \log 1.51 + (-3)$

$\qquad = .1790 + (-3)$

note: When the characteristic is negative, it's customary to keep the characteristic and mantissa separate.

We also need to perform the process shown above in reverse; i.e., we have to find *antilogarithms*:

- If x is any number and $\log y = x$, then the positive number y is called the **antilogarithm** of x.

EXAMPLE 10-29 Find y in each of the following:

$\qquad$ (a) $\log y = .3674$ $\qquad$ (b) $\log y = 2.6212$ $\qquad$ (c) $\log y = .8871 - 2$

Solution In the number on the right, identify the characteristic and mantissa. Find the mantissa in the body of Table A, determine which row and column it falls in, and record this. It will be between 1 and 10. Then adjust the decimal point to the right or left according to the size of the characteristic.

(a)
$$\log y = .3674 = .3674 + 0$$

The mantissa is .3674; the characteristic is 0. We find .3674 in the body of Table A in row 2.3 and column 3. The characteristic of 0 dictates no decimal movement. Hence the value of y is 2.33. We say that 2.33 is the antilog of .3674 exactly as .3674 is the log of 2.33.

(b)
$$\log y = 2.6212 = .6212 + 2$$

The mantissa is .6212; the characteristic is 2. From the table body we see that .6212 corresponds to 4.18. For the characteristic of 2, we move the decimal two places to the right to obtain $y = 418$. Thus 418 is the antilog of 2.6212 exactly as 2.6212 is the log of 418.

(c)
$$\log y = .8871 - 2$$

The mantissa is .8871; the characteristic is -2. From the table body we find that .8871 corresponds to 7.71. Now we move the decimal point two places to the left to get $y = .0771$. Thus .0771 is the antilog of .8871 $- 2$ exactly as .8871 $- 2$ is the log of .0771.

C. Linear interpolation

Table A gives the common logarithms for 3-digit numbers between 1 and 10. But if we wish to find the logarithm of a number made up of 4 or more digits—such as 2.322—we need to use a method of approximation called **linear interpolation**. To find a logarithm using linear interpolation, we follow these steps:

Step 1: Determine which two 3-digit numbers the number is *between.*
Step 2: Using Table A, find the logs of those two numbers.
Step 3: Determine at what proportion of the distance between the two numbers our number lies.
Step 4: Calculate the log that lies at that same proportion of the distance between the two logs. That's our answer!

EXAMPLE 10-30 Use linear interpolation to find an approximate value of log 2.322.

Solution Follow the steps outlined above.

Step 1: The number 2.322 is between the numbers 2.32 and 2.33.
Step 2: The log of 2.32 is .3655 and the log of 2.33 is .3674.
Step 3: Since $2.322 - 2.32 = .002$ and $2.33 - 2.32$ is .01, then 2.322 is $.002/.01 = \frac{2}{10}$ of the distance from 2.32 to 2.33.
Step 4: Therefore log 2.322 is $\frac{2}{10}$ of the distance from log 2.32 to log 2.33—that is, from .3655 to .3674. Since $\frac{2}{10}(.3674 - .3655) = .000\,38 \approx .0004$, we have

$$\log 2.322 \approx (\log 2.32) + .0004 \approx .3655 + .0004 = .3659$$

---

note: A hand-held calculator can usually find the logarithm of a number such as 2.322 quickly. But the technique of linear interpolation is useful because it can be extended to tables other than those of logarithms.

Linear interpolation can also be used to approximate *antilogarithms* of numbers not found in Table A. We simply use a modification of the procedure followed to approximate logarithms.

EXAMPLE 10-31 Use linear interpolation to find an approximate value of the antilogarithm of .8759.

Solution By Table A,

$$\log 7.51 = .8756 \qquad \log ?? = .8759 \qquad \log 7.52 = .8762$$

Since $.8762 - .8756 = .0006$ and $.8759 - .8756 = .0003$, it follows that $.8759$ is $.0003/.0006 = \frac{1}{2}$ of the distance from $.8756$ to $.8762$. We take the antilogarithm of $.8759$ to be $\frac{1}{2}$ of the distance from the antilogarithm of $.8756$ (i.e., 7.51) to the antilogarithm of $.8762$ (i.e., 7.52), so

$$\tfrac{1}{2}(7.52 - 7.51) = \tfrac{1}{2}(.01) = .005$$

Then the antilogarithm of $.8759$ is approximately

$$7.51 + .005 = 7.515$$

D. Using logarithms to evaluate expressions

We can often evaluate complicated expressions by using logarithms—and there's a procedure for this:

(1) Take the logarithms of both sides of the complicated expression.
(2) Apply the properties of logarithms.
(3) Perform the necessary arithmetic.
(4) Find the antilogarithm of the resulting expression.

note: In many cases hand-held calculators cannot be used to evaluate complicated expressions *directly*. But the calculator usually has log and antilog keys, which can be used to find logarithms without resorting to a table—and those logarithms can be used to find answers.

EXAMPLE 10-32 Use logarithms to evaluate

(a) $R = \sqrt[5]{40}$ (b) $T = \dfrac{(3.5)^{3/2}}{(1.36)^5}$ (c) $P = \sqrt[3]{450}\,\sqrt[4]{21.3}$

Solution Use a calculator or Table A to find the necessary logs and antilogs.

(a) Rewrite $R = \sqrt[5]{40}$ as $R = 40^{1/5}$. Then

$$\log R = \log 40^{1/5} \qquad \text{Step 1}$$
$$= \tfrac{1}{5}\log 40 \qquad \text{Step 2}$$
$$= \tfrac{1}{5}(\log 4 + \log 10)$$
$$= \tfrac{1}{5}(.6021 + 1) \qquad \text{Step 3}$$
$$= \tfrac{1}{5}(1.6021)$$
$$= .3204$$

But if $\log R = .3204$, then R is the antilog of $.3204$, which is

$$R = 2.092 \qquad \text{Step 4}$$

(b) Since $T = (3.5)^{3/2}/(1.36)^5$, then

$$\log T = \log\!\left(\frac{(3.5)^{3/2}}{(1.36)^5}\right) \qquad \text{Step 1}$$
$$= \log(3.5)^{3/2} - \log(1.36)^5 \qquad \text{Step 2}$$
$$= \tfrac{3}{2}(\log 3.5) - 5(\log 1.36)$$
$$= \tfrac{3}{2}(.5441) - 5(.1335) \qquad \text{Step 3}$$
$$= .816\,15 - .6675 = .148\,65$$

So $T = $ antilog $.148\,65$, or

$$T = 1.408 \qquad \text{Step 4}$$

(c) Rewrite $P = \sqrt[3]{450}\,\sqrt[4]{21.3}$ as $P = (450)^{1/3}(21.3)^{1/4}$. Then

$$\log P = \log(450)^{1/3} + \log(21.3)^{1/4}$$

$$= \tfrac{1}{3}(\log 450) + \tfrac{1}{4}(\log 21.3)$$

$$= \tfrac{1}{3}(2.6532) + \tfrac{1}{4}(1.3284)$$

$$= .8844 + .3321$$

$$= 1.2165$$

To find the antilog of 1.2165, we separate the number into a mantissa of $.2165$ and a characteristic of 1. Then

$$P = 1.6463 \times 10^1 = 16.463$$

EXAMPLE 10-33 Compute **(a)** $\log_2 9$ and **(b)** $\log_6 10$.

Solution Use the change of base formulas to convert the expressions to logarithms to base 10. Then use Table A or a calculator to compute the logarithms.

(a) $\log_2 9 = \dfrac{\log_{10} 9}{\log_{10} 2} = \dfrac{.9542}{.3010} = 3.17$

(b) $\log_6 10 = \dfrac{\log_{10} 10}{\log_{10} 6} = \dfrac{1}{.7782} = 1.285$

10-4. Applications of Exponential and Logarithmic Functions

There are many practical applications of exponential and logarithmic functions. They range from solving a mathematical equation to computing population growth and compound interest.

EXAMPLE 10-34 Solve for x in $8^{4x-1} = 6^{1-x}$.

Solution Take the logs of both sides to get

$$\log(8^{4x-1}) = \log(6^{1-x})$$

$$(4x - 1)\log 8 = (1 - x)\log 6$$

$$4x \log 8 - \log 8 = \log 6 - x \log 6$$

$$4x \log 8 + x \log 6 = \log 6 + \log 8$$

$$x(4 \log 8 + \log 6) = \log 6 + \log 8$$

$$x = \frac{\log 6 + \log 8}{4 \log 8 + \log 6}$$

This is the *exact* answer. To find its *approximate* value, we use a calculator or tables and compute:

$$x = \frac{.7782 + .9031}{4(.9031) + .7782} = \frac{1.6813}{4.3906} = .382\,93$$

note: Remember that logarithms are functions—except for the exact ones (e.g., $\log 1 = 0$), the numbers in log tables and in calculators are approximations.

EXAMPLE 10-35   If world population is increasing at an annual rate of 2.2%, then the exponential growth model for the population is $P(t) = p_0 e^{0.022t}$, where p_0 is an initial value of P at some starting time $t = 0$. According to this model, how long will it take for the world population to double?

Solution   Since p_0 is the beginning value, we need to find the value of time t for which the population P equals $2p_0$, i.e., when the population has doubled. We would then have

$$2p_0 = p_0 e^{.022t}$$
$$2 = e^{.022t}$$

Take the natural log of both sides:

$$\ln 2 = \ln e^{.022t} = (.022t)\ln e = (.022t) \cdot 1$$

or
$$t = \frac{\ln 2}{.022} = \frac{.6931}{.022} = 31.50$$

So, a population increasing at an annual rate of 2.2% will double in approximately 31.50 years.

EXAMPLE 10-36   Suppose a radioactive substance decays by the exponential decay model $Q = q_0 e^{-.35t}$, where q_0 is the amount of radioactive substance present at some starting time $t = 0$ and t is in months. In how many months will the original substance decay to one-half its original amount?

Solution   We need to find t in months for which $Q = .5q_0$; thus

$$.5q_0 = q_0 e^{-.35t}$$
$$.5 = e^{-.35t}$$
$$\ln .5 = \ln e^{-.35t} = (-.35t)\ln e$$
$$t = \frac{\ln .5}{-.35} = 1.98$$

So, in approximately 1.98 months, the substance will have decayed to one-half its original amount.

EXAMPLE 10-37   In a certain gas, the pressure p (in lbs/ft^3) and volume V (in ft^3) are related by the formula $pV^{1.4} = 800$. Find the pressure if the volume V is 6.52 ft^3.

Solution   Solve the equation for p and insert the given value of V to get

$$p = 800V^{-1.4} = 800(6.52)^{-1.4}$$

Now use logarithms to perform this computation:

$$\log p = \log 800 + \log(6.52)^{-1.4}$$
$$= \log 800 - 1.4\log(6.52) = 2.9031 - (1.4)(.8142) = 1.7632$$

Thus $p \doteq \text{antilog}(.7632 + 1) = 58$

If $A(t)$ denotes the amount of a quantity that is growing exponentially at the rate of r percent per year and A_0 is the original amount at time $t = 0$, then

$$A(t) = A_0\left(1 + \frac{r}{100}\right)^t$$

This formula may be used in a number of applications.

EXAMPLE 10-38　If Emily puts \$1000 into a money market account that pays interest of 11% compounded annually, how much will the account be worth in 15 years?

Solution　The given conditions are $A_0 = 1000$, $r = 11$, $t = 15$, so that

$$A(t) = 1000\left(1 + \frac{11}{100}\right)^{15} = 1000(1.11)^{15}$$

Use logarithms to compute the value of A:

$$\log A = \log[1000(1.11)^{15}] = \log 1000 + \log(1.11)^{15}$$

$$= 3 + 15\log(1.11) = 3 + 15(.0453) = 3.6795$$

$$A = \text{antilog}(.6795 + 3) = 4781$$

So, Emily will have \$4781 in her account in 15 years.

EXAMPLE 10-39　If the population of Bigtown is growing at the rate of 5% per year and 6000 people live there now, what will the population be 10 years from now?

Solution　Since $A_0 = 6000$, $r = 5$, and $t = 10$,

$$A = 6000\left(1 + \frac{5}{100}\right)^{10} = 6000(1.05)^{10}$$

Then

$$\log A = \log 6000 + 10\log(1.05)$$

$$= \log 6 + \log 1000 + 10\log(1.05) = .7782 + 3 + 10(.0212)$$

$$= 3.9902$$

$$A = \text{antilog}(.9902 + 3) = 9778$$

So there will be 9778 people in Bigtown 10 years from now.

EXAMPLE 10-40　The length L (in centimeters) of a metal rod varies with temperature T (in degrees Celsius) according to the formula $L = 60e^{.000015T}$ Find the length of the rod at $T = 100°C$.

Solution

$$L = 60e^{.000015T} = 60e^{(.000015)(100)} = 60e^{.0015}$$

$$\ln L = \ln 60 + \ln(e^{.0015}) = \ln 60 + .0015\ln e$$

$$= 4.0943 + .0015 = 4.0958$$

$$L = 60.0874$$

So, when $T = 100°C$, the length of the rod is approximately 60.0874 cm.

note:　Given $\ln L = 4.0958$, you can find L in one of two ways. On a calculator, you press the inverse function key and then the ln key—which is how we arrived at 60.0874 above. If you use a table of common logarithms, you must first change the base by the formula $\ln L = (\log_{10}L)/(\log_{10}e) \approx (\log L)/(.4343)$. Then

$$\log L = (.4343)(4.0958) \approx 1.7788$$

and

$$L = \text{antilog}(.7788 + 1) \approx 60.09$$

SUMMARY

1. An exponential function is a function of the form $f(x) = a^x$, where the base a is a positive constant not equal to 1.
2. The exponential function $y = e^x$, where $e = 2.718\ldots$, is called the natural exponential function.
3. A logarithmic function is defined with reference to an exponential function, such that

$$\log_a y = x \text{ if and only if } a^x = y$$

4. Every logarithmic function is the inverse of an exponential function.
5. Three laws of logarithms are fundamental for all work involving logarithms:

$$\log_a(xy) = \log_a x + \log_a y$$

$$\log_a\left(\frac{x}{y}\right) = \log_a x - \log_a y$$

$$\log_a x^n = n \log_a x$$

6. To change the base of a logarithm, two formulas are used:

$$\log_a x = \frac{\log_b x}{\log_b a} \quad \text{and} \quad \log_a b = \frac{1}{\log_b a}$$

7. The logarithmic function $\log_e x$ with base e is called the natural logarithmic function and is abbreviated as $\ln x$.
8. The logarithmic function $\log_{10} x$ with base 10 is called the common logarithmic function and is abbreviated as $\log x$.
9. Given $\log x = \log b + n$, the number $\log b$ is called the mantissa of the number $\log x$ and the integer n is called the characteristic of the number $\log x$.
10. If x is any number and $\log y = x$, then the positive number y is called the antilogarithm of x.

RAISE YOUR GRADES

Can you...?

☑ define and graph an exponential function
☑ use the rules of exponents to simplify expressions
☑ define the number e and give its value to four decimal places
☑ define a logarithmic function
☑ explain the relationship between an exponential function and a logarithmic function
☑ draw the graph of a logarithmic function and its associated exponential function
☑ use the definition of a logarithm and the three fundamental laws of logarithms to compute with logarithms
☑ give a number in scientific notation
☑ use a table of common logarithms
☑ define the characteristic and mantissa of a logarithm
☑ give the definition of an antilogarithm

SOLVED PROBLEMS

Exponential Functions

PROBLEM 10-1   Use the rules of exponents to simplify **(a)** $\dfrac{(5)^0(3)^{-\sqrt{12}}}{\left((2)^{\sqrt{3}}\right)^{-2}}$,   **(b)** $\left(\frac{1}{5}\right)^{\sqrt{2}}(10)^{\sqrt{8}}(17)^0$.

Solution   **(a)** $\dfrac{(5^0)(3)^{-\sqrt{12}}}{\left((2)^{\sqrt{3}}\right)^{-2}} = \dfrac{(1)(3)^{-\sqrt{12}}}{\left((2)^{\sqrt{3}}\right)^{-2}}$ $[a^0 = 1]$

$\qquad\qquad = \dfrac{\left((2)^{\sqrt{3}}\right)^2}{(3)^{\sqrt{12}}}$ $\left[a^{-x} = \dfrac{1}{a^x}\right]$

$\qquad\qquad = \dfrac{(2)^{2\sqrt{3}}}{(3)^{2\sqrt{3}}}$ $[(a^x)^y = a^{xy}]$ *Recall:* $\sqrt{12} = \sqrt{4\cdot3} = \sqrt{4}\cdot\sqrt{3} = 2\sqrt{3}$

$\qquad\qquad = \left(\dfrac{2}{3}\right)^{2\sqrt{3}}$ $\left[\left(\dfrac{a}{b}\right)^x = \dfrac{a^x}{b^x}\right]$

(b) $\left(\frac{1}{5}\right)^{\sqrt{2}}(10)^{\sqrt{8}}(17)^0 = \left(\frac{1}{5}\right)^{\sqrt{2}}(10)^{\sqrt{8}}(1)$ $[a^0 = 1]$

$\qquad = (5)^{-\sqrt{2}}(10)^{2\sqrt{2}}$ $[a^{-x} = 1/a^x]$ *Recall:* $\sqrt{8} = \sqrt{4\cdot2} = 2\sqrt{2}$

$\qquad = (5)^{-\sqrt{2}}(5)^{2\sqrt{2}}(2)^{2\sqrt{2}}$ $[(ab)^x = a^x b^x]$

$\qquad = (5)^{\sqrt{2}}(2)^{2\sqrt{2}}$ $[a^x a^y = a^{x+y}]$

$\qquad = (5)^{\sqrt{2}}(2^2)^{\sqrt{2}}$ $[(a^x)^y = a^{xy}]$

$\qquad = (5)^{\sqrt{2}}(4)^{\sqrt{2}} = (20)^{\sqrt{2}}$ $[(ab)^x = a^x b^x]$

PROBLEM 10-2   Solve the equation $x^2 2^x - 2^x = 0$.

Solution   Since 2^x is common to both terms, you can factor it out:

$$x^2 2^x - 2^x = 2^x(x^2 - 1) = 0$$

But 2^x is never equal to 0, so you can divide both sides of the equation by 2^x:

$$\frac{2^x(x^2 - 1)}{2^x} = \frac{0}{2^x}$$

$$x^2 - 1 = (x - 1)(x + 1) = 0 \qquad \text{Factor the difference of two squares.}$$

Thus $\qquad\qquad x = 1 \quad \text{or} \quad x = -1$

PROBLEM 10-3   Sketch the graph of $y = 3^{1-x} - 4$.

Solution   First revise the function to a simpler form:

$$y = 3^{1-x} - 4 = (3^1)(3^{-x}) - 4$$

$$= (3)\left(\frac{1}{3^x}\right) - 4 = 3\left(\frac{1}{3}\right)^x - 4$$

To graph $3(\frac{1}{3})^x - 4$, take the graph of $(\frac{1}{3})^x$ (shown in Figure 10-3), "stretch" it by a factor of 3, and shift it downward 4 units, as shown in Figure 10-10.

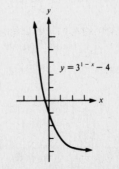

$y = 3^{1-x} - 4$

Figure 10-10

PROBLEM 10-4 Where does the graph of $y = 3^{1-x} - 4$ cross the *x*-axis? the *y*-axis?

Solution The *x*-intercept is obtained by setting $y = 0$ and solving for *x*:

$$0 = 3^{1-x} - 4$$

$$4 = 3^{1-x} = (3^1)(3^{-x}) = 3(3^{-x})$$

$$\frac{4}{3} = \frac{1}{3^x}$$

$$4(3^x) = 3$$

$$3^x = \tfrac{3}{4}$$

So
$$\log_3(3^x) = \log_3(\tfrac{3}{4})$$

$$x(\log_3 3) = \log_3(\tfrac{3}{4})$$

$$x = \log_3(\tfrac{3}{4})$$

note: By using the change of base formulas and Table A or a calculator, you'll find that the *x*-intercept is approximately $-.26$.

The *y*-intercept is obtained by setting $x = 0$ and solving for *y*:

$$y = 3^{1-0} - 4 = 3 - 4 = -1$$

See Figure 10-10.

PROBLEM 10-5 If the bacteria in a colony have unlimited nutrients and unlimited space, the bacteria increase in number according to the growth law

$$N = N_0 e^{kt}$$

where *N* is the number at any time *t*, N_0 is the number at the beginning of the experiment, and *k* is a positive constant related to the bacteria. If a particular species begins with 800 bacteria and if 5 hours later there are 2400, how many will there be after 15 hours?

Solution You are given $N_0 = 800$. Thus the growth equation is

$$N = N_0 e^{kt} = 800 e^{kt}$$

At time $t = 5$, the value of *N* is 2400. Thus

$$2400 = 800 e^{k(5)}$$

$$3 = e^{k(5)} = (e^k)^5$$

$$3^{1/5} = e^k$$

Back to the original growth equation ...

$$N = 800 e^{kt} = 800(e^k)^t = 800(3^{1/5})^t = 800(3^{t/5})$$

Since you want the value of *N* when $t = 15$ hours, put $t = 15$ into this equation:

$$N = 800(3^{t/5}) = 800(3^{15/5}) = 800(3^3) = 800(27) = 21\,600$$

So, after 15 hours, you will have 21 600 bacteria in your experimental colony.

PROBLEM 10-6 One definition of the constant *e* is the value to which the sequence $(1 + \tfrac{1}{n})^n$ goes as *n* grows unboundedly large. Verify this for some carefully chosen values of *n*.

Solution Make a chart and use a good calculator:

n	2	8	100	10 000	1 000 000	$\to \infty$
$\left(1 + \dfrac{1}{n}\right)^n$	$\left(\dfrac{3}{2}\right)^2$	$\left(\dfrac{9}{8}\right)^8$	$\left(\dfrac{101}{100}\right)^{100}$	$\left(\dfrac{10\,001}{10\,000}\right)^{10\,000}$	$\left(\dfrac{1\,000\,001}{1\,000\,000}\right)^{1\,000\,000}$	
Decimal value	2.25	2.5658	2.7048	2.7181	2.718 28	$\to e$

PROBLEM 10-7   If the population of a town is modeled by the growth equation

$$P = 21\,500e^{t/4}$$

where t is measured in years, what is the population (**a**) at time $t = 0$? (**b**) 4 years later? (**c**) 2 years earlier?

Solution

(**a**) To find P at $t = 0$, substitute $t = 0$ into the growth equation:

$$P = 21\,500e^{t/4} = 21\,500e^{0/4} = 21\,500e^0 = 21\,500$$

(**b**) Four years later $t = 4$, so

$$P = 21\,500e^{4/4} = 21\,500e^1 \simeq 21\,500(2.718\,28) = 58\,443$$

(**c**) Two years earlier $t = -2$, so

$$P = 21\,500e^{-2/4} = 21\,500e^{-1/2} \simeq 21\,500(.6065) = 13\,040$$

PROBLEM 10-8   Use a calculator to find the approximate value of

(**a**) π^π (**b**) $\pi^{\sqrt{2}}$ (**c**) $(\sqrt{2})^\pi$ (**d**) $(1.01)^{10.3}$

Solution   Use the $\boxed{y^x}$ key on the calculator:

(**a**) $\pi^\pi = 36.4622$ (**b**) $\pi^{\sqrt{2}} = 5.0475$ (**c**) $\sqrt{2}^\pi = 2.9707$ (**d**) $(1.01)^{10.3} = 1.1079$

note: It often helps to read the instructions that came with your calculator.

PROBLEM 10-9   Solve (**a**) $x^2e^x - 9e^x = 0$ for x and (**b**) $t^2e^t + te^t = 2e^t$ for t.

Solution

(**a**) Since e^x is in each term, you factor it out: (**b**)

$$t^2e^t + te^t = 2e^t$$

$$x^2e^x - 9e^x = 0$$

$$t^2e^t + te^t - 2e^t = 0$$

$$e^x(x^2 - 9) = 0$$

$$e^t(t^2 + t - 2) = 0$$

$$e^x(x - 3)(x + 3) = 0$$

$$e^t(t + 2)(t - 1) = 0$$

Then, since e^x is never zero,   Since e^t is never zero, $t = -2$ or $t = 1$.
$x = 3$ or $x = -3$.

PROBLEM 10-10   Solve the following equations for x:

(**a**) $3^{2x} = 3^{4x^2/3}$ (**b**) $(2^x)^{x+2} = 2^{3x+6}$ (**c**) $4^{\sqrt{x+3}} = 2^{2x+1}$

Solution   Use the rule that if $a^x = a^y$, then $x = y$ if $a \neq 1$:

(**a**)

$$3^{2x} = 3^{4x^2/3} \implies 2x = \frac{4x^2}{3}$$

$$6x = 4x^2$$

$$4x^2 - 6x = 0$$

$$x(4x - 6) = 0$$

$$x = 0 \quad \text{or} \quad x = \tfrac{3}{2}$$

(**b**)

$$(2^x)^{x+2} = 2^{3x+6}$$

$$2^{x^2+2x} = 2^{3x+6} \implies x^2 + 2x = 3x + 6$$

$$x^2 - x - 6 = 0$$

$$(x - 3)(x + 2) = 0$$

$$x = 3 \quad \text{or} \quad x = -2$$

(c)
$$4^{\sqrt{x+3}} = 2^{2x+1}$$

$$2^{2\sqrt{x+3}} = 2^{2x+1} \implies \quad 2\sqrt{x+3} = 2x+1$$

$$4(x+3) = (2x+1)^2$$

$$4x + 12 = 4x^2 + 4x + 1$$

$$11 = 4x^2$$

$$x^2 = 11/4$$

$$x = \sqrt{11}/2 \quad \text{or} \quad x = -\cancel{\sqrt{11}/2} \qquad [x = -\sqrt{11}/2 \text{ not valid}]$$

Logarithmic Functions

PROBLEM 10-11 Evaluate (a) $\log_2 16$, (b) $\log_5 5$, (c) $\log_{16}(\tfrac{1}{4})$.

Solution Set each expression equal to x. Then convert the logarithmic equation to an exponential equation, using $\log_a y = x$ if and only if $a^x = y$.

(a) If $x = \log_2 16$, then $2^x = 16$. Since $2^4 = 16$, $x = 4$.
(b) If $x = \log_5 5$, then $5^x = 5$. Since $5^1 = 5$, $x = 1$.
(c) If $x = \log_{16}(\tfrac{1}{4})$, then $16^x = \tfrac{1}{4}$. Since $16 = 4^2$, you have $(4^2)^x = 4^{-1}$, so $2x = -1$ or $x = -\tfrac{1}{2}$.

PROBLEM 10-12 Solve (a) $\log_8 x = \tfrac{1}{3}$, (b) $\log_3(x-3) = 2$, (c) $\log_{\sqrt{5}} x = 0$ for x.

Solution Convert each logarithmic equation to the exponential equivalent and solve:

(a) If $\log_8 x = \tfrac{1}{3}$, then $8^{1/3} = x$. Thus $x = \sqrt[3]{8} = 2$.
(b) If $\log_3(x-3) = 2$, then $3^2 = x - 3$ or $9 = x - 3$. Thus $x = 12$.
(c) If $\log_{\sqrt{5}} x = 0$, then $(\sqrt{5})^0 = x$. Thus $x = 1$.

PROBLEM 10-13 Solve (a) $\log_x 49 = 2$, (b) $\log_{x^2} 16 = 1$, (c) $\log_{x+1} 27 = 3$ for x.

Solution

(a) If $\log_x 49 = 2$, then $x^2 = 49$ or $x = 7$. (Since the base of a logarithm must be positive, you can exclude the solution $x = -7$.)
(b) If $\log_{x^2} 16 = 1$, then $(x^2)^1 = 16$ or $x = 4$ or $x = -4$. (In this case, the solution $x = -4$ is allowed, since the base is x^2, not x.)
(c) If $\log_{x+1} 27 = 3$, then $(x+1)^3 = 27$ or $x + 1 = 3$. Thus $x = 2$.

PROBLEM 10-14 Write (a) $\log_7\left(\dfrac{49}{\sqrt{7}}\right)$ and (b) $\log_3\left(\dfrac{7}{2}\right) + \log_3\left(\dfrac{4}{3}\right) - \log_3\left(\dfrac{11}{6}\right)$ in simpler logarithmic form.

Solution Use the laws of logarithms to simplify the expressions.

(a)
$$\log_7\left(\frac{49}{\sqrt{7}}\right) = \log_7 49 - \log_7\sqrt{7} \qquad [\log_a(\tfrac{x}{y}) = \log_a x - \log_a y]$$

$$= \log_7(7)^2 - \log_7(7)^{1/2}$$

$$= 2\log_7(7) - \tfrac{1}{2}\log_7(7) \qquad [\log_a x^n = n\log_a x]$$

$$= 2(1) - \tfrac{1}{2}(1)$$

$$= \tfrac{3}{2}$$

(b)
$$\log_3(\tfrac{7}{2}) + \log_3(\tfrac{4}{3}) - \log_3(\tfrac{11}{6}) = \log_3\left(\frac{(\tfrac{7}{2})(\tfrac{4}{3})}{\tfrac{11}{6}}\right) \qquad \left[\begin{matrix} \log_a(xy) = \log_a x + \log_a y; \\ \log_a(\tfrac{x}{y}) = \log_a x - \log_a y \end{matrix}\right]$$

$$= \log_3\left(\frac{\tfrac{28}{6}}{\tfrac{11}{6}}\right) = \log_3(\tfrac{28}{11})$$

PROBLEM 10-15   Solve for x in the following equations:

(a) $\log_4(x - 1) + \log_4(x + 1) = 3$ (b) $-3 \ln x = \ln 8$

Solution

(a)
$$\log_4(x - 1) + \log_4(x + 1) = 3$$
$$\log_4[(x - 1)(x + 1)] = 3 \qquad [\log_a(xy) = \log_a x + \log_a y]$$

Then
$$4^3 = x^2 - 1 \qquad \text{Convert to exponential form.}$$
$$x^2 = 64 + 1 = 65$$
$$x = +\sqrt{65}$$

(Eliminate the solution $x = -\sqrt{65}$, since both $x - 1$ and $x + 1$ would then be negative—and the logs of negative numbers are not defined.)

(b)
$$-3 \ln x = \ln 8$$
$$\ln x^{-3} = \ln 8 \qquad [\log_a x^n = n \log_a x]$$
$$x^{-3} = 8 \qquad [\log_a x = \log_a z \text{ if and only if } x = z]$$
$$x^3 = \tfrac{1}{8}$$
$$x = \tfrac{1}{2}$$

PROBLEM 10-16   Sketch the graph of $y = \log_2(x + 2)$.

Solution   The graph of $y = \log_2(x + 2)$ is simply the graph of $y = \log_2 x$ (Fig. 10-7) shifted two units to the left, as in Figure 10-11.

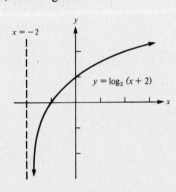

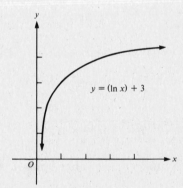

Figure 10-11   **Figure 10-12**

PROBLEM 10-17   Sketch the graph of $y = \ln x + 3$.

Solution   Notice that this is NOT $y = \ln(x + 3)$ but $y = (\ln x) + 3$. Draw the graph of $y = \ln x$ (see Fig. 10-9a) and shift it three units upward, as in Figure 10-12.

PROBLEM 10-18   Find the value(s) of x in the following equations:

(a) $\log_3 3 + \log_3(x^2 - 3x - 1) = 2$ (b) $\ln(x - 1) = \tfrac{1}{2}\ln(x + 11)$ (c) $\log_x x^{4x} = 12$

Solution

(a)
$$\log_3 3 + \log_3(x^2 - 3x - 1) = 2$$
$$\log_3[3(x^2 - 3x - 1)] = 2 \qquad [\log_a(xy) = \log_a x + \log_a y]$$

Then
$$3^2 = 3x^2 - 9x - 3 \qquad \text{Convert to exponential form.}$$
$$3x^2 - 9x - 12 = 0$$
$$3(x^2 - 3x - 4) = 0$$
$$3(x - 4)(x + 1) = 0$$

So
$$x = 4 \quad \text{or} \quad x = -1$$

(b)

$$\ln(x - 1) = \tfrac{1}{2}\ln(x + 11)$$

$$2\ln(x - 1) = \ln(x + 11) \qquad \text{Multiply both sides by 2.}$$

$$\ln(x - 1)^2 = \ln(x + 11)$$

Then

$$(x - 1)^2 = x + 11 \qquad \text{If } \ln A = \ln B, \text{ then } A = B.$$

$$x^2 - 2x + 1 = x + 11$$

$$x^2 - 3x - 10 = 0$$

$$(x - 5)(x + 2) = 0$$

So

$$x = 5 \quad \text{or} \quad x \ne -2 \qquad \ln(x - 1) = \ln(-3) \text{ is not defined.}$$

(c)

$$\log_x x^{4x} = 12$$

$$x^{12} = x^{4x}$$

$$4x = 12 \quad \text{so} \quad x = 3$$

Computation with Logarithms

PROBLEM 10-19   Evaluate **(a)** $(.0041)(258\,000)$, **(b)** $.397 \div .005\,81$, and **(c)** $(5.27 \times 10^{25}) \div (2.08 \times 10^{11})^2$, giving the result in scientific notation.

Solution

(a) $(.0041)(258\,000) = (4.1 \times 10^{-3})(2.58 \times 10^5) = (4.1)(2.58)(10^{-3} \times 10^5) = (10.578)(10^2)$

$$= 1.0578 \times 10^3$$

(b) $(.397) \div (.005\,81) = (3.97 \times 10^{-1}) \div (5.81 \times 10^{-3}) = \left(\dfrac{3.97}{5.81}\right)\left(\dfrac{10^{-1}}{10^{-3}}\right) = .6833 \times 10^2$

$$= 6.833 \times 10$$

(c) $\dfrac{5.27 \times 10^{25}}{(2.08 \times 10^{11})^2} = \dfrac{5.27 \times 10^{25}}{4.3264 \times 10^{22}} = 1.2181 \times 10^3$

PROBLEM 10-20   Use Table A in the Appendix to find approximate values for **(a)** $\log 2.73$, **(b)** $\log 6.09$, **(c)** $\log 9.0$.

Solution

(a) In the row labeled 2.7, move across to column 3 and find $\log 2.73 = .4362$.
(b) $\log 6.09 = .7846$
(c) $\log 9.0 = .9542$

PROBLEM 10-21   Use Table A to find approximate values for **(a)** $\log 41.8$, **(b)** $\log .38$, **(c)** $\log 5430$.

Solution   First write the number in scientific notation; then use the following rule:

If $x = b \times 10^n$, where $1 \le b \le 10$ and n is an integer, then $\log x = \log b + n$

(a) $\log 41.8 = \log(4.18 \times 10^1)$ **(b)** $\log .38 = \log(3.8 \times 10^{-1})$ **(c)** $\log 5430 = \log(5.43 \times 10^3)$

$\qquad = (\log 4.18) + 1 \qquad\qquad = (\log 3.8) + (-1) \qquad\qquad = (\log 5.43) + 3$

$\qquad = .6212 + 1 \qquad\qquad\quad = .5798 + (-1)) \qquad\qquad = .7348 + 3$

$\qquad = 1.6212 \qquad\qquad\qquad\qquad = 3.7348$

PROBLEM 10-22   Find the value of x for **(a)** $\log x = 1.6484$, **(b)** $\log x = .8299 - 2$, **(c)** $\log x = 4.7443$.

Solution In each case, you must find the antilogarithm of the given number.

(a) For $\log x = 1.6484$, you have a characteristic of 1 and a mantissa of .6484. Referring to the body of Table A, you see that the mantissa .6484 is the logarithm of 4.45. Adjust the decimal point placement according to the value of the characteristic—to the right if it's positive; to the left if it's negative. Here, the characteristic is positive 1, so $x = 44.5$.

(b) For $\log x = .8299 - 2$, locate the mantissa .8299 in the body of Table A and obtain the value 6.76. Since the characteristic is -2, you move the decimal point two places to the left to get $x = .0676$.

(c) For $\log x = 4.7443$, the mantissa .7443 is the logarithm of 5.55. And since the characteristic is 4, $x = 55\,500$.

PROBLEM 10-23 Find the logarithm of 15 340 000.

Solution First rewrite the number in scientific notation:

$$15\,340\,000 = 1.534 \times 10^7$$

Then, $\log(1.534 \times 10^7) = 7 + \log 1.534$. The number 1.534 has more than 3 significant digits, so you won't find its log in Table A. You'll need to use linear interpolation to find an approximation of $\log 1.534$.

Step 1: The number 1.534 is between the numbers 1.53 and 1.54.
Step 2: The log of 1.53 is .1847 and the log of 1.54 is .1875.
Step 3: Since $1.534 - 1.53 = .004$ and $1.54 - 1.53 = .01$, then 1.534 is $.004/.01 = \frac{4}{10}$ of the distance from 1.53 to 1.54.
Step 4: Therefore $\log 1.534$ is $\frac{4}{10}$ of the distance from $\log 1.53$ to $\log 1.54$—that is, from .1847 to .1875. Since $\frac{4}{10}(.1875 - .1847) = .001\,12 \approx .0011$, you have

$$\log 1.534 \approx (\log 1.53) + .0011 \approx .1847 + .0011 = .1858$$

Consequently

$$\log 15\,340\,000 = 7 + \log 1.534 \approx 7 + .1858 = 7.1858$$

PROBLEM 10-24 Find the antilogarithm of $-.6123$.

Solution First rewrite the logarithm in standard form so that the mantissa is between 0 and 1:

$$-.6123 = .3877 - 1$$

Then, by Table A,
$$\log 2.44 = .3874$$
$$\log ?? = .3877$$
$$\log 2.45 = .3892$$

Since $.3892 - .3874 = .0018$ and $.3877 - .3874 = .0003$, it follows that .3877 is $.0003/.0018 = \frac{1}{6}$ of the distance from .3874 to .3892. The antilogarithm of .3877 is $\frac{1}{6}$ of the distance from the antilogarithm of .3874 (i.e., 2.44) to the antilogarithm of .3892 (i.e., 2.45), so

$$\tfrac{1}{6}(2.45 - 2.44) = \tfrac{1}{6}(.01) \approx .0017$$

Then the antilogarithm of .3877 is approximately $2.44 + .0017 = 2.4417 \approx 2.442$. But the characteristic is -1, so the antilogarithm of $-.6123 = .3877 - 1 \approx .2442$.

PROBLEM 10-25 Use logarithms to find values for **(a)** $A = \sqrt[7]{75}$, **(b)** $P = \sqrt[3]{22.5}\sqrt[4]{8.12}$, **(c)** $(80)^{1.7} \div (6.2)^{.9}$.

Solution

(a) Rewrite $\quad\quad A = \sqrt[7]{75} = (75)^{1/7}$

Then $\quad\quad \log A = \log(75)^{1/7} = \tfrac{1}{7}\log 75 = \tfrac{1}{7}(1.8571) = .2679$

So $\quad\quad A = \text{antilog}(.2679) = 1.853$

note: If you're using Table A, you can find antilog(.2679) as follows:

$$\log 1.85 = .2672$$

$$\log A = .2679$$

$$\log 1.86 = .2695$$

Then, since $.2695 - .2672 = .0023$ and $.2679 - .2672 = .0007$, it follows that .2679 is $\frac{7}{23}$ of the distance between .2672 and .2695. So $\frac{7}{23}(1.86 - 1.85) = \frac{7}{23}(.01) = .003$ and

$$\text{antilog}(.2679) = 1.85 + .003 = 1.853$$

(b) $$P = \sqrt[3]{22.5}\sqrt[4]{8.12} = (22.5)^{1/3}(8.12)^{1/4}$$

$$\log P = \log[(22.5)^{1/3}(8.12)^{1/4}]$$

$$= \log(22.5)^{1/3} + \log(8.12)^{1/4} = \tfrac{1}{3}\log(22.5) + \tfrac{1}{4}\log(8.12) = \tfrac{1}{3}(1.3522) + \tfrac{1}{4}(.9096)$$

$$= .6781$$

$$P = \text{antilog}(.6781) = 4.7656$$

note: $\text{antilog}(.6781) = \left(\dfrac{.6781 - .6776}{.6785 - .6776}\right)(4.77 - 4.76) + 4.76 = .56(.01) + 4.76 = 4.7656$

where $\log 4.76 = .6776$ and $\log 4.77 = .6785$.

(c) $$Q = (80)^{1.7} \div (6.2)^{.9} = \frac{(80)^{1.7}}{(6.2)^{.9}}$$

$$\log Q = \log\left(\frac{(80)^{1.7}}{(6.2)^{.9}}\right) = \log(80)^{1.7} - \log(6.2)^{.9}$$

$$= 1.7(\log 80) - .9(\log 6.2) = 1.7(1.9031) - .9(.7924) = 3.235\,27 - .713\,16$$

$$\doteq 2.5221$$

$$Q = \text{antilog}(.5221 + 2) = 332.77$$

note: $\text{antilog}(.5221 + 2) = \left[\left(\dfrac{\log Q - \log 332}{\log 333 - \log 332}\right)(333 - 332) + 332\right]10^2$

$$= \left[\left(\dfrac{2.5221 - 2.5211}{2.5224 - 2.5211}\right)(1) + 332\right]10^2 = 332.77$$

PROBLEM 10-26 Use the change of base formula to find **(a)** $\log_4 23$ and **(b)** $\log_{16} 71$.

Solution Since $\log_a x = \dfrac{\log_b x}{\log_b a}$, rewrite the expressions, using $b = 10$.

(a) $$\log_4 23 = \frac{\log_{10} 23}{\log_{10} 4} = \frac{1.3617}{0.6021} = 2.2616$$

This means that $4^{2.2616} \approx 23$.

(b) $$\log_{16} 71 = \frac{\log_{10} 71}{\log_{10} 16} = \frac{1.8513}{1.2041} = 1.5375$$

Thus $16^{1.5375} \approx 71$.

note: As a check, recall that $16^{1.5} = 16^{3/2} = (16^{1/2})^3 = 4^3 = 64$—and 64 slightly less than 71. [This kind of quick mental check can be very useful—it lets you see if your answer is "reasonable" in a hurry.]

Applications of Exponential and Logarithmic Functions

PROBLEM 10-27 If you invest P dollars for n years at interest rate i compounded k times a year, then the amount A after n years is

$$A = P\left(1 + \frac{i}{k}\right)^{nk}$$

If the interest is compounded continuously, the formula is

$$A = Pe^{in}$$

Assume you have \$1000 and can invest it for 5 years at 10%. Find the value of the account if (a) interest is compounded annually, (b) interest is compounded quarterly, (c) interest is compounded continuously.

Solution

(a) Compounding annually (once a year) means $k = 1$. The other known values are $P = 1000$, $n = 5$, and $i = .10$; so

$$A = P\left(1 + \frac{i}{k}\right)^{nk} = 1000(1 + .10)^5 = 1000(1.1)^5 = 1000(1.610\,51) = \$1610.51$$

(b) Compounding quarterly (4 times a year) means $k = 4$. The other values remain the same. Thus

$$A = 1000\left(1 + \frac{.10}{4}\right)^{5 \cdot 4} = 1000(1.025)^{20} \approx 1000(1.638\,62) = \$1638.62$$

(c) Compounding continuously, you use $P = 1000$, $n = 5$, and $i = .10$ to get

$$A = Pe^{in} = 1000e^{(.10)(5)} = 1000e^{1/2} \approx 1000(1.648\,72) = \$1648.72$$

PROBLEM 10-28 Your boss gives you a sheet of paper 1 meter square. The thickness of the paper is .005 centimeters (that's .000 05 meters). You are told to cut the sheet in half and stack the resulting sheets. Then you must cut that stack of two half-sheets in half again and stack those sheets. Continue this procedure 24 times, ending up with a single stack of paper. How high is the final stack?

Solution You need to count the sheets in the final stack and multiply that number by the thickness of the paper. Make a chart to calculate the number of sheets:

Cuts	0	1	2	3	4	5	10	...	24
Sheets	1	2	4	8	2^4	2^5	2^{10}	...	2^{24}

The number of sheets in the resulting stack is $N = 2^{24}$, which you can find by using logarithms:

$$\log N = \log 2^{24} = 24 \log 2 = 24(.3010) = 7.2240$$

By Table A, $\log 1.67 = .2227$ and $\log 1.68 = .2253$. Therefore

$$N \approx \text{antilog}(.2240 + 7) \approx \left[\left(\frac{.2240 - .2227}{.2253 - .2227}\right)(1.68 + 1.67) + 1.67\right]10^7 \approx 16\,750\,000 \text{ sheets}$$

note: This logarithmic result is only approximate. The calculator answer for 2^{24} is $16\,777\,216$.

You have more than 16 million sheets, so the height in meters is

$$N(.000\,05) = (16\,750\,000)(.000\,05) = (1.675 \times 10^7)(5 \times 10^{-5}) = 8.375 \times 10^2 = 837.5 \text{ meters}$$

That's about 2747 feet—more than twice the tallest building in the world! (Make sure you get time and a half!)

PROBLEM 10-29   In problems of radioactive decay, the term "half-life" refers to the amount of time it takes for exactly half of the substance to disappear by the decay process. The decay equation is

$$N = N_0 e^{kt}$$

where k is a negative constant, t is a measure of time, and N_0 is the initial mass of the substance. The half-life of radium-226 is 1620 years. (a) Find the decay constant k. (b) If an initial sample of 1 gram is left for 100 years, how much will remain?

Solution   (a) The amount N remaining after $t = 1620$ years is $\frac{1}{2}N_0$. Thus

$$\tfrac{1}{2}N_0 = N_0 e^{k(1620)}$$

$$\tfrac{1}{2} = e^{1620k}$$

$$\ln\left(\tfrac{1}{2}\right) = \ln e^{1620k} = 1620k(\ln e) = 1620k$$

$$k = \frac{1}{1620}\ln\left(\frac{1}{2}\right) = \frac{1}{1620}(\ln 1 - \ln 2)$$

or

$$k \approx \frac{1}{1620}(0 - .6931) = -.000\,427\,8$$

(b) The decay constant k is $-.000\,427\,8$, so the decay equation is

$$N = N_0 e^{(-.000\,427\,8)t}$$

After $t = 100$ years, the number of grams remaining is

$$N = (1)e^{(-.000\,427\,8)(100)} = e^{-.042\,78}$$

Using a calculator with an $\boxed{e^x}$ key gives

$$N = .958\,12 \text{ g}$$

Supplementary Exercises

PROBLEM 10-30   Simplify the expressions (a) $3^{\sqrt{2}} \cdot 3^{\sqrt{50}}$ and (b) $(7^{\sqrt{2}})^{\sqrt{8}}$.

PROBLEM 10-31   Suppose that N is the number of rabbits in a population at t days and $N = N_0 e^{(.1)t}$, where N_0 is a constant. If 300 rabbits are present after 5 days have elapsed, how many rabbits were there originally?

PROBLEM 10-32   If G is the mass (in grams) of a particular radioactive substance present in a sample after t seconds, then its decay equation is $G = G_0 e^{-.5t}$, where G_0 is a constant. If you have 100 grams of the substance initially, how much will you have after (a) 2 seconds? (b) 6 seconds?

PROBLEM 10-33   Given $F(x) = 1 + a^x$, find $\dfrac{1}{F(x)} + \dfrac{1}{F(-x)}$.

PROBLEM 10-34   Given $S(x) = \dfrac{e^x - e^{-x}}{2}$ and $C(x) = \dfrac{e^x + e^{-x}}{2}$, find (a) $[C(x)]^2 - [S(x)]^2$, (b) $S(2x)$, (c) $2S(x)C(x)$.

PROBLEM 10-35   Graph the function $f(x) = 2^{|x|}$.

PROBLEM 10-36   Sketch $y = \log|x|$.

PROBLEM 10-37   If you put $500 in the bank at 8% interest compounded quarterly, how much will your account be worth in 10 years? [Use the formula $A = P(1 + \frac{i}{k})^{nk}$ from Problem 10-27.]

PROBLEM 10-38 If the population doubles every 16 years and today its size is 100 000, what will the size be in 48 years?

PROBLEM 10-39 Solve the equation $2^{x+2} = 7$ for x.

PROBLEM 10-40 Solve $\log(x + 3) + \log 2x^2 - \log x = 3 \log 2$ for x.

PROBLEM 10-41 If the half-life of strontium-90 is 25 years, what is its decay constant k in the decay equation $N = N_0 2^{kt}$?

PROBLEM 10-42 Solve for x in (a) $2^{x(x-1)} = 4$ and (b) $5^{x(x+4)} = \frac{1}{125}$.

PROBLEM 10-43 Write (a) $\log_2(\frac{1}{8}) = -3$ and (b) $\log_{10} 1000 = 3$ in exponential form.

PROBLEM 10-44 Use the change of base formula to help in solving $\log_{10} 2x + \log_{10} 3 = \log_5 20$.

PROBLEM 10-45 Write $2 \log \sqrt{x} - \log y^3 + 4 \log \sqrt[3]{x^2} - 5 \log \sqrt{y}$ as a single logarithmic expression.

PROBLEM 10-46 Given $\log 2 = .3010$, $\log 3 = .4771$, and $\log 5 = .6990$, find the values of (a) $\log \sqrt[3]{75}$, (b) $\log(\frac{75}{9})$, (c) $\log \sqrt{\frac{225}{243}}$.

PROBLEM 10-47 Find N if $\log N = 2.6994$.

PROBLEM 10-48 The loudness L of a sound of intensity I is measured in decibels (dB) by the formula $L = 10 \log(I/I_0)$, where I_0 is the minimum sound intensity detectable by human ears. For example, a jackhammer 8 feet away makes a sound 10^{10} times as intense as I_0, so that its loudness is

$$L = 10 \log\left(\frac{10^{10} I_0}{I_0}\right) = 10 \log 10^{10} = (10)(10) \log 10 = (10)(10)(1) = 100 \text{ dB}$$

If the sound of a jet plane at 50 feet is 10^{14} times as intense as I_0, how many decibels is that?

PROBLEM 10-49 Use Table A to find the values of (a) $\log_3 17$ and (b) $\log_{12} 33$.

PROBLEM 10-50 If you invest P dollars at 6% interest compounded continuously, how long will it take to double your money? [Use the formula $A = Pe^{in}$ from Problem 10-27.]

PROBLEM 10-51 Solve for x in (a) $2^{2-x} = 4^{1-2x}$ and (b) $4^{2x-3} = 5$.

PROBLEM 10-52 If $\log_5 5^{x^2} = \log_7 7^{.16}$, what is x?

PROBLEM 10-53 Solve the following exponential equations:

$$\text{(a) } e^{2x} - 5e^x + 4 = 0 \qquad \text{(b) } e^x + e^{-x} = 2$$

[*Hint:* In (b), multiply both sides by e^x to make the equation quadratic.)

Answers to Supplementary Exercises

10-30 (a) $3^{6\sqrt{2}}$ (b) 7^4

10-31 $300e^{-1/2}$

10-32 (a) $100e^{-1}$

     (b) $100e^{-3}$

10-33 1

10-34 (a) 1 (b) $\dfrac{e^{2x} - e^{-2x}}{2}$ (c) $\dfrac{e^{2x} - e^{-2x}}{2}$

10-35 See Figure 10-13.

10-36 See Figure 10-14.

10-37 $1104.02

10-38 800 000

10-39 $x \approx .8073$

10-40 $x = 1$

10-41 $k = -.04$

10-42 (a) $x = 2, x = -1$
(b) $x = -3, x = -1$

10-43 (a) $2^{-3} = \frac{1}{8}$
(b) $10^3 = 1000$

10-44 $x \approx 12.11$

10-45 $11 \log\left(\dfrac{x^{1/3}}{y^{1/2}}\right)$

10-46 (a) .625
(b) .9209
(c) $-.2552$

10-47 $N = 500.5$

10-48 140 dB

10-49 (a) 2.5789
(b) 1.407

10-50 $n = \dfrac{\ln 2}{.06} \approx 11\frac{1}{2}$ years

10-51 (a) $x = 0$
(b) $x \approx 2.08$

10-52 $x = \pm 0.4$

10-53 (a) $x = 0, x = \ln 4$
(b) $x = 0$

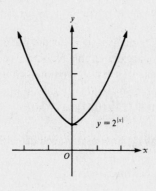

Figure 10-13

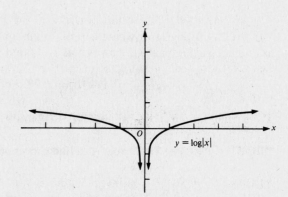

Figure 10-14

EXAM 3 (Chapters 7–10)

1. If the points $(-2, 4)$ and $(4, -6)$ are at opposite ends of a diameter of a circle, where is the center C of the circle and what is its radius r? Write the equation of the circle.

2. Find the equation of the line that passes through the points $(5, -1)$ and $(-3, 4)$.

3. Show that the lines $x + 3y = 2$ and $3x - y = 7$ are perpendicular.

4. Use graphing to find the solution set for **(a)** $2x + 3y \leq 5$ and **(b)** $-2x + y \geq 3$.

5. Consider the equations **(a)** $x = 4y - 2$ and **(b)** $y + 2x = x^2 + 2(x + 1)$. In each case, is the equation a function or a relation? What is its domain and its range? Is the equation a one-to-one function?

6. Given $f(x) = 3x^2 + x - 5$, find the expression for $\dfrac{f(x + h) - f(x)}{h}$.

7. Let $f(x) = \sqrt{2x - 5}$ and $g(x) = 3 - x^2$. Find the rule for the composite functions $(g \circ f)(x)$ and $(f \circ g)(x)$ and the domain of each composite function.

8. If $f(x) = -4x + 9$, find its inverse $f^{-1}(x)$ and graph both functions on the same axis.

9. Without graphing, find the location of the vertex of the parabola $2x = 4y^2 + y - 8$. Is this vertex a maximum value, a minimum value, or neither?

10. Find **(a)** the vertical asymptote, **(b)** the horizontal asymptote, and **(c)** the intercepts for $f(x) = \dfrac{3x^2 - 7x}{2x^2 - x - 10}$.

11. The equation $4x^2 + 9y^2 - 8x + 36y + 4 = 0$ describes an ellipse. Find its center point C, its vertices V_1, and its covertices V_2.

12. Solve for x in **(a)** $\log_x 25 = 2$ and **(b)** $2^x \cdot 8^{2x-1} = 4^{x+1}$

13. If $\log 3 = .4771$, $\log 2 = .3010$, and $\log 5 = .6990$, find **(a)** $\log 60$ and **(b)** $\log 450$ using only the rules of logarithms.

14. Find the value of $T = \log_2(\tfrac{1}{8}) + \tfrac{1}{2}(\log_7 49) + \log_{\sqrt{5}} 1$.

15. Use a table of logarithms (see Table A) to find the value for **(a)** $\sqrt[6]{83}$ and **(b)** $\sqrt[4]{13.3} \cdot \sqrt[3]{7.4}$.

Answers to Exam 3

1. $C = (1, -1); \qquad r = \sqrt{34};$
   $(x - 1)^2 + (y + 1)^2 = 34$

2. $5x + 8y = 17$

3. Slopes are $-\frac{1}{3}$ and $+3$.

4. (a) See Figure E3-1a.
   (b) See Figure E3-1b.

5. (a) Function;
       Domain all real numbers;
       range all real numbers;
       one-to-one
   (b) Function;
       domain all real numbers;
       range $y \geq 1$;
       not one-to-one

6. $6x + 3h + 1$

7. $g(f(x)) = 8 - 2x$; domain all real numbers
   $f(g(x)) = \sqrt{1 - 2x^2};$
   domain $-\dfrac{1}{\sqrt{2}} \leq x \leq \dfrac{1}{\sqrt{2}}$

8. $f^{-1}(x) = \frac{9}{4} - \frac{1}{4}x$
   See Figure E3-2.

9. $V = (-\frac{129}{32}, -\frac{1}{8})$
   Neither

10. (a) $x = \frac{5}{2}, x = -2$ (b) $y = \frac{3}{2}$ (c) $(0,0)$ and $(\frac{7}{3}, 0)$

11. $C = (1, -2)$
    $V_1 = (4, -2), (-2, -2)$
    $V_2 = (1, 0), (1, -4)$

12. (a) 5   (b) 1

13. (a) 1.7781   (b) 2.6532

14. $-3 + 1 + 0 = -2$

15. (a) 2.0887   (b) 3.7214

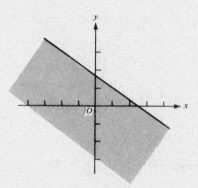

Figure E3-1a

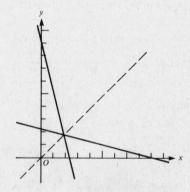

Figure E3-2

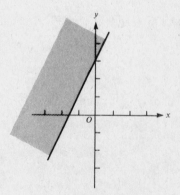

Figure E3-1b

11 SYSTEMS OF EQUATIONS AND INEQUALITIES

THIS CHAPTER IS ABOUT

☑ **Systems of Equations in Two Variables**
☑ **Linear Systems of Equations in Two Variables**
☑ **Linear Systems of Equations in Three Variables**
☑ **Systems of Inequalities in Two Variables**
☑ **Linear Programming**

11-1. Systems of Equations in Two Variables

A. Definition of a system of equations in two variables

- A **system of equations in two variables** is a collection of equations that contain two variables—usually x and y—and no other variables.

EXAMPLE 11-1 Suppose the total value of a stack of 13 nickels and dimes is $1.00. Using variables x and y, form the system of equations needed to determine how many nickels and how many dimes there are in the stack.

Solution We let x denote the number of nickels, and y the number of dimes. Then we can write the following equations in x and y:

$$\begin{cases} x + y = 13 & \text{[The total number of nickels and dimes is 13.]} \\ 5x + 10y = 100 & \text{[The total value of the nickels and dimes is 1 dollar = 100 cents.]} \end{cases}$$

We have a collection of two equations that contain variables x and y and no other variables, so this is a system of equations in two variables.

---

note: We use a *brace* to group the equations in a particular system.

EXAMPLE 11-2 Write the system of equations needed to find two numbers whose sum is 13 and whose product is 36.

Solution We call the numbers x and y and write the equations

$$\begin{cases} x + y = 13 \\ x \cdot y = 36 \end{cases}$$

- A **solution of a system of equations in two variables** is an ordered pair (a, b) with the property that if x is replaced by a and y is replaced by b in all of the equations of the system, then all of the equations become identities.

The solution of the system in Example 11-1 is the ordered pair $(x, y) = (6, 7)$, since $(6, 7)$ is a solution of both equations; i.e., $6 + 7 = 13$ and $5(6) + 10(7) = 100$. The system in Example 11-2 has $x = 4$ and $y = 9$, or $(4, 9)$, as a solution, since $4 + 9 = 13$ and $4(9) = 36$. [In this case, of course, the pair $(9, 4)$ works as well as the pair $(4, 9)$.]

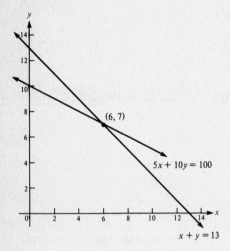

Figure 11-1

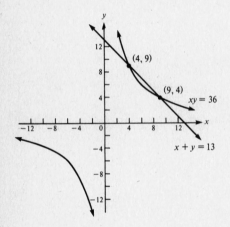

Figure 11-2

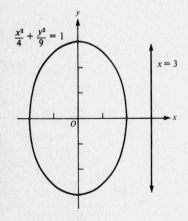

Figure 11-3

B. Solving a system of equations in two variables

Solving a system of equations in two variables means finding all the ordered pairs that are solutions of the system. And we can find the ordered pairs by *graphing* or *substitution*.

1. The graphing method

One way to solve a system of equations in two variables is to graph each equation to find if and where the graphs intersect. The points common to all of the graphs are the solution points.

EXAMPLE 11-3 Solve the system of equations $\begin{cases} x + y = 13 \\ 5x + 10y = 100 \end{cases}$.

Solution To solve the system, we graph each equation and find where the graphs intersect. The graphs are both straight lines. The first, $x + y = 13$, can be rewritten in slope-intercept form as $y = -x + 13$, which is a line with slope -1 and y-intercept 13. The second, $5x + 10y = 100$, is the line $y = -\frac{1}{2}x + 10$ with slope $-\frac{1}{2}$ and y-intercept 10. The lines, shown in Figure 11-1, cross at the point $(6, 7)$, so the point $(6, 7)$ is the solution of this system of equations.

EXAMPLE 11-4 Verify that $(4, 9)$ and $(9, 4)$ are the solutions to the system of equations $\begin{cases} x + y = 13 \\ x \cdot y = 36 \end{cases}$.

Solution These solutions can be verified by graphing. The first equation is the straight line $y = -x + 13$ with slope -1 and y-intercept 13. The second equation is a hyperbola lying in the first and third quadrants. The graphs intersect at the points $(4, 9)$ and $(9, 4)$, as shown in Figure 11-2.

EXAMPLE 11-5 Solve the system $\begin{cases} \dfrac{x^2}{4} + \dfrac{y^2}{9} = 1 \\ x = 3 \end{cases}$ by graphing.

Solution The graph of $\dfrac{x^2}{4} + \dfrac{y^2}{9} = 1$ is an ellipse with center at the origin, vertices at $(0, 3)$ and $(0, -3)$, and covertices at $(2, 0)$ and $(-2, 0)$. The graph of $x = 3$ is a vertical line drawn 3 units to the right of the origin. As Figure 11-3 shows, the graphs do not intersect at all, so there are no solutions to this system.

---

2. The substitution method

The graphing method of solving a system is not always very accurate. For example, we would have great difficulty graphically determining the solution to a system solved by $x = \frac{143}{397}$, $y = \frac{83}{591}$. But there are algebraic methods that make the task easier. A straightforward approach is the **substitution method**:

Step 1: Solve one of the given equations for one of the variables in terms of the other variable.

Step 2: Substitute the expression obtained in Step 1 into the other given equation to produce an equation in one variable only.

Step 3: Find the solution to the equation in one variable obtained in Step 2.

Step 4: Use the equation obtained in Step 1 and the solution obtained in Step 3 to find the corresponding value for the remaining variable.

Step 5: Check each solution pair in the original equations.

EXAMPLE 11-6   Use the substitution method to solve the system

$$\begin{cases} x + y = 13 \\ 5x + 10y = 100 \end{cases}$$

Solution

$$x + y = 13$$ [*Step 1:* Solve the first equation for one of the variables, say x.]

$$x = 13 - y$$

note: The decision to solve for x is an arbitrary one. We could just as easily have begun by solving for y.

$$5x + 10y = 100$$

$$5(13 - y) + 10y = 100$$ [*Step 2:* Substitute the new expression for x into the second equation.]

$$65 - 5y + 10y = 100$$

$$5y = 35$$ [*Step 3:* Solve for y.]

$$y = 7$$

$$x = 13 - y$$ [*Step 4:* Substitute the value for y into the expression for x and solve for x.]

$$= 13 - 7$$

$$= 6$$

The solution is $(6, 7)$.

$$x + y = 13 \Rightarrow 6 + 7 \overset{\checkmark}{=} 13$$ [*Step 5:* Check the solution pair in the original equations.]

$$5x + 10y = 100 \Rightarrow 5(6) + 10(7) \overset{\checkmark}{=} 100$$

EXAMPLE 11-7   Solve the system $\begin{cases} 2x - 3y = 3 \\ 3x + y = 10 \end{cases}$ using the substitution method.

Solution   Begin with the second equation, since the coefficient of y is 1:

$$3x + y = 10$$

$$y = -3x + 10$$ [*Step 1*]

Using the first equation: $2x - 3y = 3$

$$2x - 3(-3x + 10) = 3$$ [*Step 2*]

$$2x + 9x - 30 = 3$$

$$11x = 33$$

$$x = 3$$ [*Step 3*]

$$y = -3x + 10$$

$$= -3(3) + 10$$ [*Step 4*]

$$= 1$$

The solution is $(3, 1)$.

$$2x - 3y = 3 \Rightarrow 2(3) - 3(1) \overset{\checkmark}{=} 3$$ [*Step 5*]

$$3x + y = 10 \Rightarrow 3(3) + 1 \overset{\checkmark}{=} 10$$

EXAMPLE 11-8   Solve the system $\begin{cases} x^2 + y^2 = 8 \\ y - x = 4 \end{cases}$.

Solution   We begin with the second equation:

$$y - x = 4$$

$$y = x + 4 \qquad [Step\ 1]$$

Now, the first equation: $x^2 + y^2 = 8$

$$x^2 + (x + 4)^2 = 8 \qquad [Step\ 2]$$

$$x^2 + x^2 + 8x + 16 = 8$$

$$2x^2 + 8x + 8 = 0$$

$$x^2 + 4x + 4 = 0$$

$$(x + 2)(x + 2) = 0$$

$$x = -2 \qquad [Step\ 3]$$

Then $$y = x + 4$$

$$= -2 + 4 \qquad [Step\ 4]$$

$$= 2$$

The solution is $(-2, 2)$.

$$x^2 + y^2 = 8 \Rightarrow (-2)^2 + (2)^2 \overset{\leq}{=} 8 \qquad [Step\ 5]$$

$$y - x = 4 \Rightarrow 2 - (-2) \overset{\leq}{=} 4 = 4$$

EXAMPLE 11-9   Solve the system $\begin{cases} y - \sqrt{x + 1} = 0 \\ x - y = 1 \end{cases}$.

Solution $$x - y = 1$$

$$x = y + 1$$

Then, since $y - \sqrt{x + 1} = 0$, we have

$$y - \sqrt{y + 1 + 1} = 0 \qquad [\text{Substitute}]$$

$$y = \sqrt{y + 2}$$

$$y^2 = y + 2 \qquad [\text{Square both sides}]$$

$$y^2 - y - 2 = 0$$

$$(y - 2)(y + 1) = 0 \qquad [\text{Factor } y^2 - y - 2]$$

$$y = 2 \text{ or } y = -1$$

Since $x = y + 1$, when $y = 2$, then $x = 2 + 1 = 3$. When $y = -1$, then $x = -1 + 1 = 0$. So the possible solution pairs are $(3, 2)$ and $(0, -1)$.

Check:

$(3, 2)$		$(0, -1)$	
$y - \sqrt{x + 1} = 0$	$x - y = 1$	$y - \sqrt{x + 1} = 0$	$x - y = 1$
$2 - \sqrt{3 + 1} \overset{?}{=} 0$	$3 - 2 \overset{?}{=} 1$	$-1 - \sqrt{0 + 1} \overset{?}{=} 0$	$0 - (-1) \overset{?}{=} 1$
$2 \overset{\leq}{=} \sqrt{4}$	$1 \overset{\leq}{=} 1$	$-1 \neq \sqrt{1}$	$1 \overset{\leq}{=} 1$

When we check the proposed solution $(0, -1)$ by substituting 0 for x and -1 for y in the equation $y - \sqrt{x + 1} = 0$, we obtain the false statement $-1 = \sqrt{1}$. Thus we conclude that $(0, -1)$ is NOT a solution of the system and that the only solution of the system is $(3, 2)$.

note: A proposed solution of a system that is NOT a solution is called an **extraneous solution**. Extraneous solutions are likely to arise when we square both sides of an equation or multiply both sides by an expression that may be equal to zero.

EXAMPLE 11-10   Find the length and width of a rectangle whose area is 60 square inches and whose perimeter is 32 inches.

Solution   Let x be the length of the rectangle and y be the width. Since the area is 60 square inches, we have $xy = 60$; and since the perimeter is 32 inches, we have $x + y + x + y = 2x + 2y = 32$. Thus the system is

$$\begin{cases} xy = 60 \\ 2x + 2y = 32 \end{cases}$$

Now we can use the substitution method to solve the system:

$$2x + 2y = 32$$
$$2y = 32 - 2x$$
$$y = 16 - x$$

Then
$$xy = 60$$
$$x(16 - x) = 60$$
$$16x - x^2 = 60$$
$$x^2 - 16x + 60 = 0$$
$$(x - 10)(x - 6) = 0$$
$$x = 10 \text{ or } x = 6$$

The corresponding values obtained from $y = 16 - x$ are
$$y = 16 - 10 = 6 \quad \text{and} \quad y = 16 - 6 = 10$$

Thus the solutions are $(10, 6)$ or $(6, 10)$. The rectangle is 10 inches by 6 inches.

Check: Area $= xy = 10 \times 6 \overset{?}{=} 60$ Perimeter $= x + y + x + y = 10 + 6 + 10 + 6 \overset{?}{=} 32$

EXAMPLE 11-11   Show that the system $\begin{cases} 2x - y = 5 \\ -4x + 2y = 8 \end{cases}$ has no solution.

Solution
$$2x - y = 5$$
$$y = 2x - 5$$

Then
$$-4x + 2(2x - 5) = 8$$
$$-4x + 4x - 10 = 8$$
$$-10 \neq 8$$

The resulting equation $-10 \neq 8$ is impossible, so no solution exists. [A graphical analysis would show two *parallel lines*.]

EXAMPLE 11-12   Determine the values of a and b so that the line $ax + by = 1$ passes through the point $(1, -6)$ and has y-intercept 3.

Solution   If the line contains the point $(x, y) = (1, -6)$, then
$$a(1) + b(-6) = 1$$
$$a - 6b = 1$$

Since the y-intercept is 3, the line contains the point $(0, 3)$, so
$$a(0) + b(3) = 1$$
$$3b = 1$$

Thus we must solve the system $\begin{cases} a - 6b = 1 \\ 3b = 1 \end{cases}.$

From the second equation, we have $3b = 1$ or $b = \frac{1}{3}$. We substitute this value for b into the first equation:

$$a - 6b = 1$$
$$a - 6(\tfrac{1}{3}) = 1$$
$$a - 2 = 1$$
$$a = 3$$

Since the value of a is 3 and the value of b is $\frac{1}{3}$, the line $ax + by = 1$ that passes through the points $(1, -6)$ and $(0, 3)$ is given by $3x + \frac{1}{3}y = 1$ or $9x + y = 3$.

11-2. Linear Systems of Equations in Two Variables

A. Definition and classification

A *linear equation in two variables* is an equation that can be written in the form

$$ax + by = c$$

where a, b, and c are constants and a and b are not both zero. The graph of every linear equation in two variables is a straight line.

- A system that contains only linear equations in two variables is called a **linear system of equations in two variables**.

Consider first a system containing *two* linear equations in two variables. Since the graph of each equation is a straight line, there are three possible graphical situations, and for each of these situations, there is a corresponding algebraic interpretation:

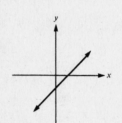

(1) Identical (coinciding) lines—infinitely many solutions

(2) Parallel lines—no solutions

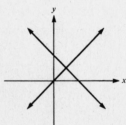

(3) Intersecting lines—one solution

Graphical Situation	Algebraic Interpretation
(1) The two lines are identical (coinciding).	(1) The system has infinitely many solutions.
(2) The two lines are parallel.	(2) The system has no solutions.
(3) The two lines intersect in exactly one point.	(3) The system has exactly one solution.

- If situation (1) occurs, then every solution of one equation of the system is also a solution of the other, and we say that the system is **consistent with infinitely many solutions**—or simply **dependent**.

- If situation (2) occurs, then the equations have NO common solutions, and we say that the system is **inconsistent**.

- If situation (3) occurs, then there is exactly ONE solution, and we say that the system is **consistent with a unique solution**—or simply **independent**.

EXAMPLE 11-13   Classify each of the following systems of equations.

(a) $\begin{cases} x + y = 1 \\ 3x + 3y = 3 \end{cases}$ (b) $\begin{cases} x + y = 4 \\ x + y = 5 \end{cases}$ (c) $\begin{cases} x + y = 2 \\ x - y = 6 \end{cases}$

Solution   In each case, we rewrite the equations in slope-intercept form

$$y = mx + b$$

(a) $\begin{cases} x + y = 1 \\ 3x + 3y = 3 \end{cases} \Rightarrow \begin{cases} y = -x + 1 \\ y = -x + 1 \end{cases}$

Since the equations of the two lines are identical, every solution of the first equation is also a solution of the second. The system is *dependent*.

(b) $\begin{cases} x + y = 4 \\ x + y = 5 \end{cases} \Rightarrow \begin{cases} y = -x + 4 \\ y = -x + 5 \end{cases}$

We now have the equations of two lines with the same slope ($m = -1$) but different y-intercepts ($b = 4$; $b = 5$). This tells us that the lines are parallel—and thus have no common solutions, since the lines never touch. This system is *inconsistent*.

(c) $\begin{cases} x + y = 2 \\ x - y = 6 \end{cases} \Rightarrow \begin{cases} y = -x + 2 \\ y = x - 6 \end{cases}$

The two equations in this system aren't identical, and they don't share the same slope. (In fact, since the product of the slopes is -1, the lines are perpendicular.) The lines must intersect in exactly one point. The system is *independent*.

B. The elimination method of solving a linear system

1. Transformations to equivalent systems

Two systems of equations are **equivalent** if they have the same solutions. When solving a system, it is often helpful to transform one or more of the equations into an equivalent equation, i.e., change the form or appearance of the equation but not its validity (solutions). The following transformations do not change the solutions to a system of equations:

(1) Interchange two equations.
(2) Multiply an equation by a nonzero constant.
(3) Add to one equation a nonzero multiple of another equation.

EXAMPLE 11-14 Create other systems equivalent to $\begin{cases} 2x - 5y = 3 \\ x + 2y = 1 \end{cases}$.

Solution Use the transformations listed above to create equivalent systems:

$\begin{cases} 2x - 5y = 3 \\ x + 2y = 1 \end{cases} \Longrightarrow \begin{cases} x + 2y = 1 \\ 2x - 5y = 3 \end{cases}$ **(1)** Interchange the equations

$\begin{cases} 2x - 5y = 3 \\ x + 2y = 1 \end{cases} \xrightarrow{2(x+2y=1)} \begin{cases} 2x - 5y = 3 \\ 2x + 4y = 2 \end{cases}$ **(2)** Multiply the second equation by 2

$\begin{cases} 2x - 5y = 3 \\ x + 2y = 1 \end{cases} \xrightarrow{+(2x+4y=2)} \begin{cases} 4x - y = 5 \\ x + 2y = 1 \end{cases}$ **(3)** Add twice the second equation to the first equation

Notice that the equations may be transformed independently—a change in one equation doesn't require a change in the other.

Sometimes a few transformations will give us an immediate solution to a system.

EXAMPLE 11-15 Create a system equivalent to $\begin{cases} 2x + y = 5 \\ 4x + 2y = 10 \end{cases}$ and solve.

Solution We create an equivalent system by multiplying the first equation by 2:

$$\begin{cases} 2x + y = 5 \\ 4x + 2y = 10 \end{cases} \Rightarrow \begin{cases} 4x + 2y = 10 \\ 4x + 2y = 10 \end{cases}$$

The system now consists of two identical equations. So any pair (a, b) that satisfies the first equation must also satisfy the second. Thus if

$$4x + 2y = 10$$
$$2y = 10 - 4x$$
$$y = 5 - 2x$$

and if x has any value a, then y has the value $5 - 2a$. The infinite collection of solution pairs is $\{(a, 5 - 2a) \mid a \text{ is a real number}\}$.

note: You can also think of this solution as a pair of coinciding lines with slope -2 and y-intercept 5.

EXAMPLE 11-16   Create a system equivalent to $\begin{cases} 3x - y = 8 \\ -9x + 3y = 5 \end{cases}$ and solve.

Solution   Multiply the first equation by 3 and add it to the second equation:

$$\begin{cases} 3x - y = 8 \\ -9x + 3y = 5 \end{cases} \Rightarrow \begin{cases} 9x - 3y = 24 \\ -9x + 3y = 5 \end{cases} \Rightarrow \begin{cases} 3x - y = 8 \\ 0x + 0y = 29 \end{cases}$$

But the equation $0x + 0y = 29$ is impossible, so the system has no solutions. The lines do not intersect—they are parallel.

---

2. The elimination method

To solve a system of equations by the **elimination method**, we use the transformation processes to *eliminate* one of the variables in order to be able to solve for the other. We follow these steps:

Step 1: Multiply one or both equations by a nonzero constant or constants. Choose the constant(s) so that the sum of one of the variable terms is zero.
Step 2: Substitute the sum of the two equations for one of the equations.
Step 3: Solve for the variable that remains (remember—one has dropped out).
Step 4: Use either equation of the original pair to solve for the remaining variable.
Step 5: Check the proposed solution(s) in the original equations.

EXAMPLE 11-17   Find solutions to the system $\begin{cases} 3x + 4y = 2 \\ 2x - y = 5 \end{cases}$.

Solution

$$\begin{cases} 3x + 4y = 2 \\ 8x - 4y = 20 \end{cases}$$ [*Step 1:* Multiply the second equation by 4.]

$$11x + 0y = 22$$ [*Step 2:* Add the equations together.]
$$11x = 22$$

$$x = 2$$ [*Step 3:* Solve for x.]

$$3x + 4y = 2$$ [*Step 4:* Go back to either of the original pair, substitute $x = 2$ into it, and solve for y.]
$$3(2) + 4y = 2$$
$$6 + 4y = 2$$
$$4y = -4$$
$$y = -1$$

The solution is $(2, -1)$.

$$3(2) + 4(-1) \overset{?}{=} 2 \qquad 2(2) - (-1) \overset{?}{=} 5 \qquad [\textit{Step 5: } \text{Check.}]$$
$$6 - 4 \overset{\checkmark}{=} 2 \qquad 4 + 1 \overset{\checkmark}{=} 5$$

EXAMPLE 11-18   Use the elimination method to solve $\begin{cases} 2x - 5y = 3 \\ -3x + y = 2 \end{cases}$.

Solution

$$\begin{cases} 2x - 5y = 3 \\ 5(-3x + y = 2) \end{cases} \Rightarrow \begin{cases} 2x - 5y = 3 \\ -15x + 5y = 10 \end{cases} \qquad [\textit{Step 1}]$$

$$\text{Add:} \quad \begin{cases} 2x - 5y = 3 \\ \underline{-15x + 5y = 10} \end{cases}$$
$$-13x = 13 \qquad [\textit{Step 2}]$$
$$x = -1 \qquad [\textit{Step 3}]$$

Then, since $x = -1$,

$$2x - 5y = 3 \qquad [\textit{Step 4}]$$
$$2(-1) - 5y = 3$$
$$-2 - 5y = 3$$
$$-5y = 5$$
$$y = -1$$

The solution is $(-1, -1)$.

Check: $2(-1) - 5(-1) \overset{?}{=} 3 \qquad -3(-1) + (-1) \overset{?}{=} 2 \qquad [\textit{Step 5}]$
$\qquad\quad -2 + 5 \overset{\checkmark}{=} 3 \qquad\qquad 3 - 1 \overset{\checkmark}{=} 2$

EXAMPLE 11-19   The sum of the digits of a certain two-digit number is 14. If the digits are reversed, the number is increased by 36. Find the number. [*Hint:* 14 is $10 + 4$, or $10 \cdot 1 + 4$, which is "1 ten plus 4 ones."]

Solution   Let the digits be m and n. Then $m + n = 14$. But the desired number is $10m + n$, and when its digits are reversed, it becomes $10n + m$. Since the reversed number is 36 more than the original, we have

$$10n + m = 10m + n + 36$$
$$(10n + m) - (10m + n) = 36$$
$$10n + m - 10m - n = 36$$
$$9n - 9m = 36$$
$$n - m = 4$$

We now have a system of equations:

$$\begin{cases} m + n = 14 \\ n - m = 4 \end{cases}$$

By adding these equations, we have

$$(m + n) + (n - m) = 18$$
$$m + n + n - m = 18$$
$$2n = 18$$
$$n = 9$$

Then since $m + n = 14$, we have $m = 5$. So the original number is $10m + n = 10(5) + 9 = 59$. The reversed number is 95, and the difference is $95 - 59 = 36$—so the answer 59 checks.

11-3. Linear Systems of Equations in Three Variables

A. Definitions

A system of equations may have more than two variables, and it also may have more than two equations. We'll restrict our discussion here to systems of three equations in three unknowns—but the solution methods we develop extend to systems with more than three variables.

- A *linear equation in three variables* is an equation that can be written in the form

$$ax + by + cz = d$$

where a, b, c, and d are constants and a, b, and c are not all zero.
- A *system* of such equations is called a **linear system of equations in three variables**.
- A *solution* of a linear system of equations in three variables is an **ordered triple** (a, b, c), such that when x is replaced by a, y is replaced by b, and z is replaced by c in all of the equations of the system, each equation becomes an identity.

The substitution and elimination methods that work for linear systems of equations in two variables also work for linear systems of equations in three variables.

note: The graph of an ordered triple (a, b, c) is a *point* in three-dimensional space. The graph of a linear equation in three variables is a *plane* in three-dimensional space.

EXAMPLE 11-20 Show that the ordered triple $(-1, 0, 2)$ is a solution of the system $\begin{cases} -4x + y - z = 2 \\ -7x + 13y - 2z = 3. \\ 4x - 2y + 3z = 2 \end{cases}$

Solution We substitute $x = -1$, $y = 0$, and $z = 2$ into each equation:

$$\begin{bmatrix} -4x + y - z = 2 \\ -4(-1) + 0 - 2 \stackrel{?}{=} 2 \\ 4 + 0 - 2 \stackrel{?}{=} 2 \\ 2 \stackrel{\checkmark}{=} 2 \end{bmatrix} \qquad \begin{bmatrix} -7x + 13y - 2z = 3 \\ -7(-1) + 13(0) - 2(2) \stackrel{?}{=} 3 \\ 7 + 0 - 4 \stackrel{?}{=} 3 \\ 3 \stackrel{\checkmark}{=} 3 \end{bmatrix}$$

$$\begin{bmatrix} 4x - 2y + 3z = 2 \\ 4(-1) - 2(0) + 3(2) \stackrel{?}{=} 2 \\ -4 - 0 + 6 \stackrel{?}{=} 2 \\ 2 \stackrel{\checkmark}{=} 2 \end{bmatrix}$$

Since each equation in the system becomes an identity when x is replaced by -1, y by 0, and z by 2, the ordered triple $(-1, 0, 2)$ is a solution of the system.

B. The substitution method

Although the substitution method for three equations in three unknowns is a little more complicated than it is for two equations in two unknowns, the basic steps are much the same:

Step 1: Use one equation to express one chosen variable in terms of the other two variables.

Step 2: In each of the remaining equations, substitute the expression obtained in Step 1 for the chosen variable, thereby obtaining two equations in two unknowns.

Step 3: Treat the remaining two equations as a linear system in two unknowns—follow the steps of the substitution method for two equations in two variables.

EXAMPLE 11-21   Solve the system $\begin{cases} x - y + 2z = 1 \\ -x - y + 3z = -2. \\ 2x - y - 4z = -3 \end{cases}$

Solution   Solve the first equation for x:

$$x - y + 2z = 1$$

$$x = 1 + y - 2z \qquad\qquad \textit{[Step 1]}$$

Substitute $x = 1 + y - 2z$ in the second and third equations:

$$-x - y + 3z = -2 \qquad\qquad 2x - y - 4z = -3$$

$$-(1 + y - 2z) - y + 3z = -2 \qquad 2(1 + y - 2z) - y - 4z = -3 \qquad \textit{[Step 2]}$$

$$-1 - y + 2z - y + 3z = -2 \qquad 2 + 2y - 4z - y - 4z = -3$$

$$-2y + 5z = -1 \qquad\qquad y - 8z = -5$$

Now we have a system of two equations in two variables:

$$\begin{cases} -2y + 5z = -1 \\ y - 8z = -5 \end{cases} \qquad\qquad \textit{[Step 3]}$$

From the second equation of this pair, we see that $y = 8z - 5$, so

$$-2(8z - 5) + 5z = -1$$

$$-16z + 10 + 5z = -1$$

$$-11z = -11$$

$$z = 1$$

Now that we have one solution value, we work back to get the other two:
Since $z = 1$,

$$y = 8z - 5 = 8(1) - 5 = 3$$

Then, since $z = 1$ and $y = 3$,

$$x = 1 + y - 2z = 1 + 3 - 2(1) = 2$$

The solution to the system is the ordered triple $(2, 3, 1)$.

C. The elimination method

The elimination method for finding the solution to a system of three linear equations in three unknowns works on the same principles that work for smaller systems. Our aim is to eliminate variables and reduce the system to **triangular form**—where the first equation has three variables, the second two variables, and the third one variable. Then we can use back-substitution to find the solution values—if any—of x, y, and z.

EXAMPLE 11-22   Solve the following system, which is given in *triangular form*:

$$4x - 2y - 3z = 8$$

$$11y - \tfrac{1}{2}z = -12$$

$$12z = 24$$

Solution Since the third equation contains only one variable, z, solve for z:

$$12z = 24$$

$$z = 2$$

Move back to the equation with two variables, replace z with 2, and solve for y:

$$11y - \tfrac{1}{2}z = -12$$

$$11y - \tfrac{1}{2}(2) = -12$$

$$11y - 1 = -12$$

$$11y = -11$$

$$y = -1$$

Finally, use these values of y and z in the top equation of the triangle and solve for x:

$$4x - 2y - 3z = 8$$

$$4x - 2(-1) - 3(2) = 8$$

$$4x + 2 - 6 = 8$$

$$4x = 12$$

$$x = 3$$

The solution is the ordered triple $(3, -1, 2)$.

---

note: There are many different graphical situations possible for linear systems of three equations in three variables. The solution of the system in Example 11-22—the ordered triple $(3, -1, 2)$—means that the three planes intersect in exactly one point, i.e., at $(3, -1, 2)$. If there is no ordered triple that satisfies all three equations—e.g., if one of the three equations is impossible—then we might have (among other possibilities) three parallel planes, or two parallel planes and a third that slices both of them. If there are infinitely many solutions—e.g., if one of the equations reduces to the identity $0 = 0$—then two possibilities are three coinciding planes, and three planes intersecting in one line.

EXAMPLE 11-23 Find the solution to the system $\begin{cases} -6x + 3y + 2z = 6 \\ 3x + 2y - z = 4 \\ 3x + y + 3z = -2 \end{cases}$.

Solution Form equivalent systems until the system is in triangular form:

$$\begin{cases} -6x + 3y + 2z = 6 \\ 3x + 2y - z = 4 \\ 3x + y + 3z = -2 \end{cases} \Rightarrow \begin{cases} -6x + 3y + 2z = 6 \\ 6x + 4y - 2z = 8 \\ 6x + 2y + 6z = -4 \end{cases}$$

[Multiply both the second and third equations by 2.]

$$\begin{cases} -6x + 3y + 2z = 6 \\ 6x + 4y - 2z = 8 \\ 6x + 2y + 6z = -4 \end{cases} \Rightarrow \begin{cases} -6x + 3y + 2z = 6 \\ 0x + 7y + 0z = 14 \\ 0x + 5y + 8z = 2 \end{cases}$$

[Add the first equation to the second; add the first equation to the third.]

Interchange the second and third equations:

$$-6x + 3y + 2z = 6$$

$$5y + 8z = 2$$

$$7y = 14$$

Now use back-substitution to find the solution:

$$7y = 14 \Rightarrow y = 2$$

$$8z + 5y = 2 \Rightarrow 8z + 5(2) = 2$$

$$8z = -8$$

$$z = -1$$

$$-6x + 3y + 2z = 6 \Rightarrow -6x + 3(2) + 2(-1) = 6$$

$$-6x + 6 - 2 = 6$$

$$-6x = 2$$

$$x = -\tfrac{1}{3}$$

The solution is the ordered triple $(-\tfrac{1}{3}, 2, -1)$.

EXAMPLE 11-24 A nursery is offering three package deals on shrubbery. They have three sizes of shrubs, classified as small, medium, and large. Deal #1 offers 1 small shrub, 2 medium shrubs, and 2 large shrubs for $22. Deal #2 offers 3 small, 4 medium, and 1 large for $28. Deal #3 offers 2 small, 1 medium, and 1 large for $14. How much is the nursery actually charging for each size of shrub?

Solution   Let $x =$ the price of a small shrub
$y =$ the price of a medium shrub
$z =$ the price of a large shrub

The three deals can be translated into the following system of equations:

$$\begin{cases} x + 2y + 2z = 22 \\ 3x + 4y + z = 28 \\ 2x + y + z = 14 \end{cases}$$

We use the elimination method to put the system into triangular form:

$$\begin{cases} x + 2y + 2z = 22 \\ 3x + 4y + z = 28 \\ 2x + y + z = 14 \end{cases} \Rightarrow \begin{cases} x + 2y + 2z = 22 \\ 3x + 4y + z = 28 \\ 0x - 3y - 3z = -30 \end{cases}$$ [Subtract twice the first equation from the third equation.]

$$\begin{cases} x + 2y + 2z = 22 \\ 3x + 4y + z = 28 \\ 0x - 3y - 3z = -30 \end{cases} \Rightarrow \begin{cases} x + 2y + 2z = 22 \\ 0x - 2y - 5z = -38 \\ -3y - 3z = -30 \end{cases}$$ [Subtract three times the first equation from the second equation.]

$$\begin{cases} x + 2y + 2z = 22 \\ -2y - 5z = -38 \\ -3y - 3z = -30 \end{cases} \Rightarrow \begin{cases} x + 2y + 2z = 22 \\ -2y - 5z = -38 \\ y + z = 10 \end{cases}$$ [Divide the third equation by -3.]

$$\begin{cases} x + 2y + 2z = 22 \\ -2y - 5z = -38 \\ y + z = 10 \end{cases} \Rightarrow \begin{cases} x + 2y + 2z = 22 \\ -2y - 5z = -38 \\ -3z = -18 \end{cases}$$ [Add the second equation to twice the third equation.]

The system is now in triangular form, so we can solve for z, then y, then x.

$$-3z = -18 \Rightarrow z = 6$$

$$-2y - 5z = -38 \Rightarrow -2y - 5(6) = -38$$

$$y = 4$$

$$x + 2y + 2z = 22 \Rightarrow x + 2(4) + 2(6) = 22$$

$$x = 2$$

The solution is the ordered triple $(2, 4, 6)$, so the small shrubs are $2, the medium shrubs are $4, and the large shrubs are $6.

11-4. Systems of Inequalities in Two Variables

A. Linear systems of inequalities in two variables

A *linear inequality in two variables* is an inequality that can be written in one of the following forms:

$$ax + by > c, \qquad ax + by < c, \qquad ax + by \geq c, \qquad ax + by \leq c$$

- A system that contains only linear inequalities in two variables is called a **linear system of inequalities in two variables**.

To find the solution set of a system of two or more linear inequalities in two variables, we graph all the inequalities on the same coordinate plane. The intersection of their solution sets (the region common to all) is the solution of the system.

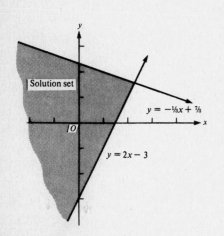

Figure 11-4

EXAMPLE 11-25 Graph the solution set of the system $\begin{cases} 2x - y \leq 3 \\ x + 3y \leq 7 \end{cases}$.

Solution First we write each inequality in slope-intercept form. Then $2x - y \leq 3$ becomes $y \geq 2x - 3$, so its graph consists of the half-plane above the line $y = 2x - 3$ and the line itself. And $x + 3y \leq 7$ becomes $y \leq -\frac{1}{3}x + \frac{7}{3}$, so its graph consists of the half-plane below the line $y = -\frac{1}{3}x + \frac{7}{3}$ and the line itself (see also Chapter 7, Section 7-5). The solution set of the system, shown in Figure 11-4, is the shaded area that is common to both half-planes.

Figure 11-5

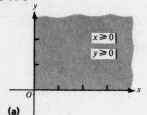

(a)

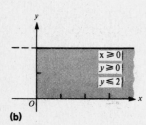

(b)

EXAMPLE 11-26 Sketch the solution set of the system

$$\begin{cases} x \geq 0 \\ y \geq 0 \\ y \leq 2 \\ x + y \geq 3 \\ 2x + y \leq 6 \end{cases}$$

Solution The first and second inequalities of the system, $x \geq 0$ and $y \geq 0$, indicate that the solution set is restricted to the first quadrant, as we see in Figure 11-5a. The third inequality, $y \leq 2$, restricts the solution set to that part of the first quadrant on or below the line $y = 2$, as shown in Figure 11-5b. The last two inequalities, $y \geq -x + 3$ and $y \leq -2x + 6$, tell us that the points in the solution set must be on or above the line $y = -x + 3$ and on or below the line $y = -2x + 6$, respectively, as shown in Figure 11-5c and d. The solution set is the area common to all the half-planes, shown in Figure 11-5e.

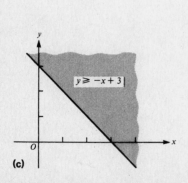

(c)

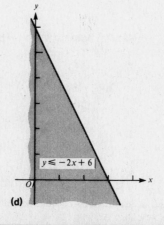

(d)

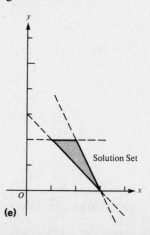

(e)

B. Systems containing second-degree inequalities in two variables

We use the same general methods to graph second-degree inequalities that we use for linear (first-degree) inequalities. We choose a test point not on the boundary—i.e., not *on* the circle, parabola, etc.—to determine which region constitutes the solution set.

EXAMPLE 11-27   Graph the solution set of the system $\begin{cases} y \geq x^2 - 2 \\ x + y \leq 2 \end{cases}$.

Solution   First we graph the parabola $y = x^2 - 2$. Choosing $(0,0)$ as the test point, we have $y \geq x^2 - 2$ or $0 \geq -2$—which is a true statement. Thus we shade the region "above" or "inside" the parabola as in Figure 11-6a. But, since the system also includes the inequality $x + y \leq 2$ or $y \leq -x + 2$, the solution set will contain only points below and on the line $y = -x + 2$, as indicated by the shaded area in Figure 11-6b. So the solution set contains those points both inside and on the parabola $y = x^2 - 2$ AND below and on the line $y - x + 2$, as in Figure 11-6c.

Figure 11-6

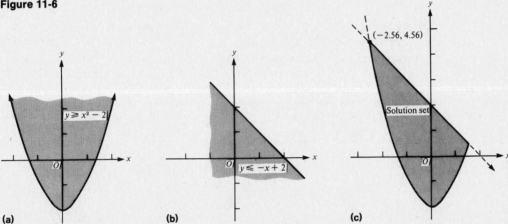

(a)   (b)   (c)

EXAMPLE 11-28   Graph the solution set of $\begin{cases} x^2 + y^2 \leq 9 \\ (x + 3)^2 + (y + 3)^2 \geq 9 \end{cases}$.

Solution   Graph each inequality and shade in the region common to both, as shown in Figure 11-7.

note: Since $(x + 3)^2 + (y + 3)^2$ is equal to or *greater than* 9, the solution set of this inequality includes those points on and *outside* the circle $(x + 3)^2 + (y + 3)^2 = 9$.

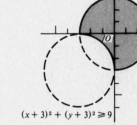

Figure 11-7

11-5. Linear Programming

A. Definitions

Linear programming allows us to combine what we know about systems of equations with what we know about systems of inequalities.

- **Linear programming** involves finding the *maximum* or *minimum* value of a linear expression whose variables are subject to a set of conditions or **constraints** given in the form of linear inequalities.
- If a problem involves two variables, the linear expression is of the form

$$ax + by$$

note: Linear programming has many business and scientific applications. In business, for example, it is often used by management to maximize profit or minimize cost subject to constraints of material cost, labor cost, shipping cost, stock maintenance, etc.

EXAMPLE 11-29   Your pizza company produces frozen pizzas in two forms—the Super, which sells for $3 wholesale, and the Magnificent, which sells for $4 wholesale. It costs you $1.50 per pie to produce the Super and $3 per pie to produce the Magnificent. Your total production costs must stay below $600 per day. Demand shows that you should not produce more than 250 pies per day and that the number of Supers produced per day must never exceed the number of Magnificents produced by more than 50 units.

Assume you want to know how many Supers and how many Magnificents you should produce per day in order to maximize (**a**) your revenue and (**b**) your profits. Set up the equations and inequalities to be analyzed in these problems.

Solution   Let $x =$ the number of Super pizzas produced per day
$y =$ the number of Magnificent pizzas produced per day

Since the Super sells for $3 and the Magnificent sells for $4, the expression $3x + 4y$ represents the total revenue per day. So the linear expression you want to maximize is

(**a**) $$ax + by = 3x + 4y \qquad \text{[Revenue]}$$

(**b**) Since it costs you $1.50 per pie to produce each Super and $3 per pie to produce each Magnificent, the expression $(3 - 1.5)x + (4 - 3)y$ or $1.5x + y$ represents the total profit per day. So the linear expression you want to maximize is

$$ax + by = 1.5x + y \qquad \text{[Profit]}$$

Next, set up the constraints in the form of linear inequalities:

$$\begin{cases} 1.5x + 3y \le 600 & \text{[Production cost constraint]} \\ x + y \le 250 & \text{[Quantity constraint]} \\ x - y \le 50 & \text{[Quantity constraint]} \\ x \ge 0 & \text{[Reality constraints:} \\ y \ge 0 & \quad \text{The number of pizzas produced can't} \\ & \quad \text{be negative.]} \end{cases}$$

The constraint conditions of a linear programming problem translate into a system of linear inequalities—which we can solve by graphing.

- A **feasible solution** is any ordered pair that satisfies all the constraints, i.e., any point that is a solution of the system of inequalities.
- The **feasible set** is the set of all feasible solutions, i.e., the shaded area representing the solution to the system of inequalities.
- A feasible solution that gives a maximum or minimum value of the objective (linear) function $ax + by$ is called an **optimal solution** of the linear programming problem.
- Since all of the constraints are linear inequalities, the graph of the feasible set is a closed polygon with boundaries of straight-line segments. The points of intersection of these line segments are called the **vertices** of the feasible set.

B. Solving linear programming problems

The foundation for solving linear programming problems is a theorem called the **Fundamental Principle of Linear Programming**:

FUNDAMENTAL PRINCIPLE OF LINEAR PROGRAMMING   The optimal solution to a linear programming problem—if one exists—occurs at one of the vertices of the feasible set.

Assuming the Fundamental Principle of Linear Programming, we follow these steps in solving linear programming problems:

Step 1: Find the region of the feasible set by graphing the system of linear inequalities determined by the constraints.

Step 2: Find the coordinates of the vertices of the feasible set. Usually, this involves solving the system of equations corresponding to the two intersecting boundary segments at each vertex.

Step 3: Calculate the value of the objective function $ax + by$ at each of the vertex points.

Step 4: If the function is to be *maximized,* choose the *largest* value found in Step 3; if the function is to be *minimized,* choose the *smallest* value found in Step 3.

EXAMPLE 11-30   Refer to the pizza production problem of Example 11-29. How many of each type of pizza should be produced per day to maximize **(a)** revenue and **(b)** profit?

Solution   Follow the steps outlined above:

(a) *Step 1:* Graph the system of inequalities

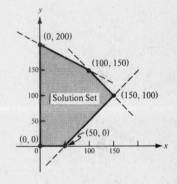

Figure 11-8

$$\begin{cases} x \geq 0 \\ y \geq 0 \\ 1.5x + 3y \leq 600 \Leftrightarrow y \leq -\tfrac{1}{2}x + 200 \\ x + y \leq 250 \Leftrightarrow y \leq -x + 250 \\ x - y \leq 50 \Leftrightarrow y \geq x - 50 \end{cases}$$

The feasible set is the shaded region common to all, as shown in Figure 11-8.

Step 2: (See Figure 11-8.) Start at the origin—the first vertex $(0,0)$—and move counterclockwise around the boundary.

The second vertex is the intersection of $y = 0$ and $y = x - 50$. This system of two equations is easy to solve, since you already have a value for y. Substituting $y = 0$ in $y = x - 50$ gives $x = 50$. Thus the second vertex is the point $(50,0)$.

The third vertex is the intersection of $y = x - 50$ and $y = -x + 250$. Using the elimination method, add the two equations to get $2y = 200$ or $y = 100$. If $y = 100$, then $x = 150$, so the third vertex is the point $(150,100)$.

The fourth vertex is the intersection of $y = -x + 250$ and $y = -\tfrac{1}{2}x + 200$. Solving this system of equations gives the fourth vertex as the point $(100,150)$.

Finally, the fifth vertex is the intersection of $y = -\tfrac{1}{2}x + 200$ and $x = 0$, which occurs at $(0,200)$.

Step 3: Calculate the value of the revenue expression $3x + 4y$ at each of the vertices:

$$(0,0) \Rightarrow 3x + 4y = 3(0) + 4(0) = 0$$

$$(50,0) \Rightarrow \qquad 3(50) + 4(0) = 150$$

$$(150,100) \Rightarrow \qquad 3(150) + 4(100) = 850$$

$$(100,150) \Rightarrow \qquad 3(100) + 4(150) = 900$$

$$(0,200) \Rightarrow \qquad 3(0) + 4(200) = 800$$

Step 4: The revenue is maximized at \$900 if you produce 100 Supers and 150 Magnificents per day.

(b) *Step 3:* Calculate the value of the profit expression $1.5x + y$ at each of the vertices:

$$(0,0) \Rightarrow \quad 1.5(0) + 1(0) = 0$$

$$(50,0) \Rightarrow \quad 1.5(50) + 1(0) = 75$$

$$(150,100) \Rightarrow 1.5(150) + 1(100) = 325$$

$$(100,150) \Rightarrow 1.5(100) + 1(150) = 300$$

$$(0,200) \Rightarrow \quad 1.5(0) + 1(200) = 200$$

Step 4: The profit is maximized at $325 if you produce 150 Supers and 100 Magnificents per day.

EXAMPLE 11-31 Find the maximum and minimum values of the linear expression $2x + y$ subject to the following constraints:

$$\begin{cases} x \geq 0, y \geq 0 \\ 2x - y \geq -2 \Leftrightarrow y \leq 2x + 2 \\ \quad x - y \leq 2 \quad \Leftrightarrow y \geq x - 2 \\ \quad x + y \leq 5 \end{cases}$$

Solution Graph the system of inequalities and shade the feasible set, as shown in Figure 11-9. The vertices occur at the points $(0,0)$, $(2,0)$, $(\frac{7}{2},\frac{3}{2})$, $(1,4)$, and $(0,2)$. Evaluate $2x + y$ at each of the vertices:

$$(0,0) \Rightarrow 2(0) + 0 = 0$$

$$(2,0) \Rightarrow 2(2) + 0 = 4$$

$$(\tfrac{7}{2},\tfrac{3}{2}) \Rightarrow 2(\tfrac{7}{2}) + \tfrac{3}{2} = \tfrac{17}{2}$$

$$(1,4) \Rightarrow 2(1) + 4 = 6$$

$$(0,2) \Rightarrow 2(0) + 2 = 2$$

Thus the maximum value occurs at $(\frac{7}{2},\frac{3}{2})$ and the minimum occurs at $(0,0)$.

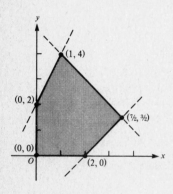

Figure 11-9

EXAMPLE 11-32 A farmer can allocate at most 400 acres of his farm for growing two kinds of peppers—green peppers and hot red peppers. Each acre of green peppers requires 3 hours of labor, while each acre of hot red peppers requires only 2 hours. The farmer has a total of 1000 hours of labor to devote to the pepper crops. The green peppers bring a profit of $20 per acre, while the red peppers bring a profit of $16 per acre. How many acres of each type of pepper should the farmer plant in order to realize maximum profit? What is the maximum profit?

Solution Let $x =$ the number of acres devoted to green peppers
$y =$ the number of acres devoted to red peppers

Then the value that must be maximized is Total Profit $= 20x + 16y$—subject to the following constraints:

$$\begin{cases} x \geq 0, y \geq 0 \\ \quad x + y \leq 400 \\ 3x + 2y \leq 1000 \end{cases}$$

From these inequalities, graph the feasible set, as shown in Figure 11-10. The vertices occur at the points $(0,0)$, $(\frac{1000}{3} \approx 333, 0)$, $(200,200)$, and $(0,400)$. Evaluating $20x + 16y$ at the vertices, you have

$$(0,0) \Rightarrow \quad 20(0) + 16(0) = 0$$

$$(333,0) \Rightarrow \quad 20(333) + 16(0) = 6660$$

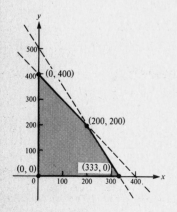

Figure 11-10

$$(200, 200) \Rightarrow 20(200) + 16(200) = 7200$$

$$(0, 400) \Rightarrow 20(0) + 16(400) = 6400$$

Thus the farmer should plant 200 acres of green peppers and 200 acres of hot red peppers in order to reach his maximum profit of $7200.

SUMMARY

1. A system of equations in two variables is a collection of equations containing two variables—usually x and y—and no other variables.
2. A system of two linear equations in two variables may have a unique solution (two intersecting lines), no solution (two parallel lines), or an infinite number of solutions (two coinciding lines).
3. You can solve a system of equations by the graphing method, the substitution method, or the elimination method.
4. To solve a system of three linear equations in three variables, you first transform the system into an equivalent system in triangular form.
5. The solution of a system of two or more linear inequalities in two variables is the intersection of the individual solution sets—i.e., the region in the coordinate plane common to all.
6. Linear programming involves finding the maximum or minimum value of a linear expression whose variables are subject to a set of conditions or constraints given in the form of linear inequalities.
7. The Fundamental Principle of Linear Programming states that the optimal solution to a linear programming problem—if one exists—occurs at one of the vertices of the feasible set.

RAISE YOUR GRADES

Can you . . . ?

☑ verify that a given ordered pair satisfies a system of equations in two variables
☑ find the graphical solution to a system of equations
☑ outline the steps in the substitution method
☑ explain the conditions under which a system has no solution or an infinite number of solutions
☑ list the algebraic operations used in transforming a system to an equivalent system
☑ use the elimination method to solve systems of equations
☑ reduce a system of three equations to an equivalent system of two equations
☑ transform a system of three linear equations to triangular form
☑ graph the solution set to a system of inequalities
☑ define a linear programming problem
☑ state the Fundamental Principle of Linear Programming and define its terms

SOLVED PROBLEMS

Systems of Equations in Two Variables

PROBLEM 11-1 Form the system of equations to represent this problem: If the sum of two numbers is 12 and their difference is 6, what are the numbers?

Solution   Let the numbers be presented by the letters x and y. Then the system is

$$\begin{cases} x + y = 12 \\ x - y = 6 \end{cases}$$

PROBLEM 11-2 Consider the system $\begin{cases} x + y = 4 \\ 2x - 3y = 3 \end{cases}$. Is $(2, 2)$ a solution? Is $(3, 1)$ a solution?

Solution   Check the proposed solution into each equation. Although $(2, 2)$ is a solution to the first equation $(2 + 2 = 4)$, it does not solve the second one since

$$2(2) - 3(2) = -2 \neq 3$$

But $(3, 1)$ is a solution since

$$3 + 1 = 4$$

and

$$2(3) - 3(1) = 3$$

PROBLEM 11-3 Graph the equations in the system of Problem 11-2 and verify the solution $(3, 1)$.

Solution   Each equation is a straight line. The first is

$$x + y = 4$$

or

$$y = -x + 4$$

The second is

$$2x - 3y = 3$$

$$3y = 2x - 3$$

or

$$y = \tfrac{2}{3}x - 1$$

The equations are graphed as in Figure 11-11. The two lines intersect at the point $(3, 1)$, so $(3, 1)$ is the solution of the system.

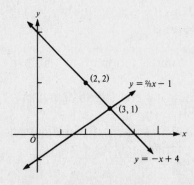

Figure 11-11

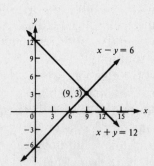

Figure 11-12

PROBLEM 11-4 Use the graphical method to find the solution of $\begin{cases} x + y = 12 \\ x - y = 6 \end{cases}$.

Solution   Each of these equations can be graphed reasonably accurately: They are both straight lines, which must intersect at the solution point if they aren't parallel or don't coincide. The first line is $y = -x + 12$, having slope -1 and y-intercept 12. The second line is $y = x - 6$, having slope $+1$ and y-intercept -6. Since their slopes are different, the lines aren't parallel; and since both their slopes and y-intercepts are different, the lines don't coincide. The graph in Figure 11-12 shows that the lines intersect at $(9, 3)$, which is the solution point.

Check: $9 + 3 \overset{\checkmark}{=} 12$; $9 - 3 \overset{\checkmark}{=} 6$

PROBLEM 11-5 Use the substitution method to find the solution to $\begin{cases} xy = 1 \\ 2x - y = -1 \end{cases}$.

Solution If you begin with the linear equation and solve it for y, you get the simplest substitution:

$$2x - y = -1$$

$$y = 2x + 1$$

Put this value of y into the original first equation:

$$xy = 1$$

$$x(2x + 1) = 1$$

$$2x^2 + x = 1$$

$$2x^2 + x - 1 = 0$$

$$(2x - 1)(x + 1) = 0$$

$$x = \tfrac{1}{2} \quad \text{or} \quad x = -1$$

Substitute these values into either of the original pair (choose the first one):

$$xy = 1 \qquad\qquad xy = 1$$

$$\tfrac{1}{2}y = 1 \qquad -1(y) = 1$$

$$y = 2 \qquad\qquad y = -1$$

So the possible solution pairs are $(\tfrac{1}{2}, 2)$ and $(-1, -1)$.

Check:

Does $(\tfrac{1}{2}, 2)$ satisfy $\{xy = 1, 2x - y = -1\}$?

$$\tfrac{1}{2} \cdot 2 \overset{\checkmark}{=} 1$$

$$2(\tfrac{1}{2}) - 2 \overset{\checkmark}{=} -1 \quad \text{Yes}$$

Does $(-1, -1)$ satisfy $\{xy = 1, 2x - y = -1\}$?

$$(-1)(-1) \overset{\checkmark}{=} 1$$

$$2(-1) - (-1) \overset{\checkmark}{=} -1 \quad \text{Yes}$$

PROBLEM 11-6 Check the solutions to the system of Problem 11-5 by graphical methods.

Solution The first equation $xy = 1$ is a hyperbola whose two parts lie in the first and third quadrants and whose center is at the origin. (This hyperbola is easily graphed by calculating a collection of pairs.) The second equation $2x - y = -1$ is a straight line having slope 2 and y-intercept 1. The two curves intersect at $(\tfrac{1}{2}, 2)$ and $(-1, -1)$, as shown in Figure 11-13—so these solutions are verified.

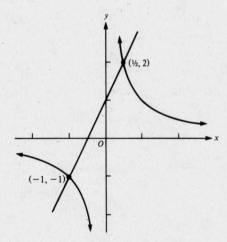

Figure 11-13

PROBLEM 11-7 **(a)** Use the substitution method to solve $\begin{cases} y &= \sqrt{x} \\ 8x - 2y + 1 = 0 \end{cases}$. **(b)** Check your answer by graphical methods.

Solution

(a) Substitute the first equation directly into the second one:

$$8x - 2y + 1 = 0$$

$$8x - 2\sqrt{x} + 1 = 0$$

$$8(\sqrt{x})^2 - 2(\sqrt{x}) + 1 = 0$$

This is a quadratic equation in the unknown $\sqrt{x}$. The quadratic formula gives

$$\sqrt{x} = \frac{-(-2) \pm \sqrt{(-2)^2 - 4(8)(1)}}{2 \cdot 8}$$

$$= \frac{2 \pm \sqrt{4 - 32}}{16}$$

$$= \frac{2 \pm \sqrt{-28}}{16}$$

This is an impossible equation because the value of $\sqrt{x}$ is a complex number—which cannot be in the real plane. Hence there are no solutions to this system.

(b) A graphical check shows that the two functions do not intersect (Figure 11-14).

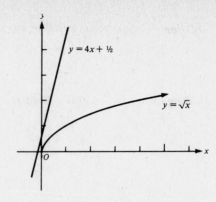

Figure 11-14

PROBLEM 11-8 If 2 apples and 5 oranges cost $1.00 and 4 apples and 3 oranges cost 88¢, how much does each apple and each orange cost?

Solution Let x and y be the cost (in cents) of each apple and each orange, respectively. Then the first condition is

$$2 \times (\text{cost of each apple}) + 5 \times (\text{cost of each orange}) = 100 \text{ cents}$$

$$2x + 5y = 100$$

The second equation is

$$4 \times (\text{cost of each apple}) + 3 \times (\text{cost of each orange}) = 88 \text{ cents}$$

$$4x + 3y = 88$$

This gives the system $\begin{cases} 2x + 5y = 100 \\ 4x + 3y = 88 \end{cases}$. Using the substitution method, solve the first equation for x:

$$2x + 5y = 100$$

$$2x = 100 - 5y$$

$$x = 50 - \tfrac{5}{2}y$$

Put this value of x into the second equation:

$$4(50 - \tfrac{5}{2}y) + 3y = 88$$

$$200 - 10y + 3y = 88$$

$$200 - 88 = 10y - 3y$$

$$112 = 7y$$

$$y = 16 \qquad \text{[Oranges are 16 cents each.]}$$

Put this value of y into the x equation:

$$x = 50 - \tfrac{5}{2}y$$

$$x = 50 - \tfrac{5}{2}(16)$$

$$x = 50 - 40 = 10 \qquad \text{[Apples are 10 cents each.]}$$

Check: $2x + 5y = 100$ $4x + 3y = 88$

$2(10) + 5(16) \overset{?}{=} 100$ $4(10) + 3(16) \overset{?}{=} 88$

$20 + 80 \overset{\checkmark}{=} 100$ $40 + 48 \overset{\checkmark}{=} 88$

Linear Systems of Equations in Two Variables

PROBLEM 11-9 Show that the system $\begin{cases} -3x + 2y = 3 \\ 8x - \tfrac{16}{3}y = 5 \end{cases}$ has no solution.

Solution Using the substitution method, find from the first equation that

$$2y = 3x + 3$$

or

$$y = \tfrac{3}{2}x + \tfrac{3}{2}$$

Putting this value into the second equation gives

$$8x - \tfrac{16}{3}(\tfrac{3}{2}x + \tfrac{3}{2}) = 5$$

$$8x - \tfrac{16}{3}\cdot\tfrac{3}{2}x - \tfrac{16}{3}\cdot\tfrac{3}{2} = 5$$

$$8x - 8x - 8 = 5$$

$$-8 \neq 5 \qquad \text{[Impossible.]}$$

Since the system of equations has produced an invalid equation, there is no solution.

PROBLEM 11-10 Verify the conclusion of Problem 11-9 by the graphing method.

Solution Write both equations in slope-intercept form. The first equation is

$$-3x + 2y = 3$$

or

$$y = \tfrac{3}{2}x + \tfrac{3}{2}$$

which is a straight line having slope $\tfrac{3}{2}$ and y-intercept $\tfrac{3}{2}$. The second equation is

$$8x - \tfrac{16}{3}y = 5$$

$$8x - 5 = \tfrac{16}{3}y$$

$$\tfrac{3}{16}(8x - 5) = y$$

or

$$y = \tfrac{3}{2}x - \tfrac{15}{16}$$

which is a straight line having slope $\tfrac{3}{2}$ and y-intercept $-\tfrac{15}{16}$. Since their slopes are the same, the lines are parallel—and thus never cross—so there is no solution to the system. (See Figure 11-15.)

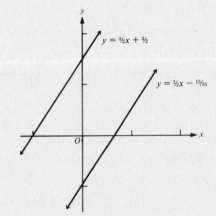

Figure 11-15

PROBLEM 11-11 Find the solution set of the system $\begin{cases} x - 4y = 2 \\ -3x + 12y = -6 \end{cases}$. Graph the result.

Solution Since $x = 4y + 2$,

$$-3x + 12y = -6$$

$$-3(4y + 2) + 12y = -6$$

$$-12y - 6 + 12y = -6$$

$$0 = 0$$

This means that the equations are the same, so their graphs coincide, as shown in Figure 11-16. To verify this and to obtain the solution set, go back to the original pair and create an equivalent pair by multiplying both sides of the first equation by -3:

$$\begin{cases} x - 4y = 2 \\ -3x + 12y = -6 \end{cases} \Leftrightarrow \begin{cases} -3x + 12y = -6 \\ -3x + 12y = -6 \end{cases}$$

Hence any number pair that satisfies the equation

$$-3x + 12y = -6 \Leftrightarrow x - 4y = 2$$

is a solution pair. If the first coordinate is c, then

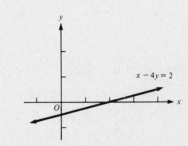

Figure 11-16

$$c - 4y = 2$$

$$c - 2 = 4y$$

$$y = \frac{c}{4} - \frac{1}{2}$$

so the solution set is $\left\{ \left(c, \frac{c}{4} - \frac{1}{2} \right) : c \text{ is a real number} \right\}$.

PROBLEM 11-12 Create other systems equivalent to the system $\begin{cases} 3x - y = 4 \\ -x + 2y = 7 \end{cases}$.

Solution Equivalent systems can be created in three ways:
(1) Interchange two equations:

$$\begin{cases} 3x - y = 4 \\ -x + 2y = 7 \end{cases} \xrightarrow{\text{Interchange}} \begin{cases} -x + 2y = 7 \\ 3x - y = 4 \end{cases}$$

(2) Multiply an equation by a nonzero constant, e.g.,

$$\begin{cases} 3x - y = 4 \\ -x + 2y = 7 \end{cases} \xrightarrow[\text{Multiply by 3}]{} \begin{cases} 3x - y = 4 \\ -3x + 6y = 21 \end{cases}$$

(3) Add to one equation a nonzero multiple of another equation, e.g.,

$$\begin{cases} 3x - y = 4 \\ -x + 2y = 7 \end{cases} \xrightarrow{\text{Multiply by 2}} \begin{cases} 6x - 2y = 8 \\ -x + 2y = 7 \end{cases} \xrightarrow[\text{equation to the second}]{\text{Add the first}} \begin{cases} 6x - 2y = 8 \\ 5x + 0y = 15 \end{cases}$$

PROBLEM 11-13 Find the solution to the system $\begin{cases} -5x + 3y = 8 \\ 2x + 6y = -2 \end{cases}$.

Solution Form an equivalent system by multiplying the first equation by -2:

$$\begin{cases} -2(-5x + 3y = 8) \\ 2x + 6y = -2 \end{cases} \Rightarrow \begin{cases} 10x - 6y = -16 \\ 2x + 6y = -2 \end{cases}$$

Now add the equations together to get

$$10x + 2x - 6y + 6y = -16 - 2$$

$$12x + 0y = -18$$

and solve for x:

$$12x = -18$$

$$x = -\frac{18}{12} = -\frac{3}{2}$$

Put this value of x into either of the original pair of equations to get

$$2(-\tfrac{3}{2}) + 6y = -2$$

$$-3 + 6y = -2$$

$$6y = 1$$

$$y = \tfrac{1}{6}$$

The solution is $(-\frac{3}{2}, \frac{1}{6})$.

Check: $-5(-\tfrac{3}{2}) + 3(\tfrac{1}{6}) \overset{?}{=} 8 \qquad 2(-\tfrac{3}{2}) + 6(\tfrac{1}{6}) \overset{?}{=} -2$

$\phantom{-5(-\tfrac{3}{2}) + 3}\tfrac{15}{2} + \tfrac{1}{2} \overset{\checkmark}{=} 8 \qquad -3 + 1 \overset{\checkmark}{=} -2$

PROBLEM 11-14 Solve the system $\begin{cases} 2x + 3y = 4 \\ -7x + 2y = -5 \end{cases}$.

Solution $\begin{cases} 7(2x + 3y = 4) \\ 2(-7x + 2y = -5) \end{cases} \Rightarrow \begin{cases} 14x + 21y = 28 \\ -14x + 4y = -10 \end{cases}$

note: The multiplier numbers are carefully chosen to make the x coefficients in the new set easy to work with.

Add:
$$14x - 14x + 21y + 4y = 28 - 10$$
$$0x + 25y = 18$$
$$y = \tfrac{18}{25}$$

Putting $y = \tfrac{18}{25}$ into one of the original pair of equations gives

$$2x + 3(\tfrac{18}{25}) = 4$$
$$2x + \tfrac{54}{25} = \tfrac{100}{25}$$
$$2x = \tfrac{46}{25}$$
$$x = \tfrac{23}{25}$$

The solution to the system is $(\tfrac{23}{25}, \tfrac{18}{25})$.

PROBLEM 11-15 At the candy store Carlos buys 5 taffy sticks and 4 lollipops for 60 cents. His sister Margarita gets 2 taffy sticks and 7 lollipops for 51 cents. What is the price of each item individually?

Solution Let x = price of a taffy stick
y = price of a lollipop

Then
$$\begin{cases} 5x + 4y = 60 & \text{[Carlos' purchase]} \\ 2x + 7y = 51 & \text{[Margarita's purchase]} \end{cases}$$

Multiply the first equation by 7 and the second equation by -4:

$$\begin{cases} 35x + 28y = 420 \\ -8x - 28y = -204 \end{cases}$$

So
$$27x + 0y = 216 \qquad \text{[Add.]}$$
$$x = \tfrac{216}{27} = 8 \qquad \text{[Taffy sticks cost 8¢ each.]}$$

and since $x = 8$,
$$5(8) + 4y = 60$$
$$40 + 4y = 60$$
$$4y = 20$$
$$y = 5 \qquad \text{[Lollipops cost 5¢ each.]}$$

PROBLEM 11-16 The sum of 3 times the tens digit plus the units digit of a given number is 10 while the sum of the tens digit and 5 times the units digit is 22. Find the number.

Solution Let t be the notation for the tens digit and u be the notation for the units digit. Then $10t + u$ is the desired two-digit number. Now set up the conditions in equivalent form:

$$\begin{cases} 3t + u = 10 \\ t + 5u = 22 \end{cases} \Rightarrow \begin{cases} -15t - 5u = -50 \\ t + 5u = 22 \end{cases}$$

Add these equations to get

$$-14t + 0u = -28$$
$$t = 2$$

But since
$$3t + u = 10 \qquad \text{[original equation]}$$

then
$$3(2) + u = 10$$
$$6 + u = 10$$
$$u = 4$$

So the number having the required properties is $10t + u = 10(2) + 4 = 24$.

PROBLEM 11-17   Find the area of the triangle in the first quadrant bounded by $x + 2y = 6$ and $2x - y = 5$. [*Hint:* $A = \frac{1}{2}$(base)(height).]

Solution   First, consider the triangle whose sides are the lines $y = -\frac{x}{2} + 3$ and $y = 2x - 5$. The base of the triangle, which is on the x-axis, must lie between the two x-intercepts, $x = \frac{5}{2}$ and $x = 6$. So the length of the base is $6 - \frac{5}{2} = \frac{7}{2}$. Then, the height must be the y-coordinate of the intersection point, which you find by solving the system

$$\begin{cases} x + 2y = 6 \\ 2x - y = 5 \end{cases} \Rightarrow \begin{cases} -2x - 4y = -12 \\ 2x - y = 5 \end{cases}$$

so that

$$-2x + 2x - 4y - y = -12 + 5$$

$$0x - 5y = -7$$

$$y = \tfrac{7}{5} \text{ units}$$

Then, the area of the triangle is $\frac{1}{2}$(base)(height), or

$$A = \tfrac{1}{2}(\tfrac{7}{2})(\tfrac{7}{5}) = \tfrac{49}{20}$$

PROBLEM 11-18   The length of a given rectangle is 3 more than twice its width. The perimeter is 36. What are its dimensions?

Solution   Let $x =$ width and $y =$ length. Then $\begin{cases} y = 2x + 3 \\ 2x + 2y = 36 \end{cases}$

Then

$$2x + 2(2x + 3) = 36$$

$$2x + 4x + 6 = 36$$

$$6x = 30$$

$$x = 5 \qquad \text{[The width is 5.]}$$

But

$$y = 2x + 3 = 2(5) + 3 = 13 \qquad \text{[The length is 13.]}$$

Linear Systems of Equations in Three Variables

PROBLEM 11-19   Find all solutions to the following linear system in three variables:

$$\begin{cases} 2x + 5y - 3z = 3 \\ 2y - 5z = 4 \\ 3z = 6 \end{cases}$$

Solution   This system is in triangular form, so it can be solved from the bottom up. The last equation gives

$$3z = 6$$

so

$$z = 2$$

Substitute $z = 2$ into the middle equation: $2y - 5(2) = 4$

$$2y - 10 = 4$$

so

$$y = 7$$

Substitute $z = 2$, $y = 7$ into the first equation: $2x + 5(7) - 3(2) = 3$

$$2x + 35 - 6 = 3$$

$$2x + 29 = 3$$

so

$$x = -13$$

The solution is the ordered triple $(x, y, z) = (-13, 7, 2)$.

PROBLEM 11-20   Show that the ordered triple (1, 2, 3) solves the system

$$\begin{cases} 2x + 2y - 4z = -6 \\ x + 2y + z = 8 \\ -2x + 3y - z = 1 \end{cases}$$

Solution   Substitute 1 for x, 2 for y, and 3 for z everywhere in the system and check the validity of each equation:

$2x + 2y - 4z = -6$	$x + 2y + z = 8$	$-2x + 3y - z = 1$
$2(1) + 2(2) - 4(3) \stackrel{?}{=} -6$	$1 + 2(2) + 3 \stackrel{?}{=} 8$	$-2(1) + 3(2) - 3 \stackrel{?}{=} 1$
$2 + 4 - 12 \stackrel{?}{=} -6$	$1 + 4 + 3 \stackrel{?}{=} 8$	$-2 + 6 - 3 \stackrel{?}{=} 1$
$-6 \stackrel{\checkmark}{=} -6$	$8 \stackrel{\checkmark}{=} 8$	$1 \stackrel{\checkmark}{=} 1$

PROBLEM 11-21   Use the substitution method to solve the system $\begin{cases} 2x - 4y + 3z = 3 \\ x - 3y + 4z = 7. \\ 2x - 2y + z = 1 \end{cases}$

Solution   Solve the third equation for z:

$$2x - 2y + z = 1$$
$$z = 1 - 2x + 2y$$

Substitute $z = 1 - 2x + 2y$ into the first two equations:

$$2x - 4y + 3z = 3$$
$$2x - 4y + 3(1 - 2x + 2y) = 3$$
$$2x - 4y + 3 - 6x + 6y = 3$$
$$-4x + 2y = 0 \qquad \text{[Keep this one.]}$$

$$x - 3y + 4z = 7$$
$$x - 3y + 4(1 - 2x + 2y) = 7$$
$$x - 3y + 4 - 8x + 8y = 7$$
$$-7x + 5y = 3 \qquad \text{[Keep this one.]}$$

Now work on the two "keepers": $\begin{cases} -4x + 2y = 0 \\ -7x + 5y = 3 \end{cases}$

The first of this pair gives

$$2y = 4x$$
$$y = 2x$$

and the second becomes

$$-7x + 5(2x) = 3$$
$$-7x + 10x = 3$$
$$x = 1 \qquad \text{[first coordinate of the solution]}$$

But since $y = 2x$, $y = 2(1) = 2$ [second coordinate of the solution]

Then $z = 1 - 2x + 2y$

$$= 1 - 2(1) + 2(2) = 1 - 2 + 4$$
$$= 3 \qquad \text{[third coordinate of the solution]}$$

The solution is the ordered triple $(1, 2, 3)$.

Not all systems problems come out with "nice-looking" solutions. But systems methods work nicely even when the the numbers become awkward.

PROBLEM 11-22 The graph of an equation of the form $y = ax^2 + bx + c$ is a parabola. Find the equation of the parabola that contains the points $(2, -3), (-2, 2), (-7, -1)$.

Solution Since the curve $y = ax^2 + bx + c$ is satisfied by the points on it, each pair given must satisfy this equation. Thus

$$-3 = a(2)^2 + b(2) + c \qquad \text{[from the point } (2, -3)\text{]}$$

$$2 = a(-2)^2 + b(-2) + c \qquad \text{[from the point } (-2, 2)\text{]}$$

$$-1 = a(-7)^2 + b(-7) + c \qquad \text{[from the point } (-7, -1)\text{]}$$

which can be written as a linear system in three variables:

$$\begin{cases} 4a + 2b + c = -3 \\ 4a - 2b + c = 2 \\ 49a - 7b + c = -1 \end{cases}$$

To solve this, eliminate the variable c first: Multiply the first equation by -1 and add it to the second equation and then independently to the third equation:

$$\begin{array}{r} -4a - 2b - c = 3 \\ 4a - 2b + c = 2 \end{array} \xrightarrow{\text{Add}} -4b = 5 \Rightarrow b = -\tfrac{5}{4}$$

$$\begin{array}{r} -4a - 2b - c = 3 \\ 49a - 7b + c = -1 \end{array} \xrightarrow{\text{Add}} 45a - 9b = 2$$

But $b = -\tfrac{5}{4}$, so
$$45a - 9(-\tfrac{5}{4}) = 2$$
$$45a = 2 - \tfrac{45}{4} = -\tfrac{37}{4}$$
$$a = -(\tfrac{37}{4})(\tfrac{1}{45}) = -\tfrac{37}{180}$$

Then from the second equation,

$$4a - 2b + c = 2$$
$$c = 2 - 4a + 2b$$
$$= 2 - 4(-\tfrac{37}{180}) + 2(-\tfrac{5}{4}) = 2 + \tfrac{37}{45} - \tfrac{5}{2}$$
$$= \tfrac{29}{90}$$

The desired values are $a = -\tfrac{37}{180}$, $b = -\tfrac{5}{4}$, and $c = \tfrac{29}{90}$, so the parabola that passes through the three given points is

$$y = -\tfrac{37}{180}x^2 - \tfrac{5}{4}x + \tfrac{29}{90}$$

PROBLEM 11-23 By producing successive equivalent systems, put the following system into triangular form and then solve it:

$$\begin{cases} -x + y \phantom{{}- 2z} = 3 \\ \phantom{-x + {}} y - 2z = 3 \\ 2x \phantom{{}+ y} + z = 5 \end{cases}$$

Solution

$$\begin{cases} -2x + 2y + 0z = 6 \\ \phantom{-2x + {}} y - 2z = 3 \\ 2x \phantom{{}+ 2y} + z = 5 \end{cases} \qquad \text{[Multiply the first equation by 2.]}$$

$$\begin{cases} -2x + 2y + 0z = 6 \\ \phantom{-2x + {}} y - 2z = 3 \\ \phantom{-2x + {}} 2y + z = 11 \end{cases} \qquad \begin{array}{l} \text{[Replace the third equation by the} \\ \text{sum of the first and third equations.]} \end{array}$$

$$\begin{cases} -2x + 2y + 0z = 6 \\ \quad\quad -2y + 4z = -6 \quad\quad \text{[Multiply the second equation by } -2.] \\ \quad\quad\quad 2y + z = 11 \end{cases}$$

Triangular form:

$$\begin{cases} -2x + 2y + 0z = 6 \\ \quad\quad -2y + 4z = -6 \quad\quad \text{[Replace the third equation by the sum of} \\ \quad\quad\quad\quad\quad 5z = 5 \quad\quad\quad \text{the second and third equations.]} \end{cases}$$

Begin the solution process from the bottom up:

Since
$$5z = 5$$
$$z = 1$$

then $\quad\quad -2y + 4z = -6 \quad$ and $\quad -2x + 2y + 0z = 6$

$\quad\quad\quad\quad -2y + 4(1) = -6 \quad\quad\quad\quad\quad -2x + 2(5) = 6$

$\quad\quad\quad\quad\quad -2y = -10 \quad\quad\quad\quad\quad\quad\quad -2x = -4$

$\quad\quad\quad\quad\quad\quad y = 5 \quad\quad\quad\quad\quad\quad\quad\quad\quad x = 2$

Thus the solution to the system is $(2, 5, 1)$.

PROBLEM 11-24 The basketball arena at U.R.T.W. (University of Retired Textbook Writers) has 2500 seats. The courtside seats sell for \$6, while the end zone seats are \$4 and the balcony seats are \$3 each. For a sell-out game the income is \$12 700. If all the non-balcony seats were sold at \$5, the gross income for a sell-out would be \$11 500. How many of each type of seat are there in the arena seating?

Solution Let $x =$ courtside seats
$\quad\quad\quad\quad y =$ end zone seats
$\quad\quad\quad\quad z =$ balcony seats

You know that $x + y + z =$ total seats $= 2500$. Also, the income figures tell you that

$$(x)(\$6) + (y)(\$4) + (z)(\$3) = \$12\,700$$

and that
$$(x)(\$5) + (y)(\$5) + (z)(\$3) = \$11\,500$$

So the system now is
$$\begin{cases} x + y + z = 2500 \quad\quad \text{(I)} \\ 6x + 4y + 3z = 12\,700 \quad\quad \text{(II)} \\ 5x + 5y + 3z = 11\,500 \quad\quad \text{(III)} \end{cases}$$

Replace this system by an equivalent system by the following operations:

(1) Replace equation II by the sum of equation II and -6 times equation I;
(2) Replace equation III by the sum of equation III and -5 times equation I:

$$\begin{matrix} & \\ (1) \Rightarrow \\ (2) \Rightarrow \end{matrix} \begin{cases} x + y + z = 2500 \\ 0x - 2y - 3z = -2300 \\ 0x + 0y - 2z = -1000 \end{cases}$$

This system is now in triangular form, so solve it:

If $\quad -2z = -1000 \Rightarrow z = 500$

then $\quad\quad -2y - 3z = -2300 \quad\quad\quad$ and $\quad\quad x + y + z = 2500$

$\quad\quad\quad -2y - 3(500) = -2300 \quad\quad\quad\quad\quad\quad x + 400 + 500 = 2500$

$\quad\quad\quad\quad\quad -2y = -2300 + 1500 = -800 \quad\quad\quad\quad\quad\quad x = 1600$

$\quad\quad\quad\quad\quad\quad y = 400$

So the arena has 1600 courtside seats, 400 end zone seats, and 500 balcony seats.

Systems of Inequalities in Two Variables

PROBLEM 11-25 Graph the solution set of the system $\begin{cases} x - 2y \le 4 \\ 3x + y \ge 3 \end{cases}$.

Solution First graph the line $x - 2y = 4$ or $y = \frac{1}{2}x - 2$ and $3x + y = 3$ or $y = -3x + 3$, as in Figure 11-17a. Solving the first inequality for y gives

$$x \le 4 + 2y$$
$$x - 4 \le 2y$$
$$\tfrac{1}{2}x - 2 \le y$$

So the solution set of the first inequality consists of all points whose y value is greater than or equal to the values on the straight line $y = \frac{1}{2}x - 2$, as shown by the half-plane graphed in Figure 11-17b. Then, solving the second inequality for y gives

$$3x + y \ge 3$$
$$y \ge -3x + 3$$

This solution set consists of all points in the plane whose y value is greater than or equal to the y values on the line $y = -3x + 3$, as shown by the half-plane graphed in Figure 11-17c. The solution set of the *system* of inequalities is the intersection of these two sets (the points of the plane that are in both sets at the same time), which is shown in Figure 11-17d.

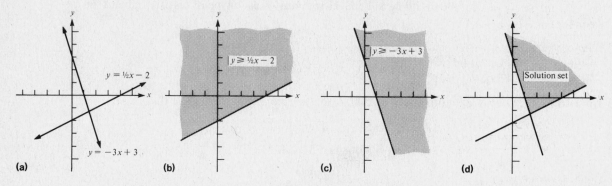

Figure 11-17

PROBLEM 11-26 Find the solution set and graph of the system

$$\begin{cases} x \ge 0 \\ y \ge 0 \\ 2x + y \le 4 \\ 2y \le 2x + 2 \end{cases}$$

Solution The first and second members of the system restrict the solution set to points in the first quadrant, including the axes. The third member, in the form $y \le -2x + 4$, includes those points on and below the line $y = -2x + 4$. The fourth member, in the form $y \le x + 1$, includes those points on and below the line $y = x + 1$. Putting these all together gives the solution set shown in Figure 11-18.

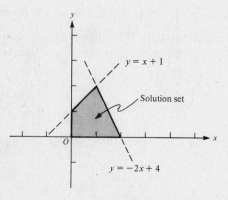

Figure 11-18

PROBLEM 11-27 Graph the set described by the system

$$\begin{cases} x \geq 0 \\ y \geq 0 \\ y \leq 3 \\ 2x - y > -1 \\ 2x + y \leq 8 \end{cases}$$

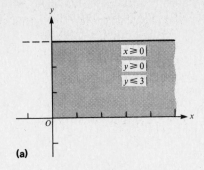

(a)

Solution The first two members restrict the set to the first quadrant, but the third member puts the set into the strip on and below $y = 3$, as shown in Figure 11-19a. The linear inequalities are analyzed independently: For $2x - y > -1$,

$$2x + 1 > y$$

so the y values are those below but not on the line $y = 2x + 1$. And for $2x + y \leq 8$,

$$y \leq -2x + 8$$

so the y values are those below and on the line $y = -2x + 8$. Putting all these together into one graph gives the solution set shown in Figure 11-19b.

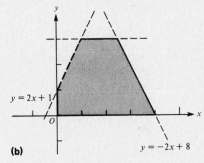

(b)

Figure 11-19

PROBLEM 11-28 Graph the set described by $\begin{cases} y \geq x^2 - 4 \\ y \leq 1 \\ x - 3y \geq -2 \end{cases}$

Solution The first member gives the points above and on the parabola $y = x^2 - 4$ (Figure 11-20a). The second member restricts the solution set to points on and below $y = 1$ (Figure 11-20b). The linear inequality is

$$x - 3y \geq -2$$
$$x + 2 \geq 3y$$
$$y \leq \tfrac{1}{3}x + \tfrac{2}{3}$$

which restricts the y values to points on and below the line $y = \tfrac{1}{3}x + \tfrac{2}{3}$. Combining all three gives the solution set shown in Figure 11-20c.

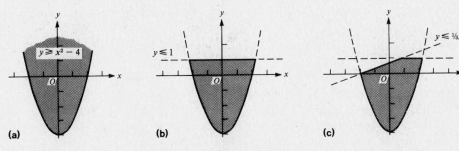

(a) (b) (c)

Figure 11-20

Linear Programming

PROBLEM 11-29 Find the values of x and y that produce the largest possible value of $P = x + 2y$, given that x and y are subject to the restrictions

$$\begin{cases} x \geq 0, \quad y \geq 0 \\ x + y \leq 6 \\ y - x \leq 3 \end{cases}$$

Solution   First, graph the feasible set from the given constraint conditions: The graph is a polygon in the first quadrant, as shown in Figure 11-21. Then find the coordinates of the vertices of this feasible set, which are $(0,0)$, $(6,0)$, $(0,3)$, $(\frac{3}{2},\frac{9}{2})$. Insert these coordinates into the objective function $P = x + 2y$:

Point	$P = x + 2y$
$(0,0)$	$P = 0 + 2(0) = 0$
$(6,0)$	$P = 6 + 2(0) = 6$
$(0,3)$	$P = 0 + 2(3) = 6$
$(\frac{3}{2},\frac{9}{2})$	$P = \frac{3}{2} + 2(\frac{9}{2}) = \frac{21}{2}$

The maximum value of P occurs at the point $(x,y) = (\frac{3}{2},\frac{9}{2})$, since $\frac{21}{2}$ is the largest value obtained.

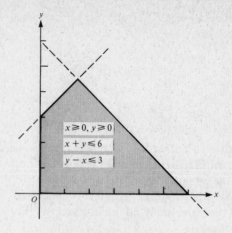

$x \geq 0, y \geq 0$

$x + y \leq 6$

$y - x \leq 3$

Figure 11-21

PROBLEM 11-30   A manufacturer of solar heating panels constructs two models, one large and one small. The cost of material for the large panel is $90 and the cost of material for the small panel is $45. Assembly of the small panel requires 2 hours of labor and assembly of the large one requires 6 hours. Labor cost is $6 per hour. Cash-flow constraints limit the allocation each day of $600 for labor and $1800 for material. Find the largest daily profit the company can make (profit equals revenue minus cost) if the small model is sold at $100 and the large model is sold at $175.

Solution   Let x denote the number of small panels and y denote the number of large panels constructed each day. The total material cost is then $45x + $90y and this must be held at or below $1800 daily. The total labor cost is $12x + $36y to be held at or below $600. The constraint conditions then are

$$x \geq 0, y \geq 0 \qquad \text{[You can't build a negative number of panels.]}$$

$$45x + 90y \leq 1800 \qquad \text{[Material constraint]}$$

$$12x + 36y \leq 600 \qquad \text{[Labor constraint]}$$

The objective function, which is to be maximized, is

$$\text{Profit} = \text{Revenue} - \text{Cost}$$

$$= \underbrace{(100x}_{\substack{\text{Sales of} \\ \text{small} \\ \text{panels}}} + \underbrace{175y)}_{\substack{\text{Sales of} \\ \text{large} \\ \text{panels}}} - (\underbrace{(45 + 12)x}_{\substack{\text{Material and} \\ \text{labor on} \\ \text{small panels}}} + \underbrace{(90 + 36)y)}_{\substack{\text{Material and} \\ \text{labor on} \\ \text{large panels}}}$$

$$= 100x + 175y - 57x - 126y$$

$$= 43x + 49y$$

The graph of the constraint conditions (Figure 11-22) is a polygon, which is the feasible set. Reduce each condition to simplest form and find the vertices, which are at $(0,0)$, $(40,0)$, $(0,\frac{50}{3})$ and at the intersection of $y = -\frac{1}{2}x + 20$ and $y = -\frac{1}{3}x + \frac{50}{3}$. Find this intersection by solving these equations simultaneously:

$$-\tfrac{1}{3}x + \tfrac{50}{3} = -\tfrac{1}{2}x + 20$$

$$(\tfrac{1}{2} - \tfrac{1}{3})x = 20 - \tfrac{50}{3} = \tfrac{60}{3} - \tfrac{50}{3}$$

$$\tfrac{1}{6}x = \tfrac{10}{3}$$

$$x = 20$$

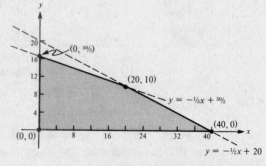

$(0, \frac{50}{3})$

$(20, 10)$

$y = -\frac{1}{3}x + \frac{50}{3}$

$(40, 0)$

$y = -\frac{1}{2}x + 20$

$(0, 0)$

Figure 11-22

Then $y = -\frac{1}{2}x + 20 = -\frac{1}{2}(20) + 20 = 10$, so the fourth vertex is at $(20, 10)$. Now test the objective function at the four vertices:

Point	Profit
$(0, 0)$	0
$(40, 0)$	$43(40) + 49(0) = \mathbf{1720}$
$(0, \frac{50}{3})$	$43(0) + 49(\frac{50}{3}) = 816.67$
$(20, 10)$	$43(20) + 49(10) = 1350$

The daily profit is maximized at $1720 by constructing 40 small panels and no large ones.

note: The high labor cost ($36 per panel) on the large panels constrains a maximum of 16 large panels if those were the only ones being made—that's the point $(0, \frac{50}{3}) \approx (0, 16)$ [$\frac{2}{3}$ of a panel is meaningless!] But the profit is low there. The low labor cost ($12 per panel) on the small panels allows, with the materials-cost constraint, 40 of these to be made per day, giving the largest possible profit level. You may also notice that pure profit per panel is $43 per small one and $49 per large one.

Supplementary Exercises

PROBLEM 11-31 Is the point $(2, -1)$ a solution to the system $\begin{cases} 2x + y = 3 \\ -x + 4y = -6 \end{cases}$?

PROBLEM 11-32 Is there a solution to the system $\begin{cases} 3x - y = 5 \\ -9x + 3y = 2 \end{cases}$?

PROBLEM 11-33 Find two numbers whose sum is 45 and whose difference is 17.

Solve the following systems.

PROBLEM 11-34 $\begin{cases} x + 4y = 7 \\ x - 5y = -2 \end{cases}$

PROBLEM 11-37 $\begin{cases} 3x + 5y - 4 = 0 \\ 2x - 3y + 10 = 0 \end{cases}$

PROBLEM 11-35 $\begin{cases} 5x - 3 = -3y \\ x + 9y = 2 \end{cases}$

PROBLEM 11-38 $\begin{cases} 8x - 12y = -20 \\ -2x + 3y = 5 \end{cases}$

PROBLEM 11-36 $\begin{cases} \dfrac{1}{x} + \dfrac{2}{y} = 2 \\ \dfrac{1}{x} - \dfrac{1}{y} = -4 \end{cases}$

PROBLEM 11-39 $\begin{cases} 3x^2 - y^2 = 4 \\ x^2 + y^2 = 12 \end{cases}$

PROBLEM 11-40 The perimeter of a rectangle is 20 in. If the length is doubled, the perimeter becomes 32 in. Find the dimensions of the original rectangle.

PROBLEM 11-41 The line $y = ax + b$ contains the points $(3, 1)$ and $(-4, -2)$. Construct the system of equations and solve it to find the slope-intercept equation of this particular line.

PROBLEM 11-42 A pile of 33 coins contains dimes and quarters. The total value is $5.70. How many of each coin are there?

PROBLEM 11-43 A chemist wants to make 100 mL of a 29% acid solution by combining a 20% solution with a 50% solution. How many milliliters of each solution should she use?

PROBLEM 11-44 The cost of tickets for the concert is $4 for adults and $2.50 for children. The income from 570 tickets sold is $2100.00. How many of each kind were sold?

PROBLEM 11-45 Solve the system $\begin{cases} x - y = 2 \\ x = y^2 - 2y + 4 \end{cases}$.

PROBLEM 11-46 Solve the system $\begin{cases} x^2 + y^2 = 4 \\ x - 2y + 6 = 0 \end{cases}$.

PROBLEM 11-47 We have three positive numbers such that one of them is twice another. Their sum is 23 and their product is 360. What are the three numbers?

Solve the following systems.

PROBLEM 11-48 $\begin{cases} 4x - 2y + z = 8 \\ y + z = 5 \\ 3z = 6 \end{cases}$

PROBLEM 11-51 $\begin{cases} x - y + z = 3 \\ y - z + x = 1 \\ z + y - x = 7 \end{cases}$

PROBLEM 11-49 $\begin{cases} x - y = 4z \\ x + y = 2z \\ y + z = 0 \end{cases}$

PROBLEM 11-52 $\begin{cases} \dfrac{1}{x} + \dfrac{1}{y} = \dfrac{5}{2} \\ \dfrac{1}{y} + \dfrac{1}{z} = 3 \\ \dfrac{1}{x} + \dfrac{1}{z} = \dfrac{3}{2} \end{cases}$

PROBLEM 11-50 $\begin{cases} x - 2y - 3z = 2 \\ -x + 4y + 13z = -4 \\ 3x - 5y - 4z = -2 \end{cases}$

PROBLEM 11-53 When the piggy bank was smashed, it contained a collection of pennies, nickels and dimes worth $4.90. Of the 116 coins, there were twice as many dimes as there were nickels. How many of each coin were there?

PROBLEM 11-54 The sum of three numbers is 37. The sum of the smaller two is three less than the largest and the largest number is four times the smallest. Find the three numbers.

PROBLEM 11-55 Manny, Moe, and Curly have a combined total of $120. Moe loses a $15 bet to Curly. Now Moe and Curly have the same amount, which is exactly half of what Manny has. How much did each stooge have before the bet?

PROBLEM 11-56 Construct the graph of the solution set to

$$\{x \geq 0,\ y \geq 0,\ x + y \leq 3,\ x - 2y \geq -2\}$$

PROBLEM 11-57 Sketch the graph for $\{y \geq 1,\ y \leq x,\ x \leq 3\}$.

PROBLEM 11-58 Find the maximum value for $C = 2x + 5y$ subject to the following constraints:

$$\begin{cases} x \geq 0 \\ y \geq 0 \\ 3y - x \geq 6 \\ x + 2y \geq 6 \end{cases}$$

PROBLEM 11-59 Find the maximum and minimum values of $P = 5x + 2y$ subject to the constraints

$$\begin{cases} y \geq 9 - \frac{3}{2}x \\ y \leq 9 + 2x \\ x \leq 6 \end{cases}$$

PROBLEM 11-60   A farmer can grow either wheat or oats on his 300 acres of land. His profit is $50 per acre on the wheat, but it requires 4 hours of labor per acre; his profit on oats is $30 per acre, which requires 2 hours of labor per acre. He figures that he has 1000 hours of labor available during the growing season to tend to these crops. How many acres of each should he plant to maximize his profit?

Answers to Supplementary Exercises

11-31   Yes

11-32   No

11-33   31 and 14

11-34 $(3, 1)$

11-35 $(\frac{1}{2}, \frac{1}{6})$

11-36 $(-\frac{1}{2}, \frac{1}{2})$

11-37 $(-2, 2)$

11-38 $\left(a, \dfrac{2a + 5}{3}\right)$
(i.e., x can be any real number a)

11-39 $(\pm 2, 2\sqrt{2})$
$(\pm 2, -2\sqrt{2})$

11-40   4 in. by 6 in.

11-41 $y = \frac{3}{7}x - \frac{2}{7}$

11-42   17 dimes, 16 quarters

11-43   30 mL of 50% solution
70 mL of 20% solution

11-44   450 adult, 120 children

11-45 $(4, 2), (3, 1)$

11-46   No solution

11-47   5, 6, 12

11-48 $(3, 3, 2)$

11-49 $(3a, -a, a)$
(z can be any real number a.)

11-50   No solution

11-51 $(2, 4, 5)$

11-52 $(2, \frac{1}{2}, 1)$

11-53   65 pennies, 17 nickels, 34 dimes

11-54   5, 12, 20

11-55   Manny, $60; Moe, $45; Curly, $15

11-56   See Figure 11-23.

11-57   See Figure 11-24.

11-58 $C = \frac{72}{5}$ at $(\frac{6}{5}, \frac{12}{5})$

11-59   Maximum, 72; minimum, 18

11-60   200 wheat, 100 oats

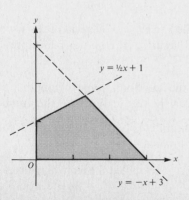

Figure 11-23

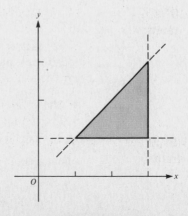

Figure 11-24

12-1. Matrix Methods for Systems of Equations

A. Matrix definitions and terminology

When we're using the elimination method on a system of equations, we're manipulating the coefficients but not the variables. If we simply don't write the variables, we're left with an ordered arrangement of numbers, which we can write in a bracketed rectangular array. For example,

$$\begin{cases} 2x + y + 6z \\ -x + 3y + 4z \end{cases} \Rightarrow \begin{bmatrix} 2 & 1 & 6 \\ -1 & 3 & 4 \end{bmatrix} \quad \text{or} \quad \begin{cases} 2x + y + 3z \\ -3x \quad\quad + 4z \\ 5x - 2y \end{cases} \Rightarrow \begin{bmatrix} 2 & 1 & 3 \\ -3 & 0 & 4 \\ 5 & -2 & 0 \end{bmatrix}$$

This array is called a *matrix*.

note: If one variable doesn't appear in a given equation, we use a zero in the appropriate position of the matrix.

• A **matrix** is a rectangular array of numbers denoted by

$$\begin{array}{c} \\ \text{Rows} \to i = 1 \\ 2 \\ 3 \\ \\ m \end{array} \overset{\displaystyle j = 1 \quad 2 \quad\quad 3 \quad\quad\quad n \leftarrow \text{Columns}}{\begin{bmatrix} a_{11} & a_{12} & a_{13} & \cdots & a_{1n} \\ a_{21} & a_{22} & a_{23} & \cdots & a_{2n} \\ a_{31} & a_{32} & a_{33} & \cdots & a_{3n} \\ \vdots & \vdots & \vdots & & \vdots \\ a_{m1} & a_{m2} & a_{m3} & \cdots & a_{mn} \end{bmatrix}}$$

where each real number a_{ij} is called an **entry** or **element** of the matrix. The location of each element a_{ij} (read "a sub i,j") in the matrix is expressed by a unique pair of subscripts:

• The first subscript i indicates the (horizontal) *row* the number is in.
• The second subscript j indicates the (vertical) *column* the number is in.

Thus a_{11} is the entry in row 1, column 1; a_{23} is the entry in row 2, column 3; and a_{mn} is the entry in row m, column n.

note: The elements $a_{11}, a_{22}, a_{33}, \ldots$ are called the **main diagonal elements** of the matrix.

EXAMPLE 12-1 (a) Write the matrix for $\begin{cases} 4x - y + 3z \\ 15y + 2z \end{cases}$. (b) Using subscript notation, identify all of the elements of the matrix.

Solution

(a)
$$A = \begin{bmatrix} 4 & -1 & 3 \\ 0 & 15 & 2 \end{bmatrix}$$

   note: We usually denote matrices by capital letters (A, B, M, etc.).

(b) Since the number 4 appears in the *first* row and the *first* column, we write $a_{11} = 4$. Likewise, $a_{12} = -1$, $a_{13} = 3$, $a_{21} = 0$, $a_{22} = 15$, and $a_{23} = 2$.

We classify a matrix by its size. An **$m \times n$ matrix**—read "m by n"—is a matrix with m rows and n columns. A matrix with the same number of rows and columns—i.e., of size $n \times n$—is called a **square matrix of order n**. A matrix that contains a single row or a single column is called a **vector**.

caution: "$m \times n$" does NOT mean "m times n."

EXAMPLE 12-2 Classify the following matrices by size:

(a) $A = [4]$ (b) $B = \begin{bmatrix} 1 & 1 \\ 3 & 2 \end{bmatrix}$ (c) $M = \begin{bmatrix} 0 & 5 & 3 & 2 \\ -2 & 4 & 5 & 6 \end{bmatrix}$ (d) $N = \begin{bmatrix} 8 \\ 7 \\ 3 \end{bmatrix}$

Solution

(a) A is a 1×1 matrix, since there is just 1 row and 1 column.
(b) B is a 2×2 matrix; it is also a square matrix of order 2.
(c) M is a 2×4 matrix.
(d) N is a 3×1 matrix; it is also a vector.

If we have a system of equations such as
$$\begin{cases} 5x - 3y - 2z = 2 \\ 4x \quad\quad - 3z = 8 \\ \quad\quad 2y + 3z = -3 \end{cases}$$

the **coefficient matrix** of the system is
$$\begin{bmatrix} 5 & -3 & -2 \\ 4 & 0 & -3 \\ 0 & 2 & 3 \end{bmatrix}$$

and the **augmented matrix** is
$$\begin{bmatrix} 5 & -3 & -2 & \vdots & 2 \\ 4 & 0 & -3 & \vdots & 8 \\ 0 & 2 & 3 & \vdots & -3 \end{bmatrix}$$

This is the matrix that represents the *system* of equations.

EXAMPLE 12-3 Form the augmented matrices for

(a) $\begin{cases} 2x - 5y = 3 \\ -x + 3y = 4 \end{cases}$ (b) $\begin{cases} x + 3z - 5 = 0 \\ 2x - 4y + 3z = 0 \\ 3y + 5z - 12 = 8 \end{cases}$

Solution

(a) The various terms of the equations must be lined up in columns according to the variable letter; the constants must be on the right of the equal sign. This system is already in the required form:

$$\begin{cases} 2x - 5y = 3 \\ -x + 3y = 4 \end{cases} \Rightarrow \left[\begin{array}{cc|c} 2 & -5 & 3 \\ -1 & 3 & 4 \end{array} \right]$$

(b) Reorganize each of the equations to line up the variables in the correct (x, y, z) place and get the constants on the right side:

$$\begin{cases} x + 3z - 5 = 0 \\ 2x - 4y + 3z = 0 \Rightarrow \\ 3y + 5z - 12 = 8 \end{cases} \begin{cases} x + 0y + 3z = 5 \\ 2x - 4y + 3z = 0 \\ 0x + 3y + 5z = 20 \end{cases}$$

Now the augmented matrix is easy to form:

$$\left[\begin{array}{ccc|c} 1 & 0 & 3 & 5 \\ 2 & -4 & 3 & 0 \\ 0 & 3 & 5 & 20 \end{array} \right]$$

B. Equivalent matrices: Elementary row operations

When we're working with systems of equations and using the elimination method, we operate on the equations. We can multiply or divide by a constant, interchange equations, replace one equation by the sum or difference of that equation and another, and/or use any combination of these operations—none of these alterations changes the system or its solution. These alterations on the equations of a system correspond exactly to the operations we are allowed to perform on a matrix. Since the rows of the matrix actually are the equations without the "frills," the **elementary row operations** are

(1) Interchange any two rows.
(2) Multiply or divide all the elements of any row by any nonzero constant.
(3) Replace one row by the sum or difference of that row and another row.
(4) Replace one row by the sum of that row and a nonzero multiple of another row.

We say that two augmented matrices are **equivalent matrices** if one of them can be derived from the other by using only the elementary row operations (1)–(4).

The notation below the arrows here is just a convenient shorthand: $R_2 \rightarrow -1R_2$ is short for "Replace row two by multiplying row two by -1"; $R_1 \rightarrow R_1 + R_2$ means "Replace row one by the sum of rows one and two"; $R_2 \rightarrow R_2 + -2R_1$ means "Replace row two by the sum of row two and -2 times row one."

caution: If you go on to study computer science (i.e., algorithmic) notation, you'll find that the symbol for "replacement" is an arrow pointing to the *left*: For example, $P \leftarrow P + Q$ would mean "P is replaced by $P + Q$."

EXAMPLE 12-4 Show that the matrices

$$M_1 = \begin{bmatrix} 2 & 3 & 5 \\ 5 & -2 & 3 \end{bmatrix} \quad \text{and} \quad M_2 = \begin{bmatrix} -3 & 5 & 2 \\ 1 & -8 & -7 \end{bmatrix}$$

are equivalent.

Solution Use operations (1)–(4) on M_1 in a succession of steps to get M_2:

$$\begin{bmatrix} 2 & 3 & 5 \\ 5 & -2 & 3 \end{bmatrix} \xrightarrow[\substack{(R_2 \rightarrow -1R_2)}]{\substack{\text{Step 1: Multiply row two} \\ \text{by } -1}} \begin{bmatrix} 2 & 3 & 5 \\ -5 & 2 & -3 \end{bmatrix} \xrightarrow[\substack{(R_1 \rightarrow R_1 + R_2)}]{\substack{\text{Step 2: Replace row one by the sum of} \\ \text{rows one and two}}}$$

$$\begin{bmatrix} -3 & 5 & 2 \\ -5 & 2 & -3 \end{bmatrix} \xrightarrow[\substack{(R_2 \rightarrow R_2 + -2R_1)}]{\substack{\text{Step 3: Replace row two by the sum of} \\ \text{row two and } -2 \text{ times row one}}} \begin{bmatrix} -3 & 5 & 2 \\ 1 & -8 & -7 \end{bmatrix}$$

C. Solving a system by matrix methods

To see how the matrix operations compare to the operations on the equations in a system, we'll do a side-by-side comparison as we reduce the system to triangular form.

EXAMPLE 12-5 (a) Reduce the following system to triangular form. (b) Solve the system.

$$\begin{cases} x + y + z = 13 \\ 3x + y - 3z = 5 \\ x - 2y + 4z = 10 \end{cases}$$

Solution

(a) Replace the third equation/row by the difference of the first and third equations/rows:

$$\begin{cases} x + y + z = 13 \\ 3x + y - 3z = 5 \\ x - 2y + 4z = 10 \end{cases} \qquad \begin{bmatrix} 1 & 1 & 1 & | & 13 \\ 3 & 1 & -3 & | & 5 \\ 1 & -2 & 4 & | & 10 \end{bmatrix}$$

$$(R_3 \to R_1 - R_3) \quad \begin{cases} x + y + z = 13 \\ 3x + y - 3z = 5 \\ 0 + 3y - 3z = 3 \end{cases} \qquad \begin{bmatrix} 1 & 1 & 1 & | & 13 \\ 3 & 1 & -3 & | & 5 \\ 0 & 3 & -3 & | & 3 \end{bmatrix}$$

Replace the second equation/row by the sum of the second equation/row and -3 times the first equation/row:

$$(R_2 \to R_2 + -3R_1) \quad \begin{cases} x + y + z = 13 \\ 0 - 2y - 6z = -34 \\ 0 + 3y - 3z = 3 \end{cases} \qquad \begin{bmatrix} 1 & 1 & 1 & | & 13 \\ 0 & -2 & -6 & | & -34 \\ 0 & 3 & -3 & | & 3 \end{bmatrix}$$

Divide the second equation/row by -2; divide the third equation/row by 3:

$$\begin{matrix} (R_2 \to -\frac{1}{2}R_2) \\ (R_3 \to \frac{1}{3}R_3) \end{matrix} \quad \begin{cases} x + y + z = 13 \\ 0 + y + 3z = 17 \\ 0 + y - z = 1 \end{cases} \qquad \begin{bmatrix} 1 & 1 & 1 & | & 13 \\ 0 & 1 & 3 & | & 17 \\ 0 & 1 & -1 & | & 1 \end{bmatrix}$$

Replace the third equation by the difference of the second and third equation:

$$(R_3 \to R_2 - R_3) \quad \begin{cases} x + y + z = 13 \\ y + 3z = 17 \\ 4z = 16 \end{cases} \qquad \begin{bmatrix} 1 & 1 & 1 & | & 13 \\ 0 & 1 & 3 & | & 17 \\ 0 & 0 & 4 & | & 16 \end{bmatrix}$$

While the equations were being put into triangular form, the augmented matrix was changed into its own "triangular form" called **echelon form**, in which all the elements below the main diagonal are zero.

(b) Continue only on the matrix now. Divide the last row by 4:

$$\begin{bmatrix} 1 & 1 & 1 & | & 13 \\ 0 & 1 & 3 & | & 17 \\ 0 & 0 & 1 & | & 4 \end{bmatrix}$$

We read the rows of the matrix as equations. From the bottom up, we see that $z = 4$, that $y + 3z = 17$ or $y + 3 \cdot 4 = 17$ or $y = 5$, and that $x + y + z = 13$ or $x + 5 + 4 = 13$ or $x = 4$. So the solution set is $(4, 5, 4)$.

EXAMPLE 12-6 Use the augmented matrix to show that the following system has infinitely many solutions and write its solution set:

$$\begin{cases} x + y - 3z = 10 \\ -5x - 8y + 8z = -48 \\ 2x + 5y + z = 18 \end{cases}$$

Solution Form the matrix first; then use the matrix row operations:

$$\begin{bmatrix} 1 & 1 & -3 & \vdots & 10 \\ -5 & -8 & 8 & \vdots & -48 \\ 2 & 5 & 1 & \vdots & 18 \end{bmatrix} \xrightarrow{R_3 \to -2R_1 + R_3} \begin{bmatrix} 1 & 1 & -3 & \vdots & 10 \\ -5 & -8 & 8 & \vdots & -48 \\ 0 & 3 & 7 & \vdots & -2 \end{bmatrix}$$

$$\begin{bmatrix} 1 & 1 & -3 & \vdots & 10 \\ -5 & -8 & 8 & \vdots & -48 \\ 0 & 3 & 7 & \vdots & -2 \end{bmatrix} \xrightarrow{R_2 \to 5R_1 + R_2} \begin{bmatrix} 1 & 1 & -3 & \vdots & 10 \\ 0 & -3 & -7 & \vdots & 2 \\ 0 & 3 & 7 & \vdots & -2 \end{bmatrix}$$

$$\begin{bmatrix} 1 & 1 & -3 & \vdots & 10 \\ 0 & -3 & -7 & \vdots & 2 \\ 0 & 3 & 7 & \vdots & -2 \end{bmatrix} \xrightarrow{R_3 \to R_2 + R_3} \begin{bmatrix} 1 & 1 & -3 & \vdots & 10 \\ 0 & -3 & -7 & \vdots & 2 \\ 0 & 0 & 0 & \vdots & 0 \end{bmatrix}$$

The identity $0 = 0$ in the last matrix shows that we have an infinite number of solutions. The equations that remain are $x + y - 3z = 10$ and $-3y - 7z = 2$. Solve the second for y in terms of z:

$$y = \frac{-2 - 7z}{3} = -\frac{2}{3} - \frac{7}{3}z$$

Then, from the first one we have

$$x = 10 + 3z - y = 10 + 3z + \frac{2}{3} + \frac{7}{3}z = \frac{32}{3} + \frac{16}{3}z$$

The infinite solution set is the set of triples

$$\{(\tfrac{32}{3} + \tfrac{16}{3}z, -\tfrac{2}{3} - \tfrac{7}{3}z, z) \mid z \text{ is any real number}\}$$

12-2. Determinants

A. Definition of a determinant

If a system of equations has the same number of variables as equations, i.e., $m = n$, then the coefficient matrix of the system is square. In this case, the system can be solved by a method that uses a special number associated with square matrices only, called the *determinant* of the matrix.

- The **determinant** of a square matrix A, such that

$$A = \begin{bmatrix} a_{11} & a_{12} & \cdots & a_{1n} \\ a_{21} & a_{22} & \cdots & a_{2n} \\ \vdots & \vdots & & \vdots \\ a_{n1} & a_{n2} & \cdots & a_{nn} \end{bmatrix}$$

is a number det A denoted by

$$\det A = \begin{vmatrix} a_{11} & a_{12} & \cdots & a_{1n} \\ a_{21} & a_{22} & \cdots & a_{2n} \\ \vdots & \vdots & & \vdots \\ a_{n1} & a_{n2} & \cdots & a_{nn} \end{vmatrix}$$

- For a 1×1 matrix $A_1 = [a_{11}]$, the **1×1 determinant** is

$$\det A_1 = |a_{11}| = a_{11}$$

- For a 2 × 2 matrix $A_2 = \begin{bmatrix} a_{11} & a_{12} \\ a_{21} & a_{22} \end{bmatrix}$, the **2 × 2 determinant** is

$$\det A_2 = \begin{vmatrix} a_{11} & a_{12} \\ a_{21} & a_{22} \end{vmatrix} = a_{11}a_{22} - a_{21}a_{12}$$

whose value is the difference of the products on the diagonals.

EXAMPLE 12-7 Find the value of the determinant of

(a) $A = [5]$ (b) $B = \begin{bmatrix} 2 & 3 \\ 1 & 5 \end{bmatrix}$ (c) $C = \begin{bmatrix} -1 & 4 \\ 2 & 6 \end{bmatrix}$

Solution

(a) Since the matrix $A = [5]$ is 1×1, the value of $\det A = 5$.

(b) This matrix is 2×2, so $\det B = \begin{vmatrix} 2 & 3 \\ 1 & 5 \end{vmatrix} = 2 \cdot 5 - 1 \cdot 3 = 10 - 3 = 7$

(c) This matrix is 2×2, so $\det C = \begin{vmatrix} -1 & 4 \\ 2 & 6 \end{vmatrix} = -1 \cdot 6 - 2 \cdot 4 = -14$

Things start getting more complicated with a 3×3 matrix.

- For a 3×3 matrix $D = \begin{bmatrix} a_{11} & a_{12} & a_{13} \\ a_{21} & a_{22} & a_{23} \\ a_{31} & a_{32} & a_{33} \end{bmatrix}$, the 3×3 determinant is

$$\det D = \begin{vmatrix} a_{11} & a_{12} & a_{13} \\ a_{21} & a_{22} & a_{23} \\ a_{31} & a_{32} & a_{33} \end{vmatrix} = a_{11}a_{22}a_{33} + a_{12}a_{23}a_{31} + a_{13}a_{21}a_{32}$$
$$- a_{31}a_{22}a_{13} - a_{32}a_{23}a_{11} - a_{33}a_{21}a_{12}$$

There's a process that will help us to find the value of a 3×3 determinant— we follow the arrows in this diagram:

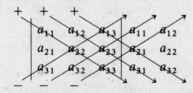

(1) Form the products of the elements on the *main diagonals* (those from upper left to lower right) and give them a plus sign.
(2) Form the products of the elements on the *minor diagonals* (those from lower left to upper right) and give them a negative sign.
(3) Add these products to get the final value.

EXAMPLE 12-8 Evaluate

(a) $\det A = |3|$ (b) $\det B = \begin{vmatrix} 2 & 7 \\ 1 & 6 \end{vmatrix}$ (c) $\det C = \begin{vmatrix} 1 & 0 & -2 \\ 4 & 3 & 1 \\ -1 & 5 & 2 \end{vmatrix}$

(d) $\det A$ if $A = \begin{bmatrix} 2 & 4 & 1 \\ 0 & -3 & -2 \\ 6 & 1 & 0 \end{bmatrix}$ (e) $\det B$ if $B = \begin{bmatrix} 3 & 0 & -1 \\ 4 & 5 & 6 \\ -2 & 1 & 3 \end{bmatrix}$

Solution

(a) $\det A = 3$

(b) $\det B = \begin{vmatrix} 2 & 7 \\ 1 & 6 \end{vmatrix} = 2 \cdot 6 - 1 \cdot 7 = 12 - 7 = 5$

(c) $\det C = \begin{vmatrix} 1 & 0 & -2 \\ 4 & 3 & 1 \\ -1 & 5 & 2 \end{vmatrix}$

$= (1 \cdot 3 \cdot 2) + (0 \cdot 1(-1)) + ((-2)4 \cdot 5) - ((-1)(3)(-2))$

$\quad - (5 \cdot 1 \cdot 1) - (2 \cdot 4 \cdot 0)$

$= 6 + 0 + (-40) - 6 - 5 - 0 = -45$

(d) $\det A = \begin{vmatrix} 2 & 4 & 1 \\ 0 & -3 & -2 \\ 6 & 1 & 0 \end{vmatrix}$

$= (2(-3)0) + (4(-2)6) + (1 \cdot 0 \cdot 1) - (6(-3)1) - (1(-2)2) - (0 \cdot 0 \cdot 4)$

$= 0 - 48 + 0 + 18 + 4 - 0 = -26$

(e) $\det B = \begin{vmatrix} 3 & 0 & -1 \\ 4 & 5 & 6 \\ -2 & 1 & 3 \end{vmatrix}$

$= (3 \cdot 5 \cdot 3) + (0 \cdot 6(-2)) + ((-1)4 \cdot 1)$

$\quad - ((-2)(5)(-1)) - (1 \cdot 6 \cdot 3) - (3 \cdot 4 \cdot 0)$

$= 45 - 0 - 4 - 10 - 18 - 0 = 13$

B. Definitions of a determinant in terms of minors and cofactors

recall: We label the elements of a matrix by their pair of subscripts; e.g., the element a_{23} is in the second row and third column.

There are two attributes of $n \times n$ (square) matrices, where $n > 1$, that allow us to find the numerical value of their determinants by using these attributes: We can evaluate the determinant of a square matrix in terms of (1) *minors* and (2) *cofactors.*

(1) If A is an $n \times n$ matrix, the minor M_{ij} of A is the determinant of the matrix formed by deleting the ith row and the jth column from A.

EXAMPLE 12-9   Form the minor M_{13} and the minor M_{22} for the matrix

$$A = \begin{bmatrix} 4 & 1 & 2 \\ 1 & -3 & 1 \\ 2 & -1 & 2 \end{bmatrix}$$

Solution   To get M_{13}, delete row one and column three from A to get the 2×2 matrix B:

$$\begin{bmatrix} 4 & 1 & 2 \\ 1 & -3 & 1 \\ 2 & -1 & 2 \end{bmatrix} \Rightarrow \begin{bmatrix} 1 & -3 \\ 2 & -1 \end{bmatrix} = B$$

Then find det B:

$$M_{13} = \det B = \begin{vmatrix} 1 & -3 \\ 2 & -1 \end{vmatrix} = 1(-1) - 2(-3) = -1 + 6 = 5$$

To get M_{22}, delete row two and column two from A to get the 2×2 matrix C:

$$\begin{bmatrix} 4 & 1 & 2 \\ 1 & -3 & 1 \\ 2 & -1 & 2 \end{bmatrix} \Rightarrow \begin{bmatrix} 4 & 2 \\ 2 & 2 \end{bmatrix} = C$$

Then $M_{22} = \det C = \begin{vmatrix} 4 & 2 \\ 2 & 2 \end{vmatrix} = 4 \cdot 2 - 2 \cdot 2 = 8 - 4 = 4$

EXAMPLE 12-10 Find the minor M_{23} for the matrix

$$A = \begin{bmatrix} 2 & 1 & 1 & 3 \\ 4 & -2 & 0 & 1 \\ -3 & 0 & 2 & 5 \\ 1 & 1 & 4 & 2 \end{bmatrix}$$

Solution Delete the second row and the third column to get the 3×3 matrix B:

$$B = \begin{bmatrix} 2 & 1 & 3 \\ -3 & 0 & 5 \\ 1 & 1 & 2 \end{bmatrix}$$

and then $M_{23} = \det B = \begin{vmatrix} 2 & 1 & 3 \\ -3 & 0 & 5 \\ 1 & 1 & 2 \end{vmatrix} = 0 + 5 - 9 - 0 - 10 + 6 = -8$

The definition of a 3×3 determinant can be restated in terms of minors:

- If M_{ij} is a minor of a 3×3 matrix A, then

$$\det A = \begin{vmatrix} a_{11} & a_{12} & a_{13} \\ a_{21} & a_{22} & a_{23} \\ a_{31} & a_{32} & a_{33} \end{vmatrix} = a_{11}M_{11} - a_{12}M_{12} + a_{13}M_{13}$$

But the definition in terms of M_{ij} has a minus sign on the second term, which we can get rid of by defining *cofactors*:

(2) If A is an $n \times n$ matrix, the **cofactor** A_{ij} of an element a_{ij} is

$$A_{ij} = (-1)^{i+j}M_{ij}$$

Then since $A_{11} = (-1)^{1+1}M_{11} = M_{11}$, $A_{12} = (-1)^{1+2}M_{12} = -M_{12}$, and $A_{13} = (-1)^{1+3}M_{13} = M_{13}$, we can write the definition of a 3×3 determinant in terms of cofactors:

- If A_{ij} is the cofactor of an element a_{ij} of a 3×3 matrix A, then

$$\det A = \begin{vmatrix} a_{11} & a_{12} & a_{13} \\ a_{21} & a_{22} & a_{23} \\ a_{31} & a_{32} & a_{33} \end{vmatrix} = a_{11}A_{11} + a_{12}A_{12} + a_{13}A_{13}$$

This form of the definition leads us to the mathematical definition of the determinant of any square matrix larger than 1×1.

The **determinant of a square matrix larger than 1 × 1** *may be found by multiplying the elements in any one row (or column) by their respective cofactors and adding the resulting products.*

note: The cofactor definition above uses the elements in the first row, so that $\det A = a_{11}A_{11} + a_{12}A_{12} + a_{13}A_{13}$. But *any* row/column can be used.

EXAMPLE 12-11   Use three different processes to find the value of det A for

$$A = \begin{bmatrix} 1 & 4 & 2 \\ 0 & 1 & -2 \\ 3 & 0 & 1 \end{bmatrix}$$

Solution   First, find the value by the diagonals product:

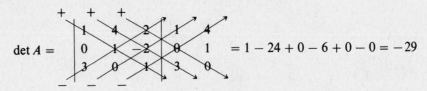

$$\det A = \begin{vmatrix} 1 & 4 & 2 \\ 0 & 1 & -2 \\ 3 & 0 & 1 \end{vmatrix} = 1 - 24 + 0 - 6 + 0 - 0 = -29$$

Second, find the value by the minors products:

$$\det A = a_{11}M_{11} - a_{12}M_{12} + a_{13}M_{13}$$

$$= 1 \begin{vmatrix} 1 & -2 \\ 0 & 1 \end{vmatrix} - 4 \begin{vmatrix} 0 & -2 \\ 3 & 1 \end{vmatrix} + 2 \begin{vmatrix} 0 & 1 \\ 3 & 0 \end{vmatrix}$$

$$= 1(1 - 0) - 4(0 - (-6)) + 2(0 - 3) = 1 - 24 - 6 = -29$$

Finally, use the cofactor method, with the first column as the basis column (this column is a good choice because it has a zero in it).

$$\det A = 1A_{11} + 0A_{21} + 3A_{31}$$

$$= 1\{(-1)^{1+1} \cdot M_{11}\} + 0 + 3\{(-1)^{3+1} \cdot M_{31}\}$$

$$= 1 \cdot 1 \cdot \begin{vmatrix} 1 & -2 \\ 0 & 1 \end{vmatrix} + 3 \cdot 1 \cdot \begin{vmatrix} 4 & 2 \\ 1 & -2 \end{vmatrix}$$

$$= 1 \cdot 1(1 - 0) + 3 \cdot 1(-8 - 2) = 1 - 24 - 6 = -29$$

Or, use the third row as a basis row:

$$\det A = 3A_{31} + 0A_{32} + 1A_{33}$$

$$= 3\{(-1)^{3+1} \cdot M_{31}\} + 0 + 1\{(-1)^{3+3} \cdot M_{33}\}$$

$$= 3 \cdot 1 \cdot \begin{vmatrix} 4 & 2 \\ 1 & -2 \end{vmatrix} + 0 + 1 \cdot 1 \cdot \begin{vmatrix} 1 & 4 \\ 0 & 1 \end{vmatrix}$$

$$= 3 \cdot 1(-8 - 2) + 1 \cdot 1(1 - 0) = -24 - 6 + 1 = -29$$

note: Since the cofactor method allows us to choose the elements of any row or column as the multipliers of the cofactors, we always try to choose the basis row or column that contains the *most zeros*. Clearly, multiplying zero times any cofactor produces a zero value, so the cofactor need not be evaluated there.

EXAMPLE 12-12   Use the cofactor method to evaluate the determinant of the 4×4 matrix

$$B = \begin{bmatrix} 2 & 0 & 1 & 3 \\ -1 & 2 & 0 & 4 \\ 0 & 2 & 5 & 1 \\ -1 & 3 & 0 & 1 \end{bmatrix}$$

Solution   The third column is the best one to use because it has two zeros, so we only have to do two cofactor evaluations:

$$\det B = 1A_{13} + 0A_{23} + 5A_{33} + 0A_{43}$$

$$= 1(-1)^{1+3}\begin{vmatrix} -1 & 2 & 4 \\ 0 & 2 & 1 \\ -1 & 3 & 1 \end{vmatrix} + 5(-1)^{3+3}\begin{vmatrix} 2 & 0 & 3 \\ -1 & 2 & 4 \\ -1 & 3 & 1 \end{vmatrix}$$

$$= 1\begin{vmatrix} -1 & 2 & 4 \\ 0 & 2 & 1 \\ -1 & 3 & 1 \end{vmatrix} + 5\begin{vmatrix} 2 & 0 & 3 \\ -1 & 2 & 4 \\ -1 & 3 & 1 \end{vmatrix}$$

Now use the process again on each of these. First expand the first determinant using basis row two:

$$0A'_{21} + 2A'_{22} + 1A'_{23} = 2(-1)^{2+2}\begin{vmatrix} -1 & 4 \\ -1 & 1 \end{vmatrix} + 1(-1)^{2+3}\begin{vmatrix} -1 & 2 \\ -1 & 3 \end{vmatrix}$$

$$= 2 \cdot 1(-1 + 4) + 1(-1)(-3 + 2)$$

$$= 6 + 1 = 7$$

Expand the second one using basis column two:

$$0A''_{12} + 2A''_{22} + 3A''_{32} = 2 \cdot (-1)^{2+2}\begin{vmatrix} 2 & 3 \\ -1 & 1 \end{vmatrix} + 3(-1)^{3+2}\begin{vmatrix} 2 & 3 \\ -1 & 4 \end{vmatrix}$$

$$= 2 \cdot 1(2 + 3) + 3(-1)(8 + 3)$$

$$= 10 - 33 = -23$$

So

$$\det B = 1(7) + 5(-23) = 7 - 115 = -108$$

C. Special operations on determinants

The evaluation of a large determinant can be time-consuming, especially if the determinant doesn't contain any zeros. Fortunately, there are a few simple operations we can do that don't change the value of a determinant but which do change the numbers in it. The following changes can aid in finding the value of a determinant:

(1) If two rows (or two columns) of a determinant are interchanged, a multiplier of -1 is introduced in the value of the determinant.

(2) If all of the elements in any row (or column) are multiplied by a nonzero constant k, the value of the new determinant is k times the value of the original determinant.

(3) If all of the elements in any row (or column) are divided by a nonzero constant c, the value of the new determinant is $1/c$ times the value of the original determinant.

(4) If a nonzero multiple of any row (or column) is added to another row (or column), then the value of the determinant is unchanged.

Two properties, which evolve from the above, are also useful to know:

(5) If all of the elements in any row (or column) of a determinant are zero, the value of the determinant is zero.

(6) If any two rows (or two columns) are identical, the value of the determinant is zero.

EXAMPLE 12-13 Evaluate the following using the operations on determinants:

$$
\text{(a)} \begin{vmatrix} 2 & 1 & 6 \\ 4 & 8 & 3 \\ 7 & 5 & 9 \end{vmatrix}
\qquad
\text{(b)} \begin{vmatrix} 3 & 6 & 2 & -2 \\ -2 & 1 & -3 & 5 \\ 1 & 4 & 1 & -2 \\ 0 & 2 & -4 & 7 \end{vmatrix}
$$

Solution   The trick is to keep changing numbers until there are enough zeros to make working the problem easy.

(a) The third column contains a common factor 3, so divide column three (C_3) by 3. To keep the value equal, put this value 3 as a multiplier in front:

$$
\begin{vmatrix} 2 & 1 & 6 \\ 4 & 8 & 3 \\ 7 & 5 & 9 \end{vmatrix}
\xrightarrow{C_3 \to \frac{1}{3}C_3}
3 \begin{vmatrix} 2 & 1 & 2 \\ 4 & 8 & 1 \\ 7 & 5 & 3 \end{vmatrix}
$$

Now multiply the second row by -2 and add it to the first row:

$$
3 \begin{vmatrix} 2 & 1 & 2 \\ 4 & 8 & 1 \\ 7 & 5 & 3 \end{vmatrix}
\xrightarrow{R_1 \to -2R_2 + R_1}
3 \begin{vmatrix} -6 & -15 & 0 \\ 4 & 8 & 1 \\ 7 & 5 & 3 \end{vmatrix}
$$

Now multiply the second row by -3 and add it to the third row; simultaneously factor -3 from row one:

$$
3 \begin{vmatrix} -6 & -15 & 0 \\ 4 & 8 & 1 \\ 7 & 5 & 3 \end{vmatrix}
\xrightarrow{R_3 \to -3R_2 + R_3;\ R_1 \to -\frac{1}{3}R_1}
(-3)(3) \begin{vmatrix} 2 & 5 & 0 \\ 4 & 8 & 1 \\ -5 & -19 & 0 \end{vmatrix}
$$

Now expand, using the third column as a basis:

$$
= (-9)(1)(-1)^{2+3} \begin{vmatrix} 2 & 5 \\ -5 & -19 \end{vmatrix} = (-9)(-1)(-38 + 25) = -117
$$

(b) Column one already contains a zero, and since it has a 1, it's easy to get more zeros. Multiply row three (R_3) by 2 and add it to row two (R_2):

$$
\begin{vmatrix} 3 & 6 & 2 & -2 \\ -2 & 1 & -3 & 5 \\ 1 & 4 & 1 & -2 \\ 0 & 2 & -4 & 7 \end{vmatrix}
\xrightarrow{R_2 \to 2(R_3) + R_2}
\begin{vmatrix} 3 & 6 & 2 & -2 \\ 0 & 9 & -1 & 1 \\ 1 & 4 & 1 & -2 \\ 0 & 2 & -4 & 7 \end{vmatrix}
$$

Now multiply row three by -3 and add it to row one:

$$
\begin{vmatrix} 3 & 6 & 2 & -2 \\ 0 & 9 & -1 & 1 \\ 1 & 4 & 1 & -2 \\ 0 & 2 & -4 & 7 \end{vmatrix}
\xrightarrow{R_1 \to -3(R_3) + R_1}
\begin{vmatrix} 0 & -6 & -1 & 4 \\ 0 & 9 & -1 & 1 \\ 1 & 4 & 1 & -2 \\ 0 & 2 & -4 & 7 \end{vmatrix}
$$

Choose your operations carefully! Don't destroy a zero in one spot as you create a zero in another spot.

To get still more zeros, do a few more operations. Add column four to column three:

$$
\begin{vmatrix} 0 & -6 & -1 & 4 \\ 0 & 9 & -1 & 1 \\ 1 & 4 & 1 & -2 \\ 0 & 2 & -4 & 7 \end{vmatrix}
\xrightarrow{C_3 \to C_4 + C_3}
\begin{vmatrix} 0 & -6 & 3 & 4 \\ 0 & 9 & 0 & 1 \\ 1 & 4 & -1 & -2 \\ 0 & 2 & 3 & 7 \end{vmatrix}
$$

Multiply column three by 2 and add it to column two:

$$\begin{vmatrix} 0 & -6 & 3 & 4 \\ 0 & 9 & 0 & 1 \\ 1 & 4 & -1 & -2 \\ 0 & 2 & 3 & 7 \end{vmatrix} \xrightarrow{C_2 \to 2(C_3) + C_2} \begin{vmatrix} 0 & 0 & 3 & 4 \\ 0 & 9 & 0 & 1 \\ 1 & 2 & -1 & -2 \\ 0 & 8 & 3 & 7 \end{vmatrix}$$

Multiply row one by -1 and add it to row four:

$$\begin{vmatrix} 0 & 0 & 3 & 4 \\ 0 & 9 & 0 & 1 \\ 1 & 2 & -1 & -2 \\ 0 & 8 & 3 & 7 \end{vmatrix} \xrightarrow{R_4 \to -1(R_1) + R_4} \begin{vmatrix} 0 & 0 & 3 & 4 \\ 0 & 9 & 0 & 1 \\ 1 & 2 & -1 & -2 \\ 0 & 8 & 0 & 3 \end{vmatrix}$$

Finally, expand using cofactors of the elements in the first column:

$$= 1(-1)^{3+1} \begin{vmatrix} 0 & 3 & 4 \\ 9 & 0 & 1 \\ 8 & 0 & 3 \end{vmatrix} \qquad \text{[The only nonzero entry in column one was used.]}$$

Now expand this determinant by cofactors of the second column:

$$= 1 \cdot 1 \cdot 3(-1)^{1+2} \begin{vmatrix} 9 & 1 \\ 8 & 3 \end{vmatrix} = 3(-1)(27 - 8) = -57$$

EXAMPLE 12-14 Show that each of the following determinants has value zero:

(a) $\begin{vmatrix} p & q \\ 4p & 4q \end{vmatrix}$ (b) $\begin{vmatrix} 2 & 1 & 6 \\ 5 & 0 & 3 \\ -10 & 0 & -6 \end{vmatrix}$ (c) $\begin{vmatrix} 4 & 2 & 1 \\ 5 & 7 & 8 \\ 4 & 2 & 1 \end{vmatrix}$

Solution

(a) This one is easy to expand by the products of diagonals:

$$\begin{vmatrix} p & q \\ 4p & 4q \end{vmatrix} = p(4q) - (4p)q = 4pq - 4pq = 0$$

(b) Expand this one using cofactors of the second column:

$$1(-1)^{1+2} \begin{vmatrix} 5 & 3 \\ -10 & -6 \end{vmatrix} = 1(-1)[-30 - (-30)] = 1(-1)(0) = 0$$

(c) The first row is the same as the third row. And by property (6), if a determinant has identical rows, its value is zero. As a reinforcement, multiply row one by -1 and add it to row three:

$$\begin{vmatrix} 4 & 2 & 1 \\ 5 & 7 & 8 \\ 4 & 2 & 1 \end{vmatrix} \xrightarrow{R_3 \to -1R_1 + R_3} \begin{vmatrix} 4 & 2 & 1 \\ 5 & 7 & 8 \\ 0 & 0 & 0 \end{vmatrix} = 0$$

The last row has all zeros. And by property (5), if a determinant has all zeros in a row, its value is zero.

Or, expand by using the cofactors of the third row:

$$0A_{31} + 0A_{32} + 0A_{33} = 0$$

12-3. Solving Linear Systems by Determinants: Cramer's Rule

By now you're probably wondering why we go through all this "tic tac toe" to find determinants. It's fun, but what *good* is it? Well, we do this because we can *use* determinants to solve systems of equations.

- If we have a system of linear equations in which the number of equations is the same as the number of unknowns, we can solve the system by a technique known as *Cramer's Rule*.

A. Cramer's Rule for two linear equations in two unknowns

Consider the system

$$\begin{cases} a_1x + b_1y = c_1 \\ a_2x + b_2y = c_2 \end{cases}$$

where the coefficient determinant $D = \begin{vmatrix} a_1 & b_1 \\ a_2 & b_2 \end{vmatrix} = a_1b_2 - a_2b_1 \neq 0$.

First, solve for x and y:

x	y
Multiply the first equation by b_2 and the second by b_1:	Multiply the first equation by a_2 and the second by a_1:
$a_1b_2x + b_1b_2y = c_1b_2$	$a_2a_1x + a_2b_1y = a_2c_1$
$a_2b_1x + b_2b_1y = c_2b_1$	$a_1a_2x + a_1b_2y = a_1c_2$
Subtract the second equation from the first:	Subtract the first equation from the second:
$(a_1b_2 - a_2b_1)x + 0y = c_1b_2 - c_2b_1$	$0x + (a_1b_2 - a_2b_1)y = a_1c_2 - a_2c_1$
or $\quad x = \dfrac{c_1b_2 - c_2b_1}{a_1b_2 - a_2b_1}$	or $\quad y = \dfrac{a_1c_2 - a_2c_1}{a_1b_2 - a_2b_1}$
so $\quad x = \dfrac{c_1b_2 - c_2b_1}{D}$	so $\quad y = \dfrac{a_1c_2 - a_2c_1}{D}$

Now, rewrite x and y in determinant form:

$$x = \frac{\begin{vmatrix} c_1 & b_1 \\ c_2 & b_2 \end{vmatrix}}{D} \quad \text{and} \quad y = \frac{\begin{vmatrix} a_1 & c_1 \\ a_2 & c_2 \end{vmatrix}}{D}$$

Finally, use the notation D_x and D_y such that

$$D_x = \begin{vmatrix} c_1 & b_1 \\ c_2 & b_2 \end{vmatrix} \quad \text{and} \quad D_y = \begin{vmatrix} a_1 & c_1 \\ a_2 & c_2 \end{vmatrix}$$

Then

$$x = \frac{D_x}{D} \quad \text{and} \quad y = \frac{D_y}{D}$$

Now we can state **Cramer's Rule** for a system of two linear equations in two unknowns: The solution(s) to the system $\begin{cases} a_1x + b_1y = c_1 \\ a_2x + b_2y = c_2 \end{cases}$, where $a_1b_2 - a_2b_1 \neq 0$, is

$$x = \frac{\begin{vmatrix} c_1 & b_1 \\ c_2 & b_2 \end{vmatrix}}{\begin{vmatrix} a_1 & b_1 \\ a_2 & b_2 \end{vmatrix}} = \frac{D_x}{D} \quad \text{and} \quad y = \frac{\begin{vmatrix} a_1 & c_1 \\ a_2 & c_2 \end{vmatrix}}{\begin{vmatrix} a_1 & b_1 \\ a_2 & b_2 \end{vmatrix}} = \frac{D_y}{D} \quad \text{where } D \neq 0$$

note: D—which *must* be nonzero—is the coefficient determinant of the original system:

D_x is obtained from D by replacing the "x-column" of D by the column of constants to the right of the equal signs;

D_y is obtained from D by replacing the "y-column" of D by the column of constants to the right of the equal sign.

EXAMPLE 12-15 Use Cramer's Rule to solve the system $\begin{cases} 2x + 5y = -4 \\ -x + 6y = -7 \end{cases}$.

Solution The coefficient determinant is

$$D = \begin{vmatrix} 2 & 5 \\ -1 & 6 \end{vmatrix} = 2 \cdot 6 - (-1)(5) = 12 + 5 = 17$$

Then

$$D_x = \begin{vmatrix} -4 & 5 \\ -7 & 6 \end{vmatrix} = -4 \cdot 6 - (-7)(5) = -24 + 35 = 11$$

and

$$D_y = \begin{vmatrix} 2 & -4 \\ -1 & -7 \end{vmatrix} = 2(-7) - (-1)(-4) = -14 - 4 = -18$$

Thus

$$x = \frac{D_x}{D} = \frac{11}{17} \quad \text{and} \quad y = \frac{D_y}{D} = \frac{-18}{17}$$

B. Cramer's Rule for three linear equations in three unknowns

Cramer's Rule also works for three equations in three unknowns. For the system

$$\begin{cases} a_1x + b_1y + c_1z = d_1 \\ a_2x + b_2y + c_2z = d_2 \\ a_3x + b_3y + c_3z = d_3 \end{cases}$$

where

$$D = \begin{vmatrix} a_1 & b_1 & c_1 \\ a_2 & b_2 & c_2 \\ a_3 & b_3 & c_3 \end{vmatrix} \neq 0$$

the unique solution is given by

$$x = \frac{\begin{vmatrix} d_1 & b_1 & c_1 \\ d_2 & b_2 & c_2 \\ d_3 & b_3 & c_3 \end{vmatrix}}{D}, \quad y = \frac{\begin{vmatrix} a_1 & d_1 & c_1 \\ a_2 & d_2 & c_2 \\ a_3 & d_3 & c_3 \end{vmatrix}}{D}, \quad z = \frac{\begin{vmatrix} a_1 & b_1 & d_1 \\ a_2 & b_2 & d_2 \\ a_3 & b_3 & d_3 \end{vmatrix}}{D}$$

so that in D_x, D_y, D_z notation

$$x = \frac{D_x}{D}, \quad y = \frac{D_y}{D}, \quad z = \frac{D_z}{D}$$

EXAMPLE 12-16 Use Cramer's Rule to solve the system $\begin{cases} 2x - y - 3z = 0 \\ 2x + y + z = 0 \\ 4x + 3y + 2z = 2 \end{cases}$.

Solution First, find the coefficient determinant of the system:

$$D = \begin{vmatrix} 2 & -1 & -3 \\ 2 & 1 & 1 \\ 4 & 3 & 2 \end{vmatrix} = 4 + (-4) + (-18) - (-12) - 6 - (-4) = -8$$

Then, find D_x, D_y, and D_z—use the cofactor method:

$$D_x = \begin{vmatrix} 0 & -1 & -3 \\ 0 & 1 & 1 \\ 2 & 3 & 2 \end{vmatrix} = 0A_{11} + 0A_{21} + 2A_{31} = 2(-1)^{3+1} \begin{vmatrix} -1 & -3 \\ 1 & 1 \end{vmatrix}$$

$$= 2 \begin{vmatrix} -1 & -3 \\ 1 & 1 \end{vmatrix} = 2(-1 + 3) = 4$$

$$D_y = \begin{vmatrix} 2 & 0 & -3 \\ 2 & 0 & 1 \\ 4 & 2 & 2 \end{vmatrix} = 0A_{12} + 0A_{22} + 2A_{32} = 2(-1)^{3+2}\begin{vmatrix} 2 & -3 \\ 2 & 1 \end{vmatrix}$$

$$= 2(-1)\begin{vmatrix} 2 & -3 \\ 2 & 1 \end{vmatrix} = -2(2+6) = -16$$

$$D_z = \begin{vmatrix} 2 & -1 & 0 \\ 2 & 1 & 0 \\ 4 & 3 & 2 \end{vmatrix} = 0A_{13} + 0A_{23} + 2A_{33} = 2(-1)^{3+3}\begin{vmatrix} 2 & -1 \\ 2 & 1 \end{vmatrix}$$

$$= 2\begin{vmatrix} 2 & -1 \\ 2 & 1 \end{vmatrix} = 2(2+2) = 8$$

So the solution is

$$x = \frac{D_x}{D} = \frac{4}{-8} = -\frac{1}{2}, \quad y = \frac{D_y}{D} = \frac{-16}{-8} = 2, \quad z = \frac{D_z}{D} = \frac{8}{-8} = -1$$

Check:

$$2(-\tfrac{1}{2}) - 2 - 3(-1) \overset{?}{=} 0 \qquad 2(-\tfrac{1}{2}) + 2 + (-1) \overset{?}{=} 0 \qquad 4(-\tfrac{1}{2}) + 3(2) + 2(-1) \overset{?}{=} 2$$

$$-1 \quad -2+3 \quad \overset{\checkmark}{=} 0 \qquad -1 \quad +2-1 \quad \overset{\checkmark}{=} 0 \qquad -2 \quad +6 \quad -2 \quad \overset{\checkmark}{=} 2$$

12-4. Matrix Products and Inverses

A. Properties of matrices

Before we obtain the process of multiplying a matrix by a matrix, we need to know the following properties of matrices:

(1) If A and B are $m \times n$ matrices, then A and B are **equal matrices**—if and only if the elements in corresponding positions in A and B are equal. Thus

EQUALITY OF MATRICES

$$\begin{bmatrix} a_{11} & a_{12} \\ a_{21} & a_{22} \end{bmatrix} = \begin{bmatrix} b_{11} & b_{12} \\ b_{21} & b_{22} \end{bmatrix}$$

if and only if $\begin{aligned} a_{11} = b_{11}, a_{12} = b_{12} \\ a_{21} = b_{21}, a_{22} = b_{22} \end{aligned}$

note: "Equivalent" matrices aren't the same as "equal" matrices.

(2) If A and B are $m \times n$ matrices, the **sum (of matrices)** A and B is the $m \times n$ matrix whose elements are the sums of the corresponding elements of A and B. Thus

SUM OF MATRICES

$$\begin{bmatrix} a_{11} & a_{12} & a_{13} \\ a_{21} & a_{22} & a_{23} \end{bmatrix} + \begin{bmatrix} b_{11} & b_{12} & b_{13} \\ b_{21} & b_{22} & b_{23} \end{bmatrix}$$
$$= \begin{bmatrix} a_{11} + b_{11} & a_{12} + b_{12} & a_{13} + b_{13} \\ a_{21} + b_{21} & a_{22} + b_{22} & a_{23} + b_{23} \end{bmatrix}$$

If A and B don't have exactly the same number of elements, the sum $A + B$ is *undefined*.

(3) If A is an $m \times n$ matrix and c is a nonzero constant, then the **scalar product** cA is the matrix whose elements are c times the corresponding element of A. Thus

SCALAR PRODUCT OF A MATRIX

$$c\begin{bmatrix} a_{11} & a_{12} \\ a_{21} & a_{22} \\ a_{31} & a_{32} \end{bmatrix} = \begin{bmatrix} ca_{11} & ca_{12} \\ ca_{21} & ca_{22} \\ ca_{31} & ca_{32} \end{bmatrix}$$

B. Products of matrices

To define the product of matrices A and B, we need to make a size check *for compatibility:*

- Two matrices A and B are **compatible** for multiplication, i.e., give the product AB, if and only if the number of columns of A is the same as the number of rows of B.

  note: If the number of columns of A does *not* equal the number of rows of B, the product is *undefined*. That is, matrix multiplication is, in general, not commutative.

Thus, we're allowed to form the product matrix AB only if A is $m \times p$ and B is $p \times n$ for any positive integers m and n: The product matrix AB will then be $m \times n$.

- If the entries of the product matrix AB are denoted c_{ij}, for the entry in the ith row and jth column, then each entry is found by

PRODUCT OF MATRICES $c_{ij} = a_{i1}b_{1j} + a_{i2}b_{2j} + a_{i3}b_{3j} + \cdots + a_{ip}b_{pj}$

This is the sum of the element-by-element products of the ith row of A with the jth column of B.

EXAMPLE 12-17   Find the product of $A = \begin{bmatrix} 2 & 1 \\ 3 & 2 \end{bmatrix}$ and $B = \begin{bmatrix} -1 & 2 \\ 3 & 1 \end{bmatrix}$.

Solution A and B are both 2×2 matrices, so AB will be a 2×2 matrix. The entries in AB, denoted by c_{ij}, are

$$c_{11} = a_{11}b_{11} + a_{12}b_{21} \Rightarrow (2)(-1) + (1)(3) = 1$$

$$c_{12} = a_{11}b_{12} + a_{12}b_{22} \Rightarrow (2)(2) + (1)(1) = 5$$

$$c_{21} = a_{21}b_{11} + a_{22}b_{21} \Rightarrow (3)(-1) + (2)(3) = 3$$

$$c_{22} = a_{21}b_{12} + a_{22}b_{22} \Rightarrow (3)(2) + (2)(1) = 8$$

So $AB = \begin{bmatrix} c_{11} & c_{12} \\ c_{21} & c_{22} \end{bmatrix} = \begin{bmatrix} 1 & 5 \\ 3 & 8 \end{bmatrix}$

EXAMPLE 12-18   Find the product **(a)** AB and **(b)** BA, where

$$A = \begin{bmatrix} 2 & 1 \\ -1 & 3 \\ 0 & 2 \end{bmatrix} \quad \text{and} \quad B = \begin{bmatrix} 1 & 0 & 3 & 1 \\ 2 & -1 & 2 & 0 \end{bmatrix}$$

Solution

(a) Since A is 3×2 and B is 2×4, they're compatible for multiplication and the product matrix AB will be 3×4.

$$c_{11} = (2)(1) + (1)(2) = 4, \qquad c_{12} = (2)(0) + (1)(-1) = -1$$

$$c_{13} = (2)(3) + (1)(2) = 8, \qquad c_{14} = (2)(1) + (1)(0) = 2$$

$$c_{21} = (-1)(1) + (3)(2) = 5, \qquad c_{22} = (-1)(0) + (3)(-1) = -3$$

$$c_{23} = (-1)(3) + (3)(2) = 3, \qquad c_{24} = (-1)(1) + (3)(0) = -1$$

$$c_{31} = (0)(1) + (2)(2) = 4, \qquad c_{32} = (0)(0) + (2)(-1) = -2$$

$$c_{33} = (0)(3) + (2)(2) = 4, \qquad c_{34} = (0)(1) + (2)(0) = 0$$

So
$$AB = \begin{bmatrix} c_{11} & c_{12} & c_{13} & c_{14} \\ c_{21} & c_{22} & c_{23} & c_{24} \\ c_{31} & c_{32} & c_{33} & c_{34} \end{bmatrix} = \begin{bmatrix} 4 & -1 & 8 & 2 \\ 5 & -3 & 3 & -1 \\ 4 & -2 & 4 & 0 \end{bmatrix}$$

(b) The matrix product BA can't be formed here; it's undefined since the number of columns of B doesn't equal the number of rows of A.

C. The identity matrix and the inverse matrix

Let n be a given positive integer. Then the **nth-order identity matrix** is the $n \times n$ matrix I_n that has a zero in every position except on the main diagonal (from upper left corner to lower right corner), where the elements are 1's. For example:

$$I_2 = \begin{bmatrix} 1 & 0 \\ 0 & 1 \end{bmatrix} \qquad I_3 = \begin{bmatrix} 1 & 0 & 0 \\ 0 & 1 & 0 \\ 0 & 0 & 1 \end{bmatrix} \qquad I_4 = \begin{bmatrix} 1 & 0 & 0 & 0 \\ 0 & 1 & 0 & 0 \\ 0 & 0 & 1 & 0 \\ 0 & 0 & 0 & 1 \end{bmatrix}$$

2nd-order	3rd-order	4th-order
identity matrix	identity matrix	identity matrix

The identity matrix, naturally enough, results in the property of identity:

- If A is any square matrix of size $m \times m$, then the product of A and its mth-order identity matrix is A:

$$AI_m = I_m A = A$$

note: In other words, the identity matrix is to matrix products what 1 is to algebraic products: $a1 = 1a = a$.

EXAMPLE 12-19   Let $A = \begin{bmatrix} 2 & 3 \\ 4 & 5 \end{bmatrix}$. Show that $AI_2 = I_2 A = A$.

Solution

$$AI_2 = \begin{bmatrix} 2 & 3 \\ 4 & 5 \end{bmatrix} \cdot \begin{bmatrix} 1 & 0 \\ 0 & 1 \end{bmatrix} = \begin{bmatrix} (2 \cdot 1 + 3 \cdot 0) & (2 \cdot 0 + 3 \cdot 1) \\ (4 \cdot 1 + 5 \cdot 0) & (4 \cdot 0 + 5 \cdot 1) \end{bmatrix} = \begin{bmatrix} 2 & 3 \\ 4 & 5 \end{bmatrix}$$

$$I_2 A = \begin{bmatrix} 1 & 0 \\ 0 & 1 \end{bmatrix} \cdot \begin{bmatrix} 2 & 3 \\ 4 & 5 \end{bmatrix} = \begin{bmatrix} (1 \cdot 2 + 0 \cdot 4) & (1 \cdot 3 + 0 \cdot 5) \\ (0 \cdot 2 + 1 \cdot 4) & (0 \cdot 3 + 1 \cdot 5) \end{bmatrix} = \begin{bmatrix} 2 & 3 \\ 4 & 5 \end{bmatrix}$$

- If a square matrix A of size $n \times n$ has $\det A \neq 0$, then A has a unique **inverse matrix** A^{-1}, with the property that the product of A and A^{-1} is the nth-order identity matrix of A:

$$AA^{-1} = A^{-1}A = I_n$$

And there are methods for finding inverses of square matrices with nonzero determinants.

note: If a square matrix has a determinant of zero, it doesn't have an inverse.

1. Inverse of a 2 × 2 matrix

If A is a 2 × 2 matrix $A = \begin{bmatrix} a & b \\ c & d \end{bmatrix}$, then

$$A^{-1} = \frac{1}{\det A} \begin{bmatrix} d & -b \\ -c & a \end{bmatrix}$$

A way to remember this is to observe that A^{-1} is the matrix derived from A by

(1) interchanging the elements on the main diagonal;
(2) changing the signs on the elements on the minor diagonal;
(3) multiplying in front by $1/\det A$.

EXAMPLE 12-20 Find the inverse of $A = \begin{bmatrix} 5 & 3 \\ -2 & -1 \end{bmatrix}$.

Solution First, find det A:

$$\det A = \begin{vmatrix} 5 & 3 \\ -2 & -1 \end{vmatrix} = 5(-1) - (-2)(3) = -5 + 6 = 1$$

Then, using A, interchange the main diagonal elements, negate the minor diagonal elements, and divide by 1:

$$A^{-1} = \frac{1}{1} \begin{bmatrix} -1 & -3 \\ -(-2) & 5 \end{bmatrix} = \begin{bmatrix} -1 & -3 \\ 2 & 5 \end{bmatrix}$$

Check: You know you're right if $AA^{-1} = I_2$:

$$AA^{-1} = \begin{bmatrix} 5 & 3 \\ -2 & -1 \end{bmatrix} \cdot \begin{bmatrix} -1 & -3 \\ 2 & 5 \end{bmatrix} = \begin{bmatrix} -5+6 & -15+15 \\ 2-2 & 6-5 \end{bmatrix} = \begin{bmatrix} 1 & 0 \\ 0 & 1 \end{bmatrix} \overset{\checkmark}{=} I_2$$

EXAMPLE 12-21 Find the inverse of $A = \begin{bmatrix} 2 & 4 \\ 1 & 5 \end{bmatrix}$.

Solution Find det A:

$$\det A = \begin{vmatrix} 2 & 4 \\ 1 & 5 \end{vmatrix} = 2 \cdot 5 - 1 \cdot 4 = 10 - 4 = 6$$

Then $$A = \begin{bmatrix} 2 & 4 \\ 1 & 5 \end{bmatrix} \Rightarrow A^{-1} = \frac{1}{6} \begin{bmatrix} 5 & -4 \\ -1 & 2 \end{bmatrix}$$

Check: $$A^{-1}A = \frac{1}{6} \begin{bmatrix} 5 & -4 \\ -1 & 2 \end{bmatrix} \cdot \begin{bmatrix} 2 & 4 \\ 1 & 5 \end{bmatrix} = \frac{1}{6} \begin{bmatrix} 5 \cdot 2 - 4 \cdot 1 & 5 \cdot 4 - 4 \cdot 5 \\ -1 \cdot 2 + 2 \cdot 1 & -1 \cdot 4 + 2 \cdot 5 \end{bmatrix}$$

$$= \frac{1}{6} \begin{bmatrix} 6 & 0 \\ 0 & 6 \end{bmatrix} = \begin{bmatrix} 1 & 0 \\ 0 & 1 \end{bmatrix} \overset{\checkmark}{=} I_2$$

---

2. Inverse of an $n \times n$ matrix, where $n \geq 3$

If A is $n \times n$ with $n \geq 3$, we find A^{-1} as follows:

(1) Rewrite the matrix A as an augmented $n \times 2n$ matrix with its identity matrix on the right:

$$\left[\begin{array}{cccc|cccc} a_{11} & a_{12} & \cdots & a_{1n} & 1 & 0 & 0 \cdots & 0 \\ a_{21} & a_{22} & \cdots & a_{2n} & 0 & 1 & 0 \cdots & 0 \\ \vdots & \vdots & & \vdots & \vdots & \vdots & \ddots & \vdots \\ a_{n1} & a_{n2} & \cdots & a_{nn} & 0 & 0 & 0 \cdots & 1 \end{array} \right]$$

(2) Perform any elementary row operations that are needed to bring this $n \times 2n$ matrix into the form

$$\begin{bmatrix} 1 & 0 & 0 \cdots 0 & e_{11} & e_{12} & \cdots & e_{1n} \\ 0 & 1 & 0 \cdots 0 & e_{21} & e_{22} & \cdots & e_{2n} \\ \vdots & \vdots & \ddots \vdots & \vdots & \vdots & & \vdots \\ 0 & 0 & 0 \cdots 1 & e_{n1} & e_{n2} & \cdots & e_{nn} \end{bmatrix}$$

so that I_n is on the left. Then A^{-1} will appear on the right. The matrix

$$A^{-1} = \begin{bmatrix} e_{11} & e_{12} & \cdots & e_{1n} \\ e_{21} & e_{22} & \cdots & e_{2n} \\ e_{n1} & e_{n2} & \cdots & e_{nn} \end{bmatrix}$$

has the property that

$$AA^{-1} = A^{-1}A = I_n$$

If the matrix A *cannot* be changed by means of elementary row operations to the special form illustrated in step (2), then matrix A has no inverse.

EXAMPLE 12-22   Find the inverse of the matrix $A = \begin{bmatrix} 1 & 0 & -2 \\ 3 & 1 & 0 \\ 2 & -1 & 2 \end{bmatrix}$.

Solution   First augment A with I_3 to get the 3×6 matrix

$$\begin{bmatrix} 1 & 0 & -2 & 1 & 0 & 0 \\ 3 & 1 & 0 & 0 & 1 & 0 \\ 2 & -1 & 2 & 0 & 0 & 1 \end{bmatrix}$$

Now, start performing elementary row operations on this 3×6 matrix. Your goal is to transform the "A part" to I_3.

$$\overbrace{\begin{bmatrix} 1 & 0 & -2 \\ 3 & 1 & 0 \\ 2 & -1 & 2 \end{bmatrix}}^{A} \ \overbrace{\begin{bmatrix} 1 & 0 & 0 \\ 0 & 1 & 0 \\ 0 & 0 & 1 \end{bmatrix}}^{I_3} \xRightarrow{R_2 \to -3R_1 + R_2} \begin{bmatrix} 1 & 0 & -2 & 1 & 0 & 0 \\ 0 & 1 & 6 & -3 & 1 & 0 \\ 2 & -1 & 2 & 0 & 0 & 1 \end{bmatrix}$$

$$\begin{bmatrix} 1 & 0 & -2 & 1 & 0 & 0 \\ 0 & 1 & 6 & -3 & 1 & 0 \\ 2 & -1 & 2 & 0 & 0 & 1 \end{bmatrix} \xRightarrow{R_3 \to -2R_1 + R_3} \begin{bmatrix} 1 & 0 & -2 & 1 & 0 & 0 \\ 0 & 1 & 6 & -3 & 1 & 0 \\ 0 & -1 & 6 & -2 & 0 & 1 \end{bmatrix}$$

$$\begin{bmatrix} 1 & 0 & -2 & 1 & 0 & 0 \\ 0 & 1 & 6 & -3 & 1 & 0 \\ 0 & -1 & 6 & -2 & 0 & 1 \end{bmatrix} \xRightarrow{R_3 \to R_2 + R_3} \begin{bmatrix} 1 & 0 & -2 & 1 & 0 & 0 \\ 0 & 1 & 6 & -3 & 1 & 0 \\ 0 & 0 & 12 & -5 & 1 & 1 \end{bmatrix}$$

$$\begin{bmatrix} 1 & 0 & -2 & 1 & 0 & 0 \\ 0 & 1 & 6 & -3 & 1 & 0 \\ 0 & 0 & 12 & -5 & 1 & 1 \end{bmatrix} \xRightarrow{R_1 \to \frac{1}{6}R_3 + R_1} \begin{bmatrix} 1 & 0 & 0 & \frac{1}{6} & \frac{1}{6} & \frac{1}{6} \\ 0 & 1 & 6 & -3 & 1 & 0 \\ 0 & 0 & 12 & -5 & 1 & 1 \end{bmatrix}$$

$$\begin{bmatrix} 1 & 0 & 0 & \frac{1}{6} & \frac{1}{6} & \frac{1}{6} \\ 0 & 1 & 6 & -3 & 1 & 0 \\ 0 & 0 & 12 & -5 & 1 & 1 \end{bmatrix} \xRightarrow{R_2 \to -\frac{1}{2}R_3 + R_2} \begin{bmatrix} 1 & 0 & 0 & \frac{1}{6} & \frac{1}{6} & \frac{1}{6} \\ 0 & 1 & 0 & -\frac{1}{2} & \frac{1}{2} & -\frac{1}{2} \\ 0 & 0 & 12 & -5 & 1 & 1 \end{bmatrix}$$

$$\begin{bmatrix} 1 & 0 & 0 & \frac{1}{6} & \frac{1}{6} & \frac{1}{6} \\ 0 & 1 & 0 & -\frac{1}{2} & \frac{1}{2} & -\frac{1}{2} \\ 0 & 0 & 12 & -5 & 1 & 1 \end{bmatrix} \xRightarrow{R_3 \to \frac{1}{12}R_3} \underbrace{\begin{bmatrix} 1 & 0 & 0 \\ 0 & 1 & 0 \\ 0 & 0 & 1 \end{bmatrix}}_{I_3} \underbrace{\begin{bmatrix} \frac{1}{6} & \frac{1}{6} & \frac{1}{6} \\ -\frac{1}{2} & \frac{1}{2} & -\frac{1}{2} \\ -\frac{5}{12} & \frac{1}{12} & \frac{1}{12} \end{bmatrix}}_{A^{-1}}$$

Thus

$$A^{-1} = \begin{bmatrix} \frac{1}{6} & \frac{1}{6} & \frac{1}{6} \\ -\frac{1}{2} & \frac{1}{2} & -\frac{1}{2} \\ -\frac{5}{12} & \frac{1}{12} & \frac{1}{12} \end{bmatrix} \text{ or } \frac{1}{12}\begin{bmatrix} 2 & 2 & 2 \\ -6 & 6 & -6 \\ -5 & 1 & 1 \end{bmatrix}$$

Check:

$$A^{-1}A = \begin{bmatrix} \frac{1}{6} & \frac{1}{6} & \frac{1}{6} \\ -\frac{1}{2} & \frac{1}{2} & -\frac{1}{2} \\ -\frac{5}{12} & \frac{1}{12} & \frac{1}{12} \end{bmatrix} \cdot \begin{bmatrix} 1 & 0 & -2 \\ 3 & 1 & 0 \\ 2 & -1 & 2 \end{bmatrix}$$

$$= \begin{bmatrix} \frac{1}{6}+\frac{3}{6}+\frac{2}{6} & 0+\frac{1}{6}-\frac{1}{6} & -\frac{2}{6}+0+\frac{2}{6} \\ -\frac{1}{2}+\frac{3}{2}-\frac{2}{2} & 0+\frac{1}{2}+\frac{1}{2} & \frac{2}{2}+0-\frac{2}{2} \\ -\frac{5}{12}+\frac{3}{12}+\frac{2}{12} & 0+\frac{1}{12}-\frac{1}{12} & \frac{10}{12}+0+\frac{2}{12} \end{bmatrix} \overset{?}{=} \begin{bmatrix} 1 & 0 & 0 \\ 0 & 1 & 0 \\ 0 & 0 & 1 \end{bmatrix}$$

12-5. Solving Linear Systems by Matrix Methods

If the number of equations in a linear system equals the number of variables and if the coefficient matrix of the system has an inverse, then the system can be solved by using that inverse.

Consider the system,

$$\begin{cases} a_1x + b_1y + c_1z = d_1 \\ a_2x + b_2y + c_2z = d_2 \\ a_3x + b_3y + c_3z = d_3 \end{cases}$$

Let the coefficient matrix be A (where $\det A \neq 0$), and let the constant matrix be D:

$$A = \begin{bmatrix} a_1 & b_1 & c_1 \\ a_2 & b_2 & c_2 \\ a_3 & b_3 & c_3 \end{bmatrix}, \qquad D = \begin{bmatrix} d_1 \\ d_2 \\ d_3 \end{bmatrix}$$

Then, define a matrix of variables:

$$X = \begin{bmatrix} x \\ y \\ z \end{bmatrix}$$

Now we can write the entire system of equations in matrix notation as

$$AX = D$$

To verify this, write out each side:

$$\begin{array}{ccccc} A & & \cdot X & = & D \\ \begin{bmatrix} a_1 & b_1 & c_1 \\ a_2 & b_2 & c_2 \\ a_3 & b_3 & c_3 \end{bmatrix} & \cdot & \begin{bmatrix} x \\ y \\ z \end{bmatrix} & = & \begin{bmatrix} d_1 \\ d_2 \\ d_3 \end{bmatrix} \end{array}$$

and multiply the 3×3 matrix by the 3×1 matrix to get two equal 3×1 matrices:

$$\begin{array}{ccc} AX & & = D \\ \begin{bmatrix} a_1x + b_1y + c_1z \\ a_2x + b_2y + c_2z \\ a_3x + b_3y + c_3z \end{bmatrix} & = & \begin{bmatrix} d_1 \\ d_2 \\ d_3 \end{bmatrix} \end{array}$$

And if these 3×1 matrices are equal, then their corresponding elements must be equal. But since these are the original three equations in the system, we now know that we can solve the system by using matrix methods:

If $\qquad AX = D$

then $\qquad A^{-1}(AX) = A^{-1}D \qquad$ [Multiply through by the inverse A^{-1}.]

$\qquad (A^{-1}A)X = A^{-1}D \qquad$ [Regroup.]

$\qquad I_nX = A^{-1}D \qquad$ [Replace $(A^{-1}A)$ by its equivalent identity matrix.]

so $\qquad X = A^{-1}D$

Hence we can find the entries of X, i.e., the values of the unknowns, if we find A^{-1} and multiply it by the known matrix D.

Caution: The inverse method works only when the *determinant of the coefficient matrix is nonzero*, i.e., only when the coefficient matrix has an inverse.

EXAMPLE 12-23   Using matrix methods, solve $\begin{cases} 3x + 2y = 8 \\ -4x + 5y = 2 \end{cases}$.

Solution

$$A = \begin{bmatrix} 3 & 2 \\ -4 & 5 \end{bmatrix}, \qquad X = \begin{bmatrix} x \\ y \end{bmatrix}, \qquad D = \begin{bmatrix} 8 \\ 2 \end{bmatrix}$$

Thus $AX = D$ is $\qquad \begin{bmatrix} 3 & 2 \\ -4 & 5 \end{bmatrix} \cdot \begin{bmatrix} x \\ y \end{bmatrix} = \begin{bmatrix} 8 \\ 2 \end{bmatrix}$

and $X = A^{-1}D = \begin{bmatrix} x \\ y \end{bmatrix} = \dfrac{1}{23}\begin{bmatrix} 5 & -2 \\ 4 & 3 \end{bmatrix} \cdot \begin{bmatrix} 8 \\ 2 \end{bmatrix} = \dfrac{1}{23}\begin{bmatrix} (5)(8) + (-2)(2) \\ (4)(8) + (3)(2) \end{bmatrix} = \begin{bmatrix} \frac{36}{23} \\ \frac{38}{23} \end{bmatrix}$

so $x = \frac{36}{23}$ and $y = \frac{38}{23}$.

Check: $\quad 3(\frac{36}{23}) + 2(\frac{38}{23}) \stackrel{?}{=} 8 \qquad -4(\frac{36}{23}) + 5(\frac{38}{23}) \stackrel{?}{=} 2$

$\qquad\qquad \frac{108}{23} + \frac{76}{23} \stackrel{?}{=} 8 \qquad\qquad -\frac{144}{23} + \frac{190}{23} \stackrel{?}{=} 2$

$\qquad\qquad\qquad 8 \stackrel{\checkmark}{=} 8 \qquad\qquad\qquad\qquad 2 \stackrel{\checkmark}{=} 2$

EXAMPLE 12-24   Using matrix methods, solve $\begin{cases} x \quad\;\; - 2z = 1 \\ 3x + y \quad\quad = -2 \\ 2x - y + 2z = 3 \end{cases}$.

Solution   First, write the system in matrix form $AX = D$:

$$\begin{bmatrix} 1 & 0 & -2 \\ 3 & 1 & 0 \\ 2 & -1 & 2 \end{bmatrix} \cdot \begin{bmatrix} x \\ y \\ z \end{bmatrix} = \begin{bmatrix} 1 \\ -2 \\ 3 \end{bmatrix}$$

We need A^{-1} now. But look back at Example 12-22, where we found exactly that!

$$A^{-1} = \begin{bmatrix} \frac{1}{6} & \frac{1}{6} & \frac{1}{6} \\ -\frac{1}{2} & \frac{1}{2} & -\frac{1}{2} \\ -\frac{5}{12} & \frac{1}{12} & \frac{1}{12} \end{bmatrix}$$

Hence $\quad X = \begin{bmatrix} x \\ y \\ z \end{bmatrix} = A^{-1}D = \begin{bmatrix} \frac{1}{6} & \frac{1}{6} & \frac{1}{6} \\ -\frac{1}{2} & \frac{1}{2} & -\frac{1}{2} \\ -\frac{5}{12} & \frac{1}{12} & \frac{1}{12} \end{bmatrix} \cdot \begin{bmatrix} 1 \\ -2 \\ 3 \end{bmatrix}$

$$= \begin{bmatrix} \frac{1}{6}(1) + \frac{1}{6}(-2) + \frac{1}{6}(3) \\ -\frac{1}{2}(1) + \frac{1}{2}(-2) + -\frac{1}{2}(3) \\ -\frac{5}{12}(1) + \frac{1}{12}(-2) + \frac{1}{12}(3) \end{bmatrix} = \begin{bmatrix} \frac{1}{3} \\ -3 \\ -\frac{1}{3} \end{bmatrix}$$

Thus $x = \frac{1}{3}$, $y = -3$, $z = -\frac{1}{3}$.

SUMMARY

1. A matrix is a rectangular array of numbers called the elements of the matrix; these elements are arranged in rows and columns, so that a matrix is referred to by its size: A matrix is said to be $m \times n$ if it has m rows and n columns.

2. Elementary row operations performed on a matrix change the numbers (elements) of the matrix to produce equivalent matrices. These operations are used to help in solving a system of equations.

3. A matrix is said to be in *echelon form* if all of the elements below the main diagonal (the lower left portion) are zero.

4. For every square ($n \times n$) matrix A there is defined a number called the determinant of the matrix:

$$\text{If } A = \begin{bmatrix} a_{11} & a_{12} & \cdots & a_{1n} \\ a_{21} & a_{22} & \cdots & a_{2n} \\ \vdots & \vdots & & \vdots \\ a_{n1} & a_{n2} & \cdots & a_{nn} \end{bmatrix}, \det A = \begin{vmatrix} a_{11} & a_{12} & \cdots & a_{1n} \\ a_{21} & a_{22} & \cdots & a_{2n} \\ \vdots & \vdots & & \vdots \\ a_{n1} & a_{n2} & \cdots & a_{nn} \end{vmatrix}$$

5. The minor M_{ij} of a square matrix A is the determinant of the matrix formed by deleting from A the ith row and the jth column.

6. The cofactor A_{ij} of an element a_{ij} of a determinant is defined by $A_{ij} = (-1)^{i+j} M_{ij}$.

7. The value of the determinant of a square matrix A may be found by adding a particular sequence of elements and cofactors:

   • The value of a 1×1 determinant $|a_{11}|$ is a_{11}.
   • The value of an $n \times n$ determinant, where $n > 1$, is found by multiplying the elements in any one row (or column) by their respective cofactors and adding the resulting products.

8. Evaluation of a determinant is made easier if the rules for operating on determinants are used to change some of the numbers in the determinant to zeros.

9. Cramer's Rule is an efficient method of solving systems of linear equations using determinants: If the number of equations in a linear system is equal to the number of variables, then (in the 3×3 case),

$$x = \frac{D_x}{D}, \qquad y = \frac{D_y}{D}, \qquad z = \frac{D_z}{D}$$

where D is the nonzero coefficient determinant of the system; and D_x, D_y, D_z are derived, respectively, from D by replacing the x, y, and z columns of the coefficient determinant by the constant column by the system.

10. Matrices of identical sizes may be added or subtracted; a matrix may be multiplied by a scalar number; and, under size requirements ($A = m \times p$; $B = p \times n$), matrices may be multiplied together ($AB = m \times n$).

11. An identity matrix I is a square matrix with the element 1 in every position on the main diagonal and the element 0 in all other positions. The identity matrix I_m has the property that $AI_m = I_m A = A$ for any $m \times m$ matrix A.

12. The inverse of a given $n \times n$ matrix A is another $n \times n$ matrix denoted A^{-1} with the property that $AA^{-1} = A^{-1}A = I_n$.

13. If a system of equations is written in matrix notation as $AX = D$, then the solution to the system is $X = A^{-1}D$.

RAISE YOUR GRADES

Can you . . . ?

☑ form the augmented matrix from a given system of linear equations
☑ define equivalent matrices
☑ use the elementary row operations to form equivalent matrices
☑ reduce a system of equations to triangular (echelon) form
☑ calculate the value of any determinant

☑ find a minor of a matrix and a cofactor of a determinant
☑ find the value of a determinant using cofactors
☑ list the operations on determinants and the consequences of applying them
☑ recognize when a determinant has value zero without a formal evaluation process
☑ solve a system of equations using Cramer's Rule
☑ explain how two matrices can be equal
☑ list the rules under which the product of two matrices is defined
☑ find the inverse of a 2 × 2 matrix in three easy steps
☑ find the inverse of a 3 × 3 matrix A by augmenting I_3 onto A and using row operations to change A to I_3
☑ change a system of equations into matrix form and solve it using the inverse

SOLVED PROBLEMS

Matrix Methods for Systems of Equations

PROBLEM 12-1 Form the coefficient matrix C and the augmented matrix A for the system of equations $\begin{cases} 2x - y = 5 \\ 4x + 2y = 3 \end{cases}$.

Solution Since the terms of the equations are already lined up, simply write down the matrices.

$$C = \begin{bmatrix} 2 & -1 \\ 4 & 2 \end{bmatrix} \quad \text{and} \quad A = \begin{bmatrix} 2 & -1 & \vdots & 5 \\ 4 & 2 & \vdots & 3 \end{bmatrix}$$

PROBLEM 12-2 Form the augmented matrix A for the system $\begin{cases} 2x \quad\ - 3z = 8 \\ \quad y + 4z = -2 \\ x - y + \ z = 4 \end{cases}$.

Solution First rewrite the equations by supplying a zero coefficient for the missing terms:

$$\begin{cases} 2x + 0y - 3z = 8 \\ 0x + \ y + 4z = -2 \\ \ x - \ y + \ z = 4 \end{cases}$$

Now

$$A = \begin{bmatrix} 2 & 0 & -3 & \vdots & 8 \\ 0 & 1 & 4 & \vdots & -2 \\ 1 & -1 & 1 & \vdots & 4 \end{bmatrix}$$

PROBLEM 12-3 Show that the matrices

$$M_1 = \begin{bmatrix} 2 & -1 & 2 \\ 3 & 4 & 1 \end{bmatrix} \quad \text{and} \quad M_2 = \begin{bmatrix} 1 & 5 & -1 \\ 0 & -11 & 4 \end{bmatrix}$$

are equivalent.

Solution Use the allowable matrix row operations on M_1 to derive M_2 from M_1:

$$\begin{bmatrix} 2 & -1 & 2 \\ 3 & 4 & 1 \end{bmatrix} \xrightarrow{R_1 \to -1R_1 + R_2} \begin{bmatrix} 1 & 5 & -1 \\ 3 & 4 & 1 \end{bmatrix} \xrightarrow{R_2 \to -3R_1 + R_2} \begin{bmatrix} 1 & 5 & -1 \\ 0 & -11 & 4 \end{bmatrix}$$

PROBLEM 12-4 Beginning with the system of two linear equations $\begin{cases} 3x - 7y = -4 \\ 6x + 2y = 8 \end{cases}$, compare the algebraic operations on the system to the corresponding elementary row operations on the augmented matrix of the system.

Solution

$$\begin{cases} 3x - 7y = -4 \\ 6x + 2y = 8 \end{cases} \qquad \begin{bmatrix} 3 & -7 & \vdots & -4 \\ 6 & 2 & \vdots & 8 \end{bmatrix}$$

Multiply the first equation/row by -2:

$$\begin{cases} -6x + 14y = 8 \\ 6x + 2y = 8 \end{cases} \qquad \begin{bmatrix} -6 & 14 & \vdots & 8 \\ 6 & 2 & \vdots & 8 \end{bmatrix}$$

Replace the second equation/row by the sum of the first and second equations/rows:

$$\begin{cases} -6x + 14y = 8 \\ 0x + 16y = 16 \end{cases} \qquad \begin{bmatrix} -6 & 14 & \vdots & 8 \\ 0 & 16 & \vdots & 16 \end{bmatrix}$$

Multiply the second equation/row by $\frac{1}{16}$:

$$\begin{cases} -6x + 14y = 8 \\ 0x + y = 1 \end{cases} \text{[Triangular form]} \qquad \begin{bmatrix} -6 & 14 & \vdots & 8 \\ 0 & 1 & \vdots & 1 \end{bmatrix} \text{[Echelon form]}$$

Multiply the second equation/row by -14 and add it to the first equation/row:

$$\begin{cases} -6x + 0y = -6 \\ 0x + y = 1 \end{cases} \qquad \begin{bmatrix} -6 & 0 & \vdots & -6 \\ 0 & 1 & \vdots & 1 \end{bmatrix}$$

Multiply the first equation/row by $-\frac{1}{6}$:

$$\begin{cases} x + 0y = 1 \\ 0x + y = 1 \end{cases} \qquad \begin{bmatrix} 1 & 0 & \vdots & 1 \\ 0 & 1 & \vdots & 1 \end{bmatrix}$$

The equations are now reduced to the solution—they read $x = 1$, $y = 1$. The matrix is also read $x = 1$, $y = 1$.

PROBLEM 12-5 Using elementary row operations, transform the following matrix to echelon form:

$$\begin{bmatrix} -1 & 4 & 2 & \vdots & 5 \\ 3 & 0 & -2 & \vdots & 1 \\ 5 & 4 & 3 & \vdots & 2 \end{bmatrix}$$

Solution

$$\begin{bmatrix} -1 & 4 & 2 & \vdots & 5 \\ 3 & 0 & -2 & \vdots & 1 \\ 5 & 4 & 3 & \vdots & 2 \end{bmatrix} \xrightarrow{R_2 \to 3R_1 + R_2} \begin{bmatrix} -1 & 4 & 2 & \vdots & 5 \\ 0 & 12 & 4 & \vdots & 16 \\ 5 & 4 & 3 & \vdots & 2 \end{bmatrix} \xrightarrow{R_3 \to 5R_1 + R_3}$$

$$\begin{bmatrix} -1 & 4 & 2 & \vdots & 5 \\ 0 & 12 & 4 & \vdots & 16 \\ 0 & 24 & 13 & \vdots & 27 \end{bmatrix} \xrightarrow{R_3 \to -2R_2 + R_3} \begin{bmatrix} -1 & 4 & 2 & \vdots & 5 \\ 0 & 12 & 4 & \vdots & 16 \\ 0 & 0 & 5 & \vdots & -5 \end{bmatrix} \xrightarrow{R_1 \to -1R_1;\ R_2 \to \frac{1}{4}R_2;\ R_3 \to \frac{1}{5}R_3}$$

$$\begin{bmatrix} 1 & -4 & -2 & \vdots & -5 \\ 0 & 3 & 1 & \vdots & 4 \\ 0 & 0 & 1 & \vdots & -1 \end{bmatrix}$$

PROBLEM 12-6 Solve the system of equations $\begin{cases} -x + 4y + 2z = 5 \\ 3x - 2z = 1 \\ 5x + 4y + 3z = 2 \end{cases}$.

Solution The matrix given in Problem 12-5 is the augmented matrix of this system of equations. So, the echelon form obtained in Problem 12-5 is equivalent to this matrix and also to the system of equations it represents. Thus, from the echelon form

$$\begin{cases} x - 4y - 2z = -5 \\ \phantom{x -{}} 3y + z = 4 \\ \phantom{x - 4y -{}} z = -1 \end{cases}$$

Solving this is easy: Start from the bottom:

$z = -1$ then $3y - 1 = 4$ then $x - 4(\frac{5}{3}) - 2(-1) = -5$

$$3y = 5 \qquad\qquad\qquad\qquad x = \frac{20}{3} - 2 - 5$$

$$y = \frac{5}{3} \qquad\qquad\qquad\qquad = -\frac{1}{3}$$

The solution to the system is $(-\frac{1}{3}, \frac{5}{3}, -1)$

Determinants

PROBLEM 12-7 Write the determinant associated with the following matrices:

$$A = \begin{bmatrix} 1 & 5 \\ -2 & 3 \end{bmatrix} \qquad B = [4] \qquad C = \begin{bmatrix} 1 & 2 & 3 \\ 4 & 5 & 6 \\ 7 & 8 & 9 \end{bmatrix}$$

Solution To form the determinant is simple:

$$\det A = \begin{vmatrix} 1 & 5 \\ -2 & 3 \end{vmatrix} \qquad \det B = |4| \qquad \det C = \begin{vmatrix} 1 & 2 & 3 \\ 4 & 5 & 6 \\ 7 & 8 & 9 \end{vmatrix}$$

PROBLEM 12-8 Find the value of the determinant for each of the following matrices:

$$A = \begin{bmatrix} 5 & 1 \\ 8 & 2 \end{bmatrix} \qquad B = \begin{bmatrix} -3 & 0 \\ 0 & 6 \end{bmatrix} \qquad C = \begin{bmatrix} 4 & 2 \\ 2 & 1 \end{bmatrix}$$

Solution For a two-by-two matrix $\begin{bmatrix} a & b \\ c & d \end{bmatrix}$ the determinant is written $\begin{vmatrix} a & b \\ c & d \end{vmatrix}$ and has value $ad - cb$. Thus

$$\det A = \begin{vmatrix} 5 & 1 \\ 8 & 2 \end{vmatrix} = 5 \cdot 2 - 8 \cdot 1 = 10 - 8 = 2$$

$$\det B = \begin{vmatrix} -3 & 0 \\ 0 & 6 \end{vmatrix} = -3 \cdot 6 - 0 \cdot 0 = -18$$

$$\det C = \begin{vmatrix} 4 & 2 \\ 2 & 1 \end{vmatrix} = 4 \cdot 1 - 2 \cdot 2 = 0$$

PROBLEM 12-9 Given the 3×3 matrix $M = \begin{bmatrix} 5 & 1 & 3 \\ -2 & 0 & 4 \\ 1 & 2 & 1 \end{bmatrix}$, find the value of $\det M$.

Solution Use the diagonals method:

$$= (5)(0)(1) + (1)(4)(1) + (3)(-2)(2) - (1)(0)(3) - (2)(4)(5) - (1)(-2)(1)$$

$$= 0 + 4 - 12 - 0 - 40 + 2 = -46$$

PROBLEM 12-10   Find the value of the determinant of the matrix

$$A = \begin{bmatrix} 1 & -1 & 2 \\ -3 & 0 & -5 \\ -2 & 4 & -3 \end{bmatrix}$$

Solution

$$= (1)(0)(-3) + (-1)(-5)(-2) + (2)(-3)(4) - (-2)(0)(2) - (4)(-5)(1) - (-3)(-3)(-1)$$

$$= 0 + (-10) + (-24) - (0) - (-20) - (-9)$$

$$= -10 - 24 + 20 + 9$$

$$= -5$$

PROBLEM 12-11   Find the minor M_{31} and the minor M_{12} for the matrix

$$A = \begin{bmatrix} 1 & 2 & 3 \\ 4 & 5 & 6 \\ 7 & 8 & 9 \end{bmatrix}$$

Solution   To get the minor M_{31}, delete row three (R_3) and column one (C_1) from A and calculate the value of the resulting determinant. To get M_{12}, delete R_1 and C_2 and calculate the value of the determinant.

$$A = \begin{bmatrix} 1 & 2 & 3 \\ 4 & 5 & 6 \\ 7 & 8 & 9 \end{bmatrix} \qquad M_{31} = \begin{vmatrix} 2 & 3 \\ 5 & 6 \end{vmatrix} = 12 - 15 = -3$$

$$A = \begin{bmatrix} 1 & 2 & 3 \\ 4 & 5 & 6 \\ 7 & 8 & 9 \end{bmatrix} \qquad M_{12} = \begin{vmatrix} 4 & 6 \\ 7 & 9 \end{vmatrix} = 36 - 42 = -6$$

PROBLEM 12-12   Find the cofactors A_{12} and A_{33} of the determinant

$$\begin{vmatrix} -1 & 2 & 1 \\ 3 & 1 & 0 \\ 0 & 4 & 2 \end{vmatrix}$$

Solution   By definition, the cofactor $A_{12} = (-1)^{1+2}M_{12} = (-1)M_{12}$ and $A_{33} = (-1)^{3+3}M_{33} = (1)M_{33}$. To get M_{12}, delete the first row and second column:

$$M_{12} = \begin{vmatrix} 3 & 0 \\ 0 & 2 \end{vmatrix} = 6 - 0 = 6$$

To get M_{33}, delete the third row and third column:

$$M_{33} = \begin{vmatrix} -1 & 2 \\ 3 & 1 \end{vmatrix} = -1 - 6 = -7$$

So $A_{12} = (-1)(6) = -6$ and $A_{33} = (1)(-7) = -7$.

PROBLEM 12-13   Using cofactors of the second column, find the value of det A where

$$A = \begin{bmatrix} 4 & 1 & 2 \\ 1 & 0 & -3 \\ 2 & 1 & 5 \end{bmatrix}$$

Solution

$$\det A = \begin{vmatrix} 4 & 1 & 2 \\ 1 & 0 & -3 \\ 2 & 1 & 5 \end{vmatrix} = a_{12}A_{12} + a_{22}A_{22} + a_{32}A_{32}$$

$$= 1(-1)^3 M_{12} + 0(-1)^4 M_{22} + 1(-1)^5 M_{32}$$

$$= -1\begin{vmatrix} 1 & -3 \\ 2 & 5 \end{vmatrix} + 0 + (-1)\begin{vmatrix} 4 & 2 \\ 1 & -3 \end{vmatrix}$$

$$= -1(5 + 6) + (-1)(-12 - 2)$$

$$= -11 + 14 = 3$$

PROBLEM 12-14   Check the result of Problem 12-13 by another method.

Solution   Do the diagonal expansion:

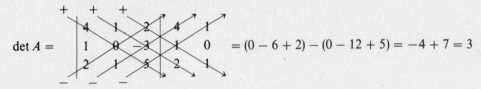

$$\det A = \qquad\qquad = (0 - 6 + 2) - (0 - 12 + 5) = -4 + 7 = 3$$

PROBLEM 12-15   Find the value of det M if

$$M = \begin{bmatrix} 1 & -1 & 0 & 1 \\ 1 & 1 & 1 & 0 \\ 0 & 2 & 1 & 0 \\ -2 & 0 & 1 & 1 \end{bmatrix}$$

Solution   The fourth column contains two zeros, so it will be a convenient one on which to base the expansion:

$$\det M = 1A_{14} + 0A_{24} + 0A_{34} + 1A_{44}$$

$$= 1(-1)^5 \begin{vmatrix} 1 & 1 & 1 \\ 0 & 2 & 1 \\ -2 & 0 & 1 \end{vmatrix} + 1(-1)^8 \begin{vmatrix} 1 & -1 & 0 \\ 1 & 1 & 1 \\ 0 & 2 & 1 \end{vmatrix}$$

$$= (-1)[(2 - 2 + 0) - (-4 + 0 + 0)] + (1)[(1 + 0 + 0) - (0 + 2 - 1)]$$

$$= (-1)(4) + (1)(0) = -4$$

PROBLEM 12-16   Using operations on the determinant of matrix M in Problem 12-15, verify its value.

Solution   First, operate on the determinant to isolate a single nonzero element in row one:

$$\det M = \begin{vmatrix} 1 & -1 & 0 & 1 \\ 1 & 1 & 1 & 0 \\ 0 & 2 & 1 & 0 \\ -2 & 0 & 1 & 1 \end{vmatrix} \xrightarrow{C_2 \to C_1 + C_2} \begin{vmatrix} 1 & 0 & 0 & 1 \\ 1 & 2 & 1 & 0 \\ 0 & 2 & 1 & 0 \\ -2 & -2 & 1 & 1 \end{vmatrix} \xrightarrow{C_4 \to -1C_1 + C_4} \begin{vmatrix} 1 & 0 & 0 & 0 \\ 1 & 2 & 1 & -1 \\ 0 & 2 & 1 & 0 \\ -2 & -2 & 1 & 3 \end{vmatrix}$$

Now expand, using cofactors of row one:

$$\det M = a_{11}A_{11} = \begin{vmatrix} 2 & 1 & -1 \\ 2 & 1 & 0 \\ -2 & 1 & 3 \end{vmatrix}$$

Then, continue to create more zeros by determinant operations. Isolate a single nonzero element in row two:

$$\begin{vmatrix} 2 & 1 & -1 \\ 2 & 1 & 0 \\ -2 & 1 & 3 \end{vmatrix} \xrightarrow{R_2 \to -1R_1 + R_2} \begin{vmatrix} 2 & 1 & -1 \\ 0 & 0 & 1 \\ -2 & 1 & 3 \end{vmatrix}$$

Now expand, using cofactors of row two:

$$\det M = (-1)^5 \begin{vmatrix} 2 & 1 \\ -2 & 1 \end{vmatrix} = (-1)[(2)(1) - (-2)(1)] = -1(2+2) = -4$$

So the value $\det M = -4$ found in Problem 12-15 is verified.

PROBLEM 12-17   Show that $\det A = 0$ if

$$A = \begin{bmatrix} 6 & 1 & -8 \\ 11 & 2 & 14 \\ -18 & -3 & 24 \end{bmatrix}$$

Solution   Use determinant operations to show that

$$\det A = \begin{vmatrix} 6 & 1 & -8 \\ 11 & 2 & 14 \\ -18 & -3 & 24 \end{vmatrix} \xrightarrow{R_3 \to 3R_1 + R_3} \begin{vmatrix} 6 & 1 & -8 \\ 11 & 2 & 14 \\ 0 & 0 & 0 \end{vmatrix} = 0$$

Since row three has all zeros, $\det A$ must be zero.

Solving Systems of Equations by Determinants: Cramer's Rule

PROBLEM 12-18   Use Cramer's Rule to solve the system $\begin{cases} 3x - 5y = 2 \\ x + 3y = 10 \end{cases}$.

Solution   Set up the three needed determinants, D, D_x, and D_y:

$$D = \begin{vmatrix} 3 & -5 \\ 1 & 3 \end{vmatrix} \qquad D_x = \begin{vmatrix} 2 & -5 \\ 10 & 3 \end{vmatrix} \qquad D_y = \begin{vmatrix} 3 & 2 \\ 1 & 10 \end{vmatrix}$$

Coefficient determinant	Replace the x-column of D by the column of constants	Replace the y-column of D by the column of constants

Find the values of D, D_x, and D_y:

$$D = (3)(3) - (1)(-5) \qquad D_x = (2)(3) - (10)(-5) \qquad D_y = (3)(10) - (1)(2)$$
$$= 9 + 5 = 14 \qquad\qquad = 6 + 50 = 56 \qquad\qquad = 30 - 2 = 28$$

Then, use Cramer's Rule:

$$x = \frac{D_x}{D} = \frac{56}{14} = 4 \qquad \text{and} \qquad y = \frac{D_y}{D} = \frac{28}{14} = 2$$

PROBLEM 12-19   Use Cramer's Rule to solve the system

$$\begin{cases} x - y + 2z = 2 \\ -x + y - z = 1 \\ 2x + y + z = 0 \end{cases}$$

Solution   Form the four needed determinants and evaluate each:

$$D = \begin{vmatrix} 1 & -1 & 2 \\ -1 & 1 & -1 \\ 2 & 1 & 1 \end{vmatrix} = (1 + 2 - 2) - [4 + (-1) + 1] = 1 - 4 = -3$$

$$D_x = \begin{vmatrix} 2 & -1 & 2 \\ 1 & 1 & -1 \\ 0 & 1 & 1 \end{vmatrix} = (2 + 0 + 2) - (0 - 2 - 1) = 4 + 3 = 7$$

$$D_y = \begin{vmatrix} 1 & 2 & 2 \\ -1 & 1 & -1 \\ 2 & 0 & 1 \end{vmatrix} = (1 - 4 + 0) - (4 + 0 - 2) = -3 - 2 = -5$$

$$D_z = \begin{vmatrix} 1 & -1 & 2 \\ -1 & 1 & 1 \\ 2 & 1 & 0 \end{vmatrix} = (0 - 2 - 2) - (4 + 1 + 0) = -4 - 5 = -9$$

So

$$x = \frac{D_x}{D} = -\frac{7}{3}, \qquad y = \frac{D_y}{D} = \frac{-5}{-3} = \frac{5}{3}, \qquad z = \frac{D_z}{D} = \frac{-9}{-3} = 3$$

Matrix Products and Inverses

PROBLEM 12-20   Let $A = \begin{bmatrix} 5 & 1 \\ -2 & 6 \end{bmatrix}$ and $B = \begin{bmatrix} 4 & -2 \\ 3 & 1 \end{bmatrix}$. Find $A + B$, $A - B$, and $4A$.

Solution   To add (or subtract) two matrices of equal size, add (or subtract) element-by-element:

$$A + B = \begin{bmatrix} 5 & 1 \\ -2 & 6 \end{bmatrix} + \begin{bmatrix} 4 & -2 \\ 3 & 1 \end{bmatrix} = \begin{bmatrix} 5+4 & 1-2 \\ -2+3 & 6+1 \end{bmatrix} = \begin{bmatrix} 9 & -1 \\ 1 & 7 \end{bmatrix}$$

$$A - B = \begin{bmatrix} 5 & 1 \\ -2 & 6 \end{bmatrix} - \begin{bmatrix} 4 & -2 \\ 3 & 1 \end{bmatrix} = \begin{bmatrix} 5-4 & 1-(-2) \\ -2-3 & 6-1 \end{bmatrix} = \begin{bmatrix} 1 & 3 \\ -5 & 5 \end{bmatrix}$$

To find the scalar product of a matrix, multiply each element of the matrix by the scalar (i.e., nonzero constant):

$$4A = 4\begin{bmatrix} 5 & 1 \\ -2 & 6 \end{bmatrix} = \begin{bmatrix} 4 \cdot 5 & 4 \cdot 1 \\ 4(-2) & 4 \cdot 6 \end{bmatrix} = \begin{bmatrix} 20 & 4 \\ -8 & 24 \end{bmatrix}$$

PROBLEM 12-21   Find the matrix products AB and BA if $A = \begin{bmatrix} 2 & 1 \\ -1 & 3 \end{bmatrix}$ and $B = \begin{bmatrix} 1 & -2 \\ 4 & 2 \end{bmatrix}$.

Solution   Both AB and BA can be found because A and B are both 2×2 matrices.

$$AB = \begin{bmatrix} 2 & 1 \\ -1 & 3 \end{bmatrix} \cdot \begin{bmatrix} 1 & -2 \\ 4 & 2 \end{bmatrix} = \begin{bmatrix} 2 \cdot 1 + 1 \cdot 4 & 2(-2) + 1 \cdot 2 \\ -1 \cdot 1 + 3 \cdot 4 & (-1)(-2) + 3 \cdot 2 \end{bmatrix} = \begin{bmatrix} 6 & -2 \\ 11 & 8 \end{bmatrix}$$

$$BA = \begin{bmatrix} 1 & -2 \\ 4 & 2 \end{bmatrix} \cdot \begin{bmatrix} 2 & 1 \\ -1 & 3 \end{bmatrix} = \begin{bmatrix} 1 \cdot 2 + (-2)(-1) & 1 \cdot 1 + (-2)3 \\ 4 \cdot 2 + 2(-1) & 4 \cdot 1 + 2 \cdot 3 \end{bmatrix} = \begin{bmatrix} 4 & -5 \\ 6 & 10 \end{bmatrix}$$

note: This example illustrates a general property: If matrices AB and BA are both defined, then in general, $AB \neq BA$ (i.e., multiplication of matrices is not commutative).

PROBLEM 12-22   Let $A = \begin{bmatrix} 2 & 0 & 1 \\ -1 & 3 & 2 \end{bmatrix}$ and $B = \begin{bmatrix} 0 & 2 & -1 \\ 1 & 2 & 0 \\ 4 & 1 & 3 \end{bmatrix}$.

Find matrix product AB and explain why BA is not defined.

Solution For a matrix product to be defined, the matrices must be *compatible*; i.e., the number of columns of the first matrix must equal the number of rows of the second matrix.

For the product AB, the matrices are compatible:

$$\underset{\underset{2\times 3}{\downarrow}}{A} \cdot \underset{\underset{3\times 3}{\downarrow}}{B} = \underset{\underset{2\times 3}{\downarrow}}{AB}$$

Three columns in
matrix A = *three*
rows in matrix B

$$\begin{bmatrix} 2 & 0 & 1 \\ -1 & 3 & 2 \end{bmatrix} \cdot \begin{bmatrix} 0 & 2 & -1 \\ 1 & 2 & 0 \\ 4 & 1 & 3 \end{bmatrix}$$

$$= \begin{bmatrix} 2\cdot 0 + 0\cdot 1 + 1\cdot 4 & 2\cdot 2 + 0\cdot 2 + 1\cdot 1 & 2(-1) + 0\cdot 0 + 1\cdot 3 \\ -1\cdot 0 + 3\cdot 1 + 2\cdot 4 & -1\cdot 2 + 3\cdot 2 + 2\cdot 1 & (-1)(-1) + 3\cdot 0 + 2\cdot 3 \end{bmatrix}$$

$$= \begin{bmatrix} 4 & 5 & 1 \\ 11 & 6 & 7 \end{bmatrix}$$

But for the product BA,

$$\underset{\underset{3\times 3}{\downarrow}}{B} \cdot \underset{\underset{2\times 3}{\downarrow}}{A}$$

Three columns in
matrix $B \neq two$
rows in matrix A

so BA is not defined.

PROBLEM 12-23 Given $A = \begin{bmatrix} 4 & 3 \\ 2 & 1 \end{bmatrix}$, find A^{-1}. Check AA^{-1}.

Solution First find the value of det A:

$$\det A = \begin{vmatrix} 4 & 3 \\ 2 & 1 \end{vmatrix} = 4 - 6 = -2$$

Then interchange the elements on the main diagonal, negate the elements on the minor diagonal, and multiply the result by $1/\det A$:

$$A^{-1} = \frac{1}{-2}\begin{bmatrix} 1 & -3 \\ -2 & 4 \end{bmatrix} = \begin{bmatrix} -\frac{1}{2} & \frac{3}{2} \\ 1 & -2 \end{bmatrix}$$

Now check—you're right if $AA^{-1} = I_n$:

$$AA^{-1} = \begin{bmatrix} 4 & 3 \\ 2 & 1 \end{bmatrix} \cdot \begin{bmatrix} -\frac{1}{2} & \frac{3}{2} \\ 1 & -2 \end{bmatrix} = \begin{bmatrix} 4(-\frac{1}{2}) + 3\cdot 1 & 4(\frac{3}{2}) + 3(-2) \\ 2(-\frac{1}{2}) + 1\cdot 1 & 2(\frac{3}{2}) + 1(-2) \end{bmatrix} = \begin{bmatrix} 1 & 0 \\ 0 & 1 \end{bmatrix} = I_2$$

PROBLEM 12-24 Can you find the inverse of $A = \begin{bmatrix} 6 & 2 \\ -3 & -1 \end{bmatrix}$?

Solution Find det A:

$$\det A = \begin{vmatrix} 6 & 2 \\ -3 & -1 \end{vmatrix} = -6 -(-6) = 0$$

This matrix A does not have an inverse, since det $A = 0$.

note: If the determinant of a square matrix A has value zero, then the matrix has no inverse. Mathematicians call this kind of matrix *singular*. If det $A \neq 0$, then an inverse does exist and the matrix is *nonsingular*.

PROBLEM 12-25   Find the inverse of the matrix $A = \begin{bmatrix} 1 & 1 & 1 \\ 0 & -1 & 2 \\ 1 & 1 & 0 \end{bmatrix}$.

Solution   First, augment A with I_3: $\left[\begin{array}{ccc:ccc} 1 & 1 & 1 & 1 & 0 & 0 \\ 0 & -1 & 2 & 0 & 1 & 0 \\ 1 & 1 & 0 & 0 & 0 & 1 \end{array}\right]$

Then, perform row operations on the augmented matrix to transform the left member to the identity matrix I_3:

$$\left[\begin{array}{ccc:ccc} 1 & 1 & 1 & 1 & 0 & 0 \\ 0 & -1 & 2 & 0 & 1 & 0 \\ 1 & 1 & 0 & 0 & 0 & 1 \end{array}\right] \xrightarrow{R_3 \to -1R_1 + R_3} \left[\begin{array}{ccc:ccc} 1 & 1 & 1 & 1 & 0 & 0 \\ 0 & -1 & 2 & 0 & 1 & 0 \\ 0 & 0 & -1 & -1 & 0 & 1 \end{array}\right]$$

$$\left[\begin{array}{ccc:ccc} 1 & 1 & 1 & 1 & 0 & 0 \\ 0 & -1 & 2 & 0 & 1 & 0 \\ 0 & 0 & -1 & -1 & 0 & 1 \end{array}\right] \xrightarrow{R_1 \to R_2 + R_1} \left[\begin{array}{ccc:ccc} 1 & 0 & 3 & 1 & 1 & 0 \\ 0 & -1 & 2 & 0 & 1 & 0 \\ 0 & 0 & -1 & -1 & 0 & 1 \end{array}\right]$$

$$\left[\begin{array}{ccc:ccc} 1 & 0 & 3 & 1 & 1 & 0 \\ 0 & -1 & 2 & 0 & 1 & 0 \\ 0 & 0 & -1 & -1 & 0 & 1 \end{array}\right] \xrightarrow{R_2 \to 2R_3 + R_2} \left[\begin{array}{ccc:ccc} 1 & 0 & 3 & 1 & 1 & 0 \\ 0 & -1 & 0 & -2 & 1 & 2 \\ 0 & 0 & -1 & -2 & 0 & 1 \end{array}\right]$$

$$\left[\begin{array}{ccc:ccc} 1 & 0 & 3 & 1 & 1 & 0 \\ 0 & -1 & 0 & -2 & 1 & 2 \\ 0 & 0 & -1 & -1 & 0 & 1 \end{array}\right] \xrightarrow{R_1 \to 3R_3 + R_1} \left[\begin{array}{ccc:ccc} 1 & 0 & 0 & -2 & 1 & 3 \\ 0 & -1 & 0 & -2 & 1 & 2 \\ 0 & 0 & -1 & -1 & 0 & 1 \end{array}\right]$$

$$\left[\begin{array}{ccc:ccc} 1 & 0 & 0 & -2 & 1 & 3 \\ 0 & -1 & 0 & -2 & 1 & 2 \\ 0 & 0 & -1 & -1 & 0 & 1 \end{array}\right] \xrightarrow{R_2 \to -1R_2;\ R_3 \to -1R_3} \left[\begin{array}{ccc:ccc} 1 & 0 & 0 & -2 & 1 & 3 \\ 0 & 1 & 0 & 2 & -1 & -2 \\ 0 & 0 & 1 & 1 & 0 & -1 \end{array}\right]$$

Now the left side is in the form I_3, so the right side is A^{-1}:

$$A^{-1} = \begin{bmatrix} -2 & 1 & 3 \\ 2 & -1 & -2 \\ 1 & 0 & -1 \end{bmatrix}$$

Check:

$$AA^{-1} = \begin{bmatrix} 1 & 1 & 1 \\ 0 & -1 & 2 \\ 1 & 1 & 0 \end{bmatrix} \begin{bmatrix} -2 & 1 & 3 \\ 2 & -1 & -2 \\ 1 & 0 & -1 \end{bmatrix}$$

$$= \begin{bmatrix} (1)(-2) + & (1)(2) + (1)(1) & (1)(1) + & (1)(-1) + (1)(0) & (1)(3) + & (1)(-2) + (1)(-1) \\ (0)(-2) + (-1)(2) + (2)(1) & (0)(1) + (-1)(-1) + (2)(0) & (0)(3) + (-1)(-2) + (2)(-1) \\ (1)(-2) + & (1)(2) + (0)(1) & (1)(1) + & (1)(-1) + (0)(0) & (1)(3) + & (1)(-2) + (0)(-1) \end{bmatrix}$$

$$= \begin{bmatrix} -2+2+1 & 1-1+0 & 3-2-1 \\ 0-2+2 & 0+1+0 & 0+2-2 \\ -2+2+0 & 1-1+0 & 3-2-0 \end{bmatrix} = \begin{bmatrix} 1 & 0 & 0 \\ 0 & 1 & 0 \\ 0 & 0 & 1 \end{bmatrix} \overset{\checkmark}{=} I_3$$

PROBLEM 12-26   Find the inverse of $A = \begin{bmatrix} 1 & 2 & 0 \\ 4 & 4 & 1 \\ 3 & 0 & 1 \end{bmatrix}$.

Solution   First write the augmented matrix: $\overbrace{}^{A}\ \overbrace{}^{I_3}$ $\left[\begin{array}{ccc:ccc} 1 & 2 & 0 & 1 & 0 & 0 \\ 4 & 4 & 1 & 0 & 1 & 0 \\ 3 & 0 & 1 & 0 & 0 & 1 \end{array}\right]$

Then perform elementary row operations:

$$
\begin{array}{cc}
\overset{A}{} & \overset{I_3}{}
\end{array}
$$

$$
\left[\begin{array}{ccc|ccc}
1 & 2 & 0 & 1 & 0 & 0 \\
4 & 4 & 1 & 0 & 1 & 0 \\
3 & 0 & 1 & 0 & 0 & 1
\end{array}\right]
\xrightarrow{R_3 \to -3R_1 + R_3; \; R_2 \to -4R_1 + R_2}
\left[\begin{array}{ccc|ccc}
1 & 2 & 0 & 1 & 0 & 0 \\
0 & -4 & 1 & -4 & 1 & 0 \\
0 & -6 & 1 & -3 & 0 & 1
\end{array}\right]
$$

$$
\left[\begin{array}{ccc|ccc}
1 & 2 & 0 & 1 & 0 & 0 \\
0 & -4 & 1 & -4 & 1 & 0 \\
0 & -6 & 1 & -3 & 0 & 1
\end{array}\right]
\xrightarrow{R_2 \to -\frac{1}{4}R_2}
\left[\begin{array}{ccc|ccc}
1 & 2 & 0 & 1 & 0 & 0 \\
0 & 1 & -\frac{1}{4} & 1 & -\frac{1}{4} & 0 \\
0 & -6 & 1 & -3 & 0 & 1
\end{array}\right]
$$

$$
\left[\begin{array}{ccc|ccc}
1 & 2 & 0 & 1 & 0 & 0 \\
0 & 1 & -\frac{1}{4} & 1 & -\frac{1}{4} & 0 \\
0 & -6 & 1 & -3 & 0 & 1
\end{array}\right]
\xrightarrow{R_1 \to -2R_2 + R_1; \; R_3 \to 6R_2 + R_3}
\left[\begin{array}{ccc|ccc}
1 & 0 & \frac{1}{2} & -1 & \frac{1}{2} & 0 \\
0 & 1 & -\frac{1}{4} & 1 & -\frac{1}{4} & 0 \\
0 & 0 & -\frac{1}{2} & 3 & -\frac{3}{2} & 1
\end{array}\right]
$$

$$
\left[\begin{array}{ccc|ccc}
1 & 0 & \frac{1}{2} & -1 & \frac{1}{2} & 0 \\
0 & 1 & -\frac{1}{4} & 1 & -\frac{1}{4} & 0 \\
0 & 0 & -\frac{1}{2} & 3 & -\frac{3}{2} & 1
\end{array}\right]
\xrightarrow{R_3 \to -2R_3}
\left[\begin{array}{ccc|ccc}
1 & 0 & 0 & 2 & -1 & 1 \\
0 & 1 & 0 & -\frac{1}{2} & \frac{1}{2} & -\frac{1}{2} \\
0 & 0 & 1 & -6 & 3 & -2
\end{array}\right]
$$

$$
\underset{I_3}{} \quad \underset{A^{-1}}{}
$$

Check:
$$
AA^{-1} =
\begin{bmatrix}
1 & 2 & 0 \\
4 & 4 & 1 \\
3 & 0 & 1
\end{bmatrix}
\cdot
\begin{bmatrix}
2 & -1 & 1 \\
-\frac{1}{2} & \frac{1}{2} & -\frac{1}{2} \\
-6 & 3 & -2
\end{bmatrix}
$$

$$
=
\begin{bmatrix}
1(2)+2(-\frac{1}{2})+0(-6) & 1(-1)+2(\frac{1}{2})+0(3) & 1(1)+2(-\frac{1}{2})+0(-2) \\
4(2)+4(-\frac{1}{2})+1(-6) & 4(-1)+4(\frac{1}{2})+1(3) & 4(1)+4(-\frac{1}{2})+1(-2) \\
3(2)+0(-\frac{1}{2})+1(-6) & 3(-1)+0(\frac{1}{2})+1(3) & 3(1)+0(-\frac{1}{2})+1(-2)
\end{bmatrix}
$$

$$
=
\begin{bmatrix}
2-1+0 & -1+1+0 & 1-1-0 \\
8-2-6 & -4+2+3 & 4-2-2 \\
6-0-6 & -3+0+3 & 3-0-2
\end{bmatrix}
=
\begin{bmatrix}
1 & 0 & 0 \\
0 & 1 & 0 \\
0 & 0 & 1
\end{bmatrix}
\overset{\checkmark}{=} I_3
$$

Solving Linear Systems by Matrix Methods

PROBLEM 12-27 Use matrix methods to solve the system $\begin{cases} 3x + y = -2 \\ 5x + 2y = 4 \end{cases}$.

Solution First obtain the coefficient matrix A, the constant matrix D, and the variable matrix X:

$$
A = \begin{bmatrix} 3 & 1 \\ 5 & 2 \end{bmatrix} \qquad
D = \begin{bmatrix} -2 \\ 4 \end{bmatrix} \qquad
X = \begin{bmatrix} x \\ y \end{bmatrix}
$$

Now the system may be written $AX = D$, or in the expanded form

$$
\begin{bmatrix} 3 & 1 \\ 5 & 2 \end{bmatrix} \cdot \begin{bmatrix} x \\ y \end{bmatrix} = \begin{bmatrix} -2 \\ 4 \end{bmatrix}
$$

Then, since $AX = D$ and $I_2 = A^{-1}A$,

$$
A^{-1}(AX) = A^{-1}(D)
$$
$$
(A^{-1}A)X = A^{-1}D
$$
$$
I_2 X = A^{-1}D
$$
$$
X = A^{-1}D
$$

Now, since $A = \begin{bmatrix} 3 & 1 \\ 5 & 2 \end{bmatrix}$, then $\det A = \begin{vmatrix} 3 & 1 \\ 5 & 2 \end{vmatrix} = 6 - 5 = 1$; so $A^{-1} = \dfrac{1}{1}\begin{bmatrix} 2 & -1 \\ -5 & 3 \end{bmatrix}$. Thus

$$
X = A^{-1}D = \begin{bmatrix} 2 & -1 \\ -5 & 3 \end{bmatrix}\begin{bmatrix} -2 \\ 4 \end{bmatrix} = \begin{bmatrix} 2(-2) + -1(4) \\ -5(-2) + 3(4) \end{bmatrix} = \begin{bmatrix} -8 \\ 22 \end{bmatrix}
$$

Therefore
$$X = \begin{bmatrix} x \\ y \end{bmatrix} = \begin{bmatrix} -8 \\ 22 \end{bmatrix} \rightarrow \begin{cases} x = -8 \\ y = 22 \end{cases}$$

Check:
$$3x + y = -2 \qquad 5x + 2y = 4$$
$$3(-8) + 22 \stackrel{?}{=} -2 \qquad 5(-8) + 2(22) \stackrel{?}{=} 4$$
$$-24 + 22 \stackrel{\checkmark}{=} -2 \qquad -40 + 44 \stackrel{\checkmark}{=} 4$$

PROBLEM 12-28 Use matrix methods to solve the system $\begin{cases} x + 2y = 3 \\ 4x + 4y + z = 1 \\ 3x + z = -2 \end{cases}$.

Solution

Let
$$A = \begin{bmatrix} 1 & 2 & 0 \\ 4 & 4 & 1 \\ 3 & 0 & 1 \end{bmatrix}, \qquad D = \begin{bmatrix} 3 \\ 1 \\ -2 \end{bmatrix}, \qquad X = \begin{bmatrix} x \\ y \\ z \end{bmatrix}$$

And the inverse of the coefficient matrix is

$$A^{-1} = \begin{bmatrix} 2 & -1 & 1 \\ -\frac{1}{2} & \frac{1}{2} & -\frac{1}{2} \\ -6 & 3 & -2 \end{bmatrix} \qquad \text{[See Problem 12-26.]}$$

Thus, since $X = A^{-1}D$,

$$X = \begin{bmatrix} x \\ y \\ z \end{bmatrix} = A^{-1}D = \begin{bmatrix} 2 & -1 & 1 \\ -\frac{1}{2} & \frac{1}{2} & -\frac{1}{2} \\ -6 & 3 & -2 \end{bmatrix} \cdot \begin{bmatrix} 3 \\ 1 \\ -2 \end{bmatrix}$$

$$= \begin{bmatrix} 2(3) + -1(1) + 1(-2) \\ -\frac{1}{2}(3) + \frac{1}{2}(1) + -\frac{1}{2}(-2) \\ -6(3) + 3(1) + -2(-2) \end{bmatrix} = \begin{bmatrix} 3 \\ 0 \\ -11 \end{bmatrix}$$

Therefore, the solution is $x = 3$, $y = 0$, $z = -11$.

Check:
$$x + 2y = 3 \qquad 4x + 4y + z = 1 \qquad 3x + z = -2$$
$$3 + 0 \stackrel{\checkmark}{=} 3 \qquad 12 + 0 - 11 \stackrel{\checkmark}{=} 1 \qquad 9 - 11 \stackrel{\checkmark}{=} -2$$

Supplementary Exercises

PROBLEM 12-29 Form the coefficient matrix C and the augmented matrix A for the system of equations $\begin{cases} 4x - 7y = 3 \\ -2x + 3y = 5 \end{cases}$.

PROBLEM 12-30 Are the matrices $\begin{bmatrix} -1 & 5 & 0 \\ 2 & 3 & 4 \end{bmatrix}$ and $\begin{bmatrix} 0 & 13 & 4 \\ 1 & 8 & 4 \end{bmatrix}$ equivalent?

PROBLEM 12-31 Using a sequence of row operations on the augmented matrix, (a) bring the system $\begin{cases} 2x - 4y = 2 \\ -x + 3y = 1 \end{cases}$ to a point where the solution may be found and (b) solve it.

PROBLEM 12-32 For the system of equations $\begin{cases} 2x - 5y + z = 5 \\ -x + 3y = -1 \\ x + y - 4z = -3 \end{cases}$,

(a) form the augmented matrix and (b) change it to echelon form.

PROBLEM 12-33 Solve the system of equations of Problem 12-32.

PROBLEM 12-34 Find the value of the determinant of the following matrices:

$$A = [-2], \qquad B = \begin{bmatrix} 4 & 1 \\ 2 & 3 \end{bmatrix}, \qquad C = \begin{bmatrix} 1 & 0 & 5 \\ -1 & 2 & 0 \\ 0 & 4 & 2 \end{bmatrix}$$

PROBLEM 12-35 Find the value of the determinant of the following matrices:

$$A = \begin{bmatrix} 1 & 4 & 3 \\ 2 & 5 & 2 \\ 3 & 0 & 1 \end{bmatrix} \qquad B = \begin{bmatrix} 1.3 & 0.9 & 0 \\ 2.2 & 1.6 & 1 \\ 7.5 & 4.6 & 0 \end{bmatrix} \qquad C = \begin{bmatrix} 2 & 8 & -4 \\ 21 & -7 & 49 \\ 6 & 24 & -12 \end{bmatrix}$$

PROBLEM 12-36 Find the minors M_{22} and M_{13} for matrix $A = \begin{bmatrix} 4 & 1 & 1 \\ -3 & 0 & 6 \\ 5 & 2 & 4 \end{bmatrix}$.

PROBLEM 12-37 Find the cofactors A_{32} and A_{11} for the determinant $\begin{vmatrix} 8 & 6 & 1 \\ 0 & 2 & 5 \\ 3 & 4 & 0 \end{vmatrix}$.

PROBLEM 12-38 Using cofactors of the third row, find the value of det A where

$$A = \begin{bmatrix} 1 & -3 & 2 \\ 5 & -2 & 4 \\ 2 & 0 & 1 \end{bmatrix}$$

PROBLEM 12-39 Find the value of det B if

$$B = \begin{bmatrix} -1 & 2 & 0 & 3 \\ 0 & 4 & 1 & 5 \\ 1 & 0 & -2 & 0 \\ 2 & 3 & 1 & -1 \end{bmatrix}$$

Hint: Form the determinant of matrix B; then perform determinant operations to produce more zeros and expand by cofactors to reduce the size. Continue to the value of det B.

PROBLEM 12-40 Use Cramer's Rule to solve the system $\begin{cases} 2x + 5y = 2 \\ 4x + 7y = 10 \end{cases}$.

PROBLEM 12-41 Use Cramer's Rule to solve the system $\begin{cases} 7x + y = 6 \\ 3x - 2y = 22 \end{cases}$.

PROBLEM 12-42 Use Cramer's Rule to solve the system $\begin{cases} 2x + y + z = 1 \\ -3x - y + 2z = 3 \\ x + 2y + 3z = -2 \end{cases}$.

PROBLEM 12-43 Find $A + B$, $C - 2A$, $A + (B + C)$, $(A + B) + C$, and AB, if

$$A = \begin{bmatrix} 2 & 5 \\ 1 & 3 \end{bmatrix}, \qquad B = \begin{bmatrix} 1 & 1 \\ 0 & 4 \end{bmatrix}, \qquad C = \begin{bmatrix} 3 & 6 \\ 2 & 0 \end{bmatrix}$$

PROBLEM 12-44 Find AB and BA, given $A = \begin{bmatrix} 1 & 0 & 4 \\ 0 & -1 & 3 \\ 2 & 2 & 1 \end{bmatrix}$ and $B = \begin{bmatrix} 0 & 2 & 1 \\ -1 & 0 & 4 \\ 3 & 1 & 2 \end{bmatrix}$.

PROBLEM 12-45 Given $A = \begin{bmatrix} 4 & 2 \\ 3 & 2 \end{bmatrix}$ and $B = \begin{bmatrix} 2 & -3 \\ 1 & -2 \end{bmatrix}$, find A^{-1} and B^{-1}. Then find AB and $(AB)^{-1}$. Now find $B^{-1}A^{-1}$. Any conclusions?

PROBLEM 12-46   Find the inverse of the matrix $A = \begin{bmatrix} 6 & -1 & 2 \\ -1 & 1 & -1 \\ 1 & 1 & -1 \end{bmatrix}$.

PROBLEM 12-47   Use matrix methods to solve the system $\begin{cases} 5a + 2b = -2 \\ a + 3b = 10 \end{cases}$.

PROBLEM 12-48   Use matrix methods to solve the system $\begin{cases} 2u - 5v = \frac{5}{2} \\ -4u + 13v = -\frac{7}{2} \end{cases}$.

PROBLEM 12-49   Use matrix methods to solve the system $\begin{cases} -x + 4y = 2 \\ 3x + 3y + 2z = 3 \\ -2x + 7y + z = 0 \end{cases}$.

PROBLEM 12-50   Use matrix methods to solve the system $\begin{cases} 6x - y + 2z = 3 \\ -x + y - z = 3 \\ x + y - z = 5 \end{cases}$.

Answers to Supplementary Exercises

12-29 $C = \begin{bmatrix} 4 & -7 \\ -2 & 3 \end{bmatrix}$, $A = \begin{bmatrix} 4 & -7 & \vdots & 3 \\ -2 & 3 & \vdots & 5 \end{bmatrix}$

12-30 Yes

12-31 (a) $\begin{bmatrix} 2 & -4 & \vdots & 2 \\ 0 & 2 & \vdots & 4 \end{bmatrix}$ (b) $(5, 2)$

12-32 (a) $\begin{bmatrix} 2 & -5 & 1 & \vdots & 5 \\ -1 & 3 & 0 & \vdots & -1 \\ 1 & 1 & -4 & \vdots & -3 \end{bmatrix}$

(b) $\begin{bmatrix} 2 & -5 & 1 & \vdots & 5 \\ 0 & 1 & 1 & \vdots & 3 \\ 0 & 0 & 2 & \vdots & 4 \end{bmatrix}$

12-33 $(4, 1, 2)$

12-34 $\det A = -2$, $\det B = 10$, $\det C = -16$

12-35 $\det A = -24$, $\det B = 0.77$, $\det C = 0$

12-36 $M_{22} = \begin{vmatrix} 4 & 1 \\ 5 & 4 \end{vmatrix} = 11$, $M_{13} = \begin{vmatrix} -3 & 0 \\ 5 & 2 \end{vmatrix} = -6$

12-37 $A_{32} = (-1)\begin{vmatrix} 8 & 1 \\ 0 & 5 \end{vmatrix} = -40$,

$A_{11} = \begin{vmatrix} 2 & 5 \\ 4 & 0 \end{vmatrix} = -20$

12-38 -3

12-39 -39

12-40 $(6, -2)$

12-41 $(2, -8)$

12-42 $(2, -5, 2)$

12-43 $A + B = \begin{bmatrix} 3 & 6 \\ 1 & 7 \end{bmatrix}$, $C - 2A = \begin{bmatrix} -1 & -4 \\ 0 & -6 \end{bmatrix}$,

$A + (B + C) = (A + B) + C = \begin{bmatrix} 6 & 12 \\ 3 & 7 \end{bmatrix}$,

$AB = \begin{bmatrix} 2 & 22 \\ 1 & 13 \end{bmatrix}$

12-44 $AB = \begin{bmatrix} 12 & 6 & 9 \\ 10 & 3 & 2 \\ 1 & 5 & 12 \end{bmatrix}$ $BA = \begin{bmatrix} 2 & 0 & 7 \\ 7 & 8 & 0 \\ 7 & 3 & 17 \end{bmatrix}$

12-45 $A^{-1} = \begin{bmatrix} 1 & -1 \\ -\frac{3}{2} & 2 \end{bmatrix}$, $B^{-1} = \begin{bmatrix} 2 & -3 \\ 1 & -2 \end{bmatrix} = B$,

$AB = \begin{bmatrix} 10 & -16 \\ 8 & -13 \end{bmatrix}$, $(AB)^{-1} = \begin{bmatrix} \frac{13}{2} & -8 \\ 4 & -5 \end{bmatrix}$;

$B^{-1}A^{-1} = \begin{bmatrix} \frac{13}{2} & -8 \\ 4 & -5 \end{bmatrix}$

$\therefore (AB)^{-1} = B^{-1}A^{-1}$

12-46 $\begin{bmatrix} 0 & -\frac{1}{2} & \frac{1}{2} \\ 1 & 4 & -2 \\ 1 & \frac{7}{2} & -\frac{5}{2} \end{bmatrix}$

12-47 $(-2, 4)$

12-48 $(\frac{5}{2}, \frac{1}{2})$

12-49 $(2, 1, -3)$

12-50 $(1, 5, 1)$

13 COMPLEX NUMBERS

THIS CHAPTER IS ABOUT

☑ **Complex Numbers Revisited**
☑ **The Arithmetic of Complex Numbers**
☑ **Complex Conjugates and Quotients**
☑ **The Complex Plane**
☑ **Equations with Complex Roots**

13-1. Complex Numbers Revisited

The study of solutions of the general polynomial equation

$$a_n x^n + a_{n-1} x^{n-1} + a_{n-2} x^{n-2} + \cdots + a_2 x^2 + a_1 x + a_0 = 0$$

where n is a positive integer and the coefficients $a_n, a_{n-1}, \ldots, a_2, a_1, a_0$ are real numbers, has interested mathematicians and scientists for centuries. Even some very simple forms such as $x^2 + 1 = 0$ have no solution in the field of real numbers. So we have to extend the field.

A. The complex number field

The **complex number field** is an extension of the real number field to include a special member i with the property that $i^2 = -1$, which provides a solution to $x^2 + 1 = 0$. This extended field has found use in many scientific fields, especially in physics and electrical and computer engineering.

The simplest complex number is the number i, defined by

$$i^2 = -1 \quad \text{or} \quad i = \sqrt{-1}$$

This number clearly solves $x^2 + 1 = 0$, since

$$i^2 + 1 = -1 + 1 = 0$$

In order to solve $x^2 + 4 = 0$ or $x^2 + x + 1 = 0$, we need to introduce **complex numbers** of the form

$$a + bi$$

where a and b are real numbers. Thus the *complex number field* may be defined as the collection of all numbers that can be written as $a + bi$ where a and b are real numbers.

- Every second-degree polynomial has a solution from the complex number field.

B. Definitions and notation

- For every complex number $a + bi$, the real number a is called the **real part Re** of the complex number and the real number b is called the **imaginary part Im** of the complex number. If we use a single letter z to refer to a complex number, such that $z = a + bi$, then the conventional notation is

$$\text{the } \textit{real part} \text{ of } z \equiv \text{Re}(z) = a$$

and

$$\text{the } \textit{imaginary part} \text{ of } z \equiv \text{Im}(z) = b$$

- A complex number of the form $0 + bi$ is called a **pure imaginary number** (or just an *imaginary number*).
- Two complex numbers $a + bi$ and $c + di$ are equal if and only if $a = c$ and $b = d$.

note: See also Chapter 3, Section 3-4.

13-2. The Arithmetic of Complex Numbers

Since we have defined a new field of numbers, the operations on these numbers must be defined.

A. Addition

To add complex numbers, we add the real and imaginary parts independently.

- The **sum of complex numbers** is defined by

$$(a + bi) + (c + di) = (a + c) + (b + d)i$$

The complex number field is closed under addition.

EXAMPLE 13-1   Find **(a)** $(2 + 3i) + (5 - 2i)$ and **(b)** $(7 - 6i) + (2 + 4i)$.

Solution

(a) $(2 + 3i) + (5 - 2i) = (2 + 5) + (3 - 2)i = 7 + i$

(b) $(7 - 6i) + (2 + 4i) = (7 + 2) + (-6 + 4)i = 9 + (-2)i = 9 - 2i$

B. Subtraction

To subtract complex numbers, we subtract the real and imaginary parts independently.

- The **difference of two complex numbers** is defined by

$$(a + bi) - (c + di) = (a - c) + (b - d)i$$

The complex number field is closed under subtraction.

EXAMPLE 13-2   Find **(a)** $(5 + 3i) - (7 + 4i)$ and **(b)** $(\frac{16}{3} + \frac{1}{3}i) - (1 + \frac{4}{3}i)$.

Solution

(a) $(5 + 3i) - (7 + 4i) = (5 - 7) + (3 - 4)i = -2 + (-1)i = -2 - i$

(b) $(\frac{16}{3} + \frac{1}{3}i) - (1 + \frac{4}{3}i) = (\frac{16}{3} - 1) + (\frac{1}{3} - \frac{4}{3})i = \frac{13}{3} + (-1)i = \frac{13}{3} - i$

C. Multiplication

- The **product of two complex numbers** is defined by

$$(a + bi)(c + di) = a(c + di) + bi(c + di)$$
$$= (ac + adi) + (bci + bdi^2)$$
$$= ac + (adi + bci) + (bd)i^2$$

Using the fact that $i^2 = -1$, we now write

$$(a + bi)(c + di) = ac + (adi + bci) + (bd)(-1)$$
$$= (ac - bd) + (ad + bc)i$$

The fact that this product is itself a complex number shows that the complex number field is closed under muitiplication.

note: DON'T memorize this formula for multiplication. In actual practice, you can use the old FOIL method for multiplying two binomials, coupled with the fact that $i^2 = -1$.

EXAMPLE 13-3   Find the following products and write the result in the form $a + bi$:

(a) $(2 + 3i)(1 + 2i)$ **(b)** $(5 - 3i)(-2 + 4i)$ **(c)** $(5 + 4i)(5 - 4i)$

Solution   Use the FOIL method.

(a)
$$(2 + 3i)(1 + 2i) = (2)(1) + (2i)(2) + (1)(3i) + (3i)(2i)$$
$$= 2 + 4i + 3i + 6i^2$$
$$= 2 + 7i - 6$$
$$= -4 + 7i$$

(b)
$$(5 - 3i)(-2 + 4i) = (5)(-2) + (5)(4i) + (-3i)(-2) + (-3i)(4i)$$
$$= -10 + 20i + 6i - 12i^2$$
$$= -10 + 26i + 12$$
$$= 2 + 26i$$

(c)
$$(5 + 4i)(5 - 4i) = (5)(5) + (5)(-4i) + (4i)(5) + (4i)(-4i)$$
$$= 25 - 20i + 20i - 16i^2$$
$$= 25 - 0i + 16$$
$$= 41$$

D. Powers

The important special products of complex numbers are

$$i^2 = i \cdot i = -1$$
$$i^3 = i^2 \cdot i = (-1)i = -i$$
$$i^4 = i^3 \cdot i = (-i)(i) = -i^2 = -(-1) = 1$$

We now obtain a repeating pattern:

$i^1 = i$	$i^5 = i^4 \cdot i = i$	$i^9 = i$
$i^2 = -1$	$i^6 = i^4 \cdot i^2 = -1$	$i^{10} = -1$
$i^3 = -i$	$i^7 = i^4 \cdot i^3 = -i$	$i^{11} = -i$
$i^4 = 1$	$i^8 = i^4 \cdot i^4 = 1$	$i^{12} = 1$

An extension of this list allows us to compute any power of i by pulling out the 4th-power parts.

EXAMPLE 13-4   Find the reduced form of **(a)** i^{15}, **(b)** i^{53}, and **(c)** $(2i)^{12}$.

Solution

(a) $i^{15} = i^{12} \cdot i^3 = (i^4 \cdot i^4 \cdot i^4) \cdot i^3 = (1 \cdot 1 \cdot 1) \cdot i^3 = i^3 = -i$

(b) $i^{53} = i^{52} \cdot i = (i^4)^{13} \cdot i = 1^{13} \cdot i = 1 \cdot i = i$

(c) $(2i)^{12} = 2^{12} \cdot i^{12} = 4096 \cdot i^{12} = 4096(i^4)^3 = 4096$

We now have the means to calculate a higher power of a complex number $a + bi$. We have

$$(a + bi)^1 = a + bi$$

$$(a + bi)^2 = (a + bi)(a + bi) = a^2 + abi + abi + b^2i^2 = (a^2 - b^2) + (2ab)i$$

$$(a + bi)^3 = a^3 + 3a^2(bi) + 3a(bi)^2 + (bi)^3$$

$$= a^3 + (3a^2b)i + (3ab^2)i^2 + b^3i^3$$

$$= a^3 + (3a^2b)i - 3ab^2 + b^3(-i)$$

$$= (a^3 - 3ab^2) + (3a^2b - b^3)i$$

Again, we do not memorize these results but use the technique suggested.

EXAMPLE 13-5 Find the complex number represented by $(2 + 3i)^3$.

Solution

$$(2 + 3i)^3 = 2^3 + 3(2)^2(3i) + 3(2)(3i)^2 + (3i)^3$$

$$= 8 + 3 \cdot 4 \cdot 3i + 3 \cdot 2 \cdot 9i^2 + 27i^3$$

$$= 8 + 36i - 54 - 27i$$

$$= -46 + 9i$$

E. Square roots

If p is any positive real number, then the equation $x^2 + p = 0$ has two solutions:

$$x = \sqrt{p}\,i \qquad \text{(since } (\sqrt{p}\,i)^2 + p = pi^2 + p = p(-1) + p = 0\text{)}$$

and $\quad x = -\sqrt{p}\,i \qquad \text{(since } (-\sqrt{p}\,i)^2 + p = pi^2 + p = -p + p = 0\text{)}$

We call $\sqrt{p}\,i$ the **principal square root of** $-p$ and write

$$\sqrt{-p} = \sqrt{p}\,i$$

Thus

$$\sqrt{-1} = \sqrt{1}\,i = i$$

$$\sqrt{-4} = \sqrt{4}\,i = 2i$$

$$\sqrt{-9} = \sqrt{9}\,i = 3i \qquad \text{etc.}$$

We are now able to express the square root of both positive and negative numbers.

13-3. Complex Conjugates and Quotients

A. Complex conjugate

- The complex number $a - bi$ is called the **complex conjugate** of the complex number $a + bi$.

The product of a complex number and its complex conjugate is a real number:

$$(a + bi)(a - bi) = a^2 - abi + abi - b^2i^2 = a^2 + 0i + b^2$$

$$= a^2 + b^2$$

EXAMPLE 13-6 What is the complex conjugate of (**a**) $5 - 7i$ and (**b**) $-2 + 8i$?

Solution (**a**) The conjugate of $5 - 7i$ is $5 + 7i$.

(**b**) The conjugate of $-2 + 8i$ is $-2 - 8i$.

B. Inverse

If $a^2 + b^2 \neq 0$, then
$$\frac{(a + bi)(a - bi)}{a^2 + b^2} = 1$$

From this it follows that, if $a + bi \neq 0$, then

INVERSE
$$\frac{1}{a + bi} = \frac{a - bi}{a^2 + b^2}$$

which is the **reciprocal of a complex number**, also called the **multiplicative inverse of a complex number**. From this definition, we see the property that

$$\frac{1}{a + bi} = \left(\frac{1}{a^2 + b^2}\right)(a - bi)$$

$$= \left(\frac{1}{(a + bi)(a - bi)}\right)(a - bi) = \left(\frac{1}{a + bi}\right)\left(\frac{a - bi}{a - bi}\right)$$

$$= \left(\frac{1}{a + bi}\right)1 \qquad \text{[The definition produces an identity.]}$$

C. Quotients

The **quotient of two complex numbers** $\dfrac{a + bi}{c + di}$ is defined as the product of the complex number $a + bi$ and the inverse of the complex number $c + di$. Thus

$$\frac{a + bi}{c + di} = (a + bi)\left(\frac{1}{c + di}\right)$$

$$= (a + bi)\left(\frac{c - di}{c^2 + d^2}\right) = \frac{(a + bi)(c - di)}{c^2 + d^2}$$

$$= \frac{ac - adi + bci - bdi^2}{c^2 + d^2} = \frac{(ac + bd) + (bc - ad)i}{c^2 + d^2}$$

$$= \frac{ac + bd}{c^2 + d^2} + \frac{(bc - ad)i}{c^2 + d^2}$$

Often it is easier to express a quotient in the form $p + qi$ if we multiply the numerator and the denominator by the complex conjugate of the denominator (see also Chapter 3, Section 3-4):

$$\frac{a + bi}{c + di} = \left(\frac{a + bi}{c + di}\right)\left(\frac{c - di}{c - di}\right) = \frac{ac + bd + (bc - ad)i}{c^2 + d^2}$$

$$= \frac{ac + bd}{c^2 + d^2} + \frac{(bc - ad)i}{c^2 + d^2}$$

We can also simplify negative powers of i by using the conjugate. Given i^{-n}, we write this as $1/i^n$. Reduce the denominator to i, -1, $-i$, or 1 and use the conjugate multiplication.

EXAMPLE 13-7 Express the following quotients in the form $p + qi$:

(a) $\dfrac{1}{i^3}$ (b) $\dfrac{2 + 3i}{1 + i}$ (c) $\dfrac{4 - 3i}{5 - 2i}$

Solution

(a) $\dfrac{1}{i^3} = \dfrac{1}{-i} = \dfrac{1}{-i} \cdot \dfrac{i}{i} = \dfrac{i}{-i^2} = \dfrac{i}{-(-1)} = \dfrac{i}{1} = i$

(b) $\dfrac{2 + 3i}{1 + i} = \left(\dfrac{2 + 3i}{1 + i}\right)\left(\dfrac{1 - i}{1 - i}\right) = \dfrac{(2 + 3i)(1 - i)}{1 - i^2} = \dfrac{(2 + 3i)(1 - i)}{2}$

$\quad = \dfrac{2 - 2i + 3i - 3i^2}{2} = \dfrac{2 + 3 + i}{2} = \dfrac{5 + i}{2} = \dfrac{5}{2} + \dfrac{1}{2}i$

(c) $\dfrac{4 - 3i}{5 - 2i} = \left(\dfrac{4 - 3i}{5 - 2i}\right)\left(\dfrac{5 + 2i}{5 + 2i}\right) = \dfrac{(4 - 3i)(5 + 2i)}{25 - 4i^2} = \dfrac{(4 - 3i)(5 + 2i)}{25 + 4}$

$\quad = \dfrac{20 + 8i - 15i - 6i^2}{29} = \dfrac{20 + 6 - 7i}{29} = \dfrac{26 - 7i}{29} = \dfrac{26}{29} - \dfrac{7}{29}i$

D. Properties of conjugation

If we let z and w stand for given complex numbers, then their complex conjugates can be denoted by $\bar{z}$ and $\bar{w}$. It follows, then, that

(1) $\bar{\bar{z}} = z$ [The conjugate of the conjugate is the original complex number.]

(2) $\overline{z + w} = \bar{z} + \bar{w}$ [The conjugate of a sum is the sum of the conjugates.]

(3) $\overline{zw} = \bar{z} \cdot \bar{w}$ [The conjugate of a product is the product of the conjugates.]

(4) $\overline{z^n} = (\bar{z})^n$ for any positive integer n [The conjugate of a power is the power of the conjugate.]

(5) $\text{Re}(z) = \dfrac{z + \bar{z}}{2}$ [The real part of a complex number may be expressed as one-half the sum of the number and its conjugate.]

$\quad\text{Im}(z) = \dfrac{z - \bar{z}}{2i}$ [The imaginary part of a complex number is $\frac{1}{2i}$ times the number minus its conjugate.]

(6) $z + \bar{z}$ is real for any complex number.

(7) $\bar{z} = z$ if and only if z is a real number.

EXAMPLE 13-8   Verify property (3), that $\overline{zw} = \bar{z} \cdot \bar{w}$.

Solution   Let $z = a + bi$ and let $w = c + di$. Then

$$zw = (a + bi)(c + di) = (ac - bd) + (ad + bc)i$$

so that $\overline{zw} = (ac - bd) - (ad + bc)i$

But also $\bar{z} \cdot \bar{w} = (a - bi)(c - di)$

$$= ac - adi - bci + bdi^2$$

$$= (ac - bd) - (ad + bc)i$$

same

13-4. The Complex Plane

A. The axes

Each complex number $a + bi$ corresponds directly to a number pair (a, b) in a rectangular coordinate system. The real part of the complex number a is measured on the horizontal axis, called the **real axis**, and the imaginary part b is measured on the vertical axis, called the **imaginary axis**. When the plane is identified with the complex numbers, we refer to it as the **complex plane**.

EXAMPLE 13-9 Identify the following points in the complex plane:

(a) $2 + i$ **(b)** $-3 + 2i$ **(c)** $4 - i$ **(d)** $-1 - 3i$

Solution See Figure 13-1.

(a) The point $2 + i$ is $(2, 1)$, which is 2 units to the right and 1 unit up from the origin, measured on the real axis and the imaginary axis, respectively.
(b) The point $-3 + 2i$ is $(-3, 2)$ in the complex plane.
(c) The point $4 - i$ is $(4, -1)$ in the complex plane.
(d) The point $-1 - 3i$ is $(-1, -3)$ in the complex plane.

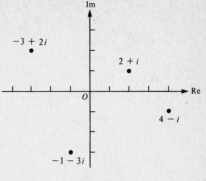

Figure 13-1

B. The absolute value

If $a + bi$ is an arbitrary complex number, then the **absolute value**, or **modulus**, of $a + bi$, denoted $|a + bi|$, is defined by

ABSOLUTE VALUE $|a + bi| = \sqrt{a^2 + b^2}$

The absolute value of a complex number may also be defined as the straight line distance of the complex number from the origin in the complex plane.

EXAMPLE 13-10 Find the absolute value of the following complex numbers:

(a) i **(b)** $3 + 4i$ **(c)** $6 - 5i$

Solution

(a) $|0 + 1 \cdot i| = \sqrt{0^2 + 1^2} = \sqrt{1} = 1$

(b) $|3 + 4i| = \sqrt{3^2 + 4^2} = \sqrt{9 + 16} = \sqrt{25} = 5$

(c) $|6 - 5i| = \sqrt{6^2 + (-5)^2} = \sqrt{36 + 25} = \sqrt{61}$

If z and w are any complex numbers, then the following properties hold:

(1) $|z| = 0$ if and only if $z = 0$ **(4)** $|zw| = |z||w|$

(2) If $z \neq 0$, then $|z| > 0$ **(5)** $\left|\dfrac{z}{w}\right| = \dfrac{|z|}{|w|}$

(3) $|z| = |-z|$ **(6)** $|z + w| \leq |z| + |w|$

EXAMPLE 13-11 Prove property (4), that $|zw| = |z||w|$, is valid.

Solution Let $z = a + bi$ and let $w = c + di$. Then

$$zw = (a + bi)(c + di) = (ac - bd) + (ad + bc)i$$

so that $|zw| = \sqrt{(ac - bd)^2 + (ad + bc)^2}$

$$= \sqrt{a^2c^2 - 2abcd + b^2d^2 + a^2d^2 + 2abcd + b^2c^2}$$

$$= \sqrt{a^2c^2 + a^2d^2 + b^2c^2 + b^2d^2}$$

But then $|z| = \sqrt{a^2 + b^2}$

and $|w| = \sqrt{c^2 + d^2}$

so that $|z||w| = \sqrt{a^2 + b^2}\sqrt{c^2 + d^2}$

$$= \sqrt{(a^2 + b^2)(c^2 + d^2)}$$

$$= \sqrt{a^2c^2 + a^2d^2 + b^2c^2 + b^2d^2}$$

same

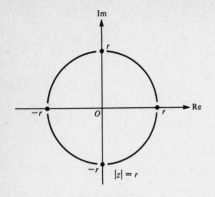

Figure 13-2

C. Geometry of complex numbers

- The set of complex numbers with a fixed absolute value constitutes a circle centered at the origin in the complex plane.

Suppose that $z = a + bi$ and $|z| = r$. Then

$$\sqrt{a^2 + b^2} = r \quad \text{or} \quad a^2 + b^2 = r^2$$

This is the same as the set $\{(a, b) \mid a^2 + b^2 = r^2\}$, which is a circle of radius r centered at $(0, 0)$. From this it follows that the expression

$$|z - z_0| = r$$

is the circle of radius r centered at the fixed point z_0. And if $z_0 = 0$, the circle is centered at the origin $(0, 0)$, as shown in Figure 13-2.

EXAMPLE 13-12   Sketch the graph of $|z + 3 - 3i| = 2$.

Solution   This is

$$|z + 3 - 3i| = 2$$
$$|z - (-3 + 3i)| = 2$$

so the center point z_0 is $-3 + 3i$ and the radius is 2, as shown in Figure 13-3.

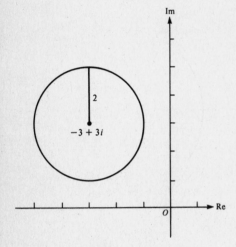

Figure 13-3

Another geometric fact for complex numbers concerns the absolute values of complex conjugates. Consider

$$z = a + bi \quad \text{and} \quad \bar{z} = a - bi$$

In the complex plane, complex conjugates have the same real part, so they are the same distance to the right or left of the origin. But their imaginary parts have opposite signs, so they go in opposite directions up and down the imaginary axis. Thus we can say that they lie on the line $x = a$ and have symmetry with respect to the x-axis, as shown in Figure 13-4. But z and $\bar{z}$ are the same distance from the origin, since

$$|z| = \sqrt{a^2 + b^2} = |\bar{z}|$$

Now we can show that

$$|z|^2 = z\bar{z}$$

To prove this, we set $z = a + bi$. Then

$$z\bar{z} = (a + bi)(a - bi) = a^2 + b^2 = (\sqrt{a^2 + b^2})^2 = |z|^2$$

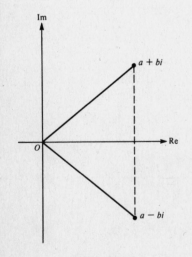

Figure 13-4

13-5. Equations with Complex Roots

A. Quadratic equations

When the discriminant $b^2 - 4ac$ of the quadratic equation $ax^2 + bx + c = 0$ is negative, the roots

$$x = \frac{-b \pm \sqrt{b^2 - 4ac}}{2a}$$

are complex numbers.

EXAMPLE 13-13   Solve the equation $x^2 + 2x + 2 = 0$.

Solution   We can see at a glance that the roots are complex numbers, because $b^2 - 4ac = 4 - 4(1)(2) = -4$. Applying the quadratic formula,

$$x = \frac{-2 \pm \sqrt{2^2 - 4(1)(2)}}{2 \cdot 1} = \frac{-2 \pm \sqrt{-4}}{2} = \frac{-2 \pm 2i}{2} = -1 \pm i$$

so the roots are $x = -1 + i$ and $x = -1 - i$. [Notice that the roots are complex conjugates.]

B. Cubic equations: Cube roots of unity

Certain cubic equations—those of the form $x^3 - 1 = 0$—may be solved using complex numbers. These equations are of special significance because one of their roots is 1, while the other two roots are complex conjugates.

EXAMPLE 13-14 Solve and graph the cubic equation $x^3 - 1 = 0$.

Solution
$$x^3 - 1 = 0$$
$$(x - 1)(x^2 + x + 1) = 0 \qquad \text{[Factor]}$$
Thus $x = 1$ or $x = \dfrac{-1 \pm \sqrt{1 - 4(1)(1)}}{2 \cdot 1} = \dfrac{-1 \pm \sqrt{-3}}{2}$

$$= -\frac{1}{2} \pm \frac{\sqrt{3}}{2} i$$

The roots are $x_1 = 1$, $x_2 = -\dfrac{1}{2} + \dfrac{\sqrt{3}}{2}i$, $x_3 = -\dfrac{1}{2} - \dfrac{\sqrt{3}}{2}i$. These roots are called the three **cube roots of unity**, because they all lie on a circle of radius 1 in the complex plane and are positioned symmetrically around the circle at angles of $360°/3 = 2\pi/3 = 120°$. See Figure 13-5.

note: Remember the radian-to-degrees conversion formula, π radians $= 180°$.

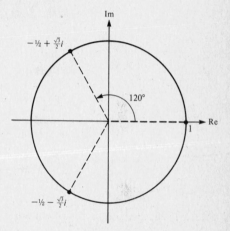

Figure 13-5

C. *n*th Roots of unity

If we want to solve the equation

$$x^4 - 1 = 0$$

for its four roots, we obtain the four **fourth roots of unity**.

EXAMPLE 13-15 Solve and graph $x^4 - 1 = 0$.

Solution
$$x^4 - 1 = 0$$
$$(x^2 - 1)(x^2 + 1) = 0$$
$$(x - 1)(x + 1)(x - i)(x + i) = 0$$

So
$$x = 1, \quad x = -1, \quad x = i, \quad x = -i$$

The graph of the four fourth roots of unity—symmetrically located at consecutive angles of $360°/4 = 90°$—is shown in Figure 13-6.

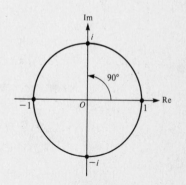

Figure 13-6

- The solutions of $x^n - 1 = 0$ (n is a positive integer) give the n **nth roots of unity**, one of which is $x = 1$ while the others are on the unit circle symmetrically located at consecutive angles of $2\pi/n = 360/n$ degrees.

These roots also come in complex conjugate pairs, so the graphing is somewhat simplified.

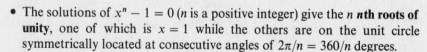

EXAMPLE 13-16 Graph the ten tenth roots of unity.

Solution We are solving $x^{10} - 1 = 0$. We know that one solution is $x = 1$, so the others are on the unit circle at $360°/10 = 36°$ apart as in Figure 13-7. Their complex notation may be found using geometry.

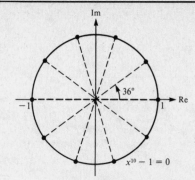

Figure 13-7

SUMMARY

1. To solve equations such as $x^2 + 1 = 0$, we extend the real number field to include $i = \sqrt{-1}$ and thus create the complex number field.
2. In a complex number $a + bi$, the real number a is called the real (Re) part and the real number b is called the imaginary (Im) part.
3. In finding the sum (or difference) of two complex numbers, we add (or subtract) the real and imaginary parts independently.
4. The complex conjugate of $a + bi$ is $a - bi$.
5. There are seven rules that govern the interaction of complex numbers and their complex conjugates: If z and w are complex numbers and $\bar{z}$ and $\bar{w}$ are their conjugates,

   (1) $\bar{\bar{z}} = z$

   (2) $\overline{z + w} = \bar{z} + \bar{w}$

   (3) $\overline{zw} = \bar{z} \cdot \bar{w}$

   (4) $\overline{z^n} = (\bar{z})^n$, where n is a positive integer

   (5) $\text{Re}(z) = \dfrac{z + \bar{z}}{2}$

   $\text{Im}(z) = \dfrac{z - \bar{z}}{2i}$

   (6) $z + \bar{z}$ is real for any complex number

   (7) $\bar{z} = z$ if and only if z is a real number

6. The complex plane is a two-dimensional graphical portrayal of the complex number system, in which the horizontal axis is the real axis and the vertical axis is the imaginary axis.
7. The modulus of a complex number $a + bi$ is the absolute value $|a + bi| = \sqrt{a^2 + b^2}$, or the distance of that complex number from the origin in the complex plane.
8. The equation $|z - z_0| = r$ is a circle of radius r centered at the point z_0 on the complex plane. [If $z_0 = 0$, the circle is centered at the origin.]
9. Solutions of the equation $x^n - 1 = 0$ for any positive integer n are called the nth roots of unity. Graphically they are symmetrically located on the unit circle with angles of $360/n$ (i.e., $2\pi/n$) between each consecutive pair.

RAISE YOUR GRADES

Can you...?

☑ explain how the complex number field is created

☑ solve $x^2 + a^2 = 0$ for any value a

☑ explain the difference between the real and imaginary parts of a complex number

☑ find the sum, difference, product, and quotient of two complex numbers

☑ identify complex numbers with points in the complex plane

☑ explain the graphical significance of the modulus of a complex number

☑ solve quadratic equations that have complex roots

☑ locate the n roots of unity of an equation $x^n - 1 = 0$.

SOLVED PROBLEMS

Complex Numbers Revisited

PROBLEM 13-1   Solve **(a)** $x^2 + 16 = 0$ and **(b)** $x^2 + 25 = 0$.

Solution

(a) $x^2 + 16 = 0$
$$x^2 = -16 = 16(-1)$$
$$x = \sqrt{16}\sqrt{-1} = \pm 4i$$

(b) $x^2 + 25 = 0$
$$x^2 = -25 = 25(-1)$$
$$x = \sqrt{25}\sqrt{-1} = \pm 5i$$

PROBLEM 13-2   Find the two roots of the quadratic equation $x^2 + 4x + 8 = 0$.

Solution   Use the quadratic formula, by which the roots of the equation $ax^2 + bx + c = 0$ are $-b \pm \sqrt{b^2 - 4ac}/2a$. For this quadratic equation $a = 1$, $b = 4$, and $c = 8$, so the roots are

$$x = \frac{-4 \pm \sqrt{4^2 - 4(1)(8)}}{2 \cdot 1} = \frac{-4 \pm \sqrt{16 - 32}}{2} = \frac{-4 \pm \sqrt{-16}}{2}$$

$$= \frac{-4 \pm \sqrt{16}\sqrt{-1}}{2} = \frac{-4 \pm 4i}{2} = -2 \pm 2i$$

Thus the roots are complex numbers, $x = -2 + 2i$ and $x = -2 - 2i$.

PROBLEM 13-3   Find the real and imaginary parts of the following complex numbers: **(a)** $z_1 = 3 + 6i$ and **(b)** $z_2 = -2 - 7i$.

Solution   For any complex number in the form $z = a + bi$, the real part of z is a and the imaginary part of z is b.

(a) The real part of $z_1 = 3 + 6i$ is 3, written $\mathrm{Re}(z_1) = 3$, and the imaginary part is 6, written $\mathrm{Im}(z_1) = 6$.
(b) $\mathrm{Re}(z_2) = -2$ and $\mathrm{Im}(z_2) = -7$.

PROBLEM 13-4   Find the value of the unknown parameters a, b, and c if the following statements are true:

$$3 + 4i = a + 4i$$
$$b - 2i = 6 + ci$$

Solution   You know that two complex numbers $a + bi$ and $c + di$ are equal if and only if $a = c$ and $b = d$. So if $3 + 4i = a + 4i$, then $a = 3$. And if $b - 2i = 6 + ci$, then $b = 6$ and $c = -2$.

The Arithmetic of Complex Numbers

PROBLEM 13-5   Given the complex numbers

$$z_1 = -1 + 2i, \qquad z_2 = 4 - 3i, \qquad z_3 = 2 - 5i, \qquad z_4 = -2 - i$$

find

(a) $z_1 + z_2$,   **(b)** $z_2 + z_3 - z_4$,   **(c)** $z_1 \cdot z_2$,   **(d)** $z_1 - (z_2 \cdot z_3) + z_4$

Solution

(a) $z_1 + z_2 = (-1 + 2i) + (4 - 3i)$

$\qquad = (-1 + 4) + (2i - 3i)$ [Add real and imaginary parts independently.]

$\qquad = 3 - i$

(b) $z_2 + z_3 - z_4 = (z_2 + z_3) - z_4$

$$= [(4 - 3i) + (2 - 5i)] - (-2 - i) = (6 - 8i) + (2 + i)$$

$$= 8 - 7i$$

(c) $z_1 \cdot z_2 = (-1 + 2i)(4 - 3i)$

$$= -4 + 3i + 8i - 6i^2 \qquad \text{[Use FOIL.]}$$

$$= -4 + 11i - 6(-1) \qquad \text{[Recall that } i^2 = \sqrt{-1^2} = -1.\text{]}$$

$$= -4 + 11i + 6$$

$$= 2 + 11i$$

(d) $z_1 - (z_2 \cdot z_3) + z_4 = (-1 + 2i) - [(4 - 3i)(2 - 5i)] + (-2 - i)$

$$= (-1 + 2i) - (8 - 20i - 6i + 15i^2) + (-2 - i)$$

$$= (-1 + 2i) - [8 - 20i - 6i + 15(-1)] + (-2 - i)$$

$$= (-1 + 2i) - (-7 - 26i) + (-2 - i)$$

$$= (-1 + 7 - 2) + (2i + 26i - i)$$

$$= 4 + 27i$$

PROBLEM 13-6   Given $z_1 = (1 + 2i)$, $z_2 = (-4 + i)$, $z_3 = (-3i)$, find **(a)** $3z_1 - 4z_3$ and **(b)** $z_1 \cdot z_2 \cdot z_3$.

Solution

(a) $3z_1 - 4z_3 = 3(1 + 2i) - 4(-3i) = 3 + 6i + 12i = 3 + 18i$

(b) $z_1 \cdot z_2 \cdot z_3 = (1 + 2i)(-4 + i)(-3i)$

$$= (-4 + i - 8i + 2i^2)(-3i) = (-6 - 7i)(-3i) = 18i + 21i^2 = -21 + 18i$$

PROBLEM 13-7   Find the reduced form for **(a)** i^{11}, **(b)** i^{33}, and **(c)** i^{372}.

Solution   Pull out the 4th-power parts:

(a) $i^{11} = i^8 \cdot i^3 = (i^4)^2 \cdot i^3 = 1^2 \cdot i^3 = i^3 = i^2 \cdot i = (-1)i = -i$

(b) $i^{33} = i^{32} \cdot i = (i^4)^8 \cdot i = 1^8 \cdot i = i$

(c) $i^{372} = i^{4 \cdot 93} = (i^4)^{93} = 1^{93} = 1$

PROBLEM 13-8   Find $\sum\limits_{n=1}^{10} i^n$. [*Hint:* See Chapter 15, Sections 15-1 and 15-2.]

Solution   The expanded form of what you want is

$$\sum_{n=1}^{10} i^n = i^1 + i^2 + i^3 + i^4 + i^5 + i^6 + i^7 + i^8 + i^9 + i^{10}$$

So $\sum\limits_{n=1}^{10} i^n = i + (-1) + (-i) + 1 + (i^4 \cdot i) + (i^4 \cdot i^2) + (i^4 \cdot i^3) + (i^4)^2 + (i^8 \cdot i) + (i^8 \cdot i^2)$

$$= i + (-1) + (-i) + 1 + i + (-1) + (-i) + 1 + i + (-1)$$

$$= i - 1$$

PROBLEM 13-9   Find the reduced form of $(1 - 2i)^3$.

Solution   Since $(1 - 2i)(1 - 2i) = 1 - 2i - 2i + 4i^2 = -3 - 4i$, we have

$$(1 - 2i)^3 = (1 - 2i)^2(1 - 2i)$$

$$= (-3 - 4i)(1 - 2i) = (-3 + 6i - 4i + 8i^2)$$

$$= -11 + 2i$$

Complex Conjugates and Quotients

PROBLEM 13-10   Find the complex conjugates of **(a)** $-2 + 4i$, **(b)** $3 - 8i$, and **(c)** $(1 - i)^2$.

Solution   The complex conjugate of $a + bi$ is $a - bi$.

(a) The conjugate of $-2 + 4i$ is $-2 - 4i$
(b) The conjugate of $3 - 8i$ is $3 + 8i$
(c) Since $(1 - i)^2 = 1 - 2i + i^2 = -2i$, the conjugate is $2i$.

PROBLEM 13-11   Remove the fraction in **(a)** $\dfrac{1}{2 - i}$ and **(b)** $\dfrac{1}{-3 + 5i}$.

Solution   Multiply both the numerator and denominator by the complex conjugate of the denominator:

(a) $\dfrac{1}{2 - i} = \left(\dfrac{1}{2 - i}\right)\left(\dfrac{2 + i}{2 + i}\right) = \dfrac{2 + i}{4 - i^2} = \dfrac{2 + i}{5} = \dfrac{2}{5} + \dfrac{1}{5}i$

(b) $\dfrac{1}{-3 + 5i} = \left(\dfrac{1}{-3 + 5i}\right)\left(\dfrac{-3 - 5i}{-3 - 5i}\right) = \dfrac{-3 - 5i}{9 - 25i^2} = \dfrac{-3 - 5i}{34} = -\dfrac{3}{34} - \dfrac{5}{34}i$

PROBLEM 13-12   Find the simplified version of **(a)** $\dfrac{-4 + i}{1 - 3i}$ and **(b)** $\dfrac{2 + 5i}{-1 + i}$.

Solution

(a) $\dfrac{-4 + i}{1 - 3i} = \left(\dfrac{-4 + i}{1 - 3i}\right)\left(\dfrac{1 + 3i}{1 + 3i}\right) = \dfrac{-4 - 12i + i + 3i^2}{1 - 9i^2} = \dfrac{-7 - 11i}{10} = \dfrac{-7}{10} - \dfrac{11}{10}i$

(b) $\dfrac{2 + 5i}{-1 + i} = \left(\dfrac{2 + 5i}{-1 + i}\right)\left(\dfrac{-1 - i}{-1 - i}\right) = \dfrac{-2 - 2i - 5i - 5i^2}{1 - i^2} = \dfrac{3 - 7i}{2} = \dfrac{3}{2} - \dfrac{7}{2}i$

PROBLEM 13-13   Find the simplified version of $\left(\dfrac{1 + i}{2 - i}\right)\left(\dfrac{3 - i}{1 + 3i}\right)$.

Solution   First multiply out the numerators and denominators; then multiply both by the conjugate of the denominator:

$$\left(\dfrac{1 + i}{2 - i}\right)\left(\dfrac{3 - i}{1 + 3i}\right) = \dfrac{3 - i + 3i - i^2}{2 + 6i - i - 3i^2} = \dfrac{4 + 2i}{5 + 5i} = \dfrac{2(2 + i)}{5(1 + i)} = \left(\dfrac{2}{5}\right)\left(\dfrac{2 + i}{1 + i}\right)$$

$$= \left(\dfrac{2}{5}\right)\left(\dfrac{2 + i}{1 + i}\right)\left(\dfrac{1 - i}{1 - i}\right) = \left(\dfrac{2}{5}\right)\left(\dfrac{2 - 2i + i - i^2}{1 - i^2}\right) = \left(\dfrac{2}{5}\right)\left(\dfrac{3 - i}{2}\right) = \dfrac{3}{5} - \dfrac{1}{5}i$$

PROBLEM 13-14   Prove the conjugation property $\overline{z + w} = \bar{z} + \bar{w}$.

Solution   Let $z = a + bi$ and $w = c + di$. Then

$$z + w = (a + c) + (b + d)i$$

and

$$\overline{z + w} = (a + c) - (b + d)i$$

But $\bar{z} = a - bi$ and $\bar{w} = c - di$, so that

$$\bar{z} + \bar{w} = (a - bi) + (c - di)$$

$$= (a + c) - (b + d)i$$

same

PROBLEM 13-15   Prove the property that $\text{Re}(z) = \dfrac{z + \bar{z}}{2}$.

Solution   Let $z = a + bi$. Then $\bar{z} = a - bi$, so that

$$\dfrac{z + \bar{z}}{2} = \dfrac{(a + bi) + (a - bi)}{2} = \dfrac{2a}{2} = a = \text{Re}(z)$$

The Complex Plane

PROBLEM 13-16 Graph each of the following points in the complex plane and find the modulus for each one:

(a) $4 + 3i$ **(b)** $-1 + i$ **(c)** $-4 + \frac{1}{2}i$

Solution The graphs are shown in Figure 13-8. ₹.)

Since the modulus of $a + bi$ is $|a + bi|$,

(a) $|4 + 3i| = \sqrt{4^2 + 3^2}$

$\qquad = \sqrt{16 + 9} = \sqrt{25} = 5$

(b) $|-1 + i| = \sqrt{(-1)^2 + 1^2}$

$\qquad = \sqrt{1 + 1} = \sqrt{2}$

(c) $|-4 + \frac{1}{2}i| = \sqrt{(-4)^2 + (\frac{1}{2})^2}$

$\qquad = \sqrt{16 + \frac{1}{4}}$

$\qquad = \sqrt{\frac{65}{4}} = \frac{1}{2}\sqrt{65}$

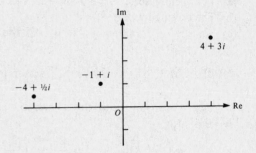

Figure 13-8

PROBLEM 13-17 Verify the property that $|z| = |-z|$.

Solution Let $z = a + bi$, so that $-z = -a - bi$. Then $|z| = \sqrt{a^2 + b^2}$ and $|-z| = \sqrt{(-a)^2 + (-b)^2} = \sqrt{a^2 + b^2}$, so $|z| = |-z|$.

PROBLEM 13-18 Verify the property that $\left|\dfrac{z}{w}\right| = \dfrac{|z|}{|w|}$.

Solution Let $z = a + bi$ and $w = c + di$. Then

$$\frac{z}{w} = \frac{a + bi}{c + di} = \left(\frac{a + bi}{c + di}\right)\left(\frac{c - di}{c - di}\right) = \frac{(ac + bd) + (bc - ad)i}{c^2 - d^2i^2}$$

$$= \frac{(ac + bd) + (bc - ad)i}{c^2 + d^2}$$

$$= \left(\frac{ac + bd}{c^2 + d^2}\right) + \left(\frac{bc - ad}{c^2 + d^2}\right)i$$

Thus

$$\left|\frac{z}{w}\right| = \sqrt{\left(\frac{ac + bd}{c^2 + d^2}\right)^2 + \left(\frac{bc - ad}{c^2 + d^2}\right)^2}$$

$$= \sqrt{\frac{a^2c^2 + 2abcd + b^2d^2 + b^2c^2 - 2abcd + a^2d^2}{(c^2 + d^2)^2}} = \frac{\sqrt{c^2(a^2 + b^2) + d^2(b^2 + a^2)}}{c^2 + d^2}$$

$$= \frac{\sqrt{(a^2 + b^2)(c^2 + d^2)}}{c^2 + d^2} = \frac{\sqrt{a^2 + b^2}\sqrt{c^2 + d^2}}{c^2 + d^2}$$

$$= \frac{\sqrt{a^2 + b^2}}{\sqrt{c^2 + d^2}}$$

And you know that

$$\frac{|z|}{|w|} = \frac{|a + bi|}{|c + di|} = \frac{\sqrt{a^2 + b^2}}{\sqrt{c^2 + d^2}}$$

So $\left|\dfrac{z}{w}\right| = \dfrac{|z|}{|w|}$.

PROBLEM 13-19   Sketch the graphs of **(a)** $|z| = 3$ and **(b)** $|z - 2 + i| = 2$.

Solution

(a) The graph of $|z| = 3$ is the locus of all points in the complex plane whose modulus is 3. This is a circle of radius 3 centered at the origin (Figure 13-9a).

(b) The graph of $|z - 2 + i| = 2$ is a circle of radius 2 centered at $(2 - i)$ since $|z - 2 + i| = |z - (2 - i)|$ (Figure 13-9b).

Figure 13-9

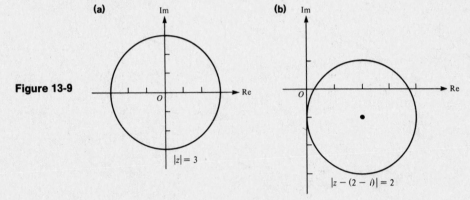

Equations with Complex Roots

PROBLEM 13-20   Solve $4x^2 - 4x + 5 = 0$.

Solution   Using the quadratic formula with $a = 4$, $b = -4$, and $c = 5$, you get

$$x = \frac{-(-4) \pm \sqrt{(-4)^2 - 4(4)(5)}}{2 \cdot 4} = \frac{4 \pm \sqrt{16 - 80}}{8} = \frac{4 \pm \sqrt{-64}}{8} = \frac{4 \pm 8i}{8} = \frac{1}{2} \pm i$$

The roots are $x = \frac{1}{2} + i$ and $x = \frac{1}{2} - i$.

PROBLEM 13-21   Is $x = \frac{1}{2} + i$ a solution to $4x^2 - 4x + 5 = 0$?

Solution   If $4x^2 - 4x + 5 = 0$, then $\frac{1}{2} + i$ is a solution if

$$4(\tfrac{1}{2} + i)^2 - 4(\tfrac{1}{2} + i) + 5 \stackrel{?}{=} 0$$

$$4(\tfrac{1}{4} + i + i^2) - 2 - 4i + 5 \stackrel{?}{=} 0$$

$$1 + 4i + 4i^2 - 2 - 4i + 5 \stackrel{?}{=} 0$$

$$1 - 4 - 2 + 5 \stackrel{?}{=} 0$$

$$0 \stackrel{\checkmark}{=} 0$$

PROBLEM 13-22   Graph the roots of

$$x^5 - 1 = 0$$

Solution   You want to find the five fifth roots of unity. By factoring you get

$$x^5 - 1 = (x - 1)(x^4 + x^3 + x^2 + x + 1) = 0$$

which gives $x = 1$ as one of the roots. Since the five roots are symmetrically located around the unit circle at consecutive angles of 360/5 degrees, you can graph the roots as in Figure 13-10.

note: You need some trigonometry to find the actual representation of the other four roots.

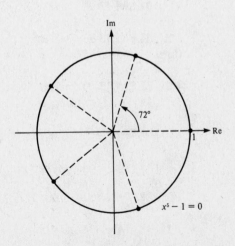

Figure 13-10

PROBLEM 13-23   Find the values of and draw the graph of the six sixth roots of unity.

Solution   The equation is $x^6 - 1 = 0$. Factor it completely:

$$(x^3 - 1)(x^3 + 1) = 0$$

$$(x - 1)(x^2 + x + 1)(x + 1)(x^2 - x + 1) = 0$$

$$(x - 1)\left[x - \left(\frac{-1 + \sqrt{3}i}{2}\right)\right]\left[x - \left(\frac{-1 - \sqrt{3}i}{2}\right)\right]$$

$$\times (x + 1)\left[x - \left(\frac{1 + \sqrt{3}i}{2}\right)\right]$$

$$\times \left[x - \left(\frac{1 - \sqrt{3}i}{2}\right)\right] = 0$$

Thus the roots are

$$x = 1, \qquad x = \frac{-1 + \sqrt{3}i}{2}, \qquad x = \frac{-1 - \sqrt{3}i}{2},$$

$$x = -1, \qquad x = \frac{1 + \sqrt{3}i}{2}, \qquad x = \frac{1 - \sqrt{3}i}{2}$$

The graph is shown in Figure 13-11, where the roots are located at consecutive angles of $360°/6 = 60°$.

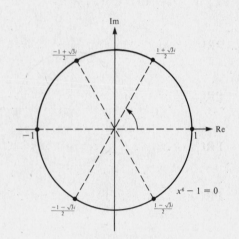

Figure 13-11

Supplementary Exercises

PROBLEM 13-24   Find the solutions to (a) $x^2 - 36 = 0$ and (b) $x^2 + 36 = 0$.

PROBLEM 13-25   Solve the quadratic equation $x^2 + 4x + 5 = 0$.

PROBLEM 13-26   Solve the equation $x^3 - 2x^2 + 5x = 0$.

PROBLEM 13-27   Find the real and imaginary parts of z_1, z_2, and $z_1 \cdot z_2$ if $z_1 = 2 + 4i$ and $z_2 = 2 - 4i$.

PROBLEM 13-28   If $x + 3i = 2 + yi$, what are x and y?

PROBLEM 13-29   Given $z_1 = -5 + i$, $z_2 = 2 + 6i$, $z_3 = -1 + 2i$, find (a) $2z_1 + 3z_2$, (b) $z_2 - 2z_3$, (c) $z_1 \cdot z_2 - z_3$, (d) $z_1(z_2 - z_3)$.

PROBLEM 13-30   Given $z_1 = -2 + 2i$, $z_2 = 2 + 2i$, $z_3 = -2 - 2i$, find (a) $z_1 + z_2 - z_3$, (b) $z_1 \cdot z_3 + (2i)z_2$, (c) $iz_1 + 3z_3 - z_2$.

PROBLEM 13-31   Find the reduced form for (a) i^{41}, (b) $i^{10} - i^4$, (c) $i^{87} - i^{42}$, (d) $i^{151} - i^{1986}$.

PROBLEM 13-32   What is the value of $i^2 + i^3 + i^4 + i^5$?

PROBLEM 13-33   Simplify $(2 + i)^3$.

PROBLEM 13-34   Simplify $(1 + 2i)^4$.

PROBLEM 13-35   Simplify $(1 - i)^5$.

PROBLEM 13-36 Find the complex conjugate of (a) $8 - 5i$, (b) $-2 - \frac{1}{3}i$, (c) $1/i$, (d) $i^3 + i$.

PROBLEM 13-37 Remove the fractions in (a) $\dfrac{1}{3 - i}$, (b) $\dfrac{2}{1 + 2i}$, (c) $\dfrac{5i}{2 + 4i}$.

PROBLEM 13-38 Find the simplified version of $\dfrac{3 + 4i}{3 - 4i}$.

PROBLEM 13-39 Find the simplified version of $\dfrac{1 - 4i}{6i + 1}$.

PROBLEM 13-40 Find the simplified version of $\dfrac{2i(1 - 4i)}{7 - i}$.

PROBLEM 13-41 Find the simplified version of $\dfrac{(5 + i)^2}{-2 + 3i}$.

PROBLEM 13-42 Find the simplified version of $\dfrac{(3 + 4i)(1 - 6i)}{(2i - 3)(1 + i^5)}$.

PROBLEM 13-43 For any complex number z, what is the value of $\dfrac{z - z}{2i}$?

PROBLEM 13-44 Locate the following points in the complex plane:
$$-2 - i, \quad -2 + i, \quad 1 - 2i, \quad 2i + 1, \quad -2i - 1$$

PROBLEM 13-45 Find the modulus of (a) $-1 - 5i$, (b) $12i - 5$, (c) $1 - \sqrt{3}i$.

PROBLEM 13-46 Let $c = \dfrac{a + bi}{a - bi} + \dfrac{a - bi}{a + bi}$. Find $\operatorname{Re}(c)$ and $\operatorname{Im}(c)$.

PROBLEM 13-47 Find the value of $\dfrac{|8 - 6i|}{(\sqrt{7} - \sqrt{3}i)(\sqrt{7} + \sqrt{3}i)}$.

PROBLEM 13-48 Expand $\left(-\dfrac{1}{2} + \dfrac{\sqrt{3}}{2}i\right)^3$.

PROBLEM 13-49 Find x if $(x - 3i)^2 = (x + i)^2$.

PROBLEM 13-50 Find the quadratic equation that has roots $2 - 3i$ and $2 + 3i$.

PROBLEM 13-51 Graph the complex relation $|z| = 4$.

PROBLEM 13-52 Graph the complex relation $|z + 3 - i| = 2$.

PROBLEM 13-53 Graph the complex relation $|z + 4| = 4$.

PROBLEM 13-54 Find the roots of $x^3 - 1 = 0$. Graph them.

PROBLEM 13-55 Find all of the solutions to $x^4 - 16 = 0$.

Answers to Supplementary Exercises

13-24 (a) $x = \pm 6$ (b) $x = \pm 6i$

13-25 $x = -2 + i, -2 - i$

13-26 $x = 0, 1 + 2i, 1 - 2i$

13-27 $\text{Re}(z_1) = 2, \quad \text{Im}(z_1) = 4$
$\text{Re}(z_2) = 2, \quad \text{Im}(z_2) = -4$
$\text{Re}(z_1 \cdot z_2) = 20, \quad \text{Im}(z_1 \cdot z_2) = 0$

13-28 $x = 2, \quad y = 3$

13-29 (a) $-4 + 20i$ (c) $-15 - 30i$
(b) $4 + 2i$ (d) $-19 - 17i$

13-30 (a) $-2 + 6i$
(b) $4 + 4i$
(c) $-10 - 10i$

13-31 (a) i (c) $1 - i$
(b) -2 (d) $1 - i$

13-32 0

13-33 $2 + 11i$

13-34 $-7 - 24i$

13-35 $-4 + 4i$

13-36 (a) $8 + 5i$
(b) $-2 + \frac{1}{3}i$
(c) i
(d) 0

13-37 (a) $\frac{3}{10} + \frac{1}{10}i$
(b) $\frac{2}{5} - \frac{4}{5}i$
(c) $1 + \frac{1}{2}i$

13-38 $-\frac{7}{25} + \frac{24}{25}i$

13-39 $-\frac{23}{37} + \frac{10}{37}i$

13-40 $\frac{27}{25} + \frac{11}{25}i$

13-41 $-\frac{18}{13} - \frac{92}{13}i$

13-42 $-\frac{121}{26} + \frac{97}{26}i$

13-43 $\text{Im}(z)$

13-44 See Figure 13-12.

13-45 (a) $\sqrt{26}$ (b) 13
(c) 2

13-46 $\text{Re}(c) = \dfrac{2(a^2 b^2)}{a^2 + b^2}$
$\text{Im}(c) = 0$

13-47 1

13-48 1

13-49 i

13-50 $x^2 - 4x + 13 = 0$

13-51 See Figure 13-13.

13-52 See Figure 13-14.

13-53 See Figure 13-15.

13-54 $x = 1$
$x = -\dfrac{1}{2} \pm \dfrac{\sqrt{3}}{2}i$
See Figure 13-16.

13-55 $2, -2, 2i, -2i$

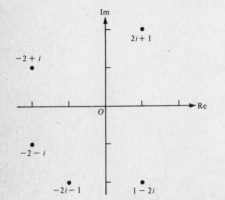

Figure 13-12

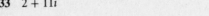

Figure 13-13

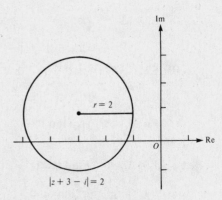

Figure 13-14

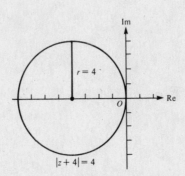

Figure 13-15

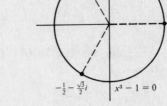

Figure 13-16

EXAM 4 (Chapters 11–13)

1. Solve $\begin{cases} 6x - y = -10 \\ -4y + 3x = 16 \end{cases}$ by the substitution method.

2. Solve the system $\begin{cases} x^2 - 4x - y = -4 \\ x - 2y = 12 \end{cases}$.

3. Solve the system $\begin{cases} x - 4y + 2z = -3 \\ -3x + y - z = 2 \\ 4x - 7y + 4z = -6 \end{cases}$.

4. Graph the solution set to $\begin{cases} x^2 + y^2 \leq 4 \\ 2y \geq 2 + x \end{cases}$.

5. You are selling flashlights and batteries door-to-door. Your empty case weighs 2 pounds, flashlights weigh $\frac{1}{3}$ pound, and batteries weigh $\frac{1}{4}$ pound each. You pack your case in the morning so its total weight is not more than 42 pounds. Each flashlight is sold with two batteries but you also sell batteries singly. Your flashlights cost you 60¢ and are sold for $1.20. The batteries cost you 20¢ and you sell them for 40¢ each. What mix of flashlights and batteries should you have in your case to maximize profit potential?

6. Use matrix methods to solve the system $\begin{cases} x - 4z = 6 \\ -3x + 2y - z = 5 \\ 2x - 6y = -26 \end{cases}$.

7. Use Cramer's Rule to solve the system $\begin{cases} 2x + \frac{1}{2}y - z = 9 \\ x - 4y - \frac{1}{5}z = -14 \\ 3y + 2z = 2 \end{cases}$.

8. Use determinant operations to evaluate $\begin{vmatrix} 2 & -1 & 5 & 0 \\ -3 & 4 & -2 & 6 \\ 0 & 2 & 1 & 1 \\ 1 & 0 & 3 & -2 \end{vmatrix}$.

9. If $A = \begin{bmatrix} 1 & 2 \\ -3 & -4 \end{bmatrix}$ and $B = \begin{bmatrix} 5 & -3 \\ -3 & 1 \end{bmatrix}$ find A^{-1}, B^{-1}, and $(AB)^{-1}$. Then verify that $(AB)^{-1} = B^{-1}A^{-1}$.

10. If $A = \begin{bmatrix} 1 & 0 & -4 \\ -3 & 2 & -1 \\ 2 & -6 & 0 \end{bmatrix}$, find A^{-1}. Then solve the system of Problem 6 using the result $X = A^{-1}C$.

11. Find the simplified form of $\dfrac{3 - 4i}{2 + 3i} + \dfrac{(1 - 6i)(5 + i)}{2 - 3i}$.

12. If $z = (2 - 5i)(i + 1)$, what is $\bar{z}$? What is $|z|$? What is $\left|\dfrac{1}{\bar{z}}\right|$?

13. Find the values of x and y so that $\dfrac{1 + ix}{2 - i} = \dfrac{3y + 4i}{1 + 3i}$.

14. Find the value of $(i^{42} - i^{27})(i - 1)^3$.

15. Graph the eight eighth roots of unity. What are the values of these roots?

Answers to Exam 4

1. $(-1, 4)$

2. No solution

3. $(-1, 2, 3)$

4. See Figure E4-1.

5. 48 flashlights, 96 batteries [*Hint*: Use $\{x \geq 0, y \geq 0, y \geq 2x, 2 + \frac{1}{3}x + \frac{1}{4}y \leq 42\}$; $P = .60x + .20y$.]

6. $(2, 5, -1)$

7. $(1, 4, -5)$

8. 43

9. $A^{-1} = \dfrac{1}{2}\begin{bmatrix} -4 & -2 \\ 3 & 1 \end{bmatrix}$, $\quad B^{-1} = -\dfrac{1}{4}\begin{bmatrix} 1 & 3 \\ 3 & 5 \end{bmatrix}$, $(AB)^{-1} = -\dfrac{1}{8}\begin{bmatrix} 5 & 1 \\ 3 & -1 \end{bmatrix}$

10. $A^{-1} = \dfrac{1}{31}\begin{bmatrix} 3 & -12 & -4 \\ 1 & -4 & -\frac{13}{2} \\ -7 & -3 & -1 \end{bmatrix}$

11. $\dfrac{103}{13} - \dfrac{42}{13}i$

12. $\bar{z} = 7 + 3i$, $\quad |z| = \sqrt{58}$, $\quad \left|\dfrac{1}{\bar{z}}\right| = \dfrac{1}{\sqrt{58}}$

13. $x = -13$, $\quad y = 6$

14. -4

15. See Figure E4-2.

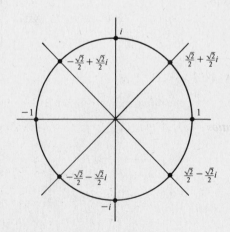

Figure E4-2

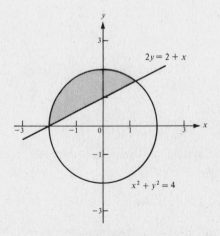

Figure E4-1

14 THEORY OF EQUATIONS

THIS CHAPTER IS ABOUT

- ☑ **General Polynomials: Definitions**
- ☑ **Division of Polynomials**
- ☑ **The Fundamental Theorem of Algebra**
- ☑ **Factoring Polynomials with Real Coefficients**

14-1. General Polynomials: Definitions

We know that we can use algebraic manipulations to solve linear equations of the form $ax + b = 0$ and quadratic equations of the form $ax^2 + bx + c = 0$. And we know that if the coefficients in quadratic equations are real numbers, we're guaranteed solutions—some of which may be complex numbers. We can now develop methods to find the solutions to the general polynomial equation

POLYNOMIAL EQUATION OF THE nth DEGREE
$$a_n x^n + a_{n-1} x^{n-1} + a_{n-2} x^{n-2} + \cdots + a_2 x^2 + a_1 x + a_0 = 0$$

where the coefficients a are real or complex constants and n is a positive integer. We call this equation the **polynomial equation of the nth degree**.

- A value $x = r$ is called a **root** (or **zero**) of the polynomial equation if it has the property that

$$a_n r^n + a_{n-1} r^{n-1} + a_{n-2} r^{n-2} + \cdots + a_2 r^2 + a_1 r + a_0 \equiv 0$$

EXAMPLE 14-1 Show that $x = 3$ is a root of the polynomial equation
$$x^3 - 7x^2 + 16x - 12 = 0$$

Solution We substitute 3 for x in the equation:
$$(3)^3 - 7(3)^2 + 16(3) - 12 \overset{?}{=} 0$$
$$27 - 7 \cdot 9 + 48 - 12 \overset{?}{=} 0$$
$$27 - 63 + 48 - 12 \overset{?}{=} 0$$
$$75 - 75 \overset{\checkmark}{=} 0 \qquad \text{[Identity]}$$

14-2. Division of Polynomials

A. Long division: A review

When polynomials in the variable x are added, subtracted, or multiplied, the result in each case is a polynomial. In the case of division this is only true *sometimes*. When the division of one polynomial by another polynomial results in a polynomial, we say that the division is EXACT.

EXAMPLE 14-2 Show that the division of $2x^2 + 5x - 3$ by $2x - 1$ is exact.

Solution

$$\frac{2x^2 + 5x - 3}{2x - 1} = \frac{(2x - 1)(x + 3)}{(2x - 1)} = x + 3$$

So the result of the division is a polynomial, $x + 3$. Thus $2x - 1$ is an exact divisor of $2x^2 + 5x - 3$.

We say that

- $D(x)$ DIVIDES $P(x)$ if there is a polynomial $Q(x)$ such that

$$P(x) = D(x) \cdot Q(x)$$

We also say that $D(x)$ is a FACTOR of $P(x)$ when $P(x)$ is DIVISIBLE by $D(x)$.

But in most cases division is not exact—and we get a nonzero remainder.

EXAMPLE 14-3 Carry out the details of

$$\text{(a) } \frac{331}{13} \qquad \text{(b) } \frac{2x^2 + 5x - 3}{2x - 1} \qquad \text{(c) } \frac{3x^3 - 2x^2 + 4x - 1}{x^2 + 3}$$

Solution

(a)

$$
\begin{array}{r}
25 \leftarrow \text{quotient} \\
\text{divisor} \longrightarrow 13 \overline{)331} \leftarrow \text{dividend} \\
-26 \\
\hline
71 \\
-65 \\
\hline
6 \leftarrow \text{remainder}
\end{array}
$$

Thus $331/13 = 25 + 6/13$.

(b)

$$
\begin{array}{r}
x + 3 \leftarrow \text{quotient} \\
\text{divisor} \longrightarrow 2x - 1 \overline{)2x^2 + 5x - 3} \leftarrow \text{dividend} \\
-(2x^2 - x) \\
\hline
6x - 3 \\
-(6x - 3) \\
\hline
0 \leftarrow \text{remainder}
\end{array}
$$

Thus $\dfrac{2x^2 + 5x - 3}{2x - 1} = x + 3 + \dfrac{0}{2x - 1} = x + 3$, or we may write $2x^2 + 5x - 3 = (2x - 1)(x + 3)$.

(c)

$$
\begin{array}{r}
3x - 2 \leftarrow \text{quotient} \\
\text{divisor} \longrightarrow x^2 + 3 \overline{)3x^3 - 2x^2 + 4x - 1} \leftarrow \text{dividend} \\
-(3x^3 + 9x) \\
\hline
-2x^2 - 5x - 1 \\
-(-2x^2 - 6) \\
\hline
-5x + 5 \leftarrow \text{remainder}
\end{array}
$$

Thus $\dfrac{3x^3 - 2x^2 + 4x - 1}{x^2 + 3} = 3x - 2 + \dfrac{-5x + 5}{x^2 + 3}$, or we may write $3x^3 - 2x^2 + 4x - 1 = (3x - 2)(x^2 + 3) + (-5x + 5)$.

note: In each case, the remainder has a lower degree than the divisor.

B. The division algorithm

We can now state the results of division formally, as the **division algorithm**:

▼ If $P(x)$ and $D(x)$ are any two polynomials of degree 1 or higher, then there are unique polynomials $Q(x)$ and $R(x)$ such that

$$P(x) = D(x) \cdot Q(x) + R(x)$$

where $R(x)$ is either zero or of degree lower than the degree of $D(x)$.

We call $D(x)$ the DIVISOR, $Q(x)$ the QUOTIENT, $P(x)$ the DIVIDEND, and $R(x)$ the REMAINDER. ▲

The division algorithm is true even if the coefficients in the polynomials are complex constants.

Division Algorithm

EXAMPLE 14-4 Divide the polynomial $4x^2 + 5ix - 2$ by $2x + 3i$.

Solution

$$
\begin{array}{r}
2x \quad - i/2 \\
2x + 3i \overline{\smash{\big)}\, 4x^2 + 5ix - 2} \\
-(4x^2 + 6ix) \\
\hline
- ix - 2 \\
-(-ix \quad\quad - 3i^2/2) \\
\hline
-2 + 3i^2/2 = -7/2
\end{array}
$$

Thus $4x^2 + 5ix - 2 = (2x + 3i)(2x - i/2) - \frac{7}{2}$

that is, $P(x)\quad = \quad D(x)\quad \cdot \quad Q(x)\quad + R(x)$

EXAMPLE 14-5 Find the quotient and remainder when $6x^4 - 3x^3 - x + 5$ is divided by $2x^2 + x$.

Solution Noting that the dividend has no term in x^2, we supply a term with a zero coefficient and then divide:

$$
\begin{array}{r}
3x^2 - 3x \;\; + \tfrac{3}{2} \longleftarrow \text{quotient} \\
\text{divisor} \longrightarrow 2x^2 + x \overline{\smash{\big)}\, 6x^4 - 3x^3 + 0x^2 - x + 5} \longleftarrow \text{dividend} \\
-(6x^4 + 3x^3) \\
\hline
-6x^3 + 0x^2 \\
-(-6x^3 - 3x^2) \\
\hline
3x^2 - x \\
-(3x^2 + \tfrac{3}{2}x) \\
\hline
-\tfrac{5}{2}x + 5 \longleftarrow \text{remainder}
\end{array}
$$

Thus $6x^4 - 3x^3 - x + 5 = (2x^2 + x)(3x^2 - 3x + \tfrac{3}{2}) + (-\tfrac{5}{2}x + 5)$

that is, $P(x)\quad = \quad D(x)\quad \cdot \quad Q(x)\quad + \quad R(x)$

C. Synthetic division

In the division process the divisor is often a linear polynomial of the form $x - c$, where c is a constant. In this case, the division algorithm takes a simpler form:

▼ Given a polynomial $P(x)$ and a first-degree binomial $x - c$, where c is a constant, there exists a unique polynomial $Q(x)$ and a unique constant R such that

$$P(x) = Q(x)(x - c) + R$$ ▲

Simplified Division Algorithm $(D(x) = x - c)$

This simpler form allows us an abbreviated method for long division.

EXAMPLE 14-6 Find the quotient and remainder when $2x^3 + x^2 + 6$ is divided by $x - 1$.

Solution

Long division:

$$
\begin{array}{r}
2x^2 + 3x\ \ + 3 \\
x - 1\,\overline{\big)\,2x^3 + \ x^2 + 0x + 6} \\
-(2x^3 - 2x^2) \\
\hline
3x^2 + 0x \\
-(3x^2 - 3x) \\
\hline
3x + 6 \\
-(3x - 3) \\
\hline
9
\end{array}
$$

Thus

$$
2x^3 + x^2 + 6 = (x - 1)\underbrace{(2x^2 + 3x + 3)}_{\text{quotient}} + \underbrace{9}_{\text{remainder}}
$$

Synthetic Division: Note that in the long division many of the numbers in vertical columns are duplicates of each other. Also note that, below the dividend, like powers of x appear in vertical columns. Thus, we can simplify the writing by eliminating all the x's and crossing out the unnecessary repetitions of the numbers:

$$
\begin{array}{r}
2x^2 + 3x\ \ + 3 \\
x - 1\,\overline{\big)\,2x^3 + 1x^2 + 0x + 6} \\
-(2x^3 - 2x^2) \\
\hline
3x^2 + 0x \\
-(3x^2 - 3x) \\
\hline
3x + 6 \\
-(3x - 3) \\
\hline
9
\end{array}
\qquad \Rightarrow \qquad
\begin{array}{r}
2\ \ \ 3\ \ \ 3 \\
-1\,\overline{\big)\,2\quad 1\quad 0\quad 6} \\
\cancel{2}\ -2 \\
\hline
3\quad 0 \\
\cancel{3}\ -3 \\
\hline
3\quad \cancel{6} \\
\cancel{3}\ -3 \\
\hline
9
\end{array}
$$

Now we push all the lower numbers left up into a second and third line:

$$
\begin{array}{r}
2\ \ \ 3\ \ \ 3 \\
-1\,\overline{\big)\,2\quad 1\quad 0\quad 6} \\
-2\quad \uparrow\quad \uparrow \\
\hline
3\quad \big| \\
-3\quad \big| \\
\hline
3\quad \big| \\
-3 \\
\hline
9
\end{array}
\qquad \Rightarrow \qquad
\begin{array}{r}
2\ \ \ 3\ \ \ 3 \\
-1\,\overline{\big)\,2\quad 1\quad\ \ 0\quad\ \ 6} \\
-2\ -3\ -3 \\
\hline
3\quad 3\quad 9
\end{array}
$$

This new result still has a repetition, because the bottom line contains the quotient (except for the front 2, which may be brought down) and the remainder. So we don't write the quotient above, but let it appear only on the bottom line. Thus

$$
\begin{array}{r}
2\ \ \ 3\ \ \ 3 \\
-1\,\overline{\big)\,2\quad 1\quad\ \ 0\quad\ \ 6} \\
-2\ -3\ -3 \\
\hline
3\quad 3\quad 9
\end{array}
\qquad \Rightarrow \qquad
\begin{array}{r}
\underline{-1}\quad 2\quad 1\quad\ \ 0\quad\ \ 6 \\
-2\ -3\ -3 \\
\hline
2\quad 3\quad 3\quad 9
\end{array}
$$

We can make one further simplification. We note that the bottom line comes from subtracting the middle line from the top line. If we change the signs on the middle line, we can add the middle line to the top line. And we compensate for this sign change by also changing the sign on the number in the divisor. Thus

$$
\begin{array}{r|rrrr}
-1 & 2 & 1 & 0 & 6 \\
   &   & -2 & -3 & -3 \\
\hline
   & 2 & 3 & 3 & 9
\end{array}
\quad \Rightarrow \quad
\begin{array}{r|rrrr}
1 & 2 & 1 & 0 & 6 \\
  &   & 2 & 3 & 3 \\
\hline
  & 2 & 3 & 3 & 9
\end{array}
$$

This final form is an example of *synthetic division*, a form in which the variable letter and the unnecessary numbers have been eliminated.

The process of **synthetic division** is outlined in the following steps:

To divide $P(x)$ by $x - c$:

(1) Use c as the divisor (or if dividing by $x + c$, use $-c$ as the divisor).
(2) Write the coefficients of the dividend horizontally, supplying a zero for any missing terms.
(3) Skip a space below the dividend, draw a horizontal line, and bring down the first dividend coefficient to its position below the line.
(4) Multiply the divisor by the first dividend coefficient below the line and put the product number above the line under the second dividend coefficient. Add the numbers in column two and put the sum under the line. Multiply the divisor by the sum in column two and put the product in column three middle line. Add. Continue the multiplication process described here until the dividend runs out.
(5) Identify the numbers: The last number under the line is the remainder; the others are the coefficients of the quotient.

EXAMPLE 14-7 Use synthetic division to find the quotient and remainder when dividing $x^3 + 5x^2 - 9x - 3$ by $x - 2$.

Solution Follow the steps.

(1) The divisor is $+2$.

$$\begin{array}{r|} 2 \\ \end{array}$$

(2) The numbers for the dividend are the coefficients.

$$\begin{array}{r|rrrr} 2 & 1 & 5 & -9 & -3 \end{array}$$

(3) Draw a horizontal line and carry down the first coefficient.

$$
\begin{array}{r|rrrr}
2 & 1 & 5 & -9 & -3 \\
\hline
  & 1
\end{array}
$$

(4) Multiply the divisor by the first coefficient: $2 \times 1 = 2$. Install the result 2 on the middle line. Add in column two.

$$
\begin{array}{r|rrrr}
2 & 1 & 5 & -9 & -3 \\
2 \times 1 & & +2 \\
\hline
  & 1 & 7
\end{array}
$$

Multiply the divisor by the total in column two: $2 \times 7 = 14$. Install the result 14 on the middle line. Add in column three.

$$
\begin{array}{r|rrrr}
2 & 1 & 5 & -9 & -3 \\
2 \times 7 & & 2 & 14 \\
\hline
  & 1 & 7 & 5
\end{array}
$$

Multiply the divisor by the total in column three: $2 \times 5 = 10$. Install the result 10 in column 4. Add in column four.

$$
\begin{array}{r|rrrr}
2 & 1 & 5 & -9 & -3 \\
2 \times 5 & & 2 & 14 & 10 \\
\hline
  & 1 & 7 & 5 & 7
\end{array}
$$

(5) Box out the final sum—it is the remainder. The other numbers provide the quotient $x^2 + 7x + 5$.

$$
\begin{array}{r|rrrr}
2 & 1 & 5 & -9 & -3 \\
  & & 2 & 14 & 10 \\
\hline
  & 1 & 7 & 5 & \boxed{7}
\end{array}
$$

Thus
$$
\underset{P(x)}{x^3 + 5x^2 - 9x - 3} = \underset{D(x) \cdot}{(x - 2)}\ \underset{Q(x)}{(x^2 + 7x + 5)} + \underset{R}{7}
$$

EXAMPLE 14-8 Use synthetic division to obtain the quotient and remainder for $4x^3 - 6x + 8$ divided by $x + 3$.

Solution   The divisor is -3 and the dividend numbers are 4, 0, -6, and 8.

$$
\begin{array}{r|rrrr}
-3 & 4 & 0 & -6 & 8 \\
\scriptstyle{-3\times 4} & & -12 & 36 & -90 \\
\scriptstyle{-3\times -12} & & & & \\
\scriptstyle{-3\times 30} & 4 & -12 & 30 & \boxed{-82} \\
\end{array}
$$

The quotient is $4x^2 - 12x + 30$ with a remainder of -82. Thus

$$4x^3 - 6x + 8 = (4x^2 - 12x + 30)(x + 3) - 82$$

EXAMPLE 14-9   Use synthetic division to find the quotient and remainder in

$$\frac{2x^4 - 5x^2 - 4x + 5}{x - 2}$$

Solution   The divisor is 2. The dividend has coefficient numbers 2, 0, -5, -4, 5.

$$
\begin{array}{r|rrrrr}
2 & 2 & 0 & -5 & -4 & 5 \\
& & 4 & 8 & 6 & 4 \\
\hline
& 2 & 4 & 3 & 2 & \boxed{9} \\
\end{array}
$$

The quotient is $2x^3 + 4x^2 + 3x + 2$ with a remainder of 9. Thus

$$2x^4 - 5x^2 - 4x + 5 = (x - 2)(2x^3 + 4x^2 + 3x + 2) + 9$$

D. The remainder theorem

If we begin with the simplified form of the division algorithm

$$P(x) = (x - c)Q(x) + R$$

and substitute $x = c$ into the equation, we get

$$P(c) = (c - c)Q(c) + R = 0 + R$$

This result is formally stated as the **remainder theorem**:

Remainder Theorem

▼ If a polynomial $P(x)$ is divided by a first-degree binomial $x - c$ and the division continues until the remainder is constant, then the value of the remainder is the value of the polynomial when x is replaced by c; i.e., $R = P(c)$. ▲

EXAMPLE 14-10   Illustrate the truth of the remainder theorem by dividing $x^3 + 3x^2 - 6x - 12$ by $x + 3$.

Solution   First use synthetic division:

$$
\begin{array}{r|rrrr}
-3 & 1 & 3 & -6 & -12 \\
& & -3 & 0 & 18 \\
\hline
& 1 & 0 & -6 & \boxed{6} \\
\end{array}
$$

The remainder is 6.
Now, by the remainder theorem,

$$P(-3) = (-3)^3 + 3(-3)^2 - 6(-3) - 12$$

$$= -27 + 27 + 18 - 12$$

$$= 6$$

The synthetic division process and the remainder theorem are always valid, even if some of the numbers are complex numbers.

EXAMPLE 14-11 Use synthetic division to divide $P(x) = x^4 - 2ix^3 + ix^2 - 4$ by $x - i$ and validate the remainder theorem for the result.

Solution

$$
\begin{array}{r|rrrrr}
i & 1 & -2i & i & 0 & -4 \\
  &   & i & 1 & i-1 & -1-i \\
\hline
  & 1 & -i & 1+i & i-1 & \boxed{-5-i}
\end{array}
$$

Thus

$$x^4 - 2ix^3 + ix^2 - 4 = (x - i)(x^3 - ix^2 + (1 + i)x + (i - 1)) + (-5 - i)$$

Then, by the remainder theorem,

$$
\begin{aligned}
P(i) &= i^4 - 2i(i^3) + i(i^2) - 4 \\
&= 1 - 2i^4 + i^3 - 4 \\
&= 1 - 2 - i - 4 \\
&= -5 - i
\end{aligned}
$$

E. The factor theorem

The remainder theorem has an important *corollary* [a result that follows from some other proven results]. This corollary is called the **factor theorem**:

▼ Let $P(x)$ be any polynomial. Then $x - c$ is a factor of $P(x)$ if and only if $P(c) = 0$.
▲

Factor Theorem

To prove the factor theorem, we use the remainder theorem. Suppose that, for a given constant c, we have $P(c) = 0$. But by the remainder theorem and the division algorithm

$$
\begin{aligned}
P(x) &= (x - c)Q(x) + P(c) \\
&= (x - c)Q(x) + 0 \\
&= (x - c)Q(x)
\end{aligned}
$$

Thus $x - c$ is a factor of $P(x)$.

Conversely, if we suppose that $x - c$ is a factor of $P(x)$, then $P(x) = (x - c)Q(x)$ for some polynomial $Q(x)$. Then

$$
\begin{aligned}
P(c) &= (c - c)Q(c) \\
&= 0 \cdot Q(c) \\
&= 0
\end{aligned}
$$

Hence $P(c)$ is zero.

EXAMPLE 14-12 Show that $x + 3$ is a factor of $P(x) = x^3 - 2x^2 - 9x + 18$. Then find the other factors.

Solution $x + 3$ is a factor of $P(x)$ if $P(-3) = 0$:

$$
\begin{aligned}
P(-3) &= (-3)^3 - 2(-3)^2 - 9(-3) + 18 \\
&= -27 - 18 + 27 + 18 \\
&= 0
\end{aligned}
$$

Use synthetic division to find the other factors and to verify that $x + 3$ is a factor.

(Show that the remainder is zero.)

$$\begin{array}{r|rrrr} -3 & 1 & -2 & -9 & 18 \\ & & -3 & 15 & -18 \\ \hline & 1 & -5 & 6 & \boxed{0} \end{array}$$

Thus $\qquad x^3 - 2x^2 - 9x + 18 = (x + 3)(x^2 - 5x + 6)$

Now factor the remaining quadratic to get

$$x^3 - 2x^2 - 9x + 18 = (x + 3)(x - 3)(x - 2)$$

EXAMPLE 14-13 Show that $x - 5$ is not a factor of $P(x) = 3x^4 - 9x^3 - 17x - 8$.

Solution Show that $P(5) \neq 0$:

$$P(5) = 3(5)^4 - 9(5)^3 - 17(5) - 8$$

$$= 3 \cdot 625 - 9 \cdot 125 - 85 - 8$$

$$= 1875 - 1125 - 85 - 8$$

$$= 1875 - 1218$$

$$= 657$$

14-3. The Fundamental Theorem of Algebra

Although we can use the factor theorem to test a specific binomial $x - c$ to see if a given polynomial $P(x)$ is divisible by $x - c$, it doesn't tell us if the polynomial actually has roots, or how to find the roots if they do exist. But we have a theorem that does—the **fundamental theorem of algebra** clears up these unanswered questions.

Fundamental Theorem of Algebra

▼ Every nonconstant polynomial has at least one zero (root) in the set of complex numbers. ▲

This theorem leads immediately to another important result called the **complete factorization theorem**:

▼ Let $P(x) = a_n x^n + a_{n-1} x^{n-1} + \cdots + a_2 x^2 + a_1 x + a_0$, where $a_n \neq 0$. Then the equation

$$P(x) = 0$$

Complete Factorization Theorem

has exactly n roots, so that $P(x)$ can be expressed as

$$P(x) = a_n(x - r_1)(x - r_2) \cdots (x - r_n)$$

where $r_1, r_2, \ldots, r_n$, which are not necessarily distinct, are the roots expressed as members of the complex number field. ▲

To see the validity of the complete factorization theorem from the fundamental theorem, begin with $P(x) = 0$:

Since, by the fundamental theorem, $P(x)$ has at least one zero, write it as

$$P(x) = (x - r_1)P_1(x) \qquad \text{[where } P_1(x) \text{ has degree } n - 1 \text{ and begins with coefficient } a_n]$$

Now $P_1(x)$ is a nonconstant polynomial, so by the fundamental theorem, $P_1(x)$ has at least one zero. Thus

$$P_1(x) = (x - r_2)P_2(x) \qquad \text{[where } P_2(x) \text{ has degree } n - 2 \text{ and begins with coefficient } a_n]$$

Hence
$$P(x) = (x - r_1)(x - r_2)P_2(x)$$

We continue by rewriting $P_2(x)$ as $(x - r_3)P_3(x)$, $P_3(x) = (x - r_4)P_4(x)$, etc.; and we finally get
$$P(x) = (x - r_1)(x - r_2)(x - r_3)\cdots(x - r_n)\cdot a_n$$

where a_n is the constant polynomial left from the last decomposition.

A. Factoring polynomials

The fundamental theorem and the complete factorization theorem are *theoretical* results—examples to show their "stuff" are not feasible. But we can say that every polynomial factors completely! And the roots mentioned are the only roots of $P(x)$. Thus we also may conclude that

- A polynomial of degree n has at most n distinct roots.

EXAMPLE 14-14 Show that the polynomial $x^4 - 3x^3 - 18x^2 + 40x$ has four distinct roots.

Solution Since x is common to each term, we can immediately factor it out:
$$x^4 - 3x^3 - 18x^2 + 40x = (x - 0)(x^3 - 3x^2 - 18x + 40)$$

Now to find additional factors, we try synthetic division with some "reasonable" trial binomials; try some $x - c$ where c is a divisor of 40. The value 2 is a reasonable starter:

$$\underline{2|} \quad 1 \quad -3 \quad -18 \quad 40$$
$$\qquad\quad 2 \quad -2 \quad -40$$
$$\overline{\qquad 1 \quad -1 \quad -20 \quad \boxed{0}}$$

Eureka! Since 2 is a root, $(x - 2)$ is a factor.

Thus
$$x^4 - 3x^3 - 18x^2 + 40x = (x - 0)(x^3 - 3x^2 - 18x + 40)$$
$$= (x - 0)(x - 2)(x^2 - x - 20)$$

The remaining factors come from the quadratic:
$$= (x - 0)(x - 2)(x - 5)(x + 4)$$

The distinct roots are 0, 2, 5, and -4.

B. Multiple roots

The roots $r_1, r_2, \ldots, r_n$ of $P(x) = 0$ do not necessarily have to be distinct.

- If a root, say r_j, appears m times in the list, we say that r_j is a root of **multiplicity m**.

For example, in the case of $P(x) = (x - 2)^3(x + 1)(x + 3)^2$, the roots 2, -1, and -3 have multiplicities 3, 1, and 2, respectively. Thus we are able to say that:

- An nth-degree polynomial has exactly n roots if we count the root of multiplicity m as m roots.

For example, the sixth-degree polynomial
$$P(x) = x^6 + x^5 - 15x^4 - 5x^3 + 70x^2 - 12x - 72$$
$$= (x - 2)^3(x + 1)(x + 3)^2$$

has six roots: 2, 2, 2, -1, -3, -3.

EXAMPLE 14-15   Form the polynomial that has roots 2, 3, and 4.

Solution   Since 2 is a root, $x - 2$ is a factor.
Since 3 is a root, $x - 3$ is a factor.
Since 4 is a root, $x - 4$ is a factor.

Thus the polynomial is

$$P(x) = (x - 2)(x - 3)(x - 4) = x^3 - 9x^2 + 26x - 24$$

We can check this by synthetic division:

$$
\begin{array}{r|rrrr}
2 & 1 & -9 & 26 & -24 \\
  &   & 2 & -14 & 24 \\
\hline
3 & 1 & -7 & 12 & \boxed{0} \\
  &   & 3 & -12 & \\
\hline
4 & 1 & -4 & \boxed{0} & \\
  &   & 4 & & \\
\hline
  & 1 & \boxed{0} & &
\end{array}
$$

14-4. Factoring Polynomials with Real Coefficients

A. Complex solutions

When a polynomial has complex coefficients, its roots may be any set of complex numbers. But if the coefficients $a_n, a_{n-1}, \ldots, a_2, a_1, a_0$ of the polynomial are all real numbers, then the complex roots, if they exist, must occur in complex conjugate pairs. If, for example, a polynomial has real coefficients and we find that $1 + i$ is a root, then automatically $1 - i$ must also be a root. And if $2 - 3i$ is a root, then $2 + 3i$ must also be a root. This is summarized in the conjugate pair theorem:

Conjugate Pair Theorem

▼If $P(x) = a_n x^n + a_{n-1} x^{n-1} + \cdots + a_2 x^2 + a_1 x + a_0$ has real coefficients and r is a complex root, then $\bar{r}$ is also a root. [*Note:* $\bar{r}$ is the notation for the complex conjugate of r.]   ▲

EXAMPLE 14-16   Let $P(x) = x^4 - 7x^3 + 10x^2 - 28x + 24$. Given that $2i$ is a root of $P(x)$, find all of the factors.

Solution   Since all the coefficients are real, the conjugate pair theorem applies—and we are assured that $-2i$ is also a root. We can use synthetic division to verify this and to find the reduced-degree quotient. Each time we find a root, we divide into the resulting quotient:

$$
\begin{array}{r|rrrrr}
2i & 1 & -7 & 10 & -28 & 24 \\
   &   & 2i & (-14i - 4) & (28 + 12i) & -24 \\
\hline
-2i & 1 & (-7 + 2i) & (-14i + 6) & 12i & \boxed{0} \\
    &   & -2i & 14i & -12i & \\
\hline
6 & 1 & -7 & 6 & \boxed{0} & \\
  &   & 6 & -6 & & \\
\hline
1 & 1 & -1 & \boxed{0} & & \\
  &   & 1 & & & \\
\hline
  & 1 & \boxed{0} & & &
\end{array}
$$

Hence

$$P(x) = x^4 - 7x^3 + 10x^2 - 28x + 24 = (x - 2i)(x + 2i)(x - 6)(x - 1)$$

is the complete factorization, which includes the complex conjugate pair of roots $2i$ and $-2i$.

caution: The conjugate-pair theorem works *only* when the coefficients of the polynomial are *real*.

It is often desirable to factor a polynomial with real coefficients into factors that are themselves polynomials with real coefficients. We can't always do this if we insist on linear factors, but we can do it if we allow the factors to be linear or quadratic polynomials. This is summarized in the following rule:

- Any polynomial with real coefficients can be factored into a product of linear polynomials or into a product of linear polynomials and quadratic polynomials with real coefficients, where the quadratic polynomials have no real roots.

EXAMPLE 14-17 Find the real factors of

$$\text{(a) } x^3 - x^2 - 19x - 5 \qquad \text{and} \qquad \text{(b) } x^3 - 2x^2 + 9x - 18$$

Solution Either we will have three linear factors or we will have one linear factor and one quadratic factor. In looking for the linear factor, we try synthetic division for which we must pick a trial root.

(a) Try 5:

$$
\begin{array}{r|rrrr}
5 & 1 & -1 & -19 & -5 \\
  &   & 5 & 20 & 5 \\
\hline
  & 1 & 4 & 1   & \boxed{0}
\end{array}
$$

Bingo! One factor is $x - 5$, and the other is $x^2 + 4x + 1$. This quotient is quadratic, so we have to test its roots by finding the discriminant:

$$D = b^2 - 4ac = (4)^2 - 4(1)(1) = 16 - 4 = 12$$

Since $D = 12$ is positive, we have two additional—linear—real roots:

$$\frac{-4 \pm \sqrt{12}}{2} = -2 \pm \sqrt{3}$$

We have all the factors, so

$$x^3 - x^2 - 19x - 5 = (x - 5)(x + 2 + \sqrt{3})(x + 2 - \sqrt{3})$$

(b) Try 2:

$$
\begin{array}{r|rrrr}
2 & 1 & -2 & 9 & -18 \\
  &   & 2 & 0 & 18 \\
\hline
  & 1 & 0 & 9 & \boxed{0}
\end{array}
$$

So $x - 2$ is a factor, and the other is $x^2 + 9$. But the real quadratic equation $x^2 = -9$ has no real roots: Its roots are the complex conjugate pair $x = 3i$ and $x = -3i$. So the given polynomial is

$$x^3 - 2x^2 + 9x - 18 = (x - 2)(x^2 + 9)$$

We can now draw the following conclusion:

- Every polynomial function of odd degree has at least one real root.

B. Integer solutions

So far it seems that solving an arbitrary polynomial equation involves luck— if a lucky guess finds a root, then synthetic division will reduce the polynomial by one degree and we can continue. And if we get to a quadratic, the quadratic formula will finish the factorization. But an intelligent way of "guessing" the roots would be a help. So we have this **theorem of integer solutions**:

Integer Solutions

▼ If $P(x) = a_n x^n + a_{n-1} x^{n-1} + \cdots + a_2 x^2 + a_1 x + a_0$ is a polynomial with integer coefficients and if k is an integer root of $P(x)$, then k is a divisor of the trailing (last) coefficient a_0. ▲

note: This theorem doesn't guarantee us integer roots—it only gives us a way to find the possible integer roots, if they exist.

EXAMPLE 14-18   Let's return to Example 14-17, in which we wished to factor (a) $P(x) = x^3 - x^2 - 19x - 5$ and (b) $x^3 - 2x^2 + 9x - 18$. What are the potential integer roots of each polynomial?

Solution

(a) The potential integer roots are those numbers that are divisors of the trailing coefficient, -5. There are four candidates—the numbers $+1$, -1, $+5$, and -5 are the only divisors of -5.

(b) The possible divisors of -18 are ± 1, ± 2, ± 3, ± 6, ± 9, and ± 18—which is a lot, but better than having to choose from *all* the integers!

C. Rational solutions

But, sometimes polynomials simply don't have integer roots. Arbitrary polynomial equations can have roots other than integer roots—and we can find these roots, too. The next easiest roots to find are those that are rational numbers, i.e., roots whose form is p/q where p and q are integers and the fraction is in lowest form. To find these roots, we use the rational root theorem:

Rational Root Theorem

▼ Let $P(x) = a_n x^n + a_{n-1} x^{n-1} + \cdots + a_2 x^2 + a_1 x + a_0 = 0$ be a polynomial equation with integer coefficients. If $r = p/q$ is a root of $P(x) = 0$, then p divides the trailing (last) coefficient a_0 and q divides the leading (first) coefficient a_n. ▲

note: Like the theorem of integer solutions, the rational root theorem doesn't guarantee that a polynomial will have rational roots—but it gives us a way to find all the possible rational roots, if they exist.

EXAMPLE 14-19   Given the polynomial equation $P(x) = 2x^4 + x^3 + x^2 + x - 1 = 0$, find the roots of the polynomial and factor completely.

Solution   The trailing coefficient is -1 and the leading coefficient is 2. If the equation has rational roots of the form p/q, then p divides -1 and q divides 2. The candidates for p are $+1$ and -1. The candidates for q are $+1$, -1, $+2$, and -2. Thus the potential roots are the proper quotients of these numbers: $(+1)/(+1)$, $(-1)/(+1)$, $(+1)/(+2)$, $(-1)/(+2)$ and repetitions of these. So we need only check 1, -1, $1/2$, and $-1/2$. Working these out gives

$$P(1) = 2(1)^4 + (1)^3 + (1)^2 + (1) - 1$$

$$= 2 + 1 + 1 + 1 - 1 = 4$$

So 1 is not a root.

$$P(-1) = 2(-1)^4 + (-1)^3 + (-1)^2 + (-1) - 1$$

$$= 2 - 1 + 1 - 1 - 1 = 0$$

So -1 is a root and $(x + 1)$ is a factor.

$$P(\tfrac{1}{2}) = 2(\tfrac{1}{2})^4 + (\tfrac{1}{2})^3 + (\tfrac{1}{2})^2 + (\tfrac{1}{2}) - 1$$
$$= \tfrac{2}{16} + \tfrac{1}{8} + \tfrac{1}{4} + \tfrac{1}{2} - 1 = 0$$

So $\tfrac{1}{2}$ is a root and $(x - \tfrac{1}{2})$ is a factor.

$$P(-\tfrac{1}{2}) = 2(-\tfrac{1}{2})^4 + (-\tfrac{1}{2})^3 + (-\tfrac{1}{2})^2 + (-\tfrac{1}{2}) - 1$$
$$= \tfrac{2}{16} - \tfrac{1}{8} + \tfrac{1}{4} - \tfrac{1}{2} - 1 = -\tfrac{5}{4}$$

So $-\tfrac{1}{2}$ is not a root.

Now we can pull out these two known factors by synthetic division, and find the remaining factors from the remaining quotient. The known roots are -1 and $\tfrac{1}{2}$. Thus

$$
\begin{array}{r|rrrrr}
-1 & 2 & 1 & 1 & 1 & -1 \\
   &   & -2 & 1 & -2 & 1 \\
\hline
\tfrac{1}{2} & 2 & -1 & 2 & -1 & \boxed{0} \\
   &   & 1 & 0 & 1 & \\
\hline
   & 2 & 0 & 2 & \boxed{0} &
\end{array}
$$

The remaining quotient is $2x^2 + 2$, so

$$2x^4 + x^3 + x^2 + x - 1 = (x + 1)(x - \tfrac{1}{2})(2x^2 + 2)$$
$$= 2(x - \tfrac{1}{2})(x + 1)(x^2 + 1)$$
$$= (2x - 1)(x + 1)(x^2 + 1)$$

This gives the real-coefficient polynomial factors. And if we use the complex numbers, then $x^2 + 1$ can be factored into $(x - i)(x + i)$, so that

$$2x^4 + x^3 + x^2 + x - 1 = (2x - 1)(x + 1)(x - i)(x + i)$$

EXAMPLE 14-20 List all the possible rational roots of $8x^4 - x^3 + 2x^2 + x - 3 = 0$.

Solution If rational roots exist, then the numerator p divides -3 and the denominator q divides 8. Thus

p could be $1, -1, 3, -3$; q could be $1, -1, 2, -2, 4, -4, 8, -8$

Consequently, the only possible rational roots are

$$1, -1, 3, -3, \tfrac{1}{2}, -\tfrac{1}{2}, \tfrac{3}{2}, -\tfrac{3}{2}, \tfrac{1}{4}, -\tfrac{1}{4}, \tfrac{3}{4}, -\tfrac{3}{4}, \tfrac{1}{8}, -\tfrac{1}{8}, \tfrac{3}{8}, -\tfrac{3}{8}$$

---

1. Variation in sign

When the list of possible rational roots gets lengthy, as in Example 14-20, we need a way to pare the list down—to make some "very smart" guesses. We use a method called *variation in sign*, which often eliminates a large group of candidates from the list.

▼ A polynomial $P(x)$ that has real-number coefficients and is arranged in descending powers of x is said to have a **variation in sign** if two consecutive terms have opposite signs. ▲

Variation in Sign

EXAMPLE 14-21 How many variations in signs are there in the following polynomials?

(a) $4x^3 + 3x^2 - 2x - 1 = 0$ (b) $8x^4 - 9x^3 + 4x^2 + 2x - 7 = 0$

(c) $x^2 + x + 4 = 0$

Solution

(a) The first sign is + and so is the second. That's not a variation. The third sign is −. That's one variation. The last sign is also −, no variation. There is one variation in sign.

(b) We go from + to − and then − to +. That's two variations. Then it stays + and then changes to −. That's another. There are three sign variations.

(c) The signs are all +. There is no sign variation.

Descartes' Rule of Signs

There's a reason for making a count of sign variation, which is summarized in **Descartes' Rule of Signs:**

▼ If a polynomial equation $P(x) = 0$ has real coefficients, then the number of positive real roots can either be equal to the number of sign variations in $P(x)$ or be less than that by an even positive integer. ▲

EXAMPLE 14-22   How many positive real roots could each of the polynomials in Example 14-21 have?

Solution

(a) In $4x^3 + 3x^2 - 2x - 1 = 0$ there is only one sign change, so there can be one positive real root.

(b) In $8x^4 - 9x^3 + 4x^2 + 2x - 7 = 0$ there are three variations in signs, so there can be either three positive real roots or one positive real root.

(c) In $x^2 + x + 4 = 0$ there are no sign variations, so there can be no positive real roots.

When we have a large collection of possible rational roots, we can often use the rule of signs to eliminate a bunch of them—especially when the conclusion is that no positive real roots exist! This throws out half of the candidates. If we find that exactly one exists, and then find it by synthetic division, we can toss out all the other candidates with positive signs.

The counterpart for real negative roots is just as easy to do:

• Information about the real negative roots of $P(x) = 0$ can be obtained by applying Descartes' Rule of Signs to $P(-x)$.

EXAMPLE 14-23   How many positive and how many negative real roots can there be for

(a) $P(x) = 3x^4 + x^3 - 4x^2 - x - 5$ (b) $Q(x) = x^4 + 3x^3 + 3x^2 + 3x + 2$

Solution

(a) In $P(x)$ we have only one sign change. Hence there exists exactly one positive real root.

Now $P(-x) = 3(-x)^4 + (-x)^3 - 4(-x)^2 - (-x) - 5$

$$= 3x^4 - x^3 - 4x^2 + x - 5$$

and here we have three sign changes. Hence we have either three negative real roots or one negative real root.

(b) In $Q(x)$ there are no sign changes, so there are no positive real roots. And in $Q(-x) = x^4 - 3x^3 + 3x^2 - 3x + 2$ there are four sign changes so there may be either four or two or zero negative real roots.

caution: Descartes' Rule of Signs applies to real roots only. For example, we have seen that $x^4 + 3x^3 + 3x^2 + 3x + 2 = 0$ has no positive real roots and either four or two or zero negative real roots. And since we know that this polynomial must have four roots, we might be tempted to assume that it must therefore have four negative roots. But

$$x^4 + 3x^3 + 3x^2 + 3x + 2 = (x + 1)(x + 2)(x + i)(x - i)$$

so there are two negative roots, -1 and -2, and two complex roots, i and $-i$. [It could also have had four complex roots, as is the case in $R(x) = (x^2 + 4)(x^2 + 9) = x^4 + 13x^2 + 36$, or in $T(x) = (x^2 + x + 1)(x^2 + 1) = x^4 + x^3 + 2x^2 + x + 1$.]

2. Intermediate value theorem

One more aid is often used to help find the rational roots from a list of potential roots: We attempt to find *bounds* between which we know roots to lie. Coupling this with the rules of signs could pinpoint a single root from a long list or at least allow us to focus on a select few. The result is called the **intermediate value theorem for polynomials**.

▼ If $P(x)$ is a polynomial with real coefficients and a and b are numbers for which $P(a)$ and $P(b)$ have opposite signs, then there is a root of $P(x)$ between a and b. ▲

Intermediate Value Theorem for Polynomials

(See Figure 14-1 for a pictorial idea of this theorem.)

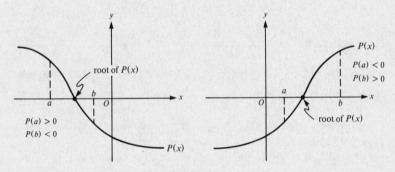

Figure 14-1

Generally, we pick a and b so that they are consecutive integers because of ease of computation. There are, however, no restrictions on the choice of a and b: We just have to find two distinct numbers for which the polynomial has opposite signs.

EXAMPLE 14-24 Let $P(x) = 16x^3 + 20x^2 - 8x - 3$. Factor $P(x)$ completely.

Solution Descartes' Rule of Signs tells us that $P(x)$ has one sign change, so it has exactly one positive real root. And $P(-x) = -16x^3 + 20x^2 + 8x - 3$ has two sign changes, so it has two or zero negative real roots.

If rational roots exist, they must have numerators that divide -3 and denominators that divide 16. This is an extensive list:

$$\pm 1, \pm 3, \pm \tfrac{1}{2}, \pm \tfrac{3}{2}, \pm \tfrac{1}{4}, \pm \tfrac{3}{4}, \pm \tfrac{1}{8}, \pm \tfrac{3}{8}, \pm \tfrac{1}{16}, \pm \tfrac{3}{16}$$

Since we have only one positive one, let's try to isolate it using the intermediate value theorem: Try $P(0)$ and $P(1)$:

$$P(0) = -3 \quad \text{and} \quad P(1) = 16 + 20 - 8 - 3 = 25$$

so the root lies between 0 and 1 because -3 and 25 have opposite signs. That eliminates 1, $\frac{3}{2}$, and 3. Next, try $P(\frac{1}{4})$:

$$P(\tfrac{1}{4}) = 16(\tfrac{1}{4})^3 + 20(\tfrac{1}{4})^2 - 8(\tfrac{1}{4}) - 3$$
$$= \tfrac{1}{4} + \tfrac{20}{16} - 2 - 3$$
$$= -\tfrac{7}{2}$$

Hence no root lies between 0 and $\frac{1}{4}$ because -3 and $-\frac{7}{2}$ have the same sign. That eliminates $\frac{1}{16}, \frac{1}{8}, \frac{3}{16}$, and $\frac{1}{4}$. The remaining candidates for the positive root are $\frac{1}{2}, \frac{3}{4}$, and $\frac{3}{8}$. Now we go to synthetic division:

$$\begin{array}{r|rrrr} \tfrac{1}{2} & 16 & 20 & -8 & -3 \\ & & 8 & 14 & 3 \\ \hline & 16 & 28 & 6 & \boxed{0} \end{array}$$

Eureka! We have the one and only positive root at $x = \frac{1}{2}$.

The quotient polynomial of lower degree is a quadratic equation

$$16x^2 + 28x + 6$$

So the quadratic formula gives the two remaining roots:

$$\frac{-28 \pm \sqrt{28^2 - 4(16)(6)}}{32} = \frac{-28 \pm \sqrt{400}}{32} = \frac{-28 \pm 20}{32} = -\frac{1}{4} \quad \text{and} \quad -\frac{3}{2}$$

Thus $-\frac{1}{4}$ and $-\frac{3}{2}$ are the negative roots, so the factors are $(x + \frac{1}{4})$ and $(x + \frac{3}{2})$. Hence the complete factorization of $P(x)$ is

$$16x^3 + 20x^2 - 8x - 3 = 16(x + \tfrac{1}{4})(x + \tfrac{3}{2})(x - \tfrac{1}{2})$$

D. Irrational roots: Methods of successive approximations

If a polynomial equation with real coefficients has roots that are *irrational numbers*, the location of these roots is sometimes very difficult to determine. Since irrational numbers have decimal expansions that are nonrepeating and nonterminating, the best we can do in finding the root (unless it is a multiple of a special symbol such as π, or e, or $\sqrt{2}$) is to get an *approximation* to it by finding the first few digits in its decimal expansion. But we don't just get one approximation: We use what we know about polynomials to get a good first approximation; then by concentrating on smaller and smaller subintervals, we get better and better approximations. The method is called—naturally enough—the **method of successive approximations**, and we can use two techniques to go about it: *bisection* and *linear interpolation*.

1. Bisection method (divide and conquer)

First, we use the rule of signs and the intermediate value theorem to locate a root between two consecutive integers a and b, so that $P(a)$ and $P(b)$ have opposite signs. Then we pick the halfway point in this interval and evaluate again. This halving lets us choose one half or the other for the root location. Then we pick the midpoint of that remaining interval and evaluate the polynomial again. This halving allows us the choice of a smaller interval for the residence of the root. Finally, we keep this process going until we have the accuracy we want.

EXAMPLE 14-25 Let $P(x) = x^3 + 2x^2 - x - 1 = 0$. Find the positive roots, if any.

Solution From Descartes' Rule of Signs, $P(x)$ has exactly one positive root. If that root is rational, it must be 1 or -1. Synthetic division will check out the possibility at 1:

$$\underline{1|}\ \ \begin{array}{rrrr} 1 & 2 & -1 & -1 \\ & 1 & 3 & 2 \\ \hline 1 & 3 & 2 & \boxed{1} \end{array}$$

The root is not at $x = 1$ so the positive root is not rational. The root must be irrational. We already know that $P(1) = 1$ (from the remainder of the synthetic division) and we can see that $P(0) = -1$, so the irrational root is between 0 and 1. Now we can divide and conquer by successive evaluations of midpoints:

$$\left.\begin{array}{l} P(0) = -1 \\ P(1) = 1 \end{array}\right\} P(\tfrac{1}{2}) = (\tfrac{1}{2})^3 + 2(\tfrac{1}{2})^2 - (\tfrac{1}{2}) - 1 = \tfrac{1}{8} + \tfrac{1}{2} - \tfrac{1}{2} - 1 = -\tfrac{7}{8}$$

$$\left.\begin{array}{l} P(\tfrac{1}{2}) = -\tfrac{7}{8} \\ P(1) = 1 \end{array}\right\} P(\tfrac{3}{4}) = (\tfrac{3}{4})^3 + 2(\tfrac{3}{4})^2 - (\tfrac{3}{4}) - 1 = \tfrac{27}{64} + \tfrac{9}{8} - \tfrac{3}{4} - 1 = -\tfrac{13}{64}$$

$$\left.\begin{array}{l} P(\tfrac{3}{4}) = -\tfrac{13}{64} \\ P(1) = 1 \end{array}\right\} P(\tfrac{7}{8}) = (\tfrac{7}{8})^3 + 2(\tfrac{7}{8})^2 - (\tfrac{7}{8}) - 1 = \tfrac{167}{512}$$

note: After we evaluate $P(\tfrac{7}{8}) = \tfrac{167}{512}$, we see that the next interval is between $P(\tfrac{3}{4})$ and $P(\tfrac{7}{8})$, not between $P(\tfrac{7}{8})$ and $P(1)$, because $P(\tfrac{7}{8})$ is positive and $P(\tfrac{3}{4})$ is negative.

$$\left.\begin{array}{l} P(\tfrac{3}{4}) = -\tfrac{13}{64} \\ P(\tfrac{7}{8}) = \tfrac{167}{512} \end{array}\right\} P(\tfrac{13}{16}) = (\tfrac{13}{16})^3 + 2(\tfrac{13}{16})^2 - (\tfrac{13}{16}) - 1 = \tfrac{181}{4096}$$

$$\left.\begin{array}{l} P(\tfrac{3}{4}) = -\tfrac{13}{64} \\ P(\tfrac{13}{16}) = \tfrac{181}{4096} \end{array}\right\} P(\tfrac{25}{32}) = (\tfrac{25}{32})^3 + 2(\tfrac{25}{32})^2 - (\tfrac{25}{32}) - 1 = \tfrac{-2743}{32\,768}$$

We have isolated the root between $\tfrac{25}{32}$ and $\tfrac{13}{16}$, i.e., between .781 25 and .8125.

Continued halving of the intervals will produce sharper results. (Clearly, a good calculator or a simple computer program would be very helpful.)

2. Linear interpolation

Suppose we know that the root of a polynomial equation lies between a and b because $P(a)$ and $P(b)$ have opposite signs. If we graph the points $(a, P(a))$ and $(b, P(b))$, they should lie on opposite sides of the x-axis. If we connect these two points with a straight line, then that line will cross the x-axis between a and b. This x-intercept becomes an approximation to the root. Repeating this process using the new point will give an even better approximation, and so on.

EXAMPLE 14-26 Find an approximation to the positive irrational root of $P(x) = x^3 - x - 3$.

Solution By the rule of signs we know there is one positive root. By inspecting the possible rational roots, ± 3 and ± 1, the positive ones may be eliminated by synthetic division:

(a)

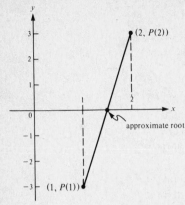

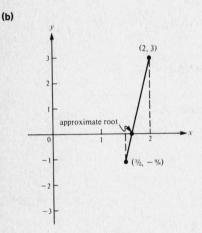

(b)

Figure 14-2

$$
\begin{array}{r|rrrr}
3\,\rule[-4pt]{0.4pt}{14pt} & 1 & 0 & -1 & -3 \\
 & & 3 & 9 & 24 \\
\hline
 & 1 & 3 & 8 & \boxed{21}
\end{array}
$$

$$
\begin{array}{r|rrrr}
1\,\rule[-4pt]{0.4pt}{14pt} & 1 & 0 & -1 & -3 \\
 & & 1 & 1 & 0 \\
\hline
 & 1 & 1 & 0 & \boxed{-3}
\end{array}
$$

We note that $P(1) = -3$ and $P(3) = 21$, so the root lies between 1 and 3. We may as well try $P(2)$:

$$
P(2) = 2^3 - 2 - 3
$$
$$
= 8 - 2 - 3 = 3
$$

so the root lies between 1 and 2, which can be graphed as in Figure 14-2a. Since the points are equidistant above and below the axis, the connecting line crosses in the middle at $(\frac{3}{2}, 0)$. Thus we try $P(\frac{3}{2})$:

$$
P(\tfrac{3}{2}) = (\tfrac{3}{2})^3 - (\tfrac{3}{2}) - 3
$$
$$
= \tfrac{27}{8} - \tfrac{9}{2} = -\tfrac{9}{8}
$$

The function at $x = \frac{3}{2}$ is at $y = P(x) = -\frac{9}{8}$ as shown in Figure 14-2b. Then, by using similar triangles we can calculate the point of crossing on the x-axis as 1.636: Let x be the distance between $(\frac{3}{2}, 0)$ and the root point, so that the distance between the root and $(2, 0)$ is $\frac{1}{2} - x$. Then

$$
\frac{\frac{9}{8}}{3} = \frac{x}{\frac{1}{2} - x}
$$

$$
x = .136
$$

So if $x = .136$, then the distance between the root and the origin is $1.5 + .136 = 1.636$.

SUMMARY

1. The division algorithm reformulates a dividend polynomial $P(x)$ into a product of a divisor polynomial $D(x)$ and a quotient polynomial $Q(x)$ with a remainder polynomial $R(x)$ added:

$$
P(x) = D(x) \cdot Q(x) + R(x)
$$

2. If the divisor is a first-degree binomial $x - c$, then the remainder is a unique constant R:

$$
P(x) = (x - c) \cdot Q(x) + R
$$

3. Synthetic division is an efficient form of polynomial division by a linear term $x - c$.
4. The remainder theorem relates the number obtained as the remainder of a polynomial division to the value obtained as a function evaluation:

$$
P(c) = (c - c)Q(c) + R = 0 + R = R
$$

5. Synthetic division and the remainder theorem remain valid even if some coefficients of the polynomials are complex numbers.
6. The factor theorem states that if a polynomial is divided by a linear term so that the remainder is zero, then the linear term is a factor; i.e., if $x - c$ is a factor of $P(x)$, then $P(c) = 0$.
7. The fundamental theorem of algebra states that every nonconstant polynomial has at least one zero (root) in the set of complex numbers.
8. The complete factorization theorem states that an nth-degree polynomial has exactly n roots and may be expressed as

$$
P(x) = a_n(x - r_1)(x - r_2) \cdots (x - r_n)
$$

Some of the roots may be repeated, but if the multiplicity of a root is m, we count it as m roots.

9. According to the conjugate pair theorem, if a polynomial with real coefficients has a complex root, then its complex conjugate is also a root.
10. Any polynomial with real coefficients can be factored into a product of linear polynomials or into a product of linear polynomials and quadratic polynomials, where the quadratic polynomials have no real roots.
11. Every polynomial function of odd degree has at least one real root.
12. The rational root theorem states that if $r = p/q$ is a rational number and is a root of $P(x) = 0$, then p divides the trailing (last) coefficient and q divides the leading (first) coefficient of $P(x)$.
13. According to Descartes' Rule of Signs, if a polynomial equation $P(x) = 0$ has real coefficients, then the number of positive real roots can either be equal to the number of sign variations in $P(x)$ or be less than that by an even positive integer.
14. The intermediate value theorem for polynomials tells us that if the sign of a polynomial function $P(x)$ changes from positive to negative or vice versa as we move through an interval of x values, the polynomial has a zero in that interval.
15. Irrational roots of polynomial functions are usually found by approximation techniques— including bisection and linear interpolation methods.

RAISE YOUR GRADES
Can you . . . ?

☑ verify a root of a polynomial equation
☑ perform the process of long division on polynomials
☑ state the division algorithm
☑ list the steps needed to do synthetic division
☑ state the remainder theorem and the factor theorem
☑ explain the fundamental theorem of algebra and how it relates to the complete factorization theorem
☑ form a polynomial that has a given collection of roots
☑ explain the significance the conjugate pair theorem
☑ create a list of the potential integer and rational roots of a polynomial equation
☑ use Descartes' Rule of Signs to predict the number of positive and/or negative roots of $P(x) = 0$
☑ isolate the intervals in which the roots of a polynomial equation lie
☑ determine an irrational root of a polynomial equation by approximation methods
☑ elaborate the difference between the bisection method and the linear interpolation method

SOLVED PROBLEMS

General Polynomials: Definitions

PROBLEM 14-1 Show that $x = -2$ is a root of the polynomial equation

$$x^3 - 7x^2 - 14x + 8 = 0$$

Solution   Substitute $x = -2$ into the equation:

$$(-2)^3 - 7(-2)^2 - 14(-2) + 8 \overset{?}{=} 0$$

$$-8 - 7(4) + 28 + 8 \overset{?}{=} 0$$

$$-8 - 28 + 28 + 8 \overset{?}{=} 0$$

$$0 \overset{\checkmark}{=} 0$$

PROBLEM 14-2   Show that $x = 2i$ is a root of the complex polynomial equation

$$x^3 - (4 + 2i)x^2 + (4 + 8i)x - 8i = 0$$

Solution   Substitute $x = 2i$ into the equation:

$$(2i)^3 - (4 + 2i)(2i)^2 + (4 + 8i)(2i) - 8i \overset{?}{=} 0$$

$$8i^3 - (4 + 2i)(4i^2) + (8i + 16i^2) - 8i \overset{?}{=} 0$$

$$-8i + (4 + 2i)(4) + 8i - 16 - 8i \overset{?}{=} 0$$

$$-8i + 16 + 8i + 8i - 16 - 8i \overset{?}{=} 0$$

$$0 \overset{\checkmark}{=} 0$$

PROBLEM 14-3   Is $\sqrt{3}$ a root of $x^4 + x^2 - 12 = 0$?

Solution $$(\sqrt{3})^4 + (\sqrt{3})^2 - 12 \overset{?}{=} 0$$

$$3^2 + 3 - 12 \overset{?}{=} 0$$

$$9 + 3 - 12 \overset{\checkmark}{=} 0 \quad \text{Yes, it is a root.}$$

Division of Polynomials

PROBLEM 14-4   Show that the division of $8x^2 + 2x - 3$ by $4x + 3$ is exact.

Solution

$$\frac{8x^2 + 2x - 3}{4x + 3} = \frac{(4x + 3)(2x - 1)}{4x + 3} = 2x - 1$$

Thus $4x + 3$ is an exact divisor of $8x^2 + 2x - 3$, or

$$P(x) \quad = \quad D(x) \cdot Q(x)$$

$$8x^2 + 2x - 3 = (4x + 3)(2x - 1)$$

PROBLEM 14-5   Use the long division process to divide

(a) $\dfrac{273}{17}$ **(b)** $\dfrac{6x^2 - 13x - 5}{2x - 5}$ **(c)** $\dfrac{5x^4 - 6x^3 - x^2 - 85x - 390}{x - 4}$ **(d)** $\dfrac{3x^2 - 2ix + 5}{x + 3i}$

Solution

(a)

$$
\begin{array}{r}
16 \quad \longleftarrow \text{quotient} \\
\text{divisor} \longrightarrow 17\,\overline{\smash{\big)}\,273} \longleftarrow \text{dividend} \\
-17 \\
\hline
103 \\
102 \\
\hline
1 \longleftarrow \text{remainder}
\end{array}
$$

Thus $\dfrac{273}{17} = 16 + \dfrac{1}{17}$.

(b)

$$\begin{array}{r} 3x \ + 1 \qquad \leftarrow \text{quotient} \\ \text{divisor} \longrightarrow 2x - 5\overline{\smash{\big)}6x^2 - 13x - 5} \leftarrow \text{dividend} \\ -(6x^2 - 15x) \\ \hline 2x - 5 \\ -(2x - 5) \\ \hline 0 \leftarrow \text{remainder} \end{array}$$

Thus $6x^2 - 13x - 5 = (2x - 5)(3x + 1)$.

(c)

$$\begin{array}{r} 5x^3 + 14x^2 + 55x \ + 135 \\ x - 4\overline{\smash{\big)}5x^4 - \ 6x^3 - \ \ x^2 - \ 85x - 390} \\ -(5x^4 - 20x^3) \\ \hline 14x^3 - \ \ x^2 \\ -(14x^3 - 56x^2) \\ \hline 55x^2 - \ 85x \\ -(55x^2 - 220x) \\ \hline 135x - 390 \\ -(135x - 540) \\ \hline 150 \end{array}$$

Thus $5x^4 - 6x^3 - x^2 - 85x - 390 = (x - 4)(5x^3 + 14x^2 + 55x + 135) + 150$.

(d)

$$\begin{array}{r} 3x \ - 11i \\ x + 3i\overline{\smash{\big)}3x^2 + \ 2ix + 5} \\ -(3x^2 + \ 9ix) \\ \hline -11ix + 5 \\ -(-11ix \ \ \ - 33i^2) \\ \hline 5 - 33i^2 \ = 38 \end{array}$$

Thus $\dfrac{3x^2 - 2ix + 5}{x + 3i} = 3x - 11i + \dfrac{38}{x + 3i}$.

PROBLEM 14-6 Illustrate the division algorithm for the division $\dfrac{x^3 - 4x^2 - 5x - 6}{x - 2}$.

Solution The division algorithm is

$$P(x) = D(x) \cdot Q(x) + R(x)$$

Here, $P(x) = x^3 - 4x^2 + 5x - 6$, $D(x) = x - 2$, and when the division is performed,

$$\begin{array}{r} x^2 - 2x \ + 1 \\ x - 2\overline{\smash{\big)}x^3 - 4x^2 + 5x - 6} \\ -(x^3 - 2x^2) \\ \hline -2x^2 + 5x \\ -(-2x^2 + 4x) \\ \hline x - 6 \\ -(x - 2) \\ \hline -4 \end{array}$$

$Q(x) = x^2 - 2x + 1$ and $R(x) = -4$.

$$\begin{array}{lcccc} \text{So} & P(x) & = \ D(x) & \cdot \ \ Q(x) & + R(x) \\ & x^3 - 4x^2 + 5x - 6 & = (x - 2)(x^2 - 2x + 1) - & & 4 \end{array}$$

PROBLEM 14-7 Find the quotient and remainder when $3x^4 - x^3 + 7x - 4$ is divided by $x^2 - x - 1$.

Solution

$$3x^2 + 2x + 5$$
$$x^2 - x - 1 \overline{\smash{\big)}\, 3x^4 - x^3 + 0x^2 + 7x - 4}$$
$$\underline{-(3x^4 - 3x^3 - 3x^2)}$$
$$2x^3 + 3x^2 + 7x$$
$$\underline{-(2x^3 - 2x^2 - 2x)}$$
$$5x^2 + 9x - 4$$
$$\underline{-(5x^2 - 5x - 5)}$$
$$14x + 1$$

Thus

$$\begin{array}{ccccccc} P(x) & = & D(x) & \cdot & Q(x) & + & R(x) \end{array}$$

$$3x^4 - x^3 + 7x - 4 = (x^2 - x - 1)(3x^2 + 2x + 5) + (14x + 1)$$

PROBLEM 14-8   Use synthetic division to divide $4x^3 - 7x^2 + 5x - 3$ by $x - 2$.

Solution   Use the five steps of the synthetic division process to divide $P(x) = 4x^3 - 7x^2 + 5x - 3$ by $x - c$, where $c = 2$.

(1) Use 2 as the divisor.

$$\underline{2|}$$

(2) Write the coefficients of the dividend horizontally.

$$\underline{2|} \quad 4 \ -7 \ \ 5 \ -3$$

(3) Skipping a space below the dividend, draw a horizontal line and bring down the first coefficient below the line.

$$\underline{2|} \quad 4 \ -7 \ \ 5 \ -3$$
$$\overline{}$$
$$4$$

(4) Multiply the divisor by the first coefficient and install the result on the middle line below the second coefficient. Add in the second column.

$$\underline{2|} \quad 4 \ -7 \ \ 5 \ -3$$
$$8$$
$$2 \times 4 \ \ 4 \quad 1$$

Repeat the multiplication and addition process until you run out of dividend.

$$\underline{2|} \quad 4 \ -7 \ \ 5 \ -3$$
$$8 \ \ 2$$
$$2 \times 1 \ \ 4 \ \ 1 \ \ 7$$

(5) Box out the remainder and identify the quotient.

$$\underline{2|} \quad 4 \ -7 \ \ 5 \ -3$$
$$8 \ \ 2 \ \ 14$$
$$2 \times 7 \ \ 4 \ \ 1 \ \ 7 \ \boxed{11}$$

Thus

$$P(x) = D(x) \cdot Q(x) + R(x)$$

$$4x^3 - 7x^2 + 5x - 3 = (x - 2)(4x^2 + x + 7) + 11$$

PROBLEM 14-9   Use synthetic division to divide

(a) $\dfrac{4x^3 - 7x^2 + 2x + 4}{x - 3}$ (b) $\dfrac{x^4 + 5x^3 - x^2 - 20x - 12}{x + 2}$ (c) $\dfrac{2x^5 + 3x^4 - 6x^3 - 15x - 4}{x + 3}$

Solution

(a)

$$\underline{3|} \quad 4 \ -7 \ \ 2 \ \ \ 4$$
$$12 \ \ 15 \ \ 51$$
$$\overline{\quad 4 \ \ \ 5 \ \ 17 \ \ \boxed{55}}$$

Thus $4x^3 - 7x^2 + 2x + 4 = (x - 3)(4x^2 + 5x + 17) + 55$

(b)

$$\underline{-2|} \quad 1 \ \ \ 5 \ -1 \ -20 \ -12$$
$$-2 \ -6 \ \ 14 \ \ \ 12$$
$$\overline{\quad 1 \ \ \ 3 \ -7 \ \ -6 \quad \boxed{0}}$$

Since the remainder is zero, the division is exact and

$$x^4 + 5x^3 - x^2 - 20x - 12 = (x + 2)(x^3 + 3x^2 - 7x - 6)$$

(c)

$$\begin{array}{r|rrrrrr} -3 & 2 & 3 & -6 & 0 & -15 & -4 \\ & & -6 & 9 & -9 & 27 & -36 \\ \hline & 2 & -3 & 3 & -9 & 12 & \boxed{-40} \end{array}$$

Hence
$$\frac{2x^5 + 3x^4 - 6x^3 - 15x - 4}{x + 3} = 2x^4 - 3x^3 + 3x^2 - 9x + 12 - \frac{40}{x + 3}$$

or
$$2x^5 + 3x^4 - 6x^3 - 15x - 4 = (x + 3)(2x^4 - 3x^3 + 3x^2 - 9x + 12) - 40$$

PROBLEM 14-10 Use the remainder theorem to find the remainder when $2x^6 + 3x^2 - 4$ is divided by $x - 1$. Verify your answer.

Solution The remainder theorem states that if $P(x)$ is divided by $x - c$, then the remainder is $P(c)$. In this problem, $P(x) = 2x^6 + 3x^2 - 4$ and $c = 1$, so the remainder is

$$P(1) = 2 \cdot 1 + 3 \cdot 1 - 4$$
$$= 5 - 4$$
$$= 1$$

You can verify this by synthetic division:

$$\begin{array}{r|rrrrrrr} 1 & 2 & 0 & 0 & 0 & 3 & 0 & -4 \\ & & 2 & 2 & 2 & 2 & 5 & 5 \\ \hline & 2 & 2 & 2 & 2 & 5 & 5 & \boxed{1} \end{array} \longleftarrow \text{remainder is 1}$$

PROBLEM 14-11 Divide the general quadratic $P(x) = ax^2 + bx + c$ by the linear divisor $x - r$. Use this result to verify the remainder theorem.

Solution Use synthetic division:

$$\begin{array}{r|ccc} r & a & b & c \\ & & ar & ar^2 + br \\ \hline & a & (ar + b) & \boxed{ar^2 + br + c} \end{array} \qquad \begin{aligned} P(r) &= a(r)^2 + b(r) + c \\ &= \underbrace{ar^2 + br + c} \end{aligned}$$

$$\underbrace{\qquad\qquad}_{\text{Same}}$$

PROBLEM 14-12 Divide $P(x) = x^3 + 3ix^2 + (1 - i)x + 2$ by $D(x) = x + 2i$, and then validate the remainder theorem.

Solution Even with complex coefficients you can use synthetic division:

$$\begin{array}{r|cccc} -2i & 1 & 3i & (1 - i) & 2 \\ & & -2i & -2i^2 & (-6i + 2i^2) \\ \hline & 1 & i & (3 - i) & \boxed{-6i} \end{array}$$

Thus
$$x^3 + 3ix^2 + (1 - i)x + 2 = (x + 2i)(x^2 + ix + (3 - i)) - 6i$$

To validate the remainder theorem you must find $P(-2i)$:

$$P(-2i) = (-2i)^3 + 3i(-2i)^2 + (1 - i)(-2i) + 2$$
$$= -8i^3 + 3i(4i^2) + (-2i + 2i^2) + 2$$
$$= 8i - 12i - 2i - 2 + 2$$
$$= -6i$$

Since $P(c) = P(-2i) = -6i = R$, the theorem is validated.

The Fundamental Theorem of Algebra

PROBLEM 14-13 Solve the equation $x^4 + 6x^3 - 17x^2 - 78x - 56 = 0$, given that $x = 4$ and $x = -2$ are roots.

Solution Since $x = 4$ is a root, then $x - 4$ is a factor. Factor it out by synthetic division.

$$
\begin{array}{r|rrrrr}
4 & 1 & 6 & -17 & -78 & -56 \\
  &   & 4 & 40 & 92 & 56 \\
\hline
  & 1 & 10 & 23 & 14 & \boxed{0}
\end{array}
$$

Since $x = -2$ is a root, then $x + 2$ is a factor. Pull it out of the remaining quotient polynomial:

$$
\begin{array}{r|rrrr}
-2 & 1 & 10 & 23 & 14 \\
   &   & -2 & -16 & -14 \\
\hline
   & 1 & 8 & 7 & \boxed{0}
\end{array}
$$

Now you have

$$
\begin{aligned}
x^4 + 6x^3 - 17x^2 - 78x - 56 &= (x - 4)(x + 2)(x^2 + 8x + 7) \\
&= (x - 4)(x + 2)(x + 7)(x + 1)
\end{aligned}
$$

So the roots are $x = 4, -2, -7$, and -1.

PROBLEM 14-14 Write the polynomial $2x^2 - 6x - 80$ in the form of the complete factorization theorem.

Solution The polynomial has a leading coefficient $a_n = 2$: Factor it out first, and then factor the rest.

$$
\begin{aligned}
2x^2 - 6x - 80 &= 2(x^2 - 3x - 40) \\
&= 2(x - 8)(x + 5)
\end{aligned}
$$

Thus
$$
P(x) = a_n(x - r_1)(x - r_2)
$$

where $r_1 = 8$ and $r_2 = -5$.

PROBLEM 14-15 Show that the polynomial $x^3 + 5x^2 - 2x - 24$ has three distinct roots.

Solution Try for factors by repeated synthetic division:

$$
\begin{array}{r|rrrr}
1 & 1 & 5 & -2 & -24 \\
  &   & 1 & 6 & 4 \\
\hline
  & 1 & 6 & 4 & \boxed{-20}
\end{array}
\longleftarrow \text{[No, } (x - 1) \text{ is not a factor.]}
$$

$$
\begin{array}{r|rrrr}
2 & 1 & 5 & -2 & -24 \\
  &   & 2 & 14 & 24 \\
\hline
  & 1 & 7 & 12 & \boxed{0}
\end{array}
\longleftarrow \text{[Yes, } (x - 2) \text{ is a factor.]}
$$

Thus
$$
x^3 + 5x^2 - 2x - 24 = (x - 2)(x^2 + 7x + 12)
$$

The further factorization is quite easy—do it by inspection:

$$
= (x - 2)(x + 4)(x + 3)
$$

So the distinct roots are $r_1 = 2, r_2 = -4$, and $r_3 = -3$.

PROBLEM 14-16 Form a polynomial that has roots $2, -1$, and 3.

Solution Since 2 is a root, $(x - 2)$ is a factor.
Since -1 is a root, $(x + 1)$ is a factor.
Since 3 is a root, $(x - 3)$ is a factor.

The factors then are $(x - 2)(x + 1)(x - 3)$ and the polynomial is

$$
x^3 - 4x^2 + x + 6
$$

PROBLEM 14-17 Check the result of Problem 14-16 using synthetic division.

Solution If the polynomial

$$P(x) = x^3 - 4x^2 + x + 6$$

has the roots, 2, -1, and 3, then the division of the polynomial and the successive quotient functions will each result in a zero remainder:

$$
\begin{array}{r|rrrr}
2\underline{|} & 1 & -4 & 1 & 6 \\
 & & 2 & -4 & -6 \\
\hline
-1\underline{|} & 1 & -2 & -3 & \boxed{0} \\
 & & & -1 & 3 \\
\hline
3\underline{|} & 1 & -3 & \boxed{0} \\
 & & & 3 \\
\hline
 & 1 & \boxed{0}
\end{array}
$$

PROBLEM 14-18 Form a polynomial with roots 2, 2, 2, and 3.

Solution Since 2 is a root of multiplicity 3 and 3 is a root of multiplicity 1, the polynomial is

$$P(x) = (x - 2)^3(x - 3) = x^4 - 9x^3 + 30x^2 - 44x + 24$$

PROBLEM 14-19 Find a quadratic function having zeros at $x = 3$ and $x = -5$ and whose graph passes through the point $(2, -21)$.

Solution If the zeros are at 3 and -5, then $(x - 3)$ and $(x + 5)$ are factors. Thus the general form of the quadratic function is

$$y = A(x - 3)(x + 5)$$

Since the function contains the point $(2, -21)$, these values can be substituted into this equation for x and y. Thus

$$-21 = A(2 - 3)(2 + 5)$$
$$-21 = A(-1)(7)$$
$$A = 3$$

The quadratic function is

$$y = 3(x - 3)(x + 5) = 3x^2 + 6x - 45$$

PROBLEM 14-20 Given $P(x) = x^4 + 5x^3 - x^2 - 20x - 12$, **(a)** determine if $x - 2$ is a factor of $P(x)$; **(b)** determine the number of factors $P(x)$ has; **(c)** find the other factors.

Solution

(a) If $x - 2$ is a factor, the remainder will be zero when you divide $P(x)$ by $x - 2$:

$$
\begin{array}{r|rrrrr}
2\underline{|} & 1 & 5 & -1 & -20 & -12 \\
 & & 2 & 14 & 26 & 12 \\
\hline
 & 1 & 7 & 13 & 6 & \boxed{0}
\end{array}
$$

Since $R = 0$ it is a factor; so

$$x^4 + 5x^3 - x^2 - 20x - 12 = (x - 2)(x^3 + 7x^2 + 13x + 6)$$

(b) Since the polynomial is of degree four, $P(x) = 0$ has four roots—by the complete factorization theorem—so $P(x)$ has four factors.

(c) The other potential rational roots must be divisors of 6. That includes $\pm 1, \pm 2, \pm 3, \pm 6$. By a trial of some of these you finally arrive at -2:

$$
\begin{array}{r|rrrr}
-2\underline{|} & 1 & 7 & 13 & 6 \\
 & & -2 & -10 & -6 \\
\hline
 & 1 & 5 & 3 & \boxed{0}
\end{array}
$$

So $x + 2$ is also a factor:

$$x^4 + 5x^3 - x^2 - 20x - 12 = (x - 2)(x + 2)(x^2 + 5x + 3)$$

The remaining quadratic can be factored using the quadratic formula:

$$x = \frac{-5 \pm \sqrt{5^2 - (4 \cdot 1 \cdot 3)}}{2} = \frac{-5 \pm \sqrt{13}}{2} = \frac{-5 + \sqrt{13}}{2} \quad \text{and} \quad \frac{-5 - \sqrt{13}}{2}$$

So $P(x)$ can be written as

$$x^4 + 5x^3 - x^2 - 20x - 12 = (x - 2)(x + 2)\left(x - \left(\frac{-5 + \sqrt{13}}{2}\right)\right)\left(x - \left(\frac{-5 - \sqrt{13}}{2}\right)\right)$$

Factoring Polynomials with Real Coefficients

PROBLEM 14-21 If it is known that $x = -i$ is a root of $P(x) = 2x^4 - 7x^3 - 2x^2 - 7x - 4$, find all of the roots.

Solution Since $P(x)$ has $-i$ as a root, then $+i$ is also a root by the conjugate pair theorem. That's two of the four roots. Now use synthetic division to pull these factors out so you can work on the remaining quotient:

$$
\begin{array}{r|rrrrr}
-i & 2 & -7 & -2 & -7 & -4 \\
& & -2i & (7i - 2i^2) & (-7i^2 + 4i) & -4i \\
\hline
i & 2 & (-7 - 2i) & (7i - 4) & 4i & \boxed{0} \Rightarrow [x = -i \text{ is a root.}] \\
& & 2i & -7i & & -4i \\
\hline
4 & 2 & -7 & -4 & \boxed{0} & \Rightarrow [x = i \text{ is a root.}] \\
& & 8 & 4 & & \\
\hline
-\frac{1}{2} & 2 & 1 & \boxed{0} & & \Rightarrow [x = 4 \text{ is a root.}] \\
& & -1 & & & \\
\hline
& 2 & \boxed{0} & & & \Rightarrow [x = -\frac{1}{2} \text{ is a root.}]
\end{array}
$$

PROBLEM 14-22 Form a polynomial with real coefficients that has $x = 1 - 2i$ and $x = -1 + i$ as roots.

Solution If a complex number is a root of a polynomial with real coefficients, then its complex conjugate is also a root. Thus $x = 1 + 2i$ and $x = -1 - i$ are also roots. The factors then are

$$
\begin{aligned}
P(x) &= \left(x - (1 - 2i)\right)\left(x - (1 + 2i)\right)\left(x - (-1 + i)\right)\left(x - (-1 - i)\right) \\
&= (x - 1 + 2i)(x - 1 - 2i)(x + 1 - i)(x + 1 + i) \\
&= (x^2 - 2x + 5)(x^2 + 2x + 2) \\
&= x^4 + 3x^2 + 6x + 10
\end{aligned}
$$

note: This is a polynomial with real coefficients for which all of the roots are complex numbers.

PROBLEM 14-23 Find the real factors of $x^3 + x^2 + 5x + 5$.

Solution This function will have either three linear factors or one linear factor and a quadratic factor not reducible to further real linear factors. Try for one linear factor first:

$$
\begin{array}{r|rrrr}
1 & 1 & 1 & 5 & 5 \\
& & 1 & 2 & 7 \\
\hline
& 1 & 2 & 7 & \boxed{12} \Rightarrow [(x - 1) \text{ is not a factor.}]
\end{array}
$$

$$
\begin{array}{r|rrrr}
-1 & 1 & 1 & 5 & 5 \\
& & -1 & 0 & -5 \\
\hline
& 1 & 0 & 5 & \boxed{0} \Rightarrow [(x + 1) \text{ is a factor.}]
\end{array}
$$

The resulting quotient is $x^2 + 5$, which is not reducible to real linear factors. Thus the factorization is

$$x^3 + x^2 + 5x + 5 = (x + 1)(x^2 + 5)$$

PROBLEM 14-24 If the polynomial $P(x) = x^4 + 5x^3 + 5x^2 - 5x - 6$ is to have integer roots, what are the available candidates?

Solution The trailing (last) coefficient is -6. If the polynomial has integer roots, they must be divisors of -6. That allows ± 1, ± 2, ± 3, ± 6.

PROBLEM 14-25 Find the roots of $P(x) = x^4 + 5x^3 + 5x^2 - 5x - 6$ from the previous problem.

Solution There are four roots: Use synthetic division to try for some:

$$
\begin{array}{r|rrrrr}
\underline{1} & 1 & 5 & 5 & -5 & -6 \\
 & & 1 & 6 & 11 & 6 \\
\hline
\underline{2} & 1 & 6 & 11 & 6 & \boxed{0} \Rightarrow [1 \text{ is a root.}] \\
 & & 2 & 16 & 54 & \\
\hline
 & 1 & 8 & 27 & \boxed{60} &
\end{array}
$$

Now, you see that 2 is not a root and neither is any number greater than 2: Since the coefficients in the new dividend polynomial are all positive, no other positive number greater than 2 will produce a zero remainder when put in for x. So try $r_2 = -2$:

$$
\begin{array}{r|rrrr}
\underline{-2} & 1 & 6 & 11 & 6 \\
 & & -2 & -8 & -6 \\
\hline
 & 1 & 4 & 3 & \boxed{0} \Rightarrow [-2 \text{ is a root.}]
\end{array}
$$

The quotient is $x^2 + 4x + 3$, which can be factored:

$$x^2 + 4x + 3 = (x + 1)(x + 3)$$

Thus the factorization is

$$P(x) = x^4 + 5x^3 + 5x^2 - 5x - 6 = (x - 1)(x + 2)(x + 1)(x + 3)$$

And the roots are $1, -2, -1, -3$.

PROBLEM 14-26 List all the possible rational roots of $P(x) = 3x^4 - 5x^2 + 2x + 2$.

Solution By the rational root theorem, the rational roots are of the form p/q where p divides 2 and q divides 3. So the candidates are ± 1, ± 2 for p and ± 1, ± 3 for q. The collection, then, is $\dfrac{\pm 1}{\pm 1}, \dfrac{\pm 1}{\pm 3}, \dfrac{\pm 2}{\pm 1}, \dfrac{\pm 2}{\pm 3}$. Removing the repetitions gives eight possible rational roots:

$$\pm 1, \quad \pm 2, \quad \pm \tfrac{1}{3}, \quad \pm \tfrac{2}{3}$$

PROBLEM 14-27 What are the possible rational roots of $P(x) = 5x^4 - 6x + 4$?

Solution The candidates for p are ± 1, ± 2, ± 4; the candidates for q are ± 1, ± 5. So the whole list is $\dfrac{\pm 1}{\pm 1}, \dfrac{\pm 1}{\pm 5}, \dfrac{\pm 2}{\pm 1}, \dfrac{\pm 2}{\pm 5}, \dfrac{\pm 4}{\pm 1}, \dfrac{\pm 4}{\pm 5}$. Removing repetitions, you have twelve possible rational roots:

$$\pm 1, \pm 2, \pm 4, \pm \frac{1}{5}, \pm \frac{2}{5}, \pm \frac{4}{5}$$

PROBLEM 14-28 List the possible rational roots of $P(x) = 3x^3 + x^2 + 21x + 7$, and then find them.

Solution The rational roots have numerators ± 1, ± 7 and denominators ± 1, ± 3. The total list is

$$\frac{\pm 1}{\pm 1}, \frac{\pm 1}{\pm 3}, \frac{\pm 7}{\pm 1}, \frac{\pm 7}{\pm 3} \Rightarrow \pm 1, \pm 7, \pm \frac{1}{3}, \pm \frac{7}{3}$$

Search this collection one at a time. But you know that if a number r is a root, then $P(r) = 0$. And by inspecting the polynomial, which has all positive coefficients, you can see that no positive number can be put in for x to make the polynomial value zero. So, eliminate all the positive candidates!

Now try some negative roots:

$$
\begin{array}{r|rrrr}
-1 & 3 & 1 & 21 & 7 \\
   &   & -3 & 2 & -23 \\
\hline
   & 3 & -2 & 23 & \boxed{-16}
\end{array} \Rightarrow [-1 \text{ is not a root.}]
$$

$$
\begin{array}{r|rrrr}
-\frac{1}{3} & 3 & 1 & 21 & 7 \\
   &   & -1 & 0 & -7 \\
\hline
   & 3 & 0 & 21 & \boxed{0}
\end{array} \Rightarrow [-\tfrac{1}{3} \text{ is a root, and the remaining quotient is } 3x^2 + 21.]
$$

Thus
$$
\begin{aligned}
3x^3 + x^2 + 21x + 7 &= (x + \tfrac{1}{3})(3x^2 + 21) \\
&= (x + \tfrac{1}{3})(3)(x^2 + 7)
\end{aligned}
$$

This is not further reducible over real factors.

PROBLEM 14-29   Count the variations in sign in each of the following polynomials:

(a) $P_1 = 5x^4 + 2x^3 + 3x^2 - x - 2$
(b) $P_2 = x^5 - 2x^3 + 5x + 5$
(c) $P_3 = -x^4 - 2x^3 + 3x^2 - x + 8$
(d) $P_4 = 8x^5 - 21x^4 - 12x^3 + 6x^2 - 2x + 8$

Solution

(a) P_1 has one variation in sign $(+ \text{ to } -)$.
(b) P_2 has two variations in sign $(+ \text{ to } - \text{ to } +)$.
(c) P_3 has three variations in sign $(- \text{ to } + \text{ to } - \text{ to } +)$.
(d) P_4 has four variations in sign $(+ \text{ to } - \text{ to } + \text{ to } - \text{ to } +)$.

PROBLEM 14-30   How many possible positive roots could each of the polynomials given in Problem 14-34 have?

Solution   By Descartes' rule of signs, the number of positive real roots equals the number of sign changes or is less than that by an even positive integer. Thus

(a) $5x^4 + 2x^3 + 3x^2 - x - 2$ has exactly one real positive root.
(b) $x^5 - 2x^3 + 5x + 5$ has either two positive real roots or no positive real roots.
(c) $-x^4 - 2x^3 + 3x^2 - x + 8$ has either three positive real roots or one positive real root.
(d) $8x^5 - 21x^4 - 12x^3 + 6x^2 - 2x + 8$ has either four or two or zero real positive roots.

PROBLEM 14-31   How many positive real roots could each of the following polynomials have?
$$
\text{(a) } P_1 = x^3 + 3x^2 + 7x + 1 \qquad \text{(b) } P_2 = -2x^4 - 3x^2 + 2x - 1
$$

Solution

(a) P_1 has no variations in sign $\Rightarrow$ No positive real roots.
(b) P_2 has two variations in sign $\Rightarrow$ Two or zero positive real roots.

PROBLEM 14-32   How many negative real roots are possible for each of the polynomials in Problem 14-31?

Solution   Apply Descartes' rule of signs to $P(-x)$:

(a)
$$
P_1(-x) = (-x)^3 + 3(-x)^2 + 7(-x) + 1
$$
$$
= -x^3 + 3x^2 - 7x + 1 \Rightarrow \text{Three sign changes}
$$
$$
\text{(three or one negative real roots)}
$$

(b)
$$
P_2(-x) = -2(-x)^4 - 3(-x)^2 + 2(-x) - 1
$$
$$
= -2x^4 - 3x^2 - 2x - 1 \Rightarrow \text{No sign changes}
$$
$$
\text{(no negative real roots)}
$$

PROBLEM 14-33   Show that the polynomial $P(x) = 2x^3 - 5x^2 + 6x - 15$ has a root between $x = 2$ and $x = 3$.

Solution   To show that $P(x)$ has a root between 2 and 3, you have to use the intermediate value theorem: You must find $P(2)$ and $P(3)$ and show that they have opposite signs. You can do this by direct substitution or by using synthetic division and the remainder theorem: Pick one—

$$P(2) = 2(2^3) - 5(2^2) + 6(2) - 15$$
$$= 2 \cdot 8 - 5 \cdot 4 + 12 - 15$$
$$= 16 - 20 + 12 - 15$$
$$= 28 - 35$$
$$= -7$$

$$\begin{array}{r|rrrr} 2 & 2 & -5 & 6 & -15 \\ & & 4 & -2 & 8 \\ \hline & 2 & -1 & 4 & \boxed{-7} \end{array}$$

— same —

$$P(3) = 2(3^3) - 5(3^2) + 6(3) - 15$$
$$= 2 \cdot 27 - 5 \cdot 9 + 18 - 15$$
$$= 54 - 45 + 18 - 15$$
$$= 72 - 60$$
$$= 12$$

$$\begin{array}{r|rrrr} 3 & 2 & -5 & 6 & -15 \\ & & 6 & 3 & 27 \\ \hline & 2 & 1 & 9 & \boxed{12} \end{array}$$

— same —

$P(2) = -7$ and $P(3) = 12$, which have opposite signs, so there is a root between $x = 2$ and $x = 3$.

PROBLEM 14-34   Isolate the intervals in which the roots of $P(x) = 8x^3 - 4x^2 - 18x + 9$ lie.

Solution   First, use Descartes' rule of signs to find out what kind of roots this thing has:

$$P(x) = 8x^3 - 4x^2 - 18x + 9 \qquad \text{[Two sign changes]}$$
$$P(-x) = -8x^3 - 4x^2 + 18x + 9 \qquad \text{[One sign change]}$$

We conclude that $P(x)$ has either two or zero positive roots and exactly one negative root.

Next, begin with a value check at $x = -1$ and move to the left until you locate the negative root:

$$\begin{array}{r|rrrr} -1 & 8 & -4 & -18 & 9 \\ & & -8 & 12 & 6 \\ \hline & 8 & -12 & -6 & \boxed{15} \end{array}$$

$$\begin{array}{r|rrrr} -2 & 8 & -4 & -18 & 9 \\ & & -16 & 40 & -44 \\ \hline & 8 & -20 & 22 & \boxed{-35} \end{array}$$

$P(-1) = 15$, $P(-2) = -35$. So the negative root is between -1 and -2.

Now, go after the potential positive roots. It's easy to see that $P(0) = 9$.

$$\begin{array}{r|rrrr} 1 & 8 & -4 & -18 & 9 \\ & & 8 & 4 & -14 \\ \hline & 8 & 4 & -14 & \boxed{-5} \end{array}$$

$P(0) = 9$
$P(1) = -5$

There is a root between 0 and 1.

$$\begin{array}{r|rrrr} 2 & 8 & -4 & -18 & 9 \\ & & 16 & 24 & 12 \\ \hline & 8 & 12 & 6 & \boxed{21} \end{array}$$

$P(1) = -5$
$P(2) = 21$

There is a root between 1 and 2.

PROBLEM 14-35   Find the roots of the polynomial of Problem 14-34.

Solution   Since $8x^3 - 4x^2 - 18x + 9$ is a cubic polynomial, you only have to find one root to be able to divide it out and solve the remaining quadratic. Since you've already eliminated the possibility of an integer root (in Problem 14-34), you can begin by looking at the candidates for rational roots: Since p divides 9 and q divides 8, you're left with

$$\pm\frac{1}{2}, \pm\frac{1}{4}, \pm\frac{1}{8}, \pm\frac{3}{2}, \pm\frac{3}{4}, \pm\frac{3}{8}, \pm\frac{9}{2}, \pm\frac{9}{4}, \pm\frac{9}{8}$$

Now, look for the positive roots:

There is a root between 1 and 2 (which you know from Problem 14-34). The only candidates in the list between 1 and 2 are $x = \frac{3}{2}$ and $x = \frac{9}{8}$. Check these out—

$$\frac{3}{2} \bigg| \begin{array}{rrrr} 8 & -4 & -18 & 9 \\ & 12 & 12 & -9 \\ \hline 8 & 8 & -6 & \boxed{0} \end{array}$$ [*Eureka!* One positive root is $x = \frac{3}{2}$.]

note: You can also choose to work with the positive root between 0 and 1—but then you'd have to try five candidates $(\frac{1}{2}, \frac{1}{4}, \frac{1}{8}, \frac{3}{4}, \frac{3}{8})$ instead of just two, which is more work.

Now you've got a foot in the door: The remaining quotient is $8x^2 + 8x - 6$, which could be solved by using the quadratic formula. But it's a good idea to practice new techniques, so start looking for the negative root:

There is a negative root between -1 and -2. Only two candidates are available, $x = -\frac{3}{2}$ and $x = -\frac{9}{8}$. Try one of these—

$$-\frac{3}{2} \bigg| \begin{array}{rrr} 8 & 8 & -6 \\ & -12 & 6 \\ \hline 8 & -4 & \boxed{0} \end{array}$$ [*Eureka* again! The negative root is $x = -\frac{3}{2}$.]

The remaining quotient is $8x - 4$, so $8x - 4 = 8(x - \frac{1}{2})$ [The other positive root is $x = \frac{1}{2}$.]

PROBLEM 14-36 Given the polynomial $P(x) = x^4 + x^3 - x^2 - 2x - 2$, find the positive roots.

Solution Descartes' Rule of Signs indicates exactly one positive real root. The rational root theorem shows that the only positive candidates are 1 and 2. Try them—

$$P(1) = 1^4 + 1^3 - 1^2 - 2 \cdot 1 - 2 \qquad P(2) = 2^4 + 2^3 - 2^2 - 2 \cdot 2 - 2$$
$$= 1 + 1 - 1 - 2 - 2 = -3 \qquad = 16 + 8 - 4 - 4 - 2 = 14$$

The remainders in both cases are nonzero—so the positive root of $P(x)$ is irrational—and there is a sign change—so the positive root is between 1 and 2. Now you have to fall back on successive approximations. The bisection method directs you to find the midpoint of this interval and then keep on halving while watching for sign variation. Begin with $x = \frac{3}{2}$:

midpoint

$$\frac{1}{2}(1 + 2) \left. \begin{array}{l} P(1) = -3 \\ P(2) = 14 \end{array} \right\} \quad P(\tfrac{3}{2}) - (\tfrac{3}{2})^4 + (\tfrac{3}{2})^3 - (\tfrac{3}{2})^2 - 2(\tfrac{3}{2}) - 2 = \tfrac{19}{16}$$

$$\frac{1}{2}(1 + \tfrac{3}{2}) \left. \begin{array}{l} P(1) = -3 \\ P(\tfrac{3}{2}) = \tfrac{19}{16} \end{array} \right\} \quad P(\tfrac{5}{4}) = (\tfrac{5}{4})^4 + (\tfrac{5}{4})^3 - (\tfrac{5}{4})^2 - 2(\tfrac{5}{4}) - 2 = -\tfrac{427}{256}$$

[$P(\tfrac{5}{4}) = -\tfrac{427}{256}$ is negative, so the next interval is
between $\tfrac{5}{4}$ and $\tfrac{3}{2}$.]

$$\frac{1}{2}(\tfrac{5}{4} + \tfrac{3}{2}) \left. \begin{array}{l} P(\tfrac{5}{4}) = -\tfrac{427}{256} \\ P(\tfrac{3}{2}) = \tfrac{19}{16} \end{array} \right\} P(\tfrac{11}{8}) = (\tfrac{11}{8})^4 + (\tfrac{11}{8})^3 - (\tfrac{11}{8})^2 - 2(\tfrac{11}{8}) - 2 = -\tfrac{1911}{4096}$$

[$P(\tfrac{11}{8}) = -\tfrac{1911}{4096}$ is negative, so the next interval is
between $\tfrac{11}{8}$ and $\tfrac{3}{2}$.]

So the value is between $x = \frac{11}{18} = 1.375$ and $x = \frac{3}{2} = 1.500$.

   The computations are getting pretty messy now, and it's easy to lose track. So draw one straight-line approximation, as in Figure 14-3. The crossing point is the "best" estimate of the root. If you measure the distance, you find the crossing point to be $\sim.0228$ units beyond $x = \frac{11}{8} = 1.375$, so the root is approximately $.0228 + 1.375 = 1.3978$.

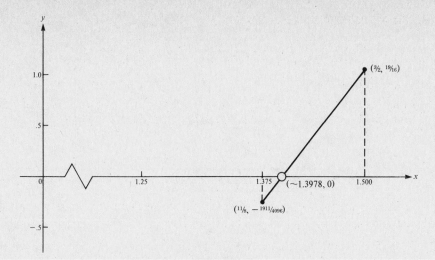

Figure 14-3

PROBLEM 14-37 Use the method of linear interpolation to find the one positive root of $P(x) = x^3 - 2x - 5 = 0$.

Solution First, find the integers between which the root lies:

$$P(0) = -5, \quad P(1) = 1 - 2 - 5 = -6, \quad P(2) = 8 - 4 - 5 = -1, \quad P(3) = 27 - 6 - 5 = 16$$

There is a sign change between $P(2) = -1$ and $P(3) = 16$, so the root is between 2 and 3. Now check $\frac{1}{2}(2 + 3) = \frac{5}{2}$:

$$P(\tfrac{5}{2}) = (\tfrac{5}{2})^3 - 2(\tfrac{5}{2}) - 5 = \tfrac{45}{8}$$

So a straight-line approximation, as in Figure 14-4a, shows the root as the point where the line connecting the points $(\frac{5}{2}, \frac{45}{8}), (2, -1)$ crosses the x-axis. Now use similar triangles to find that point:

$$\frac{x}{1} = \frac{\frac{1}{2} - x}{\frac{45}{8}} \Rightarrow \frac{45}{8}x = \frac{1}{2} - x$$

$$\frac{53}{8}x = \frac{1}{2}$$

$$x = \frac{1}{2} \cdot \frac{8}{53} = \frac{4}{53} = .07547$$

And if side x is .07547 units beyond 2, the first estimate is that the crossing point is 2.07547.

Now start a second estimate. Using a calculator, check the actual value of the function:

$$P(2.07547) = (2.07547)^3 - 2(2.07547) - 5$$

$$= 8.94024 - 4.15094 - 5 = -.2107$$

Since the function is negative there, move to the right to a convenient point, say 2.10. There

$$P(2.10) = (2.10)^3 - 2(2.10) - 5$$

$$= 9.261 - 4.20 - 5 = .061$$

Do one more straight-line approximation, as in Figure 14-4b, which is more detailed than Figure 14-4a. Then, by similar triangles, you have

$$\frac{x}{.2107} = \frac{.02453 - x}{.061}$$

$$.061x = .005168 - .2107x$$

$$x = .019022$$

Add .019022 to the previous estimate to get the new estimate, 2.09449.

(a)

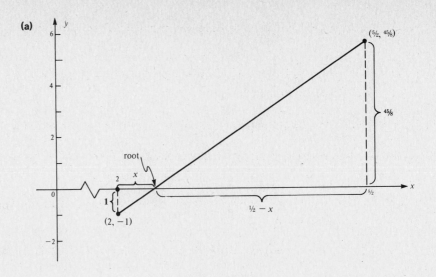

(b)

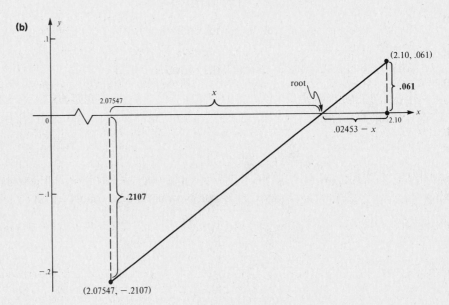

Figure 14-4

Supplementary Exercises

PROBLEM 14-38 Show that $x = 1$ is a root of $x^3 - 2x + 1 = 0$.

PROBLEM 14-39 Show that $x = -\frac{3}{2}$ is a root of $2x^3 + x^2 - 5x - 3 = 0$.

PROBLEM 14-40 Show that both $x = 2$ and $x = -3$ are roots of $2x^4 + 11x^3 + 7x^2 - 44x - 60 = 0$.

PROBLEM 14-41 Show that the division of $3x^3 - 5x^2 - 16x + 12$ by $x + 2$ is exact.

PROBLEM 14-42 Show that the division of $2x^3 + 7x^2 + 2x - 3$ by $2x - 1$ is exact. Then show the complete factorization.

PROBLEM 14-43 When $4x^3 - ix^2 + 3x - 5i$ is divided by $x - 4i$, **(a)** what is the remainder? **(b)** Check your answer by the remainder theorem.

PROBLEM 14-44 Illustrate the simplified division algorithm for $P(x) = 7x^3 - 5x^2 + x - 8$ and $D(x) = x - 2$.

PROBLEM 14-45 If the polynomial $x^5 + 4x^4 - 4x^3 - 34x^2 - 45x - 18$ has factors $(x - 3)$, $(x + 1)$, and $(x + 2)$, what are the other factors?

PROBLEM 14-46 Where does the graph of $y = -x^4 - 6x^3 - 8x^2 + 6x + 9$ cross the x-axis?

PROBLEM 14-47 Use the remainder theorem to find the value of $f(x) = 2x^6 - 4x^4 - 5x + 1$ at $x = 2$.

PROBLEM 14-48 Use the remainder theorem with synthetic division to find the value of $f(x) = x^7 - 8x^5 - x^2 + x - 8$ at $x = 3$.

PROBLEM 14-49 What is the remainder when $P(x) = ix^4 - 3x^3 + 3ix - 2$ is divided by $D(x) = x - 2i$?

PROBLEM 14-50 Find a value of k so that $x + 1$ will be a factor of $x^3 + 3x^2 + kx + 3$.

PROBLEM 14-51 Find a value of k so that $x - 2$ will be a factor of $x^3 - 4x^2 + kx + 2$.

PROBLEM 14-52 Solve the equation $x^4 + x^3 - 15x^2 + 23x - 10 = 0$, given that $x - 2$ and $x + 5$ are factors of the polynomial.

PROBLEM 14-53 Does $3x^4 + 13x^2 + 4$ have a factor of the form $x - c$ where c is a real number?

PROBLEM 14-54 Write $4x^3 - 6x^2 + 8x - 3$ in the form of the complete factorization theorem.

PROBLEM 14-55 Find a polynomial that has zeros at $x = -2$, $x = 3$, $x = 5$ and has value 3 at $x = 1$.

PROBLEM 14-56 For the function $f(x) = (4x^2 - 9)^3(5x + 3)^2$ list the zeros and their multiplicity.

PROBLEM 14-57 Find a polynomial with real coefficients with zeros $2 - i$ and $1 + 2i$.

PROBLEM 14-58 Find the real factors of $3x^4 + 11x^2 - 4$.

PROBLEM 14-59 Does $2x^4 - 3x^3 + 6x^2 - 24x + 5 = 0$ have rational roots?

PROBLEM 14-60 Find the rational roots of $9x^3 - 3x^2 - 8x + 4 = 0$.

PROBLEM 14-61 List all the possible rational roots of $3x^4 - x^3 + 5x^2 - 4 = 0$.

PROBLEM 14-62 Use synthetic division to show that $x^2 - 2x - 3$ is a factor of $x^4 - 2x^3 + x^2 - 8x - 12$.

PROBLEM 14-63 By inspection determine the remainder when $2(x + 3)^3 - 4(x + 3)^2 + 5$ is divided by $x + 3$.

PROBLEM 14-64 Find all rational roots of $3x^3 - 8x^2 + 3x + 2 = 0$.

PROBLEM 14-65 How many positive and how many negative real roots can the polynomial $6x^4 - x^3 + 23x^2 - 4x - 4 = 0$ have? Find all the roots.

PROBLEM 14-66 Given that $x^6 - 4x^5 + 13x^4 - 16x^3 + 11x^2 + 20x - 25 = 0$ has $1 - 2i$ as a root of multiplicity 2, find all of the factors.

PROBLEM 14-67 Isolate the intervals in which the roots of $2x^3 - 6x - 15 = 0$ lie.

PROBLEM 14-68 Use the bisection method to solve $x^2 - 7 = 0$. [Hint: You will be approximating the value of $\sqrt{7}$.

PROBLEM 14-69 Given $f(x) = x^3 + 2x^2 - x - 4$, find the approximate value of the zero between $x = 1$ and $x = 2$.

Answers to Supplementary Exercises

14-38
$$1 \underline{|} \quad 1 \quad 0 \quad -2 \quad 1$$
$$\phantom{1 \underline{|} \quad} 1 \quad 1 \quad -1$$
$$\overline{\phantom{1\underline{|}}\; 1 \quad 1 \quad -1 \quad \boxed{0}}$$
or $P(1) = 1^3 - 2(1) + 1 = 0$

14-39
$$-\tfrac{3}{2} \underline{|} \quad 2 \quad 1 \quad -5 \quad -3$$
$$\phantom{-\tfrac{3}{2} \underline{|} \quad} -3 \quad 3 \quad 3$$
$$\overline{\phantom{-\tfrac{3}{2}\underline{|}}\; 2 \quad -2 \quad -2 \quad \boxed{0}}$$
or
$$P(-\tfrac{3}{2}) = 2(-\tfrac{3}{2})^3 + (-\tfrac{3}{2})^2 - 5(-\tfrac{3}{2}) - 3 = 0$$

14-40
$$2 \underline{|} \quad 2 \quad 11 \quad 7 \quad -44 \quad -60$$
$$\phantom{2 \underline{|} \quad} 4 \quad 30 \quad 74 \quad 60$$
$$-3 \underline{|} \quad \overline{2 \quad 15 \quad 37 \quad 30 \quad \boxed{0}} \quad P(2) = 0$$
$$\phantom{-3 \underline{|} \quad} -6 \quad -27 \quad -30$$
$$\overline{\phantom{-3\underline{|}}\; 2 \quad 9 \quad 10 \quad \boxed{0}} \quad P(-3) = 0$$

14-41
$$-2 \underline{|} \quad 3 \quad -5 \quad -16 \quad 12$$
$$\phantom{-2 \underline{|} \quad} -6 \quad 22 \quad -12$$
$$\overline{\phantom{-2\underline{|}}\; 3 \quad -11 \quad 6 \quad \boxed{0}}$$

14-42
$$\tfrac{1}{2} \underline{|} \quad 2 \quad 7 \quad 2 \quad -3$$
$$\phantom{\tfrac{1}{2} \underline{|} \quad} 1 \quad 4 \quad 3$$
$$\overline{\phantom{\tfrac{1}{2}\underline{|}}\; 2 \quad 8 \quad 6 \quad \boxed{0}} \quad P(\tfrac{1}{2}) = 0$$
$$2(x - \tfrac{1}{2})(x + 1)(x + 3)$$

14-43 (a) $-233i$

      (b) $4(4i)^3 - i(4i)^2 + 3(4i) - 5i \overset{?}{=} -233i$
$$4(64i^2)i - i(16i^2) + 12i - 5i \overset{?}{=} -233i$$
$$-256i + 16i + 7i \overset{\checkmark}{=} -233i$$

14-44
$$P(x) = D(x) \quad \cdot \quad Q(x) + R$$
$$7x^3 - 5x^2 + x - 8 = (x - 2)(7x^2 + 9x + 19) + 30$$

14-45 $(x + 1), (x + 3)$

14-46 At $x = 1$ and $x = -1$. Touchpoint at $x = -3$.

14-47 $f(2) = R = 55$

14-48 $f(3) = R = 229$ **14-49** $-8 + 40i$

14-50 $k = 5$ **14-51** $k = 3$

14-52 $x = 1, x = 1, x = 2, x = -5$

14-53 No

14-54 $(x - \tfrac{1}{2})(4x^2 - 4x + 6)$

14-55 $\tfrac{1}{8}(x + 2)(x - 3)(x - 5)$

14-56 $-\tfrac{3}{5}$ order 2; $\tfrac{3}{2}$ order 3; $-\tfrac{3}{2}$ order 3

14-57 $x^4 - 6x^3 + 18x^2 - 30x + 25$

14-58 $(x - 2)$ and $(x + 2)$

14-59 No   **14-60** $\tfrac{2}{3}$

14-61 $\pm 1, \pm 2, \pm 4, \pm \tfrac{1}{3}, \pm \tfrac{2}{3}, \pm \tfrac{4}{3}$

14-62 $x^2 - 2x - 3 = (x - 3)(x + 1)$
$$3 \underline{|} \quad 1 \quad -2 \quad 1 \quad -8 \quad -12$$
$$\phantom{3 \underline{|} \quad} 3 \quad 3 \quad 12 \quad 12$$
$$-1 \underline{|} \quad \overline{1 \quad 1 \quad 4 \quad 4 \quad \boxed{0}}$$
$$\phantom{-1 \underline{|} \quad} -1 \quad 0 \quad -4$$
$$\overline{\phantom{-1\underline{|}}\; 1 \quad 0 \quad 4 \quad \boxed{0}}$$
$$x^4 - 2x^3 + x^2 - 8x - 12$$
$$= (x - 3)(x + 1)(x^2 + 4)$$

14-63 5   **14-64** $1, 2, -\tfrac{1}{3}$

14-65 Three or one positive real roots
One negative real root
$\tfrac{1}{2}, -\tfrac{1}{3}, \pm 2i$

14-66 $\pm 1, (1 - 2i)^2, (1 + 2i)^2$

14-67 $(2, 3)$; two roots are complex

14-68 $\sqrt{7} \cong 2.646$ **14-69** 1.266

15 SEQUENCES, SERIES, AND INDUCTION

THIS CHAPTER IS ABOUT

☑ **Number Sequences**
☑ **Partial Sums of Sequences: Series**
☑ **Arithmetic Sequences and Series**
☑ **Geometric Sequences and Series**
☑ **Mathematical Induction**

15-1. Number Sequences

- A **sequence** is a collection of numbers, called *terms*, that are arranged in a specific order.

Sequences may be *finite*, where the last term is fixed, or *infinite*, where the last term is not fixed. Thus, the collection 5, 10, 15, 20 is a finite sequence of four terms, while $1, \frac{1}{2}, \frac{1}{4}, \frac{1}{8}, \ldots$ is an infinite sequence, where the dots indicate an infinite continuation of a given pattern of computation.

A. The sequence function

We can define a sequence of numbers by means of a function:

- A sequence is the ordered range of a **sequence function** whose domain is the set $\{1, 2, 3, \ldots\}$ of all positive integers.

Thus, for example, if $f(n) = n^2$ is a sequence function, its domain is the positive integers $n = 1, 2, 3, \ldots$, and its ordered range is the sequence 1, 4, 9, 16, 25, ... In general, the function associates with each positive integer n a definite number a_n, called the **general term**, so that $f(n) = a_n$ for $n \geq 1$.

EXAMPLE 15-1 Fill in the blanks for the next three terms of each sequence given below. Then find a function to represent the infinite sequence.

(a) $\frac{1}{2}, \frac{2}{3}, \frac{3}{4}, \frac{4}{5}$, _____, _____, _____, $\cdots$

(b) 2, 4, 8, _____, _____, _____, $\cdots$

(c) 1, 4, 7, 10, 13, 16, _____, _____, _____, $\cdots$

(d) 1, 8, 27, 64, 125, _____, _____, _____, $\cdots$

Solution

(a) In each case, both the numerator and denominator increase by 1 in each successive term. Thus the sequence is

$$\frac{1}{2}, \frac{2}{3}, \frac{3}{4}, \frac{4}{5}, \frac{5}{6}, \frac{6}{7}, \frac{7}{8}, \cdots$$

And if $n = 1, n = 2, n = 3$, etc., and each positive integer is associated with the number $a_n = \dfrac{n}{n+1}$, the function is $f(n) = \dfrac{n}{n+1}$.

(b) In each term, the value is double the preceding value. Thus the sequence is

$$2, 4, 8, \underline{16}, \underline{32}, \underline{64}, \ldots$$

and $a_n = 2^n$, so the function is $f(n) = 2^n$. (Try it for $n = 1$, $n = 2$, $n = 3$, etc.)

(c) In each case, a term is always 3 larger than the preceding term (i.e., we add 3 to get another term). Thus the sequence is

$$1, 4, 7, 10, 13, 16, \underline{19}, \underline{22}, \underline{25}, \ldots$$

The function describing this sequence is $f(n) = 3n - 2$. (Try it for $n = 1$, $n = 2$, $n = 3$, etc.)

(d) The terms of this sequence are the cubes of the positive integers. Thus the sequence is

$$1, 8, 27, 64, 125, \underline{216}, \underline{343}, \underline{512}, \ldots$$

and the function is $f(n) = n^3$. (Try this one too.)

Since different people see different patterns in sequences of numbers, it is better to use the function or the formula with the general term to describe a sequence.

EXAMPLE 15-2   Write the terms of the sequence defined by **(a)** $f(n) = 5n + 3$ and **(b)** $f(n) = 2^n - 1$.

Solution   We simply let n take on its values from the domain $\mathscr{D} = \{1, 2, 3, \ldots\}$ and calculate the range by the rule of the function.

(a) From $f(n) = 5n + 3$, we find the ordered range

$$f(1) = 5 \cdot 1 + 3 = 8, \quad f(2) = 5 \cdot 2 + 3 = 13, \quad f(3) = 5 \cdot 3 + 3 = 18$$
$$f(4) = 5 \cdot 4 + 3 = 23, \quad f(5) = 5 \cdot 5 + 3 = 28, \quad f(6) = 5 \cdot 6 + 3 = 33$$

to get the sequence $8, 13, 18, 23, 28, 33, \ldots$

(b) Begin with $n = 1$ and go. The sequence is

$$1, 3, 7, 15, 31, 63, \ldots$$

EXAMPLE 15-3   Find the first four terms and the tenth term of the infinite sequence whose general term a_n is given by **(a)** $a_n = 2^{n-4}$ and **(b)** $a_n = n^2 - 3$.

Solution   The first term is a_1, the second is a_2, etc. We let n take on the values in the domain and calculate.

(a) $a_1 = 2^{1-4}, a_2 = 2^{2-4}, a_3 = 2^{3-4}, a_4 = 2^{4-4}, \ldots, a_{10} = 2^{10-4}, \ldots$

or

$$a_1 = 2^{-3}, a_2 = 2^{-2}, a_3 = 2^{-1}, a_4 = 2^0, \ldots, a_{10} = 2^6, \ldots$$

i.e., the sequence is

$$\tfrac{1}{8}, \tfrac{1}{4}, \tfrac{1}{2}, 1, \ldots, 64, \ldots$$

(b) $a_1 = 1^2 - 3, a_2 = 2^2 - 3, a_3 = 3^2 - 3, a_4 = 4^2 - 3, \ldots, a_{10} = 10^2 - 3, \ldots$

or

$$a_1 = -2, a_2 = 1, a_3 = 6, a_4 = 13, \ldots, a_{10} = 97, \ldots$$

i.e., the sequence is

$$-2, 1, 6, 13, \ldots, 97, \ldots$$

note: The general term of a sequence, usually denoted by a_n, is also denoted by enclosing the term in braces. For example, the infinite sequence for which $a_n = 2^n$, where $n \geq 1$, may be denoted by $\{a_n\}_{n=1}^{\infty} = \{2^n\}_{n=1}^{\infty}$ or the infinite sequence for which $a_n = \frac{1}{2^n}$, where $n \geq 1$, may be denoted by $\{a_n\}_{n=1}^{\infty} = \{\frac{1}{2^n}\}_{n=1}^{\infty}$.

warning: It's possible to define a sequence so that its domain is not necessarily the set of all *positive* integers: A sequence may also be defined as the range of a function whose domain is the set of all *nonnegative* integers greater than or equal to a GIVEN integer m. Thus, the domain could include 0, for example, or it might begin with 5, or 6, or .... But in this book, you can assume that the domain of a sequence function will always be $\{1, 2, 3, \ldots\}$ unless we specifically tell you otherwise.

B. Recursion formulas

Sometimes sequences are described by giving the first or the first few terms and a **recursion formula**, which is a formula that indicates how to generate the succeeding terms from one or more preceding terms.

EXAMPLE 15-4 Write out the next four terms of the infinite sequences described recursively by

(a) $a_1 = 1$, and $a_n = 4a_{n-1} - 1$ for $n > 1$
(b) $a_1 = 2$, and $a_n = 2a_{n-1} - 3$ for $n > 1$

Solution We use the given value a_1 to find the first required term, a_2, by the recursion formula. Then we use a_2 to calculate a_3, and so on up to a_5.

(a) Given $a_1 = 1$. So

$$a_2 = 4a_{2-1} - 1 = 4a_1 - 1 = 4 \cdot 1 - 1 = 3$$

Then
$$a_3 = 4a_{3-1} - 1 = 4a_2 - 1 = 4 \cdot 3 - 1 = 11$$

$$a_4 = 4a_{4-1} - 1 = 4a_3 - 1 = 4 \cdot 11 - 1 = 43$$

$$a_5 = 4a_{5-1} - 1 = 4a_4 - 1 = 4 \cdot 43 - 1 = 171$$

The sequence is 1, 3, 11, 43, 171, ...

(b) Given $a_1 = 2$. So

$$a_2 = 2a_{2-1} - 3 = 2a_1 - 3 = 2 \cdot 2 - 3 = 1$$

Then
$$a_3 = 2a_{3-1} - 3 = 2a_2 - 3 = 2 \cdot 1 - 3 = -1$$

$$a_4 = 2a_{4-1} - 3 = 2a_3 - 3 = 2(-1) - 3 = -5$$

$$a_5 = 2a_{5-1} - 3 = 2a_4 - 3 = 2(-5) - 3 = -13$$

The sequence is 2, 1, -1, -5, -13, ...

C. Important recursive formulas:
The Fibonacci sequence and factorials

There is a famous sequence that was introduced by Leonardo Fibonacci about AD 1200. This sequence, which has been studied for centuries, arose from the following problem.

Suppose that one pair (defined as male and female) of adult rabbits is the starting pair of a rabbit colony. Further suppose that each pair of mature rabbits bears a new male/female pair every month, and that a rabbit reaches maturity after its second month. Assuming that no rabbits die, how many pairs of mature rabbits will there be at the end of n months, where n is any positive integer?

EXAMPLE 15-5 Starting with Fibonacci's problem, (a) create a formula that will solve the problem and (b) solve the problem for $n = 12$.

Solution We know that there is 1 pair of adult rabbits in the 1st month. And since rabbits don't mature until after their 2nd month, we still have one pair of adults in the 2nd month. But adults produce babies, so in the following months, baby rabbits (0–1 month old) will become juvenile rabbits (1–2 months old), juveniles will become adults (2+ months old), adults will still be adults—and all those adults will produce babies, ... The only thing to do now is to make a table:

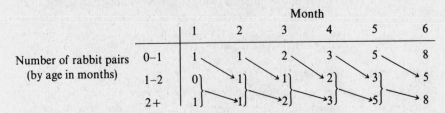

To find the number of adults in any month, we add the number of pairs that were adult in the preceding month to the number of pairs that become adult in the month of interest. Thus, in the 3rd month, there was already 1 pair of adults (from the second month) and 1 pair of juveniles (who were babies in the 1st month and juveniles in the 2nd month) who grew up to become another pair of adults: $1 + 1 = 2$. Then, in the 4th month, 2 previous adult pairs + 1 new adult pair is 3 adult pairs; 5th month, $3 + 2 = 5$; and so on, as the table shows. [Simple, once you get the hang of it!]

(a) Now we have a sequence and a pattern, so we can write the sequence by using a recursion formula called the **Fibonacci sequence**:

FIBONACCI SEQUENCE
If $n \geq 1$ and a_n is the number of mature rabbit pairs at the end of n months, then

$$a_1 = 1, a_2 = 1, a_3 = 2, \quad \text{and} \quad a_n = a_{n-1} + a_{n-2} \quad \text{for } n_2 \geq 3$$

(b) We use the Fibonacci sequence to calculate the number of adult pairs after 12 months, a_{12}. Picking up where the table left off,

$$a_7 = a_6 + a_5 = 8 + 5 = 13$$

$$a_8 = a_7 + a_6 = 13 + 8 = 21$$

$$a_9 = a_8 + a_7 = 21 + 13 = 34$$

$$a_{10} = a_9 + a_8 = 34 + 21 = 55$$

$$a_{11} = a_{10} + a_9 = 55 + 34 = 89$$

$$a_{12} = a_{11} + a_{10} = 89 + 55 = 144$$

note: In 1724, good old Bernoulli found a specific formula for the Fibonacci sequence:

$$a_n = \frac{1}{\sqrt{5}}\left[\left(\frac{1}{2} + \frac{\sqrt{5}}{2}\right)^n - \left(\frac{1}{2} - \frac{\sqrt{5}}{2}\right)^n\right]_{n=1}^{\infty}$$

Can you verify that the 12th member of the sequence is 144?

There is another important sequence that can be defined recursively by

$$a_1 = 1, \quad \text{and} \quad a_n = na_{n-1} \quad \text{for } n > 1$$

The value of a_n is called **n factorial**, denoted by $n!$. It is found for $n \geq 1$, by

$$n! = n(n-1)(n-2)\cdots 3\cdot 2\cdot 1 \quad \text{or} \quad n! = n(n-1)!$$

The terms of the sequence $n!$ are

$a_1 = 1! = 1$ (Given)

$a_2 = 2! = 2(2-1) = 2\cdot 1 = \mathbf{2}$ $(=2a_{2-1})$

$a_3 = 3! = 3(3-1)(3-2) = 3\cdot 2\cdot 1 = \mathbf{6}$ $(=3a_{3-1})$

$a_4 = 4! = 4(4-1)(4-2)(4-3) = 4\cdot 3\cdot 2\cdot 1 = \mathbf{24}$ $(=4a_{4-1})$

$a_5 = 5! = 5(5-1)(5-2)(5-3)(5-4) = 5\cdot 4\cdot 3\cdot 2\cdot 1 = \mathbf{120}$

$a_n = n! = n(n-1)(n-2)\cdots 3\cdot 2\cdot 1 = na_{n-1}$

In other words, for $n \geq 1$, $n!$ is the product of all the integers from 1 to n INCLUSIVE.

One more point: We also define $0! = 1$ (even though it doesn't make sense in the formula $n! = n(n-1)!$).

note: There's a broader way to define n factorial if we assume that the domain is $\mathscr{D} = \{0, 1, 2, 3, \ldots\}$, so that

$$a_0 = 1, \quad \text{and} \quad a_{n+1} = (n+1)a_n \quad \text{for } n \geq 0$$

or $$(n+1)! = (n+1)n!$$

By this system, $a_0 = 1$ is given, so $0! = 1$ and $1! = 1\cdot 0! = 1\cdot 1 = 1$.

EXAMPLE 15-6   Calculate the values of **(a)** $6!$, **(b)** $\dfrac{8!}{5!}$, and **(c)** $\dfrac{4!}{(4-2)!}$.

Solution   Just remember that $n!$ is the product of all the integers from 1 to n inclusive.

(a) $6! = 6\cdot 5\cdot 4\cdot 3\cdot 2\cdot 1 = 720$

(b) $\dfrac{8!}{5!} = \dfrac{8\cdot 7\cdot 6\cdot \cancel{5}\cdot \cancel{4}\cdot \cancel{3}\cdot \cancel{2}\cdot \cancel{1}}{\cancel{5}\cdot \cancel{4}\cdot \cancel{3}\cdot \cancel{2}\cdot \cancel{1}} = 8\cdot 7\cdot 6 = 336$

(c) $\dfrac{4!}{(4-2)!} = \dfrac{4!}{2!} = \dfrac{4\cdot 3\cdot 2\cdot 1}{2\cdot 1} = 4\cdot 3 = 12$

15-2. Partial Sums of Sequences

A. Summation notation

We often want to find the sum of the first n terms of a sequence. To express such sums, called **series**, we use summation notation:

- Given an infinite sequence

$$\{a_i\} = a_1, a_2, a_3, \ldots, a_n, \ldots$$

the notation $\sum\limits_{i=1}^{n} a_i$, called the nth partial sum, represents the sum of the first n terms of the sequence, where the set of integers $i = \{1, 2, 3, \ldots, n\}$ is the domain of summation and i is called the index of summation. Thus

$$\sum_{i=1}^{n} a_i = a_1 + a_2 + a_3 + \cdots + a_{n-1} + a_n$$

note: The numbers above and below the $\sum$ tell us where to start and where to leave off in summing the terms of the sequence: $i = 1$ tells us to start with the first term and n tells us to end with the nth term.

Interpret $\sum\limits_{i=1}^{n} a_i$ as follows:

"Sum all the values of a_i as the index i runs through the integers from 1 to n."

Thus, we really have—but don't write—

$$\sum_{i=1}^{i=n} a_i$$

EXAMPLE 15-7 Find the following partial sums:

(a) $\displaystyle\sum_{i=1}^{4} i^2$ **(b)** $\displaystyle\sum_{j=1}^{5} (2j)$ **(c)** $\displaystyle\sum_{k=1}^{6} (k^2 - 2)$

Solution

(a) The sequence for the index i is the integers from 1 through 4 inclusive, so we have to sum the squares of the integers from $i = 1$ through $i = 4$. The sum is

$$\sum_{i=1}^{4} i^2 = 1^2 + 2^2 + 3^2 + 4^2 = 1 + 4 + 9 + 16 = 30$$

(b) We are summing the first five even integers. The sum is

$$\sum_{j=1}^{5} (2j) = (2 \cdot 1) + (2 \cdot 2) + (2 \cdot 3) + (2 \cdot 4) + (2 \cdot 5)$$
$$= 2 + 4 + 6 + 8 + 10 = 30$$

(c) We need the sum of the first six members of the sequence $\{k^2 - 2\}$. The sum is

$$\sum_{k=1}^{6} (k^2 - 2) = (1^2 - 2) + (2^2 - 2) + (3^2 - 2)$$
$$+ (4^2 - 2) + (5^2 - 2) + (6^2 - 2)$$
$$= -1 + 2 + 7 + 14 + 23 + 34 = 79$$

In general summation notation, the sum does not need to begin with the index $i = 1$. If, for example, we have $i = 3$ and $n = 5$, as in $\displaystyle\sum_{i=3}^{5} a_i$, we begin with the value of the third term a_3, add the value of the fourth term a_4, and then add the value of the fifth—and final—term a_5.

EXAMPLE 15-8 Evaluate **(a)** $\displaystyle\sum_{i=4}^{7} i^2$ and **(b)** $\displaystyle\sum_{k=3}^{6} [k^2(k + 2)]$.

Solution

(a) The terms of the partial sum begin with the term corresponding to $i = 4$, which is $4^2 = 16$, progress through the succeeding integers, and end at the term corresponding to $i = 7$. Thus we have

$$\sum_{i=4}^{7} i^2 = 4^2 + 5^2 + 6^2 + 7^2 = 126$$

(b) Here we begin with the term corresponding to $k = 3$, include the terms corresponding to $k = 4$ and $k = 5$, and end with the term corresponding to $k = 6$. We have

$$\sum_{k=3}^{6} [k^2(k + 2)] = 3^2(3 + 2) + 4^2(4 + 2) + 5^2(5 + 2) + 6^2(6 + 2)$$
$$= 9 \cdot 5 + 16 \cdot 6 + 25 \cdot 7 + 36 \cdot 8 = 604$$

B. Sum formulas

If we need the sum of a sequence with many terms, the computations can be long and tedious, so we often find a formula that will give us the result without requiring that we calculate and add every single term.

EXAMPLE 15-9 Find $S = \sum\limits_{i=1}^{200} i$ (i.e., find the sum of the first 200 integers).

Solution We want $S = 1 + 2 + 3 + 4 + \cdots + 199 + 200$. Boring! Clearly, we need a special formula for problems of this type. This plan will do it!

Let $\quad S = \quad 1 + \quad 2 + \quad 3 + \quad 4 + \cdots + 197 + 198 + 199 + 200$

Also let $\quad S = 200 + 199 + 198 + 197 + \cdots + \quad 4 + \quad 3 + \quad 2 + \quad 1$

Add each column:

$$2S = 201 + 201 + 201 + 201 + \cdots + 201 + 201 + 201 + 201$$

How many of these 201's are there? Exactly 200, because $n = 200$. So

$$2S = 201(200) \quad \text{and} \quad S = \tfrac{1}{2}(201)(200) = 20\,100$$

Example 15-9 is a special case of the formula

$$\sum_{i=1}^{n} i = \frac{1}{2}(n + 1)(n) \tag{15.1}$$

That is, the sum of all integers from 1 to n is the product of the number of terms, n, in the sequence and the average of the first and last terms, $(n + 1)/2$.

There are also some general properties of sums that can be expressed by formulas:

- $\sum\limits_{i=1}^{n} (c) = nc$, where c is a fixed constant

- $\sum\limits_{i=1}^{n} (a_i \pm b_i) = \sum\limits_{i=1}^{n} a_i \pm \sum\limits_{i=1}^{n} b_i$

- $\sum\limits_{i=1}^{n} (ca_i) = c\left(\sum\limits_{i=1}^{n} a_i\right)$, for all constants c

Finally, observe that, in summation notation, a general polynomial

$$P(x) = a_n x^n + a_{n-1} x^{n-1} + a_{n-2} x^{n-2} + \cdots + a_2 x^2 + a_1 x + a_0$$

may be written as

$$P(x) = \sum_{i=0}^{n} a_i x^i$$

15-3. Arithmetic Sequences and Series

A. Definition of an arithmetic sequence

- An **arithmetic sequence** (or **arithmetic progression**) is a sequence in which successive terms differ by the same number, so that the difference between any term and its predecessor is constant. We express this by the formula

ARITHMETIC SEQUENCE $\qquad a_{k+1} - a_k = d \quad \text{for } k \geq 1$

where d is the common difference.

An arithmetic sequence can also be written as

$$a_1, a_1 + d, a_1 + 2d, a_1 + 3d, \ldots, a_1 + (n - 1)d, \ldots$$

Thus the nth term of an arithmetic sequence is

$$a_n = a_1 + (n - 1)d$$

EXAMPLE 15-10   Fill in the blanks for the terms of the following sequences:

(a) 9, 6, 3, 0, _____, _____, ...

(b) 2, 9, 16, 23, _____, _____, ...

(c) 5, 5.5, 6, 6.5, 7, _____, _____, ...

Solution   These are arithmetic sequences, so we can write more terms by finding the common difference and adding.

(a) The common difference is -3, so the sequence is

$$9, 6, 3, 0, \underline{-3}, \underline{-6}, \ldots$$

(b) The common difference is 7, so the sequence is

$$2, 9, 16, 23, \underline{30}, \underline{37}, \ldots$$

(c) The common difference is .5, so the sequence is

$$5, 5.5, 6, 6.5, 7, \underline{7.5}, \underline{8}, \ldots$$

EXAMPLE 15-11   Find the 10th term for each sequence in Example 15-10.

Solution

(a) We know that $a_1 = 9$ and $d = -3$. If we want the 10th term, we have $n = 10$, so

$$a_n = a_1 + (n - 1)d$$
$$a_{10} = 9 + (10 - 1)(-3) = 9 + (9)(-3) = 9 - 27 = -18$$

(b) Here, $a_1 = 2$, $d = 7$, and $n = 10$, so

$$a_{10} = 2 + (10 - 1)(7) = 2 + (9)(7) = 65$$

(c) Here, $a_1 = 5$, $d = .5$, and $n = 10$, so

$$a_{10} = 5 + (10 - 1)(.5) = 5 + (9)(.5) = 5 + 4.5 = 9.5$$

EXAMPLE 15-12   The 10th term of an arithmetic sequence is 20 and the 16th term is 32. Find the common difference, the 1st term, and a general formula for the nth term.

Solution   We know a_{10} and a_{16}, so we set up a system of equations to find d:

$$a_{16} = a_1 + (16 - 1)d = 32$$
$$\underline{a_{10} = a_1 + (10 - 1)d = 20} \qquad \text{Subtract } a_{10} \text{ from } a_{16}.$$
$$0 + 15d - 9d = 32 - 20$$
$$d = 2$$

Once we have the value of d, we can substitute it into one of our equations for a specific a_n, and solve for a_1:

$$20 = a_1 + (10 - 1)2$$
$$a_1 = 2$$

And once we have both a_1 and d, we can find the formula for this arithmetic sequence:

$$a_n = a_1 + (n - 1)d$$
$$= 2 + (n - 1)2 = 2 + 2n - 2 = 2n$$

This is the sequence of even integers, $\{2n\}$.

B. Partial sums of arithmetic sequences: Arithmetic series

If $a_1, a_2, a_3, \ldots$ is an arithmetic sequence with common difference d, then the nth partial sum of the sequence, or the **arithmetic series**, is given by

$$S_n = \sum_{i=1}^{n} a_i = a_1 + (a_1 + d) + (a_1 + 2d) + \cdots + (a_1 + (n-1)d)$$

$$= (a_1 + a_1 + a_1 + \cdots + a_1) + (d + 2d + 3d + \cdots + (n-1)d)$$

$$= na_1 + d(1 + 2 + 3 + \cdots + (n-1))$$

By using formula (15.1) we can substitute the sum of the first $n-1$ integers. Notice that $n-1$ here replaces the n of formula (15.1). Thus

$$S_n = na_1 + d\left(\frac{n(n-1)}{2}\right) \qquad \text{(15.2a)}$$

An alternative form is obtained as follows:

$$S_n = na_1 + d\left(\frac{n(n-1)}{2}\right) = na_1 + nd\left(\frac{n-1}{2}\right)$$

$$= \frac{n}{2}(2a_1 + (n-1)d) = \frac{n}{2}(a_1 + a_1 + (n-1)d)$$

and since $a_n = a_1 + (n-1)d$,

$$S_n = \frac{n}{2}(a_1 + a_n) \qquad \text{(15.2b)}$$

In summary,

PARTIAL SUM OF AN ARITHMETIC SEQUENCE	$S_n = na_1 + d\left(\dfrac{n(n-1)}{2}\right) = \dfrac{n}{2}(a_1 + a_n)$	**(15.2)**

EXAMPLE 15-13 Find the sum of the even integers from 2 through 200.

Solution This is an arithmetic sequence with $a_1 = 2$, $a_n = 200$, and a total of $n = 100$ terms. Thus

$$S_n = \frac{n}{2}(a_1 + a_n)$$

$$S_{100} = \left(\tfrac{100}{2}\right)(2 + 200) = \left(\tfrac{100}{2}\right)(202) = 10\,100$$

EXAMPLE 15-14 If the first two terms of an arithmetic sequence are 4 and 7, what is the sum of the first 50 terms?

Solution We have $a_1 = 4$, $d = 3$, and $n = 50$. Thus

$$S_n = na_1 + d\left(\frac{n(n-1)}{2}\right)$$

$$S_{50} = 50(4) + 3\left(\frac{50 \cdot 49}{2}\right) = 200 + 3675 = 3875$$

EXAMPLE 15-15 Find the sum of all multiples of 3 between 11 and 83.

Solution The multiples of 3 between 11 and 83 are 12, 15, 18, ..., 81. Their sum is $12 + 15 + 18 + \cdots + 81$. This is an arithmetic sequence with common difference 3. The first term is 12 and the last (nth term) is 81. Since $a_n = a_1 + (n-1)d$, we have $81 = 12 + (n-1)(3)$ or $81 - 12 = 3(n-1)$, so $n = 24$. Then

$$S_{24} = \tfrac{24}{2}(12 + 81) = 12(93) = 1116$$

15-4. Geometric Sequences and Series

A. Definition of a geometric sequence

In an arithmetic sequence, we progress from one term to the next by ADDING the same number. We now consider sequences in which we MULTIPLY each term by a fixed number to find the next term.

- A **geometric sequence** (or **geometric progression**) is a sequence in which, after the first term, the quotient of any term with the preceding term is the same nonzero number. This quotient is called the **common ratio** r of the geometric sequence.

EXAMPLE 15-16 What is the common ratio in the following geometric sequences? Fill in the blanks.

(a) 2, 6, 18, 54, _____ , _____ , $\cdots$

(b) $5, \frac{5}{2}, \frac{5}{4}, \frac{5}{8},$ _____ , _____ , $\cdots$

(c) 9, 3, 1, _____ , _____ , _____ , $\cdots$

Solution

(a) To obtain the common ratio, we divide the 2nd term by the 1st term to get $r = \frac{6}{2} = 3$. Check it for the 3rd term divided by the 2nd term: $r = \frac{18}{6} = 3$. Now we know that any term is 3 times its predecessor. The sequence is

$$2, 6, 18, 54, \underline{162}, \underline{486}, \cdots$$

(b) The common ratio is

$$r = \frac{\frac{5}{2}}{5} = \frac{1}{2} \qquad \text{Check it:} \quad r = \frac{\frac{5}{4}}{\frac{5}{2}} = \frac{5}{4} \cdot \frac{2}{5} = \frac{1}{2}$$

The sequence is

$$5, \frac{5}{2}, \frac{5}{4}, \frac{5}{8}, \frac{5}{16}, \frac{5}{32}, \cdots$$

(c) The common ratio is

$$r = \frac{3}{9} = \frac{1}{3}$$

The sequence is

$$9, 3, 1, \frac{1}{3}, \frac{1}{9}, \frac{1}{27}, \cdots$$

If a geometric sequence has first term a_1 and common ratio r, then it has the following terms:

$$a_1 \qquad \text{(1st term)}$$
$$a_2 = a_1 r \qquad \text{(2nd term)}$$
$$a_3 = (a_1 r)(r) = a_1 r^2 \qquad \text{(3rd term)}$$
$$a_4 = (a_1 r^2)(r) = a_1 r^3 \qquad \text{(4th term)}$$
$$\vdots$$
$$a_n = a_1 r^{n-1} \qquad \text{(nth term)}$$

Note also that

$$r = \frac{a_{n+1}}{a_n} \qquad \text{for any } n \geq 1$$

EXAMPLE 15-17 For the sequence that begins with 3, 12, 48, ..., find the common ratio and the 6th term. What is the formula for the nth term?

Solution

$$r = \frac{12}{3} = \frac{48}{12} = 4$$

The first term is $a_1 = 3$. Then the 6th term is

$$a_6 = a_1 r^5 = 3(4)^5 = 3072$$

and the formula for the nth term is

$$a_n = (3)(4)^{n-1}$$

EXAMPLE 15-18 If the first term of a geometric sequence is 6 and the common ratio $r = -\frac{1}{3}$, what is the 5th term and what is the formula for the nth term?

Solution

$$a_5 = a_1 r^4 = 6(-\tfrac{1}{3})^4 = 6(-1)^4(\tfrac{1}{3})^4 = 6(1)(\tfrac{1}{81}) = \tfrac{2}{27}$$

Then
$$a_n = 6(-\tfrac{1}{3})^{n-1}$$

B. Partial sums of geometric sequences: Geometric series

The sum S_n of the first n terms of a geometric sequence, called a **geometric series**, may be written as

$$\sum_{j=1}^{n} ar^{j-1} = S_n = a_1 + a_1 r + a_1 r^2 + a_1 r^3 + \cdots + a_1 r^{n-1}$$

$$= a_1(1 + r + r^2 + r^3 + \cdots + r^{n-1})$$

If $r = 1$, then all the terms of the series are the same and the formula is simple:

$$S_n = a_1 + a_1 + a_1 + \cdots + a_1 = na_1$$

If $r \neq 1$, the terms of the series are different, but they can be manipulated algebraically to give a formula:

$$(1 - r)S_n = (1 - r)a_1(1 + r + r^2 + r^3 + \cdots + r^{n-1})$$

$$= a_1[(1 - r)(1 + r + r^2 + r^3 + \cdots + r^{n-1})]$$

$$= a_1[(1 + r + r^2 + r^3 + \cdots + r^{n-1})$$

$$- (r + r^2 + r^3 + \cdots + r^{n-1} + r^n)]$$

$$= a_1(1 - r^n)$$

Thus
$$S_n = \frac{a_1(1 - r^n)}{1 - r}$$

In summary,

$$S_n = a_1 + a_1 r + a_1 r^2 + a_1 r^3 + \cdots + a_1 r^{n-1}$$

PARTIAL SUM OF A GEOMETRIC SEQUENCE
$$= \begin{cases} na_1 & \text{if } r = 1 \\ a_1\left(\dfrac{1 - r^n}{1 - r}\right) = a_1\left(\dfrac{r^n - 1}{r - 1}\right) & \text{if } r \neq 1 \end{cases}$$

(15.3)

note: It is more convenient to use the form $S_n = a_1\left(\dfrac{1 - r^n}{1 - r}\right)$ when $r < 1$ and to use the form $S_n = a_1\left(\dfrac{r^n - 1}{r - 1}\right)$ when $r > 1$.

EXAMPLE 15-19   Find the sum S_{12} of the first 12 terms of the geometric sequence $1, \frac{1}{2}, \frac{1}{4}, \ldots$.

Solution   The first term is 1, so we have $a_1 = 1$. Then the common ratio is

$$r = \frac{\frac{1}{2}}{1} = \frac{\frac{1}{4}}{\frac{1}{2}} = \frac{1}{2}$$

Since we want the first 12 terms, $n = 12$. Thus

$$S_{12} = a_1 \left(\frac{1 - r^n}{1 - r} \right) = (1) \left(\frac{1 - (\frac{1}{2})^{12}}{1 - \frac{1}{2}} \right) = \frac{1 - \frac{1}{4096}}{\frac{1}{2}} = 1.999\,51$$

EXAMPLE 15-20   Bright Young Lad and his father strike a deal. Lad promises to do all of the following tasks for two weeks: mow the lawn, wash the dishes, and weed the garden. His father agrees to pay him as follows: 1 cent for the first day, 2 cents for the second day, 4 cents for the third day, 8 cents for the fourth day, etc.—double the pay each consecutive day. How much will Lad earn in two weeks? What would his total pay be for a 30-day deal?

Solution   Lad's total pay in dollars is

$$S_{14} = .01 + .02 + .04 + .08 + \cdots + \boxed{\text{14th day's pay}}$$

Here $a_1 = .01$ and $r = .02/.01 = 2$. We want the sum for $n = 14$. Thus

$$S_{14} = a_1 \left(\frac{r^{14} - 1}{r - 1} \right) = (.01) \left(\frac{2^{14} - 1}{2 - 1} \right) = (.01)(16\,383) = \$163.83$$

Now the 30-day deal would result in a sum

$$S_{30} = a_1 \left(\frac{r^{30} - 1}{r - 1} \right) = (.01) \left(\frac{2^{30} - 1}{2 - 1} \right) = \$10\,737\,418.23$$

That's close to 11 million dollars! (Nice work if you can get it.)

EXAMPLE 15-21   A culture of bacteria increases by 25% each hour. If the culture originally contains 1000 bacteria, how many are there after 12 hours?

Solution   The populations in successive hours are

$$\text{Time zero} \longrightarrow 1000$$
$$\text{After 1 hour} \longrightarrow 1000 + .25(1000) = 1250$$
$$\text{After 2 hours} \longrightarrow 1250 + .25(1250) = 1562.5$$

This is a geometric sequence of sums $S_1, S_2, \ldots$, where the terms of the sequence are found from the series:

$$S_1 = 1000$$
$$S_2 = S_1 + .25S_1 = (1.25)S_1$$
$$S_3 = S_2 + .25S_2 = (1.25)S_2 = (1.25)^2 S_1 \qquad \text{[note: Here, } r = 1.25\text{]}$$
$$S_4 = S_3 + .25S_3 = (1.25)S_3 = (1.25)^3 S_1$$
$$\vdots$$
$$S_n = (1.25)^{n-1} S_1$$

Then, since the terms of the sequence are already sums, all we have to do is find S_n: after 12 hours, the bacteria count is

$$S_{12} = (1.25)^{11}(1000) = \left(\frac{5}{4} \right)^{11} (1000) = \left(\frac{5^{11}}{4^{11}} \right)(1000) \simeq 11\,642$$

C. Sum of an infinite geometric sequence: Infinite geometric series

Is it possible to add infinitely many numbers in a geometric sequence? Does it make sense to try to sum $\frac{1}{2} + \frac{1}{4} + \frac{1}{8} + \frac{1}{16} + \frac{1}{32} + \cdots$?

Suppose you are taking a walk. You go $\frac{1}{2}$ mile and stop for a rest. Then you walk $\frac{1}{4}$ mile and rest again. Then you walk $\frac{1}{8}$ mile, then $\frac{1}{16}$ mile, etc. However many times you walk a distance that is one-half the distance that you walked last, you can never get farther than 1 mile from your starting point. Thus we might say that

$$\frac{1}{2} + \frac{1}{4} + \frac{1}{8} + \frac{1}{16} + \cdots = 1$$

How can infinitely many numbers add up to only 1?

The partial sum of this geometric sequence is

$$S_n = a_1\left(\frac{1 - r^n}{1 - r}\right) = \left(\frac{1}{2}\right)\left(\frac{1 - (\frac{1}{2})^n}{1 - \frac{1}{2}}\right) = 1 - \left(\frac{1}{2}\right)^n$$

If n is a very large number, then $(\frac{1}{2})^n$ is a very small number, so that the amount we subtract from 1 is very small. Then, if we let n grow unboundedly large, $(\frac{1}{2})^n$ approaches zero and the sum approaches 1. For instance,

$$(\tfrac{1}{2})^{50} \simeq 0.000\,000\,000\,000\,000\,88$$

so that

$$1 - (\tfrac{1}{2})^{50} = 0.999\,999\,999\,999\,999\,12$$

and that is *pretty close* to 1!

We may conclude that if $|r| < 1$, the sum of an infinite number of terms in a geometric sequence may be calculated and

SUM OF AN INFINITE GEOMETRIC SERIES $\quad S_\infty = a_1\left(\dfrac{1}{1 - r}\right) \quad$ for $|r| < 1$ **(15.4)**

EXAMPLE 15-22 Find the following sums:

(a) $3 - \dfrac{3}{2} + \dfrac{3}{4} - \dfrac{3}{8} + \dfrac{3}{16} - \dfrac{3}{32} + \cdots = \displaystyle\sum_{j=1}^{\infty} 3\left(-\frac{1}{2}\right)^{j-1}$ **(b)** $\displaystyle\sum_{j=1}^{\infty}\left(\frac{2}{3}\right)^{j}$

Solution

(a) This is an infinite geometric series, where $a_1 = 3$ and $r = -\frac{1}{2}$, so

$$S_\infty = a_1\left(\frac{1}{1 - r}\right) = 3\left(\frac{1}{1 - (-\frac{1}{2})}\right) = 3\left(\frac{1}{\frac{3}{2}}\right) = 2$$

note: If we write out the sequence of partial sums, we get

$S_1 = 3$

$S_2 = 3 - \frac{3}{2} = \frac{3}{2}$

$S_3 = \frac{3}{2} + \frac{3}{4} = \frac{9}{4}$ Notice how the successive sum

$S_4 = \frac{9}{4} - \frac{3}{8} = \frac{15}{8}$ values get closer and closer to 2.

$S_5 = \frac{15}{8} + \frac{3}{16} = \frac{33}{16}$

$\vdots \qquad\qquad \vdots$

(b) We first write out a few terms to see what we have:

$$\sum_{j=1}^{\infty}\left(\frac{2}{3}\right)^{j} = \left(\frac{2}{3}\right)^{1} + \left(\frac{2}{3}\right)^{2} + \left(\frac{2}{3}\right)^{3} + \cdots = \frac{2}{3} + \frac{4}{9} + \frac{8}{27} + \cdots$$

Aha! This is an infinite geometric series, where $a_1 = \frac{2}{3}$ and $r = \frac{2}{3}$. Thus

$$S_\infty = a_1\left(\frac{1}{1-r}\right) = \frac{2}{3}\left(\frac{1}{1-\frac{2}{3}}\right) = \frac{2}{3}\left(\frac{1}{\frac{1}{3}}\right) = \left(\frac{2}{3}\right)3 = 2$$

D. Repeating decimals

We are familiar with the expressions $\frac{1}{3} = .333\,333\ldots$ and $\frac{2}{3} = .666\,666\ldots$. Clearly, these are repeating decimals—but what *are* they, and what is $2.131\,313\,131\,3\ldots$ or $1.555\,555\ldots$? You guessed it—we can use the ideas developed here to answer these questions.

Let's start with a simple one, $\frac{1}{3} = .333\,333\ldots$, which allows us to write

$$.333\,333\ldots = .3 + .03 + .003 + .0003 + .000\,03 + .000\,003 + \cdots$$

$$= \frac{3}{10} + \frac{3}{100} + \frac{3}{1000} + \frac{3}{10\,000} + \frac{3}{100\,000} + \frac{3}{1\,000\,000} + \cdots$$

This is an infinite geometric series with first term $a_1 = \frac{3}{10}$ and ratio $\frac{1}{10}$. Thus its sum is

$$S_\infty = \frac{3}{10}\left(\frac{1}{1-\frac{1}{10}}\right) = \frac{3}{10}\left(\frac{1}{\frac{9}{10}}\right) = \frac{3}{10}\cdot\frac{10}{9} = \frac{1}{3}$$

So

- A **repeating decimal** is an infinite geometric series whose sum is a rational number, or common fraction.

EXAMPLE 15-23   Write the following repeating decimal numbers as common fractions: **(a)** $1.555\,555\ldots$; **(b)** $2.131\,313\,13\ldots$.

Solution

(a) Here, write the integer 1 as a term, then start the series with $a_1 = \frac{5}{10}$.

$$1.555\,555\ldots = 1 + .5 + .05 + .005 + .0005 + \cdots$$

$$= 1 + \frac{5}{10} + \frac{5}{100} + \frac{5}{1000} + \frac{5}{10\,000} + \cdots$$

$$= 1 + \left(\frac{5}{10}\cdot\frac{1}{1-\frac{1}{10}}\right) = 1 + \left(\frac{5}{10}\cdot\frac{1}{\frac{9}{10}}\right) = 1 + \left(\frac{5}{10}\cdot\frac{10}{9}\right)$$

$$= 1 + \frac{5}{9} = \frac{14}{9}$$

(b) Here, the repeating unit is 13, and because it covers two decimal places $r = \frac{1}{100}$.

$$2.131\,313\,13\ldots = 2 + .13 + .0013 + .000\,013 + \cdots$$

$$= 2 + \frac{13}{100} + \frac{13}{10\,000} + \frac{13}{1\,000\,000} + \cdots$$

$$= 2 + \left(\frac{13}{100}\cdot\frac{1}{1-\frac{1}{100}}\right) = 2 + \left(\frac{13}{100}\cdot\frac{1}{\frac{99}{100}}\right) = 2 + \frac{13}{99} = \frac{211}{99}$$

15-5. Mathematical Induction

In Section 15-2 we used algebra to derive the formula

$$\sum_{i=1}^{n} i = 1 + 2 + 3 + \cdots + n = \frac{n(n+1)}{2}$$

And there are many similar formulas for summing numbers, such as

$$\sum_{i=1}^{n} i^2 = 1^2 + 2^2 + 3^2 + \cdots + n^2 = \frac{n(n+1)(2n+1)}{6}$$

$$\sum_{i=1}^{n} \left(\frac{1}{i(i+1)} \right) = \frac{1}{1 \cdot 2} + \frac{1}{2 \cdot 3} + \frac{1}{3 \cdot 4} + \frac{1}{4 \cdot 5} + \cdots + \frac{1}{n(n+1)} = \frac{n}{n+1}$$

$$\sum_{i=1}^{n} (2i-1)^2 = 1^2 + 3^2 + 5^2 + \cdots + (2n-1)^2 = \frac{n(2n-1)(2n+1)}{3}$$

We can establish the validity of these formulas by a different method, called the PRINCIPLE (or AXIOM) OF MATHEMATICAL INDUCTION, in which we use particular cases to prove general formulas.

To see how inductive proofs work, consider the sum of the odd positive integers:

$$\sum_{i=1}^{n} (2i-1) = S = 1 + 3 + 5 + \cdots + (2n-1)$$

First look at the partial sums and observe the pattern of the results:

$$S_1 = \mathbf{1}$$
$$S_2 = 1 + 3 = \mathbf{4}$$
$$S_3 = 4 + 5 = \mathbf{9}$$
$$S_4 = 9 + 7 = \mathbf{16}$$
$$S_5 = 16 + 9 = \mathbf{25}$$

We note immediately that the results are the squares of the integers and we see the pattern

$$S_n = n^2$$

Thus we could conclude that

$$1 + 3 + 5 + 7 + 9 + \cdots + (2n-1) = n^2$$

But at this point we don't know if this result is true for ALL n. We can't sum all the odd positive integers because there's an infinite number of them. But we can prove the truth of this statement for all n if we do two things:

(1) Show that the formula is true for $n = 1$.
(2) Show that if the formula is true for $n = k$, where k is any positive integer, then it is also true for $n = k + 1$.

EXAMPLE 15-24   Use mathematical induction to show that

$$1 + 3 + 5 + \cdots + (2n-1) = n^2$$

Solution

(1) *Show that the formula is true for $n = 1$:* If $n = 1$, the only term on the left side of the equation is $2(1) - 1 = 1$. And since the only term on the right is $1^2 = 1$, we must show that $1 = 1$. This needs no proof: It is true.
(2) *Show that if the formula is true for $n = k$, then it is true for $n = k + 1$:* If $n = k$, the formula is

$$1 + 3 + 5 + \cdots + (2k-1) = k^2$$

We assume the truth of this statement. From this we must show the truth of the statement

$$1 + 3 + 5 + \cdots + (2k-1) + (2(k+1)-1) = (k+1)^2$$

or
$$[1 + 3 + 5 + \cdots + (2k - 1)] + (2k + 1) = (k + 1)^2$$

But we've assumed that the bracketed quantity $[1 + 3 + 5 + \cdots + (2k - 1)]$ is equal to k^2, so now we need to show that

$$[k^2] + (2k + 1) = (k + 1)^2$$

This is easy, since the left side is factorable:

$$k^2 + 2k + 1 = (k + 1)(k + 1) = (k + 1)^2$$
$$(k + 1)^2 = (k + 1)^2$$

Thus the formula is valid for all n.

Now let's proceed a little more formally.

- A **statement** is a mathematical sentence [or formula or equation] that is either true or false. If a statement depends on a value n, we often designate the statement by S_n.

Thus the formula in Example 15-24 can be written as the statement

$$S_n:\ 1 + 3 + 5 + \cdots + (2n - 1) = n^2$$

Using this notation, we can state the **principle of mathematical induction**:

For each positive integer n let a statement S_n be given. If we can show that

(1) S_1 is true and
(2) the truth of S_k implies the truth of S_{k+1} for any positive integer k,

then S_n is true for all n.

We often refer to step (1) as the *verification* step or the starting step, and to step (2) as the *inductive* step.

note: The principle of mathematical induction is sometimes referred to as the "domino principle." Picture an infinite collection of dominos all lined up for a fall. In order to conclude that all of the dominos will eventually fall down, we need to see the first domino in the string fall down (this is the verification step) *and* we must be assured that, when the kth domino in the string falls down, the next one, the $(k + 1)$st one, will also fall down (this is the inductive step).

warning: BOTH conditions (1) and (2) must hold for the proof to be valid. Also, note that in step (2), we do not prove that S_{k+1} is true—we only prove that IF S_k is true, THEN S_{k+1} is true.

EXAMPLE 15-25 Prove that

$$1^2 + 2^2 + 3^2 + \cdots + n^2 = \frac{n(n + 1)(2n + 1)}{6}$$

Solution Designate the statement:

$$S_n:\ 1^2 + 2^2 + 3^2 + \cdots + n^2 = \frac{n(n + 1)(2n + 1)}{6}$$

(1) Show the truth of S_1:

$$S_1 \text{ is } 1^2 = \frac{1(1 + 1)(2 + 1)}{6} = \frac{1 \cdot 2 \cdot 3}{6} = 1$$

Hence
$$S_1:\ 1^2 = 1 \qquad S_1 \text{ is true.}$$

(2) We assume the truth of S_k, i.e., that

$$S_k: \ 1^2 + 2^2 + 3^2 + \cdots + k^2 = \frac{k(k+1)(2k+1)}{6}$$

Using this, we must show the truth of

$$S_{k+1}: \ 1^2 + 2^2 + 3^2 + \cdots + k^2 + (k+1)^2$$

$$= \frac{(k+1)[(k+1)+1][2(k+1)+1]}{6}$$

$$= \frac{(k+1)(k+2)(2k+3)}{6}$$

Now we replace the quantity equal to S_k in S_{k+1} and manipulate the result to show an identity:

$$S_{k+1}: \ [1^2 + 2^2 + 3^2 + \cdots + k^2] + (k+1)^2 \overset{?}{=} \frac{(k+1)(k+2)(2k+3)}{6}$$

$$\frac{k(k+1)(2k+1)}{6} + (k+1)^2 \overset{?}{=} \frac{(k+1)(k+2)(2k+3)}{6}$$

$$\frac{k(k+1)(2k+1)}{6} + \frac{(k+1)6(k+1)}{6} \overset{?}{=} \frac{(k+1)(k+2)(2k+3)}{6}$$

$$\left(\frac{k+1}{6}\right)[k(2k+1) + 6(k+1)] \overset{?}{=} \frac{(k+1)(k+2)(2k+3)}{6}$$

$$\left(\frac{k+1}{6}\right)[2k^2 + 7k + 6] \overset{?}{=} \frac{(k+1)(k+2)(2k+3)}{6}$$

$$\left(\frac{k+1}{6}\right)(2k+3)(k+2) \overset{\checkmark}{=} \frac{(k+1)(k+2)(2k+3)}{6}$$

$$\therefore \ S_{k+1} \text{ is true.}$$

So S_n is true for all n.

EXAMPLE 15-26 Prove that

$$\frac{1}{1 \cdot 2} + \frac{1}{2 \cdot 3} + \frac{1}{3 \cdot 4} + \cdots + \frac{1}{n(n+1)} = \frac{n}{n+1}$$

Solution

$$S_n: \ \frac{1}{1 \cdot 2} + \frac{1}{2 \cdot 3} + \frac{1}{3 \cdot 4} + \cdots + \frac{1}{n(n+1)} = \frac{n}{n+1}$$

(1) Show S_1 is true. Using $n = 1$ in S_n gives

$$S_1: \ \frac{1}{1 \cdot 2} = \frac{1}{2} \qquad \text{This is true.}$$

(2) We will assume the truth of S_k:

$$S_k: \ \frac{1}{1 \cdot 2} + \frac{1}{2 \cdot 3} + \frac{1}{3 \cdot 4} + \cdots + \frac{1}{k(k+1)} = \frac{k}{k+1}$$

Using this, we must show the validity of S_{k+1}:

$$S_{k+1}: \ \frac{1}{1 \cdot 2} + \frac{1}{2 \cdot 3} + \frac{1}{3 \cdot 4} + \cdots + \frac{1}{k(k+1)} + \frac{1}{(k+1)(k+2)} \overset{?}{=} \frac{k+1}{k+2}$$

We rewrite S_{k+1} as

$$S_{k+1}: \underbrace{\left[\frac{1}{1 \cdot 2} + \frac{1}{2 \cdot 3} + \frac{1}{3 \cdot 4} + \cdots + \frac{1}{k(k+1)}\right]}_{S_k} + \frac{1}{(k+1)(k+2)} \overset{?}{=} \frac{k+1}{k+2}$$

$$\left[\frac{k}{k+1}\right] + \frac{1}{(k+1)(k+2)} \overset{?}{=} \frac{k+1}{k+2}$$

$$\frac{k(k+2)}{(k+1)(k+2)} + \frac{1}{(k+1)(k+2)} \overset{?}{=} \frac{k+1}{k+2}$$

$$\frac{k^2 + 2k + 1}{(k+1)(k+2)} \overset{?}{=} \frac{k+1}{k+2}$$

$$\frac{(k+1)^2}{(k+1)(k+2)} \overset{?}{=} \frac{k+1}{k+2}$$

$$\frac{k+1}{k+2} \overset{\checkmark}{=} \frac{k+1}{k+2}$$

$$\therefore S_{k+1} \text{ is true.}$$

Thus S_n is true for all n.

EXAMPLE 15-27 Prove that

$$S_n: 1^2 + 3^2 + 5^2 + \cdots + (2n-1)^2 = \frac{n(2n-1)(2n+1)}{3}$$

Solution

(1)
$$S_1: 1^2 \overset{?}{=} \frac{1 \cdot (2 \cdot 1 - 1)(2 \cdot 1 + 1)}{3}$$

$$1 \overset{\checkmark}{=} \frac{1 \cdot 1 \cdot 3}{3} \qquad S_1 \text{ is true.}$$

(2) Assume

$$S_k: 1^2 + 3^2 + 5^2 + \cdots + (2k-1)^2 = \frac{k(2k-1)(2k+1)}{3}$$

Then

$$S_{k+1}: 1^2 + 3^2 + 5^2 + \cdots + (2k-1)^2 + (2k+1)^2 \overset{?}{=} \frac{(k+1)(2k+1)(2k+3)}{3}$$

Rewrite S_{k+1} as

$$S_{k+1}: \underbrace{\left[1^2 + 3^2 + 5^2 + \cdots + (2k-1)^2\right]}_{S_k} + (2k+1)^2 \overset{?}{=} \frac{(k+1)(2k+1)(2k+3)}{3}$$

$$\frac{k(2k-1)(2k+1)}{3} + \frac{3(2k+1)^2}{3} \overset{?}{=} \frac{(k+1)(2k+1)(2k+3)}{3}$$

$$\left(\frac{2k+1}{3}\right)[k(2k-1) + 3(2k+1)] \overset{?}{=} \frac{(k+1)(2k+1)(2k+3)}{3}$$

$$\left(\frac{2k+1}{3}\right)[2k^2 + 5k + 3] \overset{?}{=} \frac{(k+1)(2k+1)(2k+3)}{3}$$

$$\left(\frac{2k+1}{3}\right)(2k+3)(k+1) \overset{\checkmark}{=} \frac{(k+1)(2k+1)(2k+3)}{3}$$

$$\therefore S_{k+1} \text{ is true.}$$

Thus S_n is true for all n.

EXAMPLE 15-28 Use mathematical induction to show that

$$1^3 + 2^3 + 3^3 + \cdots + n^3 = \left[\frac{n(n+1)}{2}\right]^2$$

Solution

$$S_n: \ 1^3 + 2^3 + 3^3 + \cdots + n^3 = \left[\frac{n(n+1)}{2}\right]^2$$

(1)

$$S_1: \ 1^3 \stackrel{?}{=} \left[\frac{1 \cdot 2}{2}\right]^2$$

$$1^3 \stackrel{\checkmark}{=} 1^2 \qquad S_1 \text{ is true.}$$

(2) Write the statement S_k, which we assume to be true in the induction step:

$$S_k: \ 1^3 + 2^3 + 3^3 + \cdots + k^3 = \left[\frac{k(k+1)}{2}\right]^2$$

Now write S_{k+1}:

$$S_{k+1}: \ \underbrace{[1^3 + 2^3 + 3^3 + \cdots + k^3]}_{S_k} + (k+1)^3 \stackrel{?}{=} \left[\frac{(k+1)(k+2)}{2}\right]^2$$

$$S_{k+1}: \ \left[\frac{k(k+1)}{2}\right]^2 + \frac{4(k+1)^3}{4} \stackrel{?}{=} \left[\frac{(k+1)(k+2)}{2}\right]^2$$

$$\frac{(k+1)^2}{4}[k^2 + 4(k+1)] \stackrel{?}{=} \left[\frac{(k+1)(k+2)}{2}\right]^2$$

$$\frac{(k+1)^2}{4}[k^2 + 4k + 4] \stackrel{?}{=} \left[\frac{(k+1)(k+2)}{2}\right]^2$$

$$\frac{(k+1)^2}{4}(k+2)^2 \stackrel{?}{=} \left[\frac{(k+1)(k+2)}{2}\right]^2$$

$$\left[\frac{(k+1)(k+2)}{2}\right]^2 \stackrel{\checkmark}{=} \left[\frac{(k+1)(k+2)}{2}\right]^2$$

So S_n is true for all n.

There are some statements that appear to be true, but can be shown otherwise by induction.

EXAMPLE 15-29 Prove that

$$S_n: \ 1 + 5 + 9 + \cdots + (4n - 3) = n(2n - 1) + (n - 1)$$

is not true for all n.

Solution

(1)

$$S_1: \ 4 \cdot 1 - 3 \stackrel{?}{=} 1(2 - 1) + (1 - 1)$$

$$1 \stackrel{\checkmark}{=} 1 \qquad S_1 \text{ is true.}$$

(2) Assume that S_k is true.

If

$$S_k: \ 1 + 5 + 9 + \cdots + (4k - 3) = k(2k - 1) + (k - 1)$$

then

$$S_{k+1}: \quad \underbrace{[1 + 5 + 9 + \cdots + (4k - 3)]}_{S_k} + (4k + 1) = (k + 1)(2k + 1) + k$$

$$S_{k+1}: \quad [k(2k - 1) + (k - 1)] + (4k + 1) \overset{?}{=} 2k^2 + 4k + 1$$

$$2k^2 - k + k - 1 + 4k + 1 \overset{?}{=} 2k^2 + 4k + 1$$

$$2k^2 + 4k \neq 2k^2 + 4k + 1$$

$$\therefore \; S_{k+1} \text{ cannot be true.}$$

So S_n is not true for all n.

SUMMARY

1. A sequence is a collection of numbers that are arranged in a specific order.
2. A sequence function is a function whose domain is the set of positive integers; the ordered range of a sequence function is a sequence.
3. A recursion formula is a formula that indicates how to generate successive terms of a sequence from one or more preceding terms of the sequence.
4. In an arithmetic sequence, successive terms differ by a common difference d.
5. In a geometric sequence, the ratio r of a term to its predecessor is constant for all pairs.
6. The sum of the first n terms of a sequence is the nth partial sum of the sequence. Such sums are expressed in summation notation.
7. The sums of arithmetic and geometric sequences can be obtained from formulas:

arithmetic sequence:
$$\sum_{i=1}^{n} a_i = S_n = na_1 + d\left(\frac{n(n-1)}{2}\right) = \frac{n(a_1 + a_n)}{2}$$

geometric sequence:
$$\sum_{j=1}^{n} ar^{j-1} = S_n = \begin{cases} na_1 & \text{if } r = 1 \\ a_1\left(\dfrac{1-r^n}{1-r}\right) = a_1\left(\dfrac{r^n-1}{r-1}\right) & \text{if } r \neq 1 \end{cases}$$

8. Under one condition it is possible to sum an infinite number of terms of a geometric sequence and obtain a finite sum: If the common ratio $|r| < 1$, $S_\infty = a_1\left(\dfrac{1}{1-r}\right)$.
9. A repeating decimal number may be converted to a rational number by means of geometric series.
10. The principle of mathematical induction is a method of proof used to show the validity of a statement or formula for all values of its variable n.

RAISE YOUR GRADES
Can you ... ?

☑ define a sequence and a sequence function
☑ describe a recursion formula for a sequence
☑ calculate the value of a factorial
☑ explain the use of summation notation
☑ define an arithmetic sequence and a geometric sequence and explain the difference between them
☑ find the sum of an infinite geometric series
☑ convert a repeating decimal to a rational number
☑ explain the principle of mathematical induction

SOLVED PROBLEMS

Number Sequences

PROBLEM 15-1 Fill in the blanks in the following sequences:

(a) $\frac{1}{2}, \frac{3}{5}, \frac{5}{8}, \frac{7}{11},$ ____, ____, ____, ...

(c) $-1, \frac{1}{2}, -\frac{1}{4}, \frac{1}{8}, -\frac{1}{16},$ ____, ____, ____, ...

(b) $1, \frac{3}{4}, \frac{5}{9}, \frac{7}{16},$ ____, ____, ...

(d) 2, 5, 10, 17, 26, 37, ____, ____, ____, ...

Solution

(a) The numerators increase by 2 each time, while the denominator increases by 3:

$$\frac{1}{2}, \frac{3}{5}, \frac{5}{8}, \frac{7}{11}, \frac{9}{14}, \frac{11}{17}, \frac{13}{20}, \cdots$$

(b) The numerators are successive odd numbers, while the denominators are successive squares:

$$1, \frac{3}{4}, \frac{5}{9}, \frac{7}{16}, \frac{9}{25}, \frac{11}{36}, \cdots$$

(c) The terms alternate in sign and each term is half the previous term:

$$-1, \frac{1}{2}, -\frac{1}{4}, \frac{1}{8}, -\frac{1}{16}, \frac{1}{32}, -\frac{1}{64}, \frac{1}{128}, \cdots$$

(d) These integers are each larger by 1 than the square of an integer. The sequence is

$$2, 5, 10, 17, 26, 37, \underline{50}, \underline{65}, \underline{82}, \cdots$$

PROBLEM 15-2 Write an expression for the general term a_n for each sequence in Problem 15-1.

Solution

(a) $a_n = \dfrac{2n-1}{3n-1}$ for $n = 1, 2, 3, \ldots$

(c) $a_n = (-1)^n \dfrac{1}{2^{n-1}} = \dfrac{(-1)^n}{2^{n-1}}$ for $n = 1, 2, 3, \ldots$

(b) $a_n = \dfrac{2n-1}{n^2}$ for $n = 1, 2, 3, \ldots$

(d) $a_n = n^2 + 1$ for $n = 1, 2, 3, \ldots$

PROBLEM 15-3 If a sequence is defined by the function

$$f(n) = (-1)^n \frac{n(n+2)}{3^n}$$

where the domain is the set of positive integers, what are the first five terms of the sequence?

Solution Put $n = 1, n = 2, \ldots$, into the function $f(n) = (-1)^n \dfrac{n(n+2)}{3^n}$:

n	1	2	3	4	5
$f(n)$	$(-1)^1 \dfrac{1 \cdot 3}{3^1} = -1$	$(-1)^2 \dfrac{2 \cdot 4}{3^2} = \dfrac{8}{9}$	$(-1)^3 \dfrac{3 \cdot 5}{3^3} = -\dfrac{15}{27}$	$(-1)^4 \dfrac{4 \cdot 6}{3^4} = \dfrac{24}{81}$	$(-1)^5 \dfrac{5 \cdot 7}{3^5} = -\dfrac{35}{243}$

PROBLEM 15-4 Write the first six terms of the sequence if **(a)** $f(n) = n^2 - n$ and **(b)** $f(n) = (-1)^{n-1} x^{n+1}$.

Solution Generate the terms by using the positive integers: $1^2 - 1 = 0$, $2^2 - 2 = 2$, etc., and $(-1)^{1-1} x^{1+1} = x^2$, $(-1)^{2-1} x^{2+1} = -x^3$, etc.

(a) $0, 2, 6, 12, 20, 30, \ldots$

(b) $x^2, -x^3, x^4, -x^5, x^6, -x^7, \ldots$

PROBLEM 15-5   Write the first five terms and the eleventh term of the sequence for which

(a) $a_n = (-1)^n \dfrac{2^n}{2^n + 1}$ (b) $a_n = \dfrac{1}{n} - \dfrac{1}{2n}$

Solution   In each case use $n = 1, 2, 3, 4, 5$, and $n = 11$.

(a) $-\frac{2}{3}, \frac{4}{5}, -\frac{8}{9}, \frac{16}{17}, -\frac{32}{33}, \ldots, -\frac{2048}{2049}, \ldots$

(b) $1 - \frac{1}{2}, \frac{1}{2} - \frac{1}{4}, \frac{1}{3} - \frac{1}{6}, \frac{1}{4} - \frac{1}{8}, \frac{1}{5} - \frac{1}{10}, \ldots, \frac{1}{11} - \frac{1}{22}, \ldots$ or $\frac{1}{2}, \frac{1}{4}, \frac{1}{6}, \frac{1}{8}, \frac{1}{10}, \ldots, \frac{1}{22}, \ldots$

PROBLEM 15-6   Write the first six terms of the following recursively defined sequences:

(a) $a_1 = -5, a_n = a_{n-1} + 3$ (b) $a_1 = 4, a_n - a_{n-1} = -2$

Solution   Write the given first term, find the second term from the first, the third from the second, etc.

(a) $-5, -5 + 3, -5 + 3 + 3, -5 + 3 + 3 + 3, \ldots$ or $-5, -2, 1, 4, 7, 10, \ldots$

(b) $4, 4 - 2, 4 - 2 - 2, 4 - 2 - 2 - 2, \ldots$ or $4, 2, 0, -2, -4, -6, \ldots$

PROBLEM 15-7   The Fibonacci sequence is defined by the recursion formula

$$a_1 = 1, a_2 = 1, a_3 = 2, \quad \text{and} \quad a_n = a_{n-1} + a_{n-2} \quad \text{for} \quad n \geq 3$$

Using this as a basis, find a formula that sums the first n terms of the Fibonacci sequence.

Solution   Since $a_n = a_{n-1} + a_{n-2}$ for $n > 2$, you can replace n by $n + 2$ and rewrite the formula as

$$a_{n+2} = a_{n+1} + a_n \quad \text{for } n > 0$$

Then solve for a_n to get

$$a_n = a_{n+2} - a_{n+1}$$

Using this formula, add the first five terms of the sequence:

$$a_1 = a_3 - a_2$$
$$a_2 = a_4 - a_3$$
$$a_3 = a_5 - a_4$$
$$a_4 = a_6 - a_5$$
$$a_5 = a_7 - a_6$$
$$\overline{}$$
$$a_1 + a_2 + a_3 + a_4 + a_5 = a_7 - a_2 = a_7 - 1$$

This suggests the general formula

$$a_1 + a_2 + a_3 + \cdots + a_n = a_{n+2} - 1$$

PROBLEM 15-8   Verify the final formula of Problem 15-7 by using $n = 8$.

Solution   The first ten terms of the Fibonacci sequence are $1, 1, 2, 3, 5, 8, 13, 21, 34, 55$. And the sum of the first eight is $1 + 1 + 2 + 3 + 5 + 8 + 13 + 21 = 54$.

Then $a_{n+2} - 1 = a_{10} - 1 = 55 - 1 = 54$

PROBLEM 15-9   Find (a) $5! - 3!$, (b) $\dfrac{8!}{6!}$, (c) $\dfrac{7!}{4!3!}$, (d) $\dfrac{n!(n-1)!}{[(n+1)!]^2}$.

Solution

(a) $5! - 3! = (5 \cdot 4 \cdot 3 \cdot 2 \cdot 1) - (3 \cdot 2 \cdot 1) = 120 - 6 = 114$

(b) $\dfrac{8!}{6!} = \dfrac{8 \cdot 7 \cdot 6 \cdot 5 \cdot 4 \cdot 3 \cdot 2 \cdot 1}{6 \cdot 5 \cdot 4 \cdot 3 \cdot 2 \cdot 1} = 8 \cdot 7 = 56$

(c) $\dfrac{7!}{4!3!} = \dfrac{7 \cdot 6 \cdot 5 \cdot 4 \cdot 3 \cdot 2 \cdot 1}{4 \cdot 3 \cdot 2 \cdot 1 \cdot 3 \cdot 2 \cdot 1} = \dfrac{7 \cdot 6 \cdot 5}{3 \cdot 2 \cdot 1} = 7 \cdot 5 = 35$

(d) $\dfrac{n!(n-1)!}{[(n+1)!]^2} = \dfrac{n(n-1)!(n-1)!}{[(n+1)(n)(n-1)!]^2} = \dfrac{1}{n(n+1)^2}$ [Remember that $n! = n(n-1)!$]

Partial Sums of Sequences: Series

PROBLEM 15-10   Find the value of the partial sums

(a) $\displaystyle\sum_{j=1}^{4} (j^2 + 1)$ (b) $\displaystyle\sum_{i=1}^{5} \frac{1}{i}$ (c) $\displaystyle\sum_{t=1}^{4} (t^2 - 3t)$

Solution

(a) $\displaystyle\sum_{j=1}^{4} (j^2 + 1) = (1^2 + 1) + (2^2 + 1) + (3^2 + 1) + (4^2 + 1) = 34$

(b) $\displaystyle\sum_{i=1}^{5} \frac{1}{i} = \frac{1}{1} + \frac{1}{2} + \frac{1}{3} + \frac{1}{4} + \frac{1}{5} = \frac{60}{60} + \frac{30}{60} + \frac{20}{60} + \frac{15}{60} + \frac{12}{60} = \frac{137}{60}$

(c) $\displaystyle\sum_{t=1}^{4} (t^2 - 3t) = (1^2 - 3 \cdot 1) + (2^2 - 3 \cdot 2) + (3^2 - 3 \cdot 3) + (4^2 - 3 \cdot 4)$
$$= -2 + (-2) + 0 + 4 = 0$$

PROBLEM 15-11   Write each sum in sigma notation:

(a) $\frac{1}{3} + \frac{1}{6} + \frac{1}{9} + \frac{1}{12} + \frac{1}{15}$ (b) $1 + 4 + 9 + 16 + 25 + 36 + \cdots + 400$

Solution

(a) There are five terms, each of which is a reciprocal of a multiple of 3. Thus

$$\frac{1}{3} + \frac{1}{6} + \frac{1}{9} + \frac{1}{12} + \frac{1}{15} = \sum_{i=1}^{5} \left(\frac{1}{3i}\right)$$

(b) Each term is the square of each successive integer. Therefore the last index value is $\sqrt{400} = 20$, which means

$$1 + 4 + 9 + 16 + 25 + 36 + \cdots + 400 = \sum_{n=1}^{20} (n^2)$$

PROBLEM 15-12   Find (a) $\displaystyle\sum_{n=1}^{30} n$, (b) $\displaystyle\sum_{k=1}^{6} 5$, and (c) $\displaystyle\sum_{n=1}^{5} nx^n$.

Solution

(a) $\displaystyle\sum_{n=1}^{30} n = 1 + 2 + 3 + \cdots + 29 + 30 = \frac{30 \cdot 31}{2} = 465$

(b) $\displaystyle\sum_{k=1}^{6} 5 = 5 + 5 + 5 + 5 + 5 + 5 = 6 \cdot 5 = 30$

(c) $\displaystyle\sum_{n=1}^{5} nx^n = 1x^1 + 2x^2 + 3x^3 + 4x^4 + 5x^5$

Arithmetic Sequences and Series

PROBLEM 15-13   For the following arithmetic sequences, find the common difference d and then the general formula for the nth term.

(a) $6, 13, 20, 27, \ldots$ (b) $3, \frac{3}{2}, 0, -\frac{3}{2}, -3, \ldots$ (c) $-1, -\frac{1}{3}, \frac{1}{3}, 1, \ldots$

Solution Find the common difference d by subtracting any term from its predecessor. Then substitute the appropriate values in $a_n = a_1 + (n - 1)d$ to find the general formula for the nth term.

(a) $d = 13 - 6 = 20 - 13 = 27 - 20 = \cdots = 7$

$a_n = 6 + (n - 1)7 = 7n - 1$

(b) $d = \frac{3}{2} - 3 = 0 - \frac{3}{2} = -\frac{3}{2} - 0 = \cdots = -\frac{3}{2}$

$a_n = 3 + (n - 1)(-\frac{3}{2}) = -\frac{3}{2}n + \frac{9}{2}$

(c) $d = -\frac{1}{3} - (-1) = \frac{1}{3} - (-\frac{1}{3}) = \cdots = \frac{2}{3}$

$a_n = -1 + (n - 1)(\frac{2}{3}) = \frac{2}{3}n - \frac{5}{3}$

PROBLEM 15-14 If the first element of an arithmetic sequence is 8 and the second element is 2, what is the eighth element?

Solution $a_1 = 8$ and $a_2 = 2$, so $d = 2 - 8 = -6$. Then

$$a_8 = a_1 + (8 - 1)d = 8 + 7(-6) = -34$$

PROBLEM 15-15 The eighth element of an arithmetic sequence is 31 and the fifth element is 19. What is the sum of the first element and the tenth element?

Solution First, solve for d.

$$a_8 = a_1 + (8 - 1)d = a + 7d = 31$$
$$a_5 = a_1 + (5 - 1)d = a + 4d = 19 \qquad \text{Subtract}$$
$$\overline{}$$
$$7d - 4d = 31 - 19$$
$$3d = 12$$
$$d = 4$$

Then substitute d into one of the equations above and solve for a_1:

$$a_1 + 7(4) = 31$$
$$a_1 = 3$$

Then
$$a_{10} = a_1 + (10 - 1)d = 3 + (9)(4) = 39$$

and
$$a_1 + a_{10} = 3 + 39 = 42$$

PROBLEM 15-16 If the first three elements of an arithmetic sequence are $\frac{1}{4}$, $\frac{1}{3}$, and $\frac{5}{12}$, how many terms will it take to get to the element with the value 3?

Solution You know that $a_1 = \frac{1}{4}$ and $d = \frac{1}{3} - \frac{1}{4} = \frac{5}{12} - \frac{1}{3} = \frac{1}{12}$. Then substitute these values into $a_n = a_1 + (n - 1)d$, set $a_n = 3$, and solve the equation for n:

$$3 = \frac{1}{4} + (n - 1)\frac{1}{12}$$
$$n = [(13 - \frac{1}{4})12] + 1 = 34$$

PROBLEM 15-17 Find the sum of all the positive even integers that have two digits.

Solution The positive even integers are an arithmetic sequence, so you can calculate a partial sum of them by using formula (15.2b). But first, find the number of terms n being added:

$$a_n = a_1 + (n - 1)d$$
$$n = \frac{a_n - a_1}{d} + 1 = \frac{98 - 10}{2} + 1 = 45$$

Now you can find the sum:

$$S_n = \frac{n}{2}(a_1 + a_n) = \frac{45}{2}(10 + 98) = 2430$$

PROBLEM 15-18   Find the sum of the odd integers between 30 and 80.

Solution   These integers form an arithmetic sequence with $a_1 = 31$, $d = 2$, and $a_n = 79$. Since $a_n = a_1 + (n-1)d$,

$$n = \frac{a_n - a_1}{d} + 1 = \frac{79 - 31}{2} + 1 = 25$$

Thus the sum is

$$S_n = \frac{n}{2}(a_1 + a_n) = \frac{25}{2}(31 + 79) = 1375$$

PROBLEM 15-19   A supermarket display of canned soup has 20 cans on the bottom row and the row above it contains 2 fewer cans. If each succeeding row also contains 2 fewer cans than the row below, how many cans are in the display?

Solution   You need to sum a sequence in which the first term is 20, the last term is 2 (the top row has 2 cans), and $d = -2$. Thus

$$n = \frac{a_n - a_1}{d} + 1 = \frac{2 - 20}{-2} + 1 = 10$$

So there are $n = 10$ rows of cans containing S_n cans:

$$S_n = \tfrac{10}{2}(20 + 2) = 110 \text{ cans}$$

Geometric Sequences and Series

PROBLEM 15-20   Fill in the blanks for the missing terms in the geometric sequences.

(a) $\frac{1}{2}, 1, 2, 4, \underline{\hspace{1em}}, \underline{\hspace{1em}}, \underline{\hspace{1em}}, \dots$

(b) $9, 3, 1, \frac{1}{3}, \underline{\hspace{1em}}, \underline{\hspace{1em}}, \underline{\hspace{1em}}, \dots$

(c) $4, 0.4, 0.04, \underline{\hspace{1em}}, \underline{\hspace{1em}}, \dots$

Solution   In each sequence, find the common ratio $r = \dfrac{a_k}{a_{k-1}}$.

(a) $r = \dfrac{1}{\frac{1}{2}} = \dfrac{2}{1} = \dfrac{4}{2} = \dots = 2$. So the sequence is $\frac{1}{2}, 1, 2, 4, \underline{8}, \underline{16}, \underline{32}, \dots$.

(b) $r = \dfrac{3}{9} = \dfrac{1}{3} = \dfrac{\frac{1}{3}}{1} = \dots = \dfrac{1}{3}$. So the sequence is $9, 3, 1, \frac{1}{3}, \underline{\frac{1}{9}}, \underline{\frac{1}{27}}, \underline{\frac{1}{81}}, \dots$.

(c) $r = \dfrac{0.4}{4} = \dfrac{0.04}{0.4} = \dots = \dfrac{1}{10}$. So the sequence is $4, 0.4, 0.04, \underline{0.004}, \underline{0.0004}, \dots$.

PROBLEM 15-21   A geometric sequence begins as $1, 3, 9, 27, \dots$. What is the 10th term?

Solution   Here $a_1 = 3$ and $r = 3$, and since $a_n = a_1 r^{n-1}$,

$$a_{10} = (1)(3)^9 = 19\,683$$

PROBLEM 15-22   Find the common ratio in a geometric sequence whose first term is 4 and whose sixth term is 972.

Solution   Substitute $n = 6$, $a_1 = 4$, and $a_6 = 972$ into $a_n = a_1 r^{n-1}$. Then solve for r:

$$r^5 = \frac{972}{4} = 243$$

$$r = 3$$

PROBLEM 15-23   Find the tenth term of a geometric sequence whose second term is 6 and whose sixth term is $\frac{2}{27}$.

Solution   If $a_2 = 6$ for $n = 2$, and $a_6 = \frac{2}{27}$ for $n = 6$, then

$$\left.\begin{array}{l} 6 = a_1 r^1 \\ \frac{2}{27} = a_1 r^5 \end{array}\right\} \qquad \frac{a_1 r^5}{a_1 r} = r^4 = \frac{\frac{2}{27}}{6} = \frac{2}{27} \cdot \frac{1}{6} = \frac{1}{81}$$

Thus

$$r = \sqrt[4]{\frac{1}{81}} = \frac{1}{3}$$

Now find a_1:

$$a_n = a_1 r^{n-1} \qquad \text{so} \qquad a_1 = \frac{a_n}{r^{n-1}}$$

Substitute r and one of the given values of n and a_n:

$$a_1 = \frac{6}{\left(\frac{1}{3}\right)^{2-1}} = 6 \cdot 3 = 18$$

Finally, calculate a_n for $n = 10$:

$$a_{10} = a_1 r^{n-1} = 18\left(\frac{1}{3}\right)^{10-1} = \frac{18}{19\,683} = \frac{2}{2187}$$

PROBLEM 15-24   Find the sum of the first 10 terms of the sequence $1, \frac{1}{2}, \frac{1}{4}, \frac{1}{8}, \dots$.

Solution   This is a geometric sequence, so you can find a partial sum of it by applying formula (15.3). Clearly, $r = \frac{1}{2}$, and thus

$$S_{10} = a_1\left(\frac{r^{10} - 1}{r - 1}\right)$$

$$= 1\left(\frac{\left(\frac{1}{2}\right)^{10} - 1}{\frac{1}{2} - 1}\right) = \frac{\frac{1}{1024} - 1}{-\frac{1}{2}} = \frac{-\frac{1023}{1024}}{-\frac{1}{2}} = \frac{1023}{512} \simeq 1.998\,05$$

PROBLEM 15-25   If the first three terms of a sequence are $5, -\frac{10}{3}$, and $\frac{20}{9}$, what is the sum of the first five terms?

Solution   Use formula (15-3) and substitute $n = 5$, $a_1 = 5$, and $r = \dfrac{-\frac{10}{3}}{5} = -\frac{2}{3}$:

$$S_5 = a_1\left(\frac{r^5 - 1}{r - 1}\right) = 5\left(\frac{\left(-\frac{2}{3}\right)^5 - 1}{-\frac{2}{3} - 1}\right) = \frac{275}{81} \simeq 3.3951$$

PROBLEM 15-26   Find the sum of the infinite geometric series

(a) $1 - \frac{1}{3} + \frac{1}{9} - \frac{1}{27} + \cdots$ (b) $\frac{3}{5} + \frac{9}{25} + \frac{27}{125} + \cdots$ (c) $\frac{16}{5} + \frac{4}{5} + \frac{1}{5} + \frac{1}{20} + \cdots$

Solution   First find the common ratio r, and then use formula (15-4):

$$S_\infty = a_1\left(\frac{1}{1 - r}\right)$$

(a) $r = -\dfrac{1}{3}$ and $a_1 = 1$ so $S_\infty = 1\left(\dfrac{1}{1 - \left(-\frac{1}{3}\right)}\right) = \dfrac{1}{\frac{4}{3}} = \dfrac{3}{4}$

(b) $r = \dfrac{3}{5}$ and $a_1 = \dfrac{3}{5}$ so $S_\infty = \dfrac{3}{5}\left(\dfrac{1}{1 - \frac{3}{5}}\right) = \dfrac{3}{5} \cdot \dfrac{1}{\frac{2}{5}} = \dfrac{3}{2}$

(c) $r = \dfrac{1}{4}$ and $a_1 = \dfrac{16}{5}$ so $S_\infty = \dfrac{16}{5}\left(\dfrac{1}{1 - \frac{1}{4}}\right) = \dfrac{16}{5} \cdot \dfrac{1}{\frac{3}{4}} = \dfrac{16}{5} \cdot \dfrac{4}{3} = \dfrac{64}{15}$

PROBLEM 15-27 For the geometric series in Problem 15-26b, $\frac{3}{5} + \frac{9}{25} + \frac{27}{125} + \frac{81}{625} + \cdots$, write out the first seven terms of the sequence of partial sums. Compare these to the limiting value found above, $S_\infty = \frac{3}{2}$.

Solution

$$S_1 = \frac{3}{5} = .600\,00$$

$$S_2 = \frac{3}{5} + \frac{9}{25} = \frac{24}{25} = .960\,00$$

$$S_3 = \frac{24}{25} + \frac{27}{125} = \frac{147}{125} = 1.176\,00$$

$$S_4 = \frac{147}{125} + \frac{81}{625} = \frac{816}{625} = 1.305\,60$$

$$S_5 = \frac{816}{625} + \frac{243}{3125} = \frac{4323}{3125} = 1.383\,36$$

$$S_6 = \frac{4323}{3125} + \frac{729}{15\,625} = \frac{22\,344}{15\,625} = 1.430\,02$$

$$S_7 = \frac{22\,344}{15\,625} + \frac{2187}{78\,125} = \frac{113\,907}{78\,125} = 1.458\,01$$

$$\vdots \qquad\qquad\qquad \vdots$$

$$S_\infty \qquad\qquad\qquad 1.500\,00$$

The sequence of partial sums is "slowing down" and approaching the sum of the infinite series $\frac{3}{2}$.

PROBLEM 15-28 Write **(a)** $0.818\,181\,81\ldots$ and **(b)** $2.142\,424\,242\ldots$ as common fractions.

Solution

(a) This number is the sum of the infinite series

$$S_\infty = .81 + .0081 + .000\,081 + .000\,000\,81 + \cdots$$

$$= \frac{81}{100} + \frac{81}{10\,000} + \frac{81}{1\,000\,000} + \frac{81}{100\,000\,000} + \cdots$$

This is a geometric series with ratio $r = \frac{1}{100}$ and $a_1 = \frac{81}{100}$. Thus

$$S_\infty = a_1\left(\frac{1}{1-r}\right) = \frac{81}{100} \cdot \frac{1}{1 - \frac{1}{100}} = \frac{81}{100} \cdot \frac{100}{99} = \frac{81}{99} = \frac{9}{11}$$

(b) This number is the sum of the infinite series

$$S_\infty = 2 + \frac{1}{10} + (.042 + .000\,42 + .000\,004\,2 + \cdots)$$

$$= \frac{21}{10} + \left(\frac{42}{1000} + \frac{42}{100\,000} + \frac{42}{10\,000\,000} + \cdots\right)$$

$$= \frac{21}{10} + \left(\frac{42}{1000} \cdot \frac{1}{1 - \frac{1}{100}}\right) = \frac{21}{10} + \frac{42}{1000} \cdot \frac{100}{99} = \frac{21}{10} + \frac{42}{990} = \frac{2121}{990} = \frac{707}{330}$$

PROBLEM 15-29 Find the sums of the series

(a) $20 - 2 + 0.2 - 0.02 + \cdots$ **(b)** $1 + (1.08)^{-1} + (1.08)^{-2} + \cdots$

Solution

(a) This series has a ratio of $-\frac{1}{10}$, so its sum is

$$S_\infty = a_1\left(\frac{1}{1-r}\right) = 20\left(\frac{1}{1 - (-\frac{1}{10})}\right) = 20\left(\frac{1}{\frac{11}{10}}\right) = \frac{200}{11} \simeq 18.181\,818\ldots$$

(b) This series has a ratio of $(1.08)^{-1}$, so its sum is

$$S_\infty = 1\left(\frac{1}{1 - (1.08)^{-1}}\right) = \frac{1}{1 - \frac{1}{1.08}} = \frac{27}{2}$$

PROBLEM 15-30 A rubber ball is dropped from a height of 100 feet. Every time the ball rebounds, it climbs to height that is one-half of the distance it just fell. **(a)** What height does the ball achieve after the fourth bounce? **(b)** What is the total distance the ball can travel?

Solution

(a) Since each bounce results in a height that is exactly half the distance from which the ball just fell, the heights the ball can achieve form a geometric sequence where $a_1 = \frac{100}{2} = 50$ and $r = \frac{1}{2}$. Thus

$$a_4 = a_1 r^3 = 50(\tfrac{1}{2})^3 = 6.25 \text{ feet}$$

(b) The distances the ball travels in various parts of the path are 100 ft down, then 50 ft up and 50 ft down, 25 ft up and 25 ft down, etc. Thus the total distance D is

$$D = 100 + [2(50) + 2(25) + \cdots] = 100 + [100 + 50 + \cdots]$$

Now, after the first term, you have the sum of an infinite geometric series, where $a_1 = 100$ and $r = \frac{1}{2}$. So

$$D = 100 + S_\infty = 100 + \left(\frac{a_1}{1-r}\right) = 100 + \frac{100}{1-\frac{1}{2}} = 300 \text{ feet}$$

PROBLEM 15-31   A certain automobile depreciates 10% of its value at the beginning of each year. If the original cost is \$16 000, what is its value after 10 years?

Solution   The sequence of values is

$$\text{New:} \quad 16\,000$$

$$\text{After 1 year:} \quad 16\,000 - 1600 = 14\,400 = (.9)(16\,000)$$

$$\text{After 2 years:} \quad 14\,400 - 1440 = 12\,960 = (.81)(16\,000)$$

$$\text{After 3 years:} \quad 12\,960 - 1296 = 11\,664 = (.729)(16\,000)$$

$$\vdots$$

$$\text{After } n \text{ years:} \quad (.9)^n(16\,000)$$

Thus after 10 years the value is

$$(.9)^{10}(16\,000) = (.348\,678)(16\,000) = \$5578.85$$

PROBLEM 15-32   A certain Christmas Club calls for a \$10 deposit at the end of January and each month after that it requires 1.5 times the amount of the previous month. How much is in the account on December 1?

Solution   The total dollars D after 11 months will be

$$D = 10 + \tfrac{3}{2}(10) + \tfrac{3}{2}(\tfrac{3}{2} \cdot 10) + \tfrac{3}{2}[\tfrac{3}{2}(\tfrac{3}{2} \cdot 10)] + \cdots$$

$$= 10 + \tfrac{3}{2}(10) + (\tfrac{3}{2})^2(10) + (\tfrac{3}{2})^3(10) + \cdots + (\tfrac{3}{2})^{10}(10)$$

The sum from the 2nd term through the 11th term is a geometric series with ratio $\frac{3}{2}$, so you can apply formula (15-3):

$$D = 10 + a_1\left(\frac{1-r^n}{1-r}\right) = 10 + 15\left(\frac{1-(\tfrac{3}{2})^{10}}{1-\tfrac{3}{2}}\right) = 10 + 1699.95 = \$1709.95$$

Mathematical Induction

PROBLEM 15-33   Use the principle of mathematical induction to prove that

$$3 + 7 + 11 + \cdots + (4n - 1) = n(2n + 1)$$

is true for all n.

Solution   Write the statement:

$$S_n: \ 3 + 7 + 11 + \cdots + (4n - 1) = n(2n + 1)$$

(1) Show that the statement S_n is true for $n = 1$:

$$S_1: \ 3 = 1(2 \cdot 1 + 1) = 1(3) = 3 \qquad \text{True}$$

(2) Show that if the statement is true for $n = k$, then it is true for $n = k + 1$. Assume that the statement S_k is true; i.e.,

$$S_k: \ 3 + 7 + 11 + \cdots + (4k - 1) = k(2k + 1)$$

Now show the truth of S_{k+1}:

$$S_{k+1}: \ 3 + 7 + 11 + \cdots + (4k - 1) + (4(k + 1) - 1) \overset{?}{=} (k + 1)(2(k + 1) + 1)$$

$$\underbrace{[3 + 7 + 11 + \cdots + (4k - 1)]}_{S_k} + (4k + 3) \overset{?}{=} (k + 1)(2k + 3)$$

$$k(2k + 1) + (4k + 3) \overset{?}{=} (k + 1)(2k + 3)$$

$$2k^2 + k + 4k + 3 \overset{?}{=} (k + 1)(2k + 3)$$

$$2k^2 + 5k + 3 \overset{?}{=} (k + 1)(2k + 3)$$

$$(2k + 3)(k + 1) \overset{\checkmark}{=} (k + 1)(2k + 3) \qquad \text{True}$$

PROBLEM 15-34 Use mathematical induction to prove that

$$\frac{1}{1 \cdot 3} + \frac{1}{3 \cdot 5} + \frac{1}{5 \cdot 7} + \cdots + \frac{1}{(2n - 1)(2n + 1)} = \frac{n}{2n + 1}$$

is true for all n.

Solution The statement is

$$S_n: \ \frac{1}{1 \cdot 3} + \frac{1}{3 \cdot 5} + \frac{1}{5 \cdot 7} + \cdots + \frac{1}{(2n - 1)(2n + 1)} = \frac{n}{2n + 1}$$

(1) For $n = 1$

$$S_1: \ \frac{1}{3} = \frac{1}{(2 \cdot 1 - 1)(2 \cdot 1 + 1)} = \frac{1}{1 \cdot 3} = \frac{1}{3} \qquad \text{True}$$

(2) Assume that S_n is true for $n = k$:

$$S_k: \ \frac{1}{1 \cdot 3} + \frac{1}{3 \cdot 5} + \frac{1}{5 \cdot 7} + \cdots + \frac{1}{(2k - 1)(2k + 1)} = \frac{k}{2k + 1}$$

Now show that statement S_{k+1} is true:

$$S_{k+1}: \ \left[\frac{1}{1 \cdot 3} + \frac{1}{3 \cdot 5} + \frac{1}{5 \cdot 7} + \cdots + \frac{1}{(2k - 1)(2k + 1)} \right] + \frac{1}{(2k + 1)(2k + 3)} \overset{?}{=} \frac{k + 1}{2k + 3}$$

$$\frac{k}{2k + 1} + \frac{1}{(2k + 1)(2k + 3)} \overset{?}{=} \frac{k + 1}{2k + 3}$$

$$\frac{k(2k + 3)}{(2k + 1)(2k + 3)} + \frac{1}{(2k + 1)(2k + 3)} \overset{?}{=} \frac{k + 1}{2k + 3}$$

$$\frac{2k^2 + 3k + 1}{(2k + 1)(2k + 3)} \overset{?}{=} \frac{k + 1}{2k + 3}$$

$$\frac{(2k + 1)(k + 1)}{(2k + 1)(2k + 3)} \overset{?}{=} \frac{k + 1}{2k + 3}$$

$$\frac{k + 1}{2k + 3} \overset{\checkmark}{=} \frac{k + 1}{2k + 3} \qquad \text{True}$$

PROBLEM 15-35 Prove that

$$S_n: \ (1 + x)^n \geq nx + 1 \quad \text{for } x \geq 0$$

is true for all natural numbers n.

Solution

(1) Prove the truth of S_1:

$$S_1: (1 + x)^1 \geq 1 \cdot x + 1 \quad \text{for } x \geq 0$$

or
$$1 + x \geq x + 1 \quad \text{for } x \geq 0 \qquad \text{True}$$

(2) Assume that S_k holds:

$$S_k: (1 + x)^k \geq kx + 1 \quad \text{for } x \geq 0$$

Now use this to show the truth of S_{k+1}:

$$S_{k+1}: (1 + x)^{k+1} \geq (k + 1)x + 1 \quad \text{for } x \geq 0$$

Now do a little algebra: First, you know that $(1 + x)^{k+1} = (1 + x)^k(1 + x)$. And if you replace $(1 + x)^k$ with $kx + 1$, you make the right side of this equation smaller because you assumed that $(1 + x)^k \geq kx + 1$. Thus it is also true that

$$(1 + x)^{k+1} \geq (kx + 1)(1 + x)$$

Therefore

$$(1 + x)^{k+1} \geq kx + kx^2 + 1 + x$$
$$\geq kx + x + kx^2 + 1$$
$$\geq (k + 1)x + kx^2 + 1$$

Next, if you subtract the positive term kx^2 from the right side, you make that side even smaller; so now it is true that

$$(1 + x)^{k+1} \geq (k + 1)x + 1 \quad \text{for } x \geq 0$$

But this is exactly what you wanted to show! Thus the induction is finished, and the statement is true for all natural numbers n.

PROBLEM 15-36 Use mathematical induction to show that $n^3 - n + 3$ is divisible by 3 for all positive integers n.

Solution The statement is

$$S_n: n^3 - n + 3 \text{ is divisible by 3 for all } n$$

(1) Show that S_1 is true.

$$S_1: 1^3 - 1 + 3 \text{ is divisible by 3}$$

or
$$3 \text{ is divisible by 3} \qquad \text{True}$$

(2) Assume the truth of S_k:

$$k^3 - k + 3 \text{ is divisible by 3}$$

You may write this algebraically as

$$S_k: k^3 - k + 3 = m \cdot 3 \qquad (m \text{ is an integer})$$

Now show that S_{k+1} is true; i.e., show that

$$S_{k+1}: (k + 1)^3 - (k + 1) + 3 \text{ is divisible by 3}$$

Rewrite this statement as

$$k^3 + 3k^2 + 3k + 1 - k - 1 + 3 \text{ is divisible by 3}$$

or
$$(k^3 - k + 3) + 3k^2 + 3k \text{ is divisible by 3}$$

The first term is $m \cdot 3$ (from S_k above), so you may rewrite the statement again as

$$m \cdot 3 + 3k^2 + 3k \text{ is divisible by 3}$$

Every term now contains a 3, so the expression is divisible by 3. Thus the statement S_n is always true.

Supplementary Exercises

PROBLEM 15-37 Fill in the missing terms in each sequence:

(a) $\frac{1}{3}, \frac{2}{3}, 1, \frac{4}{3}, \underline{\quad}, \underline{\quad}, \underline{\quad}, \cdots$

(b) $\frac{1}{2}, \frac{1}{3}, \frac{1}{6}, 0, \underline{\quad}, \underline{\quad}, \underline{\quad}, \cdots$

(c) $-2, 1, 6, 13, 22, \underline{\quad}, \underline{\quad}, \underline{\quad}, \cdots$

PROBLEM 15-38 Find a function to express each sequence in Problem 15-37.

PROBLEM 15-39 Write the first four terms of $\{4 - \frac{2}{3}n\}$.

PROBLEM 15-40 If $a_1 = 6$ and $a_n = a_{n-1} + \frac{5}{2}$, find $a_5 + a_6 + a_7$.

PROBLEM 15-41 In $\{2^n - 2n\}$, find a_6 and a_{10}.

PROBLEM 15-42 Find the 12th term of $\left\{6 + \dfrac{n}{2}\right\}$.

PROBLEM 15-43 Find the general formula for any term in the sequence $1, \frac{1}{2}, \frac{1}{6}, \frac{1}{24}, \frac{1}{120}, \ldots$.

PROBLEM 15-44 List the first six terms of $\left\{1 + \dfrac{(-1)^n}{3^n}\right\}$.

PROBLEM 15-45 Evaluate (a) $\dfrac{5!}{2!3!}$ and (b) $\dfrac{6!}{4!2!}$.

PROBLEM 15-46 Write

  (a) $2 \times 4 \times 6 \times 8 \times \cdots \times (2n)$ and (b) $1 \times 3 \times 5 \times 7 \times 9 \times \cdots \times (2n - 1)$

in factorial notation.

PROBLEM 15-47 Use sigma (summation) notation to express

(a) $\frac{1}{2} + \frac{2}{3} + \frac{3}{4} + \frac{4}{5} + \frac{5}{6}$ (b) $1 + 4 + 7 + 10 + 13 + 16$ (c) $1^3 + 2^3 + 3^3 + 4^3 + 5^3$

PROBLEM 15-48 Find the value of each sum:

(a) $\displaystyle\sum_{n=1}^{5} 7$ (b) $\displaystyle\sum_{k=1}^{6} (k^2 - 2k)$ (c) $\displaystyle\sum_{j=3}^{7} (j^2 + 2)$

PROBLEM 15-49 Find the value of the sum $\displaystyle\sum_{i=1}^{50} \left(\dfrac{1}{i+1} - \dfrac{1}{i}\right)$.

PROBLEM 15-50 Use a calculator to find the value of $\displaystyle\sum_{j=1}^{13} \left(\dfrac{1}{2^j}\right)$.

PROBLEM 15-51   For each given arithmetic sequence, find the 5th term and the formula for the nth term:

(a) $a_1 = 2, d = -3$ (b) $a_1 = 4, d = -\frac{2}{3}$ (c) $a_1 = 2\sqrt{2}, d = 2\sqrt{2}$

PROBLEM 15-52   If an arithmetic sequence begins with 12 and has $d = -\frac{4}{3}$, what is its 10th term?

PROBLEM 15-53   An arithmetic sequence has a common difference of 6 and the sum of the first four terms is zero. (a) What is the first term? (b) What is the sum of ten terms?

PROBLEM 15-54   In 10 weeks a youngster must save $85. She begins by putting away $4 the first week, but wishes to increase the amount saved each week by a constant amount. What should the constant amount be?

PROBLEM 15-55   The 2nd term of an arithmetic sequence is -3 and the 10th term is 13. What is the 4th term and what is the 25th term?

PROBLEM 15-56   If a geometric sequence has $a_1 = 5$ and $r = -\frac{1}{2}$, what is a_{12}?

PROBLEM 15-57   Find the 6th term and the nth term if a given geometric series begins with $\frac{1}{3}$ and has common ratio $\frac{2}{3}$.

PROBLEM 15-58   Find the value of the geometric series $\sum_{j=1}^{8} \left(\frac{5}{2^j} \right)$.

PROBLEM 15-59   The number of bacteria in a certain culture doubles each hour. If the culture contains 200 bacteria now, how many will it have in eight hours?

PROBLEM 15-60   If a ball is dropped from 60 feet and rebounds to $\frac{2}{3}$ of its previous height at each bounce, how high is it after the third bounce? How far does it travel in its entire path?

PROBLEM 15-61   Find the sum of the odd integers between 40 and 90.

PROBLEM 15-62   There are 28 logs in the bottom layer of a wood pile, 27 in the second, 26 in the third, etc. The top layer has 8 logs. How many logs are in the pile?

PROBLEM 15-63   A soap box racer is cruising down a hill. It covers 2 feet in the first second but increases the distance by 5 feet in each succeeding second. If it takes 18 seconds to reach the finish line, how far does it travel?

PROBLEM 15-64   Find the nth term in the geometric sequence $5, -3, \frac{9}{5}, \ldots$.

PROBLEM 15-65   The number of bowling pins needed to form a triangular set of pins generates an interesting sequence. This is the placement for the first four cases:

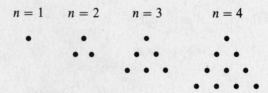

(a) What are the next three values for n? (b) What is the general formula for the nth member of this sequence?

PROBLEM 15-66   If a geometric sequence has $\sqrt{2}$ and -2 for its first and second terms, respectively, what is its sixth term and what is the sum of the first ten terms?

PROBLEM 15-67 Find the sum of the series

(a) $1 - \frac{1}{4} + \frac{1}{16} - \frac{1}{64} + \cdots$ (b) $\frac{1}{5} + \frac{2}{15} + \frac{4}{45} + \cdots$ (c) $100 - 40 + 16 \cdots$

PROBLEM 15-68 Find the rational number equivalent to (a) $3.141\,414\,14\ldots$ and (b) $0.617\,617\,617\ldots$.

PROBLEM 15-69 Find the sum of the series (a) $2.2 + .22 + .022 + .0022 + \cdots$ and (b) $5 + (1.05)^{-1} + (1.05)^{-2} + \cdots$.

Answers to Supplementary Exercises

15-37 (a) $\frac{5}{3}, 2, \frac{7}{3}$

   (b) $-\frac{1}{6}, -\frac{1}{3}, -\frac{1}{2}$

   (c) $33, 46, 61$

15-38 (a) $\frac{1}{3} + \frac{n}{3}$ for $n = 0, 1, 2, \ldots$

   (b) $\frac{1}{2} - \frac{n}{6}$ for $n = 0, 1, 2, \ldots$

   (c) $n^2 - 3$ for $n = 1, 2, 3, \ldots$

15-39 $\frac{10}{3}, \frac{8}{3}, \frac{6}{3}, \frac{4}{3}$ **15-40** $\frac{111}{2}$

15-41 $a_6 = 52, \quad a_{10} = 1004$

15-42 12   **15-43** $\frac{1}{n!}$

15-44 $\frac{2}{3}, \frac{10}{9}, \frac{26}{27}, \frac{82}{81}, \frac{242}{243}, \frac{730}{729}$

15-45 (a) 10   (b) 15

15-46 (a) $2^n \cdot n!$ (b) $\frac{(2n)!}{2^n \cdot n!}$

15-47 (a) $\sum_{i=1}^{5} \left(\frac{i}{i+1} \right)$

   (b) $\sum_{j=0}^{5} (3j + 1)$

   (c) $\sum_{n=1}^{5} n^3$

15-48 (a) 35   (b) 49   (c) 145

15-49 $-1 + \frac{1}{51}$ **15-50** $.999\,877\,6$

15-51 (a) $a_5 = -10, \quad a_n = 5 - 3n$

   (b) $a_5 = \frac{4}{3}, \quad a_n = \frac{14 - 2n}{3}$

   (c) $a_5 = 10\sqrt{2}, \quad a_n = (2\sqrt{2})n$

15-52 0

15-53 (a) -9,   (b) 180

15-54 \$1

15-55 $a_4 = 1, \quad a_{25} = 43$

15-56 $-\frac{5}{2048}$

15-57 $a_6 = \frac{32}{729}, \quad a_n = \frac{2^{n-1}}{3^n}$

15-58 $\frac{1275}{256}$

15-59 51 200

15-60 $\frac{160}{9}$ ft,   300 ft

15-61 1625

15-62 378

15-63 801 ft

15-64 $a_n = \frac{(-3)^{n-1}}{5^{n-2}}$

15-65 (a) 15, 21, 28

   (b) $a_n = \frac{n(n+1)}{2}$

15-66 $a_6 = -8, \quad S_{10} = \frac{31\sqrt{2}}{\sqrt{2} - 1}$

15-67 (a) $\frac{4}{5}$ (b) $\frac{3}{5}$ (c) $\frac{1000}{14}$

15-68 (a) $\frac{311}{99}$ (b) $\frac{617}{999}$

15-69 (a) 2.4444...   (b) 25

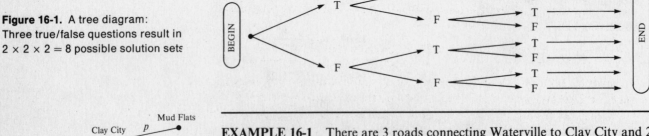

16 COUNTING TECHNIQUES

THIS CHAPTER IS ABOUT

☑ **A Counting Principle**
☑ **Permutations and Combinations**
☑ **The Binomial Theorem**

16-1. A Counting Principle

Counting is not always as easy as 1, 2, 3, ... When we're counting results [things, people, whatever] in a multistage process, we need some special techniques, from which we can define a counting principle.

A. Tree diagrams

Suppose you are required to take a quiz that consists of 3 true/false questions. You haven't studied the material, so you must guess on every question. In how many different ways could you form an answer set? That is, how would you count the number of possible answers?

One way to find out is to construct a **tree diagram**, which is a graphical depiction of the possible solution sets in a multistage counting process. Figure 16-1 shows its basic construction. From "begin" to "end" there are 8 paths, which represent the 8 possible solution sets. Follow the paths to find the solution sets. That is: The first question presents you with 2 choices, true (T) or false (F). For each of these responses, you have 2 additional choices T or F for the second question, which gives you a total of 4 choices on the first two questions: TT, TF, FT, or FF. For each of these 4 choices you have 2 additional choices on the third question, for a total of $2 \times 2 \times 2 = 8$ choices: TTT, TTF, TFT, TFF, FTT, FTF, FFT, FFF.

Figure 16-1. A tree diagram: Three true/false questions result in $2 \times 2 \times 2 = 8$ possible solution sets

Figure 16-2

EXAMPLE 16-1 There are 3 roads connecting Waterville to Clay City and 2 roads connecting Clay City to Mud Flats. A truck driver starting in Waterville must make a delivery in Clay City and end up in Mud Flats. How many different routes can she take?

Solution Construct a tree diagram, as in Figure 16-2. Label the 3 routes from Waterville to Clay City a, b, c, and the 2 routes from Clay City to Mud Flats p, q. Then follow the paths of the routes, which are easily listed by road pairs: ap, aq, bp, bq, cp, cq, for a total of $3 \times 2 = 6$ ways.

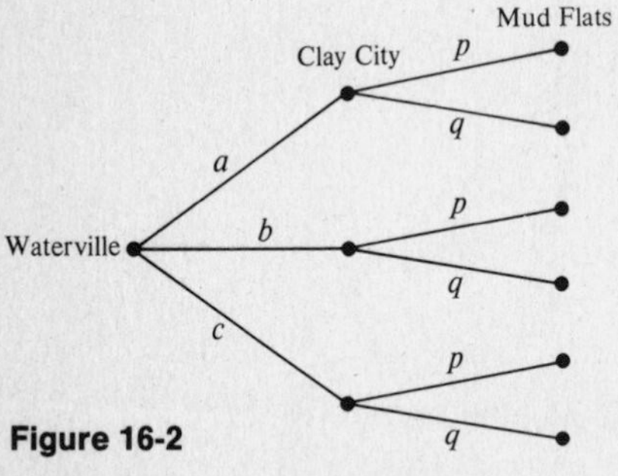

B. The multiplication rule

The tree diagram is just a picture of a basic technique that may be applied to situations in which counting is required.

EXAMPLE 16-2   A hiking club with 8 members needs to choose a chairperson, secretary, and treasurer for the group. In how many different ways can this be done?

Solution   The tree diagram would be overly cumbersome here, but we can still use its construction technique to solve the problem. At step 1, the chairperson can be any of the 8 persons. So this step has 8 choices. Once that office has been filled, any one of the remaining 7 persons could be chosen as secretary. For each of the original 8 choices there are now 7 additional possibilities for a total of $8 \times 7 = 56$ ways of electing the first two officers. Now, for each of those 56 choices any one of the remaining 6 people could be chosen for the treasurer's post. That gives $8 \times 7 \times 6 = 336$ ways of choosing the 3 officers from among the 8 club members.

If the items we are counting are selected in stages, and if we know how many choices we have at each stage, we multiply the choice numbers of the various stages to get the total number of ways of doing the counting. The formal statement of this technique is called the **Multiplication Rule**:

MULTIPLICATION RULE   Suppose N objects are to be selected in a sequence of events. If the first event can occur in k_1 ways, the second in k_2 ways, the third in k_3 ways, and so forth, then the number of ways all N of the events occur is

$$k_1 \times k_2 \times k_3 \times \cdots \times k_N \text{ ways}$$

The multiplication rule is, in fact, a **counting principle**.

EXAMPLE 16-3   Many years ago the automobile license plates in Michigan consisted of 2 letters followed by 4 digits (e.g., MN 5216). How many different letter/digit groupings could be produced in this pattern?

Solution   Consider the 6 entries to be selected in 6 stages. The first entry, a letter of the alphabet, may be chosen in 26 ways. The second entry may also be chosen in 26 ways (assuming there is no restriction on the letters' being the same). The next 4 stages involve digits. A digit may be chosen from 0, 1, 2, 3, 4, 5, 6, 7, 8, 9, so there are 10 choices for each of the 4-digit entries. Thus a 6-entry plate may be constructed in

$$26 \times 26 \times 10 \times 10 \times 10 \times 10 = 6\,760\,000 \text{ ways}$$

EXAMPLE 16-4   Not long ago the number of cars in Michigan was approaching 6.5 million and rising. The state officials needed to change the format for the plates to avoid repetition of letter/number combinations. Now the license plates have 3 letters and 3 numbers. How many different plates are now possible?

Solution   For each of the 3 letters there are 26 possible choices, and for each of the 3 digits there are 10 possible choices. The total number of ways in which plates can be constructed now is

$$26 \times 26 \times 26 \times 10 \times 10 \times 10 = 17\,576\,000 \text{ ways}$$

EXAMPLE 16-5   At a choose-it-yourself sandwich bar you have the following choices:

Type of bread: Whole wheat, rye, sourdough, white
Type of spread: Butter, mayonnaise, mustard
Type of meat: Ham, beef, turkey, salami
Type of cheese: American, Swiss
Extras (pick 1):   Lettuce, tomato, sliced pickles, chopped olives

How many visits could you make and have a different sandwich each time?

Solution There are 5 main choices, so this is a 5-stage process, and the number of choices at each stage must be counted. The total number of different sandwiches is

$$4 \times 3 \times 4 \times 2 \times 4 = 384$$

16-2. Permutations and Combinations

A. Simple permutations

We use a special case of the multiplication rule when we want to count the number of possible ordered arrangements of a given collection of items.

EXAMPLE 16-6 Suppose we have 4 balls—red, blue, green, and yellow—which we want to arrange in a row. How many different arrangements are possible?

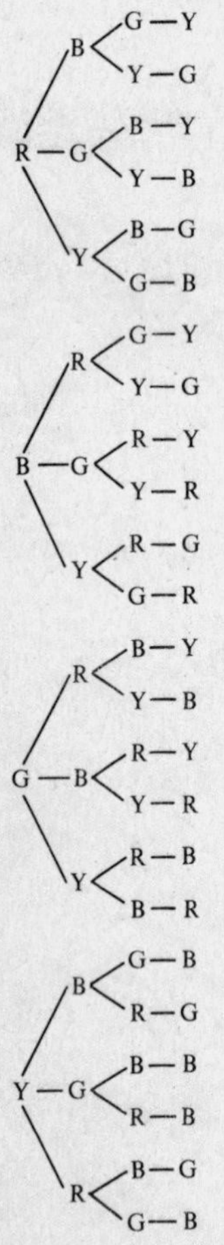

Solution We may pick any of the 4 colored balls for the first position. Then, there are 3 choices left for the second position, 2 choices for the third, and 1 choice for the fourth. Thus the total number of arrangements is $4 \times 3 \times 2 \times 1 = 24$ arrangements (see Figure 16-3).

- An ordered arrangement of a set of objects is called a **permutation** of those objects.

EXAMPLE 16-7 In how may different ways could 6 people line up to board a bus?

Solution We can think of this as if we were filling 6 waiting spots with 6 different people. We can choose any of the 6 people to put in spot 1, which leaves 5 choices for spot 2, 4 for spot 3, etc. Thus the total number of ordered arrangements, or permutations, is

$$6 \times 5 \times 4 \times 3 \times 2 \times 1 = 6! = 720 \text{ ways}$$

These two examples illustrate a general law:

- The number of permutations of n objects taken n at a time is $n!$, denoted

$$_nP_n = n! \tag{16.1}$$

   note: The phrase "n objects taken n at a time" means that we use ALL the objects, n of n.

But, if we want to choose only r of the n original distinguishable objects and put these r objects into an arrangement, then we are finding the permutation of n objects taken r at a time, denoted by $_nP_r$.

EXAMPLE 16-8 If we are given the letters in the word DEPOT, how many 3-letter code groupings (not necessarily words) could we form?

$4 \times 3 \times 2 \times 1 = 24$

Figure 16-3

Solution We are to fill three slots, ___ ___ ___, with the letters from DEPOT. Any one letter could go first. Thus there are 5 choices for the first slot. Then, of the remaining letters, we have 4 choices for the second slot and 3 choices for the third slot. Hence there are

$$5 \times 4 \times 3 = 60 \text{ total choices}$$

That is, we have the permutation of 5 things taken 3 at a time.

- The general formula for computing the number of permutations of n objects taken r at a time is

$$_nP_r = n(n-1)(n-2)\cdots(n-r+2)(n-r+1) \tag{16.2a}$$

EXAMPLE 16-9 If we use the digits 1, 2, 3, 4, 5, 6, 7 in forming various 4-digit numbers, how many numbers can be formed if no repetitions are allowed?

Solution We have 7 objects that we are taking 4 at a time and we are concerned about their arrangement; i.e., this is the permutation of $n = 7$ objects taken $r = 4$ at a time. We can do this logically: We know that we have 7 choices for the first digit, 6 choices for the second digit, 5 for the third digit, and 4 for the final digit. Thus there are $7 \times 6 \times 5 \times 4 = 840$ different 4-digit numbers. Or, we can use formula (16-2a) for $_7P_4$, which is the product of the digits from $n = 7$ going DOWN to $(n - r + 1) = (7 - 4 + 1) = 4$. Hence

$$_7P_4 = 7(7-1)(7-2)(7-3) = 7 \times 6 \times 5 \times 4 = 840$$

EXAMPLE 16-10 Suppose we have the same conditions as in Example 16-9 except that repetition of the digits is allowed. Now how many numbers can be formed?

Solution We go back to the counting procedure for multistage processes and use the multiplication rule. The choices are 7-fold for each of the four slots, so we would have

$$7 \times 7 \times 7 \times 7 = 2401 \text{ numbers}$$

The formula for the number of permutations of n objects taken r at a time may be algebraically rewritten as follows:

$$_nP_r = n(n-1)(n-2)\cdots(n-r+2)(n-r+1)$$

$$= n(n-1)(n-2)\cdots(n-r+2)(n-r+1)\left(\frac{(n-r)(n-r-1)\cdots3\cdot2\cdot1}{(n-r)(n-r-1)\cdots3\cdot2\cdot1}\right)$$

or

NUMBER OF PERMUTATIONS (*r* of *n*)
$$_nP_r = \frac{n(n-1)(n-2)\cdots3\cdot2\cdot1}{(n-r)(n-r-1)\cdots3\cdot2\cdot1} = \frac{n!}{(n-r)!} \tag{16.2b}$$

EXAMPLE 16-11 Evaluate (a) $_4P_2$, (b) $_6P_3$, and (c) $_5P_5$.

Solution We need only put the numbers into the formula $_nP_r = \dfrac{n!}{(n-r)!}$:

(a) $_4P_2 = \dfrac{4!}{(4-2)!} = \dfrac{4!}{2!} = \dfrac{4\cdot3\cdot2\cdot1}{2\cdot1} = 4\cdot3 = 12$

(b) $_6P_3 = \dfrac{6!}{(6-3)!} = \dfrac{6!}{3!} = \dfrac{6 \cdot 5 \cdot 4 \cdot 3 \cdot 2 \cdot 1}{3 \cdot 2 \cdot 1} = 6 \cdot 5 \cdot 4 = 120$

(c) $_5P_5 = \dfrac{5!}{(5-5)!} = \dfrac{5!}{0!} = \dfrac{5!}{1} = 5! = 5 \cdot 4 \cdot 3 \cdot 2 \cdot 1 = 120$

note: Remember that 0! is defined as 1.

EXAMPLE 16-12   A softball team is made up of 9 players. Find the number of different batting orders the manager can make if

(a) The first 4 batters are fixed in the line-up.
(b) Only the pitcher, who bats last, is fixed in the line-up.

Solution

(a) Four slots are already filled. The manager must permute the last 5 players into the remaining 5 slots. This is

$$_5P_5 = \frac{5!}{(5-5)!} = \frac{5!}{0!} = 5! = 120$$

(b) Only 1 slot is filled, the other 8 players get permuted into the 8 remaining slots. This is

$$_8P_8 = 8! = 40\,320$$

EXAMPLE 16-13   If 10 basketball teams are in a tournament, in how many ways could the potential first, second, and third place teams be listed?

Solution   This is an ordered arrangement of 10 items taken 3 at a time. This is

$$_{10}P_3 = \frac{10!}{(10-3)!} = \frac{10!}{7!} = \frac{10 \cdot 9 \cdot 8 \cdot 7 \cdot 6 \cdot 5 \cdot 4 \cdot 3 \cdot 2 \cdot 1}{7 \cdot 6 \cdot 5 \cdot 4 \cdot 3 \cdot 2 \cdot 1} = 10 \cdot 9 \cdot 8 = 720$$

EXAMPLE 16-14   Find the number of ways the first 4 cards can be dealt from a deck of 52 cards.

Solution   We have 52 items and we are choosing an ordered set of 4. We have

$$_{52}P_4 = \frac{52!}{(52-4)!} = \frac{52!}{48!} = 52 \cdot 51 \cdot 50 \cdot 49 = 6\,497\,400$$

B. Permutations with complications

So far we've assumed that the objects we were arranging were distinguishable one from the other. But an arrangement can contain some objects that are exactly alike. If we have some objects that are not distinguishable, we have fewer arrangements.

EXAMPLE 16-15   Dennis' boat carries 3 identical blue flags as well as 1 red flag and 1 green flag. If he displays all 5 flags on a vertical pole, how many distinguishable arrangements are possible?

Solution   If the flags were all different colors, he could make $_5P_5 = 5! = 120$ different arrangements. But the 3 blue flags can themselves be put into $3! = 3 \times 2 \times 1 = 6$ different arrangements without anyone's being able to tell them apart. For example, BRBBG could be displayed in 6 different ways by switching the blue ones around ($B_1RB_2B_3G, B_2RB_1B_3G$, etc.). Hence we need to divide out these repetitions. We divide by 3! to get

$$\frac{5!}{3!} = \frac{5 \cdot 4 \cdot 3 \cdot 2 \cdot 1}{3 \cdot 2 \cdot 1} = 5 \cdot 4 = 20 \quad \text{different arrangements}$$

In general, we have the following rule:

- If in a collection of N objects there are k_1 of one kind, k_2 of another kind, k_3 of a third kind, etc., then the number of distinguishable permutations of the N objects is

$$_N P_{k_i} = \frac{N!}{(k_1)!(k_2)! \cdots (k_i)!} \qquad (16.3)$$

EXAMPLE 16-16   We have 3 black disks, 2 white disks, and 1 orange disk. We want to form a line of 6 disks. How many distinguishable permutations could be formed?

Solution   We have $N = 6$, $k_1 = 3$, $k_2 = 2$, $k_3 = 1$. Thus we have

$$\frac{N!}{k_1! k_2! k_3!} = \frac{6!}{3! 2! 1!} = \frac{6 \cdot 5 \cdot 4 \cdot 3 \cdot 2 \cdot 1}{3 \cdot 2 \cdot 1 \cdot 2 \cdot 1 \cdot 1} = 60 \text{ distinguishable lines}$$

EXAMPLE 16-17   How many distinguishable 9-letter groupings can be formed from the letters of TENNESSEE?

Solution   We have 9 letters, but there are 2 N's, 2 S's, and 4 E's besides the lonesome T. Thus we have

$$\frac{9!}{2! 2! 4! 1!} = \frac{9 \cdot 8 \cdot 7 \cdot 6 \cdot 5 \cdot 4 \cdot 3 \cdot 2 \cdot 1}{2 \cdot 1 \cdot 2 \cdot 1 \cdot 4 \cdot 3 \cdot 2 \cdot 1 \cdot 1} = 9 \cdot 4 \cdot 7 \cdot 3 \cdot 5 = 3780 \text{ groupings}$$

C. Combinations

When a baseball manager makes out his batting line-up, he assigns specific positions in the line-up to specific players. This is an ordered arrangement, or permutation. But sometimes the order of the persons (or objects) is not important. For the basketball coach to pick 5 starters from among the 10 players does not involve ordering. To pick a set of 3 volunteers from a group of 9 people does not involve ordering them. If we need to select 4 books from an assortment of 15 books for our Book-of-the-Month premium, we are not concerned about the order of the books in the selection. These "unordered" selections are called *combinations*.

- When we select subsets of r objects from a set of n objects, without regard to order, we are forming **combinations** of n objects taken r at a time, denoted $_n C_r$.

Thus the basketball coach needs combinations of 10 things taken 5 at a time, $_{10} C_5$. The volunteers chosen from a group of people make up combinations of 9 things taken 3 at a time, $_9 C_3$. The books chosen from an assortment are combinations of 15 things taken 4 at a time, $_{15} C_4$.

The number of permutations of n things taken r at a time, $_n P_r$, takes into account the order in which the r things are selected. But we are not concerned with order in combinations, so we divide out the ordered sets to find the number of unordered sets. Since each combination corresponds to $r!$ permutations of the r objects such that $_n P_r = (r!)(_n C_r)$, we get

NUMBER OF COMBINATIONS *(r of n)*

$$_n C_r = \frac{_n P_r}{r!} = \frac{\dfrac{n!}{(n-r)!}}{r!} = \frac{n!}{r!(n-r)!} \qquad (16.4)$$

EXAMPLE 16-18   Calculate the combinations in **(a)** $_{10} C_5$, **(b)** $_9 C_3$, and **(c)** $_{15} C_4$.

Solution

(a) $_{10}C_5 = \dfrac{10!}{5!(10-5)!} = \dfrac{10!}{5!5!} = \dfrac{10\cdot9\cdot8\cdot7\cdot6}{5\cdot4\cdot3\cdot2\cdot1} = 2\cdot3\cdot2\cdot7\cdot3 = 252$

(b) $_9C_3 = \dfrac{9!}{3!(9-3)!} = \dfrac{9!}{3!6!} = \dfrac{9\cdot8\cdot7}{3\cdot2\cdot1} = 3\cdot4\cdot7 = 84$

(c) $_{15}C_4 = \dfrac{15!}{4!11!} = \dfrac{15\cdot14\cdot13\cdot12}{4\cdot3\cdot2\cdot1} = 15\cdot7\cdot13 = 1365$

EXAMPLE 16-19 A child has a penny (P), a nickel (N), a dime (D), a quarter (Q), and a half-dollar (H). How many different 2-coin payouts can be made? List the payouts.

Solution The order of the two coins is not important. So we need to find the combinations of 5 things taken 2 at a time:

$$_5C_2 = \frac{5!}{2!3!} = \frac{5\cdot4\cdot3\cdot2\cdot1}{2\cdot1\cdot3\cdot2\cdot1} = \frac{5\cdot4}{2\cdot1} = 10 \text{ two-coin payouts}$$

PN PD PQ PH ND NQ NH DQ DH QH

EXAMPLE 16-20 Suppose we construct a 12-sided polygon by picking 12 points on a circle and connecting them consecutively. Then, defining a diagonal as a line segment joining two nonconsecutive vertices, we draw all its diagonals. How many diagonals could we draw?

Solution The total number of lines that could be put in as diagonals for n points is $_nC_2$. But if we include all the possibilities, we are including lines between adjacent points, which are not diagonals but are edges. So we deduct the number of edges, which is $n = 12$. Thus for n points we have $_nC_2 - n$. For a 12-sided polygon we have

$$_{12}C_2 - 12 = \frac{12!}{2!10!} - 12 = \frac{12\cdot11}{2} - 12 = 66 - 12 = 54 \text{ diagonals}$$

In Section 16-3 we'll learn how to find the terms of the expansion of the binomial expression $(x + y)^n$ for positive integer n. The coefficients in this expansion are called **binomial coefficients**, and they turn out to be $_nC_k$ for various values of k. In that context they are written using the notation

$$_nC_k = \binom{n}{k} = \frac{n!}{k!(n-k)!}$$

EXAMPLE 16-21 Find the value of (a) $\binom{6}{2}$ and (b) $\binom{9}{6}$.

Solution

(a) $\binom{6}{2} = {_6C_2} = \dfrac{6!}{2!4!} = \dfrac{6\cdot5}{2\cdot1} = 15$

(b) $\binom{9}{6} = {_9C_6} = \dfrac{9!}{6!3!} = \dfrac{9\cdot8\cdot7}{3\cdot2\cdot1} = 3\cdot4\cdot7 = 84$

EXAMPLE 16-22 Dear Aunt Emily offers her cat Brunhilde a choice of any 4 of 6 different flavors of Kitty Treats—tuna, liver, beef, egg, milk, and cheese. In how many ways can the choices be made?

Solution   In any problem involving choosing we always need to decide whether the order of the choices is important or immaterial. If order is important, we have a permutation problem; if the order of the choices is immaterial, we have a combination problem. Here, Brunhilde is not concerned over the order in which the treats are chosen—she just picks 'em and starts crunchin'. So we have to determine the number of combinations of 6 things taken 4 at a time:

$$_6C_4 = \frac{6!}{4!2!} = \frac{6 \cdot 5}{2 \cdot 1} = 15$$

note: From Examples 16-21 and 16-22 you may have observed that

$$_6C_2 = \frac{6!}{2!4!} = 15 = \frac{6!}{4!2!} = {_6C_4}$$

Thus

$$\binom{6}{2} = \binom{6}{4}$$

EXAMPLE 16-23   A student has 4 English books, 3 math books, and 5 psychology books. In how many ways can these books be arranged on a shelf if the books in the same subject must be kept together?

Solution   Here, order is very important, since an interchange of just two of the books on the shelf results in a different arrangement. Thus we have permutation problems to solve. First look at the subject groups. The 4 English books can be permuted in $_4P_4 = 4! = 24$ ways. For each of these ways the math books are permuted in $_3P_3 = 3! = 6$ ways. And for each of these 24×6 ways for English and math we have $_5P_5 = 5! = 120$ ways for the psychology books. So far we have

$$_4P_4 \times {_3P_3} \times {_5P_5} = 24 \times 6 \times 120 = 17\,280 \text{ ways}$$

But the books, although kept in groups by subject matter, may have permutations within groups, e.g.,

| Eng | Math | Psych | or | Psych | Eng | Math | or ···

There are $_3P_3 = 3! = 6$ ways of doing this. Thus our final answer is

$$6(24 \times 6 \times 120) = 6 \times 17\,280 = 103\,680 \text{ arrangements}$$

EXAMPLE 16-24   The PTA executive board, consisting of 9 women and 7 men, must have a subcommittee consisting of exactly 3 women and 3 men. In how many ways can this subcommittee be selected?

Solution   Since the order in which the people are selected is not at all important, we have a combination problem. We have two nonoverlapping sets (men and women), so we work on each segment separately. We must choose 3 women from the available 9 women. This is done in $_9C_3$ ways. For each of these ways we can now choose the 3 men from the available 7 in $_7C_3$ ways. The total number of subcommittees then is

$$_9C_3 \cdot {_7C_3} = \frac{9!}{3!6!} \cdot \frac{7!}{3!4!} = \frac{9 \cdot 8 \cdot 7}{3 \cdot 2 \cdot 1} \cdot \frac{7 \cdot 6 \cdot 5}{3 \cdot 2 \cdot 1} = (3 \cdot 4 \cdot 7)(7 \cdot 5) = 2940 \text{ ways}$$

16-3. The Binomial Theorem

A. Finding and using the binomial theorem

The binomial theorem is a formula to find the expansion of the product $(x + y)^n$ for positive integer n. In order to obtain the formula, we'll first look at a few simple cases and examine the patterns that emerge:

$$(x + y)^1 = \qquad\qquad\qquad x + y$$
$$(x + y)^2 = \qquad\qquad\qquad x^2 + 2xy + y^2$$
$$(x + y)^3 = \qquad\qquad\quad x^3 + 3x^2y + 3xy^2 + y^3$$
$$(x + y)^4 = \qquad\quad x^4 + 4x^3y + 6x^2y^2 + 4xy^3 + y^4$$
$$(x + y)^5 = x^5 + 5x^4y + 10x^3y^2 + 10x^2y^3 + 5xy^4 + y^5$$

From these cases we make some observations:

(1) The number of terms in the expansion of $(x + y)^n$ is $n + 1$.
(2) The first term is always x^n; the last term is always y^n.
(3) The coefficient of the second term is always n.
(4) In every term, the sum of the exponents on x and y is n.
(5) The exponents on x begin at n on the left and decrease in steps of 1 to an exponent of 0 at the right end. The exponents on y begin at 0 on the left end and increase in steps of 1 to an exponent of n on the right end.
(6) The sequence of coefficients in any expansion read from left to right is the same as the sequence of coefficients read from right to left.

(7) The coefficient of the first term is $1 = {}_nC_0 = \dbinom{n}{0}$.

The coefficient of the second term is $n = {}_nC_1 = \dbinom{n}{1}$.

The coefficient of the third term is $\dfrac{n(n-1)}{2} = {}_nC_2 = \dbinom{n}{2}$.

The coefficient of the fourth term is $\dfrac{n(n-1)(n-2)}{3 \cdot 2 \cdot 1} = {}_nC_3 = \dbinom{n}{3}$.

$$\vdots$$

The coefficient of the last term is $1 = {}_nC_n = \dbinom{n}{n}$.

And we know the notation for binomial coefficients,

$$_nC_k = \binom{n}{k} = \frac{n!}{k!(n-k)!}$$

so that, for example, we write

$$_nC_2 = \binom{n}{2} = \frac{n!}{2!(n-2)!} \quad \text{and} \quad {}_nC_3 = \binom{n}{3} = \frac{n!}{3!(n-3)!}$$

These observations allow us to form a conclusion, which is called the **Binomial Theorem**:

$$(x + y)^n = \binom{n}{0}x^n + \binom{n}{1}x^{n-1}y + \binom{n}{2}x^{n-2}y^2 + \cdots + \binom{n}{n-1}xy^{n-1} + \binom{n}{n}y^n$$

$$\textbf{(16.5)}$$

or

BINOMIAL THEOREM

$$(x + y)^n = \sum_{k=0}^{n} \binom{n}{k}x^{n-k}y^k$$

EXAMPLE 16-25 Find the binomial expansion of $(x + y)^6$.

Solution We substitute into formula (16-5) using $n = 6$:

$$(x+y)^6 = \binom{6}{0}x^6 + \binom{6}{1}x^5y + \binom{6}{2}x^4y^2 + \binom{6}{3}x^3y^3 + \binom{6}{4}x^2y^4 + \binom{6}{5}xy^5 + \binom{6}{6}y^6$$

We have computed some of these coefficients before; e.g., $\binom{6}{2} = \binom{6}{4} = 15$. The others are

$$\binom{6}{0} = \binom{6}{6} = \frac{6!}{6!0!} = \frac{6!}{6!1} = 1$$

$$\binom{6}{1} = \binom{6}{5} = \frac{6!}{5!1!} = 6$$

$$\binom{6}{3} = \frac{6!}{3!3!} = \frac{6 \cdot 5 \cdot 4}{3 \cdot 2 \cdot 1} = 5 \cdot 4 = 20$$

Then

$$(x + y)^6 = x^6 + 6x^5y + 15x^4y^2 + 20x^3y^3 + 15x^2y^4 + 6xy^5 + y^6$$

note: All seven of the observations made above hold for this example.

Suppose we want to pick out a specific term of the expansion, say the ith term, where $i = 1, 2, 3, \ldots$. Observe that the exponent on y is 1 less than the term number and the lower number in the coefficient notation is the same as the exponent on y. The exponent sum is still n and the upper number in the coefficient notation is n:

$$(x+y)^n = \overbrace{\binom{n}{0}x^ny^0}^{\substack{\text{1st}\\\text{term}}} + \overbrace{\binom{n}{1}x^{n-1}y^1}^{\substack{\text{2nd}\\\text{term}}} + \overbrace{\binom{n}{2}x^{n-2}y^2}^{\substack{\text{3rd}\\\text{term}}} + \overbrace{\binom{n}{3}x^{n-3}y^3}^{\substack{\text{4th}\\\text{term}}} + \overbrace{\binom{n}{4}x^{n-4}y^4}^{\substack{\text{5th}\\\text{term}}} + \cdots$$

(16.6)

Hence the ith term is

ith TERM OF A BINOMIAL EXPANSION $\qquad \binom{n}{i-1}x^{n-(i-1)}y^{i-1}$

EXAMPLE 16-26 What is the 5th term in the expansion of $(x + y)^{12}$?

Solution Let $i = 5$. We have $n = 12$. Then the 5th term is

$$\binom{n}{i-1}x^{n-(i-1)}y^{i-1} = \binom{12}{5-1}x^{12-5+1}y^{5-1}$$

$$= \binom{12}{4}x^8y^4 = \frac{12!}{4!8!}x^8y^4 = \frac{12 \cdot 11 \cdot 10 \cdot 9}{4 \cdot 3 \cdot 2 \cdot 1}x^8y^4$$

$$= 495x^8y^4$$

EXAMPLE 16-27 Find the expansion of $(a^2 + 3b^2)^4$.

Solution Here we need to substitute a^2 for x and $(3b^2)$ for y everywhere in the expansion of $(x + y)^4$, so we expand $(x + y)^4$ first:

$$(x + y)^4 = \binom{4}{0}x^4 + \binom{4}{1}x^3y + \binom{4}{2}x^2y^2 + \binom{4}{3}xy^3 + \binom{4}{4}y^4$$

$$= x^4 + 4x^3y + 6x^2y^2 + 4xy^3 + y^4$$

Then we do the substitution:

$$(a^2 + 3b^2)^4 = (a^2)^4 + 4(a^2)^3(3b^2) + 6(a^2)^2(3b^2)^2 + 4(a^2)(3b^2)^3 + (3b^2)^4$$

$$= a^8 + 4a^6(3b^2) + 6a^4(9b^4) + 4a^2(27b^6) + 81b^8$$

$$= a^8 + 12a^6b^2 + 54a^4b^4 + 108a^2b^6 + 81b^8$$

EXAMPLE 16-28   Find the expansion of $(a - 2b)^5$.

Solution   We need to substitute a for x and $-2b$ for y in the expansion of $(x + y)^5$ after expanding.

If $\qquad (x + y)^5 = x^5 + 5x^4y + 10x^3y^2 + 10x^2y^3 + 5xy^4 + y^5$

then

$$(a - 2b)^5 = a^5 + 5(a)^4(-2b) + 10(a)^3(-2b)^2 + 10(a)^2(-2b)^3$$

$$+ 5(a)(-2b)^4 + (-2b)^5$$

$$= a^5 + 5a^4(-2b) + 10a^3(4b^2) + 10a^2(-8b^3) + 5a(16b^4) + (-32b^5)$$

$$= a^5 - 10a^4b + 40a^3b^2 - 80a^2b^3 + 80ab^4 - 32b^5$$

EXAMPLE 16-29   Find the expansion of $(1 - t^2)^6$.

Solution   We need to use the expansion of $(x + y)^6$ and substitute $x = 1$ and $y = -t^2$.

If $\qquad (x + y)^6 = x^6 + 6x^5y + 15x^4y^2 + 20x^3y^3 + 15x^2y^4 + 6xy^5 + y^6$

then

$$(1 - t^2)^6 = (1)^6 + 6(1)^5(-t^2) + 15(1)^4(-t^2)^2 + 20(1)^3(-t^2)^3$$

$$+ 15(1)^2(-t^2)^4 + 6(1)(-t^2)^5 + (-t^2)^6$$

$$= 1 - 6t^2 + 15t^4 - 20t^6 + 15t^8 - 6t^{10} + t^{12}$$

B. Pascal's triangle

The binomial coefficients $\begin{pmatrix} n \\ k \end{pmatrix}$ have a significant property in their position in a special number display. If we write the coefficients $\begin{pmatrix} n \\ k \end{pmatrix}$ for $k = 0, 1, 2, 3, \ldots$, on the nth line, we get **Pascal's Triangle**:

Line 0						1					$(x + y)^0$
Line 1					1		1				$(x + y)^1$
Line 2				1		2		1			$(x + y)^2$
Line 3			1		3		3		1		$(x + y)^3$
Line 4		1		4		6		4		1	$(x + y)^4$
Line 5	1		5		10		10		5		1	$(x + y)^5$
•		1		6		15	20	15		6		1	•
•	1		7		21		35	35		21		7		1	•

In Pascal's Triangle, the numbers in line n are the coefficients in the expansion $(x + y)^n$. And this display *shows* some properties of these coefficients:

- Any number (*except* 1) in line n can be obtained by adding together the two numbers closest to it in the line $n - 1$ above it.

- The sum of the numbers in any row is 2^n.
- The triangle is symmetric with respect to a vertical line down the middle.

All of these properties are provable—in fact, we'll do one. This property is sometimes called "**rabbit ears**": Draw the "rabbit ears" from a number, say the first 10 in line 5. The sum of the two numbers at the ends of the "ears," $4 + 6$, is the chosen number, 10.

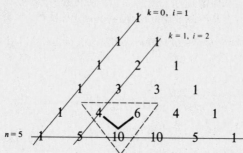

Proof: Pick any number in line n and let it be the ith one from the left end. Then its coefficient designation is $\binom{n}{k} = \binom{n}{i-1}$. Thus the first 10 above is in line 5, position $i = 3$, so its designation is $\binom{5}{3-1} = \binom{5}{2}$. Then the first 4 above is in line $n - 1 = 4$ and position $i = 2$, so its designation is $\binom{4}{1}$; and 6 is in line $n - 1 = 4$ and position $i = 3$, so its designation is $\binom{4}{2}$. Now using these coefficient designations, we can set up our rabbit ears property as an equation:

$$\binom{n}{k} = \binom{n-1}{k-1} + \binom{n-1}{k}$$

We express each of these in factorials:

$$\binom{n}{k} = \frac{n!}{(n-k)!k!}$$

$$\binom{n-1}{k-1} = \frac{(n-1)!}{(n-k)!(k-1)!}$$

$$\binom{n-1}{k} = \frac{(n-1)!}{(n-1-k)!k!}$$

Then

$$\frac{n!}{(n-k)!k!} = \frac{(n-1)!}{(n-k)!(k-1)!} + \frac{(n-1)!}{(n-1-k)!k!}$$

$$= \frac{(n-1)!}{(n-k)!(k-1)!}\left(\frac{k}{k}\right) + \frac{(n-1)!}{(n-1-k)!k!}\left(\frac{n-k}{n-k}\right)$$

$$= \frac{(n-1)!k}{(n-k)!k!} + \frac{(n-1)!(n-k)}{(n-k)!k!}$$

$$= \frac{(n-1)!}{k!}\left[\frac{k}{(n-k)!} + \frac{n-k}{(n-k)!}\right]$$

$$= \frac{(n-1)!n}{(n-k)!k!} \qquad \text{[Remember that } n! = n(n-1)!]$$

$$= \frac{n!}{(n-k)!k!} \qquad \text{QED}$$

SUMMARY

1. The counting principle allows multistage processes to be evaluated for a number of possible occurrences:

   - If stage 1 of an N-stage process may be done in k_1 ways and stage 2 following it may be done in k_2 ways, etc., then the combination of stages may be done in $k_1 \times k_2 \times \cdots \times k_N$ ways.
   - A tree diagram is a graphical way of depicting a multistage process.

2. If a set of objects n is put into an ordered arrangement, the ordering is called a permutation P. The number of permutations of n objects taken n at a time is

   $$_nP_n = n!$$

   The number of permutations of n objects taken r at a time is

   $$_nP_r = \frac{n!}{(n-r)!}$$

   The number of distinguishable permutations of N objects taken k_i at a time, where k_1 is one kind, k_2 is another kind, etc., is

   $$_NP_{k_i} = \frac{N!}{(k_1)!(k_2)!\cdots(k_i)!}$$

3. If an arrangement of a set of n objects is made without regard to ordering it is called a combination. The number of combinations of n objects taken r at a time is

   $$_nC_r = \frac{n!}{r!(n-r)!}$$

4. The number of permutations of n objects taken r at a time is more than the number of combinations of n objects taken r at a time: The connection is $_nP_r = (r!)(_nC_r)$.

5. The coefficients in the expansion of the power of a binomial are found by evaluating combination values:

   If $\binom{n}{k} = {}_nC_k = \dfrac{n!}{k!(n-k)!}$, then

   $$(x+y)^n = \binom{n}{0}x^n + \binom{n}{1}x^{n-1}y + \binom{n}{2}x^{n-2}y^2 + \cdots + \binom{n}{n-1}xy^{n-1} + \binom{n}{n}y^n$$

   These same coefficients may be found by observing the correct row of Pascal's triangle.

RAISE YOUR GRADES

Can you...?

☑ form the tree diagram for a counting problem
☑ use the multiplication rule for counting objects
☑ decide, in a given problem, whether the permutation technique or the combination technique should be used
☑ find the number of permutations of n objects taken r at a time
☑ find the number of distinguishable permutations of N objects if some of the objects are identical
☑ find the number of combinations of n objects taken r at a time
☑ relate the number of permutations to the number of combinations of a collection of objects

☑ use the Binomial Theorem to find the binomial expansion $(x + y)^n$
☑ explain the correlation between combinations and binomial coefficients
☑ construct Pascal's triangle and explain how the rows are formed and their
 significance

SOLVED PROBLEMS

A Counting Principle

PROBLEM 16-1 You are preparing to compete in the league championship track meet and you are allowed to pick exactly 1 running event and 1 field event. You have 3 specialties in running—the 100-meter, 200-meter, and 400-meter races—and 2 specialties in the field events—the long jump (LJ) and the triple jump (TJ). How many different event selections do you have?

Solution Form a tree diagram, as in Figure 16-4, and count the possible selections. There are six possible selections.

Or, use the multiplication rule: If you break the problem down into choices at the various stages, you get

Total choices = $3 \times 2 = 6$

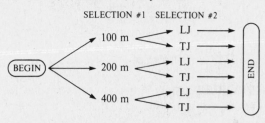

Figure 16-4

PROBLEM 16-2 In how many ways can you form a 3-digit number using only the digits 4, 5, 6, 7, 8 with no repetitions?

Solution You have 5 choices for the first digit. Once this digit has been chosen, you have 4 choices for the second digit; then 3 choices for the third digit. So you can make $5 \times 4 \times 3 = 60$ number choices.

PROBLEM 16-3 In how many different ways could you arrange the letters in the word WORD without repetitions?

Solution You have to fill 4 slots, each with a different letter. You can pick any letter to start, so you have 4 choices for the first slot. Then there are 3 choices for the second slot, 2 choices for the third slot, and 1 choice (the only letter left) for the final slot. So you have

$$\underline{4} \times \underline{3} \times \underline{2} \times \underline{1} = 24 \text{ choices}$$

PROBLEM 16-4 You toss one coin and roll one 6-sided die. How many different outcomes are possible? List them.

Solution There are 2 outcomes, H and T, for the coin toss and 6 outcomes, 1 through 6, for the die roll, so there are $2 \times 6 = 12$ outcomes. They are

$$\{H1, H2, H3, H4, H5, H6, T1, T2, T3, T4, T5, T6\}$$

PROBLEM 16-5 If an ID number has 1 letter followed by 4 digits, how many ID numbers could be created?

Solution There are 26 choices for the letter and 10 choices for each digit, so the total is

$$26 \times 10 \times 10 \times 10 \times 10 = 260\,000$$

PROBLEM 16-6   If 7 horses run a race, how many possibilities are there for the first 3 finishes, win, place, and show?

Solution   Any one of the 7 horses could win. Any one of the remaining 6 horses could place second, and any one of the remaining 5 could "show" the crowd that she is third. Thus there are $7 \times 6 \times 5 = 210$ finish orders.

PROBLEM 16-7   Your casual wardrobe consists of 4 different pairs of jeans, 5 different sweatshirts, and 3 different pairs of shoes. How many outfits are possible?

Solution $4 \times 5 \times 3 = 60$ choices.

Permutations and Combinations

PROBLEM 16-8   Evaluate (a) $_7P_2$, (b) $_8P_4$, and (c) $_6P_1$.

Solution   You know that $_nP_r = \dfrac{n!}{(n-r)!}$, so

(a) $_7P_2 = \dfrac{7!}{(7-2)!} = \dfrac{7!}{5!} = \dfrac{7 \cdot 6 \cdot 5 \cdot 4 \cdot 3 \cdot 2 \cdot 1}{5 \cdot 4 \cdot 3 \cdot 2 \cdot 1} = 7 \cdot 6 = 42$

(b) $_8P_4 = \dfrac{8!}{(8-4)!} = \dfrac{8!}{4!} = 8 \cdot 7 \cdot 6 \cdot 5 = 1680$

(c) $_6P_1 = \dfrac{6!}{(6-1)!} = \dfrac{6!}{5!} = 6$

PROBLEM 16-9   A bookstore owner has just bought 8 different bestsellers at a hefty discount, and he wants to display 5 of them on his bargain shelf. How many different displays could he make?

Solution   Because changing the order of the books is the only way to make a different display, this is a permutation problem. This is the permutation of 8 objects taken 5 at a time, $_8P_5$. Thus

$$_8P_5 = \frac{8!}{(8-5)!} = \frac{8!}{3!} = 8 \cdot 7 \cdot 6 \cdot 5 \cdot 4 = 6720 \text{ ways}$$

PROBLEM 16-10   You have 6 different coins (a penny, a nickel, a dime, a quarter, a half-dollar, and an odd shilling). (a) If you arrange some subset of 4 of them in a row, how many different arrangements are possible? (b) How many different arrangements are possible if you use all six of the coins?

Solution

(a) You're concerned with order, so you want to permute 6 objects taken 4 at a time:

$$_6P_4 = \frac{6!}{(6-4)!} = \frac{6!}{2!} = 6 \cdot 5 \cdot 4 \cdot 3 = 360 \text{ arrangements}$$

(b) $_6P_6 = 6! = 720$ arrangements

PROBLEM 16-11   How many 3-letter sets of initials (repetitions not allowed) are possible?

Solution   You are permuting 3 objects from 26 objects:

$$_{26}P_3 = \frac{26!}{(26-3)!} = \frac{26!}{23!} = 26 \cdot 25 \cdot 24 = 15\,600 \text{ sets}$$

PROBLEM 16-12   How many ways are there to order the 5 letters P, Q, R, S, T?

Solution $_5P_5 = 5! = 120$ ways

PROBLEM 16-13   How many different letter arrangements can you make with the letters in MISSISSIPPI?

Solution   If the letters were all different, you would have $N! = 11!$ permutations. But the letter I occurs 4 times, the letter S occurs 4 times, and P occurs twice, so $k_1 = 4$, $k_2 = 4$, and $k_3 = 2$. Thus there are

$$\frac{N!}{k_1!k_2!k_3} = \frac{11!}{4!4!2!} = \frac{11 \cdot 10 \cdot 9 \cdot 8 \cdot 7 \cdot 6 \cdot 5 \cdot 4 \cdot 3 \cdot 2 \cdot 1}{4 \cdot 3 \cdot 2 \cdot 1 \cdot 4 \cdot 3 \cdot 2 \cdot 1 \cdot 2 \cdot 1}$$

$$= 11 \cdot 10 \cdot 9 \cdot 7 \cdot 5$$

$$= 34\,650 \text{ distinguishable permutations}$$

PROBLEM 16-14   How many distinguishable permutations can you make with the letters in COLLEGE?

Solution   There are 7 objects, but there are two L's and two E's. Thus there are

$$\frac{7!}{2!2!} = \frac{7 \cdot 6 \cdot 5 \cdot 4 \cdot 3 \cdot 2 \cdot 1}{2 \cdot 1 \cdot 2 \cdot 1} = 7 \cdot 6 \cdot 5 \cdot 3 \cdot 2 = 1260 \text{ distinguishable permutations}$$

PROBLEM 16-15   When you're dealing a poker hand, you don't care about the order in which the cards are dealt. How many different hands of 5 cards can you deal from a standard 52-card deck?

Solution   Since order is not important, this is a combination problem. Since there are 52 objects taken 5 at a time, there are

$$_nC_r = \frac{n!}{r!(n - r)!}$$

$$_{52}C_5 = \frac{52!}{5!(52 - 5)!} = \frac{52!}{5!47!} = \frac{52 \cdot 51 \cdot 50 \cdot 49 \cdot 48}{5 \cdot 4 \cdot 3 \cdot 2 \cdot 1} = 2\,598\,960 \text{ different hands}$$

PROBLEM 16-16   How many different subcommittees of 3 persons can be chosen from a committee of 10 persons?

Solution   Since the subcommittees can be made up of any 3 persons without regard to order, the number of different subcommittees is

$$_{10}C_3 = \frac{10!}{3!7!} = \frac{10 \cdot 9 \cdot 8}{3 \cdot 2 \cdot 1} = 10 \cdot 3 \cdot 4 = 120$$

PROBLEM 16-17   You have 8 kinds of flowers and you need to make up bouquets consisting of 3 different kinds of flowers. How many different kinds of bouquets can you make?

Solution

$$_8C_3 = \frac{8!}{3!5!} = \frac{8 \cdot 7 \cdot 6}{3 \cdot 2 \cdot 1} = 56 \text{ bouquets}$$

PROBLEM 16-18   How many different basketball teams of 5 girls can a coach form from a squad of 15 girls?

Solution   Since the coach is not choosing positions, she doesn't distinguish among the team members; i.e., order is not important. Thus there are

$$_{15}C_5 = \frac{15!}{5!10!} = \frac{15 \cdot 14 \cdot 13 \cdot 12 \cdot 11}{5 \cdot 4 \cdot 3 \cdot 2 \cdot 1} = 3003 \text{ teams}$$

PROBLEM 16-19   How many ways are there of getting exactly 3 heads in the toss of 5 coins?

Solution   There are

$$_5C_3 = \frac{5!}{3!2!} = \frac{5 \cdot 4}{2 \cdot 1} = 10 \text{ ways of getting 3 heads}$$

(out of a total of $2^5 = 32$ ways the coins could fall).

PROBLEM 16-20 In how many ways can a committee of 3 Democrats and 4 Republicans be chosen from a group of 5 Democrats and 8 Republicans?

Solution The 3 Democrats are chosen from the available 5 in $_5C_3$ ways. The 4 Republicans are chosen from the available 8 in $_8C_4$ ways. Thus the total number of ways is

$$_5C_3 \cdot {}_8C_4 = \frac{5!}{3!2!} \cdot \frac{8!}{4!4!} = \frac{5 \cdot 4}{2 \cdot 1} \cdot \frac{8 \cdot 7 \cdot 6 \cdot 5}{4 \cdot 3 \cdot 2 \cdot 1} = 10 \cdot 70 = 700 \text{ ways}$$

PROBLEM 16-21 In a league of 10 softball teams, how many games will be played in a season if each team plays each other team once?

Solution There are 10 teams and each game involves 2 teams, so

$$_{10}C_2 = \frac{10}{2!8!} = \frac{10 \cdot 9}{2 \cdot 1} = 45 \text{ games}$$

PROBLEM 16-22 Pepe's Pizza Palace offers 3 choices of salad, 12 kinds of pizza, and 4 different drinks. How many different selections can you make for a salad, a pizza, and a drink?

Solution This is a 3-stage counting problem. You have 3 salad choices. Then, for each salad choice, you have 12 choices for pizza. Then for each of these two-item selections, you have 4 choices for the drink. The total number of selections is $3 \times 12 \times 4 = 144$.

PROBLEM 16-23 Izzy's Ice Cream Shoppe carries 31 flavors. How many different double-dip ice cream cones could you make, assuming that the dips are of different flavors and you don't care which flavor is on top?

Solution There are two ways to approach this problem. By the counting principle, you have 31 choices for the first flavor and 30 choices for the second flavor, which gives you a total of $31 \times 30 = 930$ different cones. But if you use this total, you are double-counting the cones with the flavors flipped (e.g., chocolate atop strawberry is the same as strawberry atop chocolate for all practical purposes). So you must divide by 2 to get $930/2 = 465$ cones.

Approaching the question as a combination problem, you are choosing 2 flavors from among the 31 available, so there are

$$_{31}C_2 = \frac{31!}{2!29!} = \frac{31 \cdot 30}{2 \cdot 1} = 31 \cdot 15 = 465 \text{ cones}$$

The Binomial Theorem

PROBLEM 16-24 Working with the notation for binomial coefficients, evaluate

(a) $\binom{5}{2}$ (b) $\binom{8}{6}$ (c) $\binom{12}{11}$ (d) $\binom{4}{4}$

Solution By definition $\binom{n}{k} = {}_nC_k = \frac{n!}{k!(n-k)!}$, so

(a) $\binom{5}{2} = \frac{5!}{2!3!} = \frac{5 \cdot 4}{2 \cdot 1} = 10$ (c) $\binom{12}{11} = \frac{12!}{11!1!} = 12$

(b) $\binom{8}{6} = \frac{8!}{6!2!} = \frac{8 \cdot 7}{2 \cdot 1} = 28$ (d) $\binom{4}{4} = \frac{4!}{4!0!} = 1$

PROBLEM 16-25 Create the expansion of $(x + y)^7$.

Solution Use the expanded formula (16-5) from the Binomial Theorem, with $n = 7$:

$$(x + y)^n = \binom{n}{0}x^n + \binom{n}{1}x^{n-1}y + \binom{n}{2}x^{n-2}y^2 + \cdots + \binom{n}{n-1}xy^{n-1} + \binom{n}{n}y^n$$

$$(x + y)^7 = \binom{7}{0}x^7 + \binom{7}{1}x^6y + \binom{7}{2}x^5y^2 + \binom{7}{3}x^4y^3 + \binom{7}{4}x^3y^4 + \binom{7}{5}x^2y^5 + \binom{7}{6}xy^6 + \binom{7}{7}y^7$$

Then find the coefficients for each term in the expansion by using the formula

$$\binom{n}{k} = \frac{n!}{k!(n-k)!}$$

The coefficients of the first and seventh terms are particularly easy to find:

$$\binom{7}{0} = \binom{7}{7} = \frac{7!}{7!} = 1$$ [The coefficient of the first term in the expansion of $(x+y)^n$ is always 1. By the symmetry property the coefficient of the last term will also be 1.]

And the coefficients of the second and sixth terms are just as easy to find:

$$\binom{7}{1} = \binom{7}{6} = \frac{7!}{6!} = 7$$ [The coefficient of the 2nd term is always n; and since the left-to-right sequence is the same as the right-to-left, the coefficient of the 6th term in a 7-term expansion also equals n.]

Now you only have to calculate the coefficients of the third and fourth terms, since by the symmetry property the coefficient of the fifth term will be the same as that of the third:

$$\binom{7}{2} = \frac{7!}{2!(7-2)!} = \frac{7!}{2!5!} = 7 \cdot 3 = 21 = \binom{7}{5}$$

$$\binom{7}{3} = \frac{7!}{3!(7-3)!} = \frac{7!}{3!4!} = 7 \cdot 5 = 35$$

Thus $(x+y)^7 = x^7 + 7x^6y + 21x^5y^2 + 35x^4y^3 + 35x^3y^4 + 21x^2y^5 + 7xy^6 + y^7$

PROBLEM 16-26 What is the ninth term of $(2a-b)^{14}$?

Solution The ith term of $(x+y)^n$ is

$$\binom{n}{i-1}x^{n-i+1}y^{i-1}$$

So if $i = 9$, $n = 14$, $x = 2a$, $y = -b$, the solution is

$$\binom{14}{8}(2a)^6(-b)^8 = \frac{14!}{8!6!}(2)^6a^6b^8 = 7 \cdot 3 \cdot 11 \cdot 3(64)a^6b^8 = 192\,192a^6b^8$$

PROBLEM 16-27 Find the expansion of $(a^3 - 2b)^5$.

Solution Begin with the expansion of $(x+y)^5$, then substitute a^3 for x and $(-2b)$ for y: Since

$$(x+y)^5 = \binom{5}{0}x^5 + \binom{5}{1}x^4y + \binom{5}{2}x^3y^2 + \binom{5}{3}x^2y^3 + \binom{5}{4}xy^4 + \binom{5}{5}y^5$$

$$= x^5 + 5x^4y + 10x^3y^2 + 10x^2y^3 + 5xy^4 + y^5$$

Then

$$(a^3 - 2b)^5 = (a^3)^5 + 5(a^3)^4(-2b) + 10(a^3)^3(-2b)^2 + 10(a^3)^2(-2b)^3 + 5(a^3)(-2b)^4 + (-2b)^5$$

$$= a^{15} - 10a^{12}b + 40a^9b^2 - 80a^6b^3 + 80a^3b^4 - 32b^5$$

PROBLEM 16-28 Use the binomial expansion to find the value $(1.02)^{15}$, accurate to three decimal places.

Solution First, note that $1.02 = 1 + .02$. Then, since the first four terms of $(x+y)^{15}$ are

$$(x+y)^{15} = x^{15} + 15x^{14}y + 105x^{13}y^2 + 455x^{12}y^3 + \cdots$$

You have

$$(1 + .02)^{15} \simeq 1^{15} + 15(1)^{14}(.02) + 105(1)^{13}(.02)^2 + 455(1)^{12}(.02)^3 + \cdots$$

$$= 1 + 15(.02) + 105(.0004) + 455(.000\,008) = 1 + .30 + .042 + .003\,64 + \cdots$$

$$= 1.345\,64\ldots \simeq 1.346$$

note: Using a calculator to get the accurate result, you find $(1.02)^{15} = 1.34583 \simeq 1.346$. Not bad!

PROBLEM 16-29   Find the expansion of $(1 - 2i)^4$.

Solution   Using $(x + y)^4$ with $x = 1$ and $y = -2i$, you have

$$(x + y)^4 = x^4 + 4x^3y + \left(\frac{4!}{2!2!}\right)x^2y^2 + 4xy^3 + y^4$$

$$= x^4 + 4x^3y + 6x^2y^2 + 4xy^3 + y^4$$

Then $(1 - 2i)^4 = 1^4 + 4(1)^3(-2i) + 6(1)^2(-2i)^2 + 4(1)(-2i)^3 + (-2i)^4$

$$= 1 - 8i + 24i^2 - 32i^3 + 16i^4 = 1 - 8i - 24 + 32i + 16$$

$$= -7 + 24i$$

PROBLEM 16-30   Find the expansion of $(2x^{1/2} - y^{1/2})^6$.

Solution   First write out $(a + b)^6$, then replace a by $(2x^{1/2})$ and b by $(-y^{1/2})$:

$$(a + b)^6 = a^6 + 6a^5b + \left(\frac{6!}{2!4!}\right)a^4b^2 + \left(\frac{6!}{3!3!}\right)a^3b^3 + \left(\frac{6!}{4!2!}\right)a^2b^4 + 6ab^5 + b^6$$

$$= a^6 + 6a^5b + 15a^4b^2 + 20a^3b^3 + 15a^2b^4 + 6ab^5 + b^6$$

$$(2x^{1/2} - y^{1/2})^6 = (2x^{1/2})^6 + 6(2x^{1/2})^5(-y^{1/2}) + 15(2x^{1/2})^4(-y^{1/2})^2 + 20(2x^{1/2})^3(-y^{1/2})^3$$

$$+ 15(2x^{1/2})^2(-y^{1/2})^4 + 6(2x^{1/2})(-y^{1/2})^5 + (-y^{1/2})^6$$

$$= 64x^3 - 192x^{5/2}y^{1/2} + 240x^2y - 160x^{3/2}y^{3/2} + 60xy^2 - 12x^{1/2}y^{5/2} + y^3$$

PROBLEM 16-31   Prove that $_nC_0 + {}_nC_1 + {}_nC_2 + \cdots + {}_nC_n = 2^n$.

Solution   Write the general formula for $(x + y)^n$:

$$(x + y)^n = {}_nC_0x^n + {}_nC_1x^{n-1}y + {}_nC_2x^{n-2}y^2 + \cdots + {}_nC_ny^n$$

Now set $x = 1$, $y = 1$ to get

$$(1 + 1)^n = {}_nC_0(1)^n + {}_nC_1(1)^{n-1}(1) + {}_nC_2(1)^{n-2}(1)^2 + \cdots + {}_nC_n(1)^n$$

or

$$2^n = {}_nC_0 + {}_nC_1 + {}_nC_2 + \cdots + {}_nC_n$$

Try it on sample values:

$$n = 2:   {}_2C_0 + {}_2C_1 + {}_2C_2 = 1 + 2 + 1 = 4 = 2^2$$

$$n = 3:   {}_3C_0 + {}_3C_1 + {}_3C_2 + {}_3C_3 = 1 + 3 + 3 + 1 = 8 = 2^3$$

$$n = 4:   {}_4C_0 + {}_4C_1 + {}_4C_2 + {}_4C_3 + {}_4C_4 = 1 + 4 + 6 + 4 + 1 = 16 = 2^4$$

Supplementary Exercises

PROBLEM 16-32   How many seven-digit telephone numbers can be formed if neither 0 nor 1 can be used as the first or second digit?

PROBLEM 16-33   Given the digits 1 through 9, you wish to form four-digit numbers whose digits are all different. How many are there?

PROBLEM 16-34   Of the above four-digit numbers, how many are less than 5000?

PROBLEM 16-35   Of the above four-digit numbers, how many are odd?

PROBLEM 16-36 If six people enter a bus and there are eight vacant seats, in how many ways can they be seated?

PROBLEM 16-37 Find n if $_nP_5 = 6 \cdot {}_nP_4$.

PROBLEM 16-38 The Greek alphabet has 24 letters. How many three-letter fraternity names could be formed if (a) repetitions are not permitted, (b) repetitions are permitted?

PROBLEM 16-39 $_nC_{(n-r)} = {}_nC_r$: True or False?

PROBLEM 16-40 From an ordinary deck of playing cards, how many ways are there of drawing three black cards?

PROBLEM 16-41 How many distinct numbers greater than 6000 without repeated digits can be formed with 1, 2, 3, 8, 9?

PROBLEM 16-42 In how many ways can you divide 9 books between two of your friends if one is to receive 5 books and the other 4 books?

PROBLEM 16-43 A post office keeps supplies of 8 different airmail stamps. In how many ways can a person buy 3 different stamps?

PROBLEM 16-44 How many seven-letter code words may be formed from the letters POPOVER?

PROBLEM 16-45 If 4 red, 3 black, 6 green, and 3 yellow disks are to be arranged in a row, what is the number of possible arrangements of the disks by color?

PROBLEM 16-46 In how many ways can a panel of 8 judges reach a majority decision?

PROBLEM 16-47 If twelve motorcyclists enter a race, in how many ways could the first three place winners be listed?

PROBLEM 16-48 A class of 22 students wants to elect a president, secretary, and treasurer. In how many ways can this be done?

PROBLEM 16-49 If a family has 6 children, in how many different orders could 4 boys and 2 girls have been born?

PROBLEM 16-50 If you invest $\$A$ at $p\%$ compounded annually, then the amount after n years is $A(1 + \frac{p}{100})^n$. Suppose you invest \$1000 at 10%. How much will your investment be worth after 5 years?

PROBLEM 16-51 Expand (a) $(3x + y)^4$, (b) $(x^2 + 2y^2)^3$, (c) $(5w - 4v)^4$, (d) $(a^3 - b^2)^5$.

PROBLEM 16-52 Find the first four terms of $(2a^2 + 3b^2)^9$.

PROBLEM 16-53 Find the fifth term of the expansion of $(5u^2 - 7v)^8$.

PROBLEM 16-54 Prove that if everyone in Smalltown, USA—population 25 000—has 3 initials, two people must have the same initials.

PROBLEM 16-55 Find n if $_nP_6 = 720$.

PROBLEM 16-56 In how many ways can 5 people line up to buy bus tickets?

PROBLEM 16-57 Find the number of ways three men and three women can use one restroom one at a time.

PROBLEM 16-58 Expand $(2 + 3i)^4$.

PROBLEM 16-59 If an automobile manufacturer makes four body models in six colors with three different engines, how many cars does a dealer need to stock to have one of each kind on hand?

Answers to Supplementary Exercises

16-32 64×10^5

16-33 3024

16-34 1344

16-35 1680

16-36 20 160

16-37 10

16-38 (a) 12 144   (b) 13 824

16-39 True

16-40 15 600

16-41 5760

16-42 126

16-43 56

16-44 1260

16-45 33 633 600

16-46 93

16-47 1320

16-48 9240

16-49 $_6C_2 = 15$

16-50 \$1610.51

16-51 (a) $81x^4 + 108x^3y + 54x^2y^2 + 12xy^3 + y^4$

(b) $x^6 + 6x^4y^2 + 12x^2y^4 + 8y^6$

(c) $625w^4 - 2000w^3v + 2400w^2v^2 - 1280wv^3 + 256v^4$

(d) $a^{15} - 5a^{12}b^2 + 10a^9b^4 - 10a^6b^6 + 5a^3b^8 - b^{10}$

16-52 $512a^{18} + 6912a^{16}b^2 + 41\,472a^{14}b^4 + 145\,152a^{12}b^6 + \cdots$

16-53 $105\,043\,750u^8v^4$

16-54 17 576 different initial combinations

16-55 $n = 6$

16-56 120

16-57 20

16-58 $-119 - 120i$

16-59 72

EXAM 5 (Chapters 14–16)

1. Find the remainder and quotient when $4x^4 - 7x^3 + x^2 - 2x + 3$ is divided by $x^2 - 4$.

2. Using synthetic division, find the factors of $P(x) = 2x^4 - 11x^3 - 8x^2 + 59x + 30$.

3. Given that $x^5 - x^4 + 24x^3 + 8x^2 + 135x + 153 = 0$ has roots $1 - 4i$ and $3i$, find all the roots.

4. What are the possible rational roots of $P(x) = 5x^3 - 6x^2 + 7x - 6$?

5. Using Descartes' Rule of Signs, determine the possible number of positive and negative real roots for

   (a) $P(x) = 2x^4 + 3x^3 - x^2 - x + 5$ and   (b) $Q(x) = x^5 - 2x^4 - x^2 + x - 3$

6. Show that the equation $T(x) = x^3 + 3x - 2 = 0$ has exactly one positive irrational root.

7. If you start your Christmas Club account at the local bank with a \$100 deposit and add a \$20 deposit each week for half of a year, what is the formula for the balance after n weeks?

8. Find the sum of all positive multiples of 3 that are less than 300.

9. Find the sum of the finite series $5 + \frac{10}{3} + \frac{20}{9} + \frac{40}{27} + \cdots$.

10. Use mathematical induction to prove that
$$1 \cdot 3 + 2 \cdot 4 + 3 \cdot 5 + \cdots + n(n + 2) = \tfrac{1}{6}n(n + 1)(2n + 7)$$
is true for all n.

11. You have one copy of each of the following textbooks: math, history, Spanish, chemistry, and psychology. In how many ways can you arrange these on your bookshelf?

12. If your club has 20 members, in how many ways could you choose a president, a vice president, and a secretary from the membership?

13. How many sets of 4 volunteers are possible from a group of 13 people?

14. What is the fourth term of the binomial expansion of $(2a + b)^7$?

15. Without using a calculator, prove that $(1.015)^{20} > 1.34$.

Answers to Exam 5

1. $R = -30x + 71$; $Q = 4x^2 - 7x + 17$

2. $(x + 2)(x - 3)(x - 5)(2x + 1)$

3. $1 \pm 4i, \pm 3i, -1$

4. $\pm 1, \pm 2, \pm 3, \pm 6, \pm \frac{1}{5}, \pm \frac{2}{5}, \pm \frac{3}{5}, \pm \frac{6}{5}$

5. (a) 2 or 0 positive; 2 or 0 negative
   (b) 3 or 1 positive; no negative

6. $T(0) = -2,$ $T(1) = 2$

7. $B = 100 + 20n$

8. 14 850

9. 15

11. 120

12. $20 \cdot 19 \cdot 18 = 6840$

13. $_{13}C_4 = 715$

14. $\binom{7}{3}(2a)^4(b)^3 = 35 \cdot 16a^4 \cdot b^3 = 560a^4b^3$

FINAL EXAM

Part 1

1. Solve for x in $3(x - 4) + 7 = 6x - 4(x + 2)$.

2. Find the x values that solve $|2x - 5| < 9$.

3. Use the quadratic formula to solve **(a)** $6x^2 - 7x - 3 = 0$ and **(b)** $2x^2 + 3x + 13 = 0$.

4. Find two consecutive odd integers whose product is 195.

5. Solve the equation $27^x = 3^{x + \sqrt{5}}$ for x.

6. Given $f(x) = \dfrac{1}{2}x - \dfrac{2}{3}$, find **(a)** $f^{-1}(x)$ and **(b)** $\dfrac{f(x + h) - f(x)}{h}$.

7. Reduce to triangular form and solve $\begin{cases} x + 5y - 2z = 0 \\ -4x + y + 4z = 9 \\ 3x + 7y + 3z = 19 \end{cases}$.

8. Use Cramer's Rule to solve $\begin{cases} 4x - 2y = 3 \\ x + 2y = -5 \end{cases}$.

9. Find all of the roots of $P(x) = 2x^5 - 5x^4 - 37x^3 + 103x^2 - 39x + 108$, given that $x = i$ is one of the roots.

10. Use mathematical induction to prove that $1 + 4 + 4^2 + \cdots + 4^{n-1} = \frac{1}{3}(4^n - 1)$ for every positive interger n.

Part 2

Simplify the following expressions. **11.** $\dfrac{-1 + \dfrac{1}{1 - x}}{1 - \dfrac{1}{1 + x}}$ **12.** $\dfrac{(a^2 b^{-3} c^{-1})^4}{(a^0 b^{-5} c^4)^{-3}}$

13. Susan's father, John, is four times as old as Susan. In 6 years John's age will be 10 years more than double Susan's age at that time. **(a)** How old is Susan now? **(b)** John?

14. Find the values of x that satisfy $x^2 - 2x - 15 < 0$.

15. If $f(x) = \dfrac{1}{1 - x}$, does $f\left(\dfrac{1}{x}\right) = -xf(x)$?

16. Find the intercepts and the asymptotes for $f(x) = \dfrac{x^2 - 4x + 3}{x^3 - 3x^2 - 10x}$.

17. Solve for x in $\ln(2x - 1) - \ln(3x + 1) = -2$.

18. Using the inverse of the coefficient matrix, solve $\begin{cases} 3x - 4y = 2 \\ -2x + y = -3 \end{cases}$.

19. Simplify $(5 + 12i)(\overline{1 + i}) + 25\left(\dfrac{2 - i}{3 + 4i}\right)$ to the form $a + bi$.

20. If -3 is a root of multiplicity two of $P(x) = x^4 + 11x^3 + 42x^2 + 63x + 27$, what are the other roots?

Part 3

21. Factor each of the following expressions:

        (a) $3x^2 - 2x - 1$　　**(b)** $x^3 + 27$　　**(c)** $7x^2 - 53x - 24$

22. Solve $y^2 = 8y + 4$ for y by completing the square.

23. **(a)** Find the vertex V of the parabola $y = x^2 - 4x + 9$. **(b)** What is its axis of symmetry?

24. Find the equation of the line passing through the points $(4, -2)$ and $(-3, 7)$.

25. Solve the equation $y = \ln(x - \sqrt{x^2 - 1})$ for x.

26. Find the sum of the infinite series $2 + \frac{4}{3} + \frac{8}{9} + \frac{16}{27} + \cdots$.

27. Use the binomial formula to obtain the expansion of $(x + 3a)^4$.

28. Find the maximum and minimum values of $P = .33x + .19y$ subject to $\{x \geq 0, \ y \geq 0, \ 2x - y \leq 5, 2x + 3y \leq 9\}$.

29. Find the equation of the parabola that passes through the points $(1, 2), (-1, 14)$, and $(4, -1)$.

30. If a room has 9 seats, in how many ways can a teacher assign seats to 6 students?

Answers to Final Exam

1. $x = -3$

2. $-2 < x < 7$

3. (a) $x = -\dfrac{1}{3}, x = \dfrac{3}{2}$

    (b) $x = -\dfrac{3}{4} \pm i\dfrac{\sqrt{95}}{4}$

4. 13 and 15

5. $x = \frac{1}{2}\sqrt{5}$

6. (a) $f^{-1} = 2x + \frac{4}{3}$

    (b) $\frac{1}{2}$

    $(1, 1, 3)$

8. $x = -0.4, y = -2.3$

9. $i, -i, 3, 4, -\frac{9}{2}$

11. $\dfrac{1 + x}{1 - x}$

12. $a^8 b^3 c^8$

13. (a) 8   (b) 32

14. $-3 < x < 5$

15. Yes

16. Intercepts: $x = 3, x = 1$
    Asymptotes: $x = 0, x = 5, x = -2, y = 0$

17. $x = \dfrac{1 + e^2}{2e^2 - 3}$

18. $X = \begin{bmatrix} 2 \\ 1 \end{bmatrix}$

19. $19 - 4i$

20. $-\dfrac{5}{2} \pm \dfrac{\sqrt{13}}{2}$

21. (a) $(3x + 1)(x - 1)$

    (b) $(x + 3)(x^2 - 3x + 9)$

    (c) $(7x + 3)(x - 8)$

22. $y = 4 \pm \sqrt{20}$

23. (a) $V = (2, 5)$ (b) $x = 2$

24. $9x + 7y = 22$

25. $x = \frac{1}{2}(e^y + e^{-y})$

26. $S = \dfrac{a}{1 - r} = \dfrac{2}{1 - \frac{2}{3}} = 6$

27. $x^4 + 12x^3 a + 54x^2 a^2 + 108xa^3 + 81a^4$

28. Minimum is 0 at $(0, 0)$
    Maximum is 1.18 at $(3, 1)$

29. $y = x^2 - 6x + 7$

30. $60\,480$

APPENDIX A: Common Logarithms

	0	1	2	3	4	5	6	7	8	9
1.0	.0000	.0043	.0086	.0128	.0170	.0212	.0253	.0294	.0334	.0374
1.1	.0414	.0453	.0492	.0531	.0569	.0607	.0645	.0682	.0719	.0755
1.2	.0792	.0828	.0864	.0899	.0934	.0969	.1004	.1038	.1072	.1106
1.3	.1139	.1173	.1206	.1239	.1271	.1303	.1335	.1367	.1399	.1430
1.4	.1461	.1492	.1523	.1553	.1584	.1614	.1644	.1673	.1703	.1732
1.5	.1761	.1790	.1818	.1847	.1875	.1903	.1931	.1959	.1987	.2014
1.6	.2041	.2068	.2095	.2122	.2148	.2175	.2201	.2227	.2253	.2279
1.7	.2304	.2330	.2355	.2380	.2405	.2430	.2455	.2480	.2504	.2529
1.8	.2553	.2577	.2601	.2625	.2648	.2672	.2695	.2718	.2742	.2765
1.9	.2788	.2810	.2833	.2856	.2878	.2900	.2923	.2945	.2967	.2989
2.0	.3010	.3032	.3054	.3075	.3096	.3118	.3139	.3160	.3181	.3201
2.1	.3222	.3243	.3263	.3284	.3304	.3324	.3345	.3365	.3385	.3404
2.2	.3424	.3444	.3464	.3483	.3502	.3522	.3541	.3560	.3579	.3598
2.3	.3617	.3636	.3655	.3674	.3692	.3711	.3729	.3747	.3766	.3784
2.4	.3802	.3820	.3838	.3856	.3874	.3892	.3909	.3927	.3945	.3962
2.5	.3979	.3997	.4014	.4031	.4048	.4065	.4082	.4099	.4116	.4133
2.6	.4150	.4166	.4183	.4200	.4216	.4232	.4249	.4265	.4281	.4298
2.7	.4314	.4330	.4346	.4362	.4377	.4393	.4409	.4425	.4440	.4456
2.8	.4472	.4487	.4502	.4518	.4533	.4548	.4564	.4579	.4594	.4609
2.9	.4624	.4639	.4654	.4669	.4683	.4698	.4713	.4728	.4742	.4757
3.0	.4771	.4786	.4800	.4814	.4829	.4843	.4857	.4871	.4885	.4900
3.1	.4914	.4928	.4942	.4955	.4969	.4983	.4997	.5011	.5024	.5038
3.2	.5051	.5065	.5079	.5092	.5105	.5119	.5132	.5145	.5159	.5172
3.3	.5185	.5198	.5211	.5224	.5237	.5250	.5263	.5276	.5289	.5302
3.4	.5315	.5328	.5340	.5353	.5366	.5378	.5391	.5403	.5416	.5428
3.5	.5441	.5453	.5465	.5478	.5490	.5502	.5514	.5527	.5539	.5551
3.6	.5563	.5575	.5587	.5599	.5611	.5623	.5635	.5647	.5658	.5670
3.7	.5682	.5694	.5705	.5717	.5729	.5740	.5752	.5763	.5775	.5786
3.8	.5798	.5809	.5821	.5832	.5843	.5855	.5866	.5877	.5888	.5899
3.9	.5911	.5922	.5933	.5944	.5955	.5966	.5977	.5988	.5999	.6010
4.0	.6021	.6031	.6042	.6053	.6064	.6075	.6085	.6096	.6107	.6117
4.1	.6128	.6138	.6149	.6159	.6170	.6180	.6191	.6201	.6212	.6222
4.2	.6232	.6243	.6253	.6263	.6274	.6284	.6294	.6304	.6314	.6325
4.3	.6335	.6345	.6355	.6365	.6375	.6385	.6395	.6405	.6415	.6425
4.4	.6435	.6444	.6454	.6464	.6474	.6484	.6493	.6503	.6513	.6522
4.5	.6532	.6542	.6551	.6561	.6571	.6580	.6590	.6599	.6609	.6618
4.6	.6628	.6637	.6646	.6656	.6665	.6675	.6684	.6693	.6702	.6712
4.7	.6721	.6730	.6739	.6749	.6758	.6767	.6776	.6785	.6794	.6803
4.8	.6812	.6821	.6830	.6839	.6848	.6857	.6866	.6875	.6884	.6893
4.9	.6902	.6911	.6920	.6928	.6937	.6946	.6955	.6964	.6972	.6981
5.0	.6990	.6998	.7007	.7016	.7024	.7033	.7041	.7050	.7059	.7067
5.1	.7076	.7084	.7093	.7101	.7110	.7118	.7126	.7135	.7143	.7152
5.2	.7160	.7168	.7177	.7185	.7193	.7202	.7210	.7218	.7226	.7235
5.3	.7243	.7251	.7259	.7267	.7275	.7284	.7292	.7300	.7308	.7316
5.4	.7324	.7332	.7340	.7348	.7356	.7364	.7372	.7380	.7388	.7396

Common Logarithms (Continued)

	0	1	2	3	4	5	6	7	8	9
5.5	.7404	.7412	.7419	.7427	.7435	.7443	.7451	.7459	.7466	.7474
5.6	.7482	.7490	.7497	.7505	.7513	.7520	.7528	.7536	.7543	.7551
5.7	.7559	.7566	.7574	.7582	.7589	.7597	.7604	.7612	.7619	.7627
5.8	.7634	.7642	.7649	.7657	.7664	.7672	.7679	.7686	.7694	.7701
5.9	.7709	.7716	.7723	.7731	.7738	.7745	.7752	.7760	.7767	.7774
6.0	.7782	.7789	.7796	.7803	.7810	.7818	.7825	.7832	.7839	.7846
6.1	.7853	.7860	.7868	.7875	.7882	.7889	.7896	.7903	.7910	.7917
6.2	.7924	.7931	.7938	.7945	.7952	.7959	.7966	.7973	.7980	.7986
6.3	.7993	.8000	.8007	.8014	.8021	.8028	.8035	.8041	.8048	.8055
6.4	.8062	.8069	.8075	.8082	.8089	.8096	.8102	.8109	.8116	.8122
6.5	.8129	.8136	.8142	.8149	.8156	.8162	.8169	.8176	.8182	.8189
6.6	.8195	.8202	.8209	.8215	.8222	.8228	.8235	.8241	.8248	.8254
6.7	.8261	.8267	.8274	.8280	.8287	.8293	.8299	.8306	.8312	.8319
6.8	.8325	.8331	.8338	.8344	.8351	.8357	.8363	.8370	.8376	.8382
6.9	.8388	.8395	.8401	.8407	.8414	.8420	.8426	.8432	.8439	.8445
7.0	.8451	.8457	.8463	.8470	.8476	.8482	.8488	.8494	.8500	.8506
7.1	.8513	.8519	.8525	.8531	.8537	.8543	.8549	.8555	.8561	.8567
7.2	.8573	.8579	.8585	.8591	.8597	.8603	.8609	.8615	.8621	.8627
7.3	.8633	.8639	.8645	.8651	.8657	.8663	.8669	.8675	.8681	.8686
7.4	.8692	.8698	.8704	.8710	.8716	.8722	.8727	.8733	.8739	.8745
7.5	.8751	.8756	.8762	.8768	.8774	.8779	.8785	.8791	.8797	.8802
7.6	.8808	.8814	.8820	.8825	.8831	.8837	.8842	.8848	.8854	.8859
7.7	.8865	.8871	.8876	.8882	.8887	.8893	.8899	.8904	.8910	.8915
7.8	.8921	.8927	.8932	.8938	.8943	.8949	.8954	.8960	.8965	.8971
7.9	.8976	.8982	.8987	.8993	.8998	.9004	.9009	.9015	.9020	.9025
8.0	.9031	.9036	.9042	.9047	.9053	.9058	.9063	.9069	.9074	.9079
8.1	.9085	.9090	.9096	.9101	.9106	.9112	.9117	.9122	.9128	.9133
8.2	.9138	.9143	.9149	.9154	.9159	.9165	.9170	.9175	.9180	.9186
8.3	.9191	.9196	.9201	.9206	.9212	.9217	.9222	.9227	.9232	.9238
8.4	.9243	.9248	.9253	.9258	.9263	.9269	.9274	.9279	.9284	.9289
8.5	.9294	.9299	.9304	.9309	.9315	.9320	.9325	.9330	.9335	.9340
8.6	.9345	.9350	.9355	.9360	.9365	.9370	.9375	.9380	.9385	.9390
8.7	.9395	.9400	.9405	.9410	.9415	.9420	.9425	.9430	.9435	.9440
8.8	.9445	.9450	.9455	.9460	.9465	.9469	.9474	.9479	.9484	.9489
8.9	.9494	.9499	.9504	.9509	.9513	.9518	.9523	.9528	.9533	.9538
9.0	.9542	.9547	.9552	.9557	.9562	.9566	.9571	.9576	.9581	.9586
9.1	.9590	.9595	.9600	.9605	.9609	.9614	.9619	.9624	.9628	.9633
9.2	.9638	.9643	.9647	.9652	.9657	.9661	.9666	.9671	.9675	.9680
9.3	.9685	.9689	.9694	.9699	.9703	.9708	.9713	.9717	.9722	.9727
9.4	.9731	.9736	.9740	.9745	.9750	.9754	.9759	.9763	.9768	.9773
9.5	.9777	.9782	.9786	.9791	.9795	.9800	.9805	.9809	.9814	.9818
9.6	.9823	.9827	.9832	.9836	.9841	.9845	.9850	.9854	.9859	.9863
9.7	.9868	.9872	.9877	.9881	.9886	.9890	.9894	.9899	.9903	.9908
9.8	.9912	.9917	.9921	.9926	.9930	.9934	.9939	.9943	.9948	.9952
9.9	.9956	.9961	.9965	.9969	.9974	.9978	.9983	.9987	.9991	.9996

INDEX

Abscissa, 145
Absolute-value equations and inequalities, 83–85
Absolute value function, 182
Addition, of polynomials, 10–11
Algebraic operation, rules of, 77
Antilogarithm, 258
Arithmetic sequence, 409–11
Associative law, 3
Asymptotes, 219–22
Axioms of equality, 3

Binomial Theorem, 443–47

Cartesian coordinate system, 145–46
Characteristic, 258
Circles, graphing, 223–24
Closure law, 3
Cofactors, 320–23
Combinations, 441–43
Common denominator, 33–34
Common factor, 16–17
Commutative law, 3
Complete factorization theorem, 376–77
Completing the square, 101–3, 209
Complex numbers, 58–61
Conic sections, 222–29
Constant function, 181
Counting principle, 436–38
Cramer's Rule, 325–28

Decimals, repeating, 416
Degree, classification by, 9–10
Degree-comparison test, 220
Descartes' Rule of Signs, 382–83
Determinants
    definition of, 318–23
    solving linear systems by, 325–28
    special operations on, 323–25
Discriminant, 104
Distance formulas, 147–48
Distributive law, 3
Division algorithm, 371
Domain, 176, 179, 185

Elementary row operations, 316
Ellipses, graphing, 224–27
Equations
    absolute-value, 83–84
    equivalent, 77
    exponential, 250
    first-degree, in one variable, 76–80
    first-order, word problems for, 124–26
    general equation of a conic section, 229
    logarithmic, 255–56
    in two variables, graphing, 149–50
    quadratic, 99–104, 127–29
    quadratic-type, 104–8
    systems of, in two variables, 279–84
Equivalent equations, 77
Equivalent matrices, 316
Exponential equations, rules for solving, 250
Exponential functions, 182, 247–50

Exponents
    integral, 49–53
    rational, 57–58
    rules of, 248–49

Factoring
    polynomials, 13–17, 377, 378–86
    solving quadratic equations by, 99
Factorization formulas, 14–16
Factor theorem, 375–76
Fibonacci Sequence, 405–6
Field properties, 3–4
First-degree equations
    in one variable, 76–80
    in two variables, graphing, 151–58
First-degree inequalities
    in one variable, 80–83
    in two variables, 158–59
First-order equations, word problems for, 124–26
First-order inequalities, word problems for, 126–27
FOIL, 12–13
Function notation, 179–80
Functions, 176–82
    composite, 185–86
    exponential, 247–50, 261–64
    graphing, 182–85
    inverse, 186–90
    logarithmic, 250–54, 261–64
    polynomial, higher-degree, 214–16
    quadratic, 207–13
    special, 181–84
Fundamental Principle of Linear Programming, 294–95
Fundamental Theorem of Algebra, 376–78

Geometric sequence, 412–15
Graphing, 145–60
    asymptotes and, 219–22
    conic sections, 222–29
    logarithmic functions, 252–55
    quadratic functions, 209–13

Horizontal line test, 181
Hyperbolas, graphing, 227–28

Identity function, 181
Identity law, 3
Inequalities
    absolute-value, 84–85
    first degree, in two variables, 158–59
    first-order, word problems for, 126–27
    in one variable, 80–83
    quadratic, 108–10
    quadratic, word problems for, 129–30
Integers, 2
Integral exponents, 49–53
Intercepts, 152, 211–12
Intermediate value theorem for polynomials, 383–84
Interval notation, 80
Intervals, bounded and unbounded, 80–81

Inverse law, 3
Inverse matrices, 330–33
Irrational numbers, 2–3
Irrational roots, approximation of, 384–86

Least common denominator, 34–35
Linear equations, 77–80
    in three variables, systems of, 288–91
    in two variables, systems of, 284–88
    systems of, solving by determinants, 325–28
    systems of, solving by matrix methods, 333–34
    in two variables, 151–58
Linear function, 182
Linear inequalities in two variables, systems of, 292–93
Linear interpolation, 259–60
Linear programming, 293–97
Linear systems
    solving by determinants, 325–28
    solving by matrix methods, 333–34
Lines, parallel and perpendicular, 158
Lines, straight
    finding the midpoint of, 148–49
    properties of, 151–54
Logarithmic equations, solving, 255–56
Logarithmic functions, 250–54
Logarithms
    computation with, 256–61
    to evaluate expressions, 260–61
    laws of, 254–55
Long division, 369–70

Mantissa, 258
Mathematical induction, 416–22
Matrices
    definitions, 314–16
    equivalent, 316
    identity, 330
    inverse, 330–33
    products of, 329–30
    properties of, 328
    solving systems by, 317–18
Minors, 320–23
Monomial functions, graphing, 214–16
Multiple roots, 377–78
Multiplication of polynomials, 11–13
Multiplication Rule, 437–38

Natural numbers, 2
n factorial, 406–7
Numbers
    integers, 2
    irrational, 2–3
    natural, 2
    rational, 2
    real, 2–3
Number sequences, 403–7

One-to-one function, 181
Ordered pair, 145